The Cambridge Guide to the Solar System

Richly illustrated with full-color images, this book is a comprehensive, up-to-date description of the planets, their moons, and recent exoplanet discoveries.

The second edition of this classic reference is brought up-to-date with the fascinating new discoveries made during recent years from 12 new solar system missions. Representative examples include water on the Moon; widespread volcanism on Mercury's previously unseen half; vast buried glaciers on Mars; geysers on Saturn's active water moon Enceladus; lakes of methane and ethane on Saturn's moon Titan; the encounter with asteroid Itokawa; and an encounter and sample return from comet Wild 2. The book is further enhanced by hundreds of striking new images of the planets and moons.

Written at an introductory level appropriate for high-school and undergraduate students, it provides fresh insights that appeal to anyone with an interest in planetary science. A website hosted by the author contains all of the images in the book with an overview of their importance. A link to this can be found at www.cambridge.org/solarsystem/.

KENNETH R. LANG is a Professor of Astronomy at Tufts University. He is a well-known author and has published 25 books. *The Cambridge Encyclopedia of the Sun* (Cambridge University Press, 2001) was recommended by the *Library Journal* as one of the best reference books published that year. He has extensive teaching experience, and has served as a Visiting Senior Scientist at NASA Headquarters.

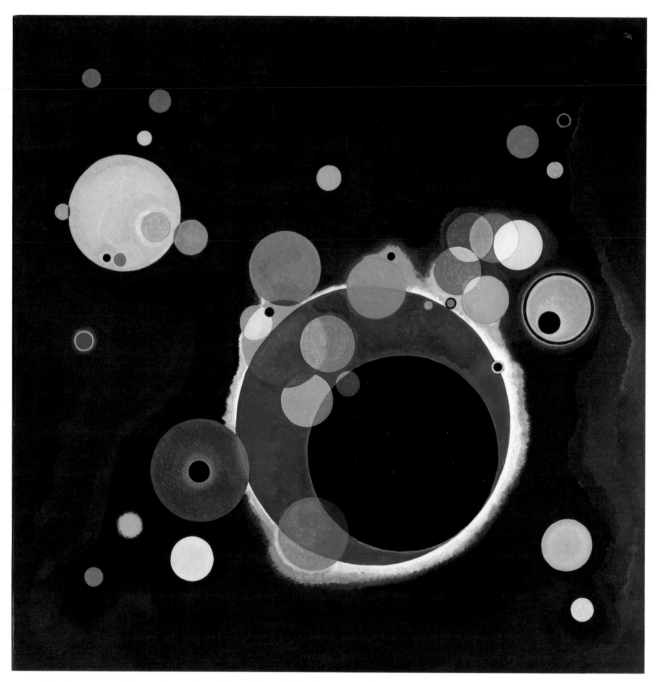

Several Circles. January–February 1926. The artist Vasily Kandinsky (1866–1944) seems to capture the essence of our space-age exploration of previously unseen worlds in this cosmic and harmonious painting. According to Kandinsky, "The circle is the synthesis of the greatest oppositions. It combines the concentric and the eccentric in a single form and in equilibrium. Of the three primary forms, it points most clearly to the fourth dimension." (Courtesy of the Solomon R. Guggenheim Museum, New York City, New York.)

The Cambridge Guide to the

Solar System

Second Edition

Kenneth R. Lang
Tufts University, Medford, Massachusetts, USA

CAMBRIDGE
UNIVERSITY PRESS

CAMBRIDGE UNIVERSITY PRESS
Cambridge, New York, Melbourne, Madrid, Cape Town,
Singapore, São Paulo, Delhi, Tokyo, Mexico City

Cambridge University Press
The Edinburgh Building, Cambridge CB2 8RU, UK

Published in the United States of America by Cambridge University Press, New York

www.cambridge.org
Information on this title: www.cambridge.org/solarsystem

First published 2003
Second edition 2011

Printed in the United Kingdom at the University Press, Cambridge

A catalogue record for this publication is available from the British Library

ISBN 978-0-521-19857-8 Hardback

Contents

Part 3 The giant planets, their satellites and their rings: worlds of liquid, ice and gas

Part 4 Remnants of creation: small worlds in the solar system

Focus elements

Tables

Preface to the second edition

The second edition of *The Cambridge Guide to the Solar System* brings this comprehensive description of the planets and moons up to date, by extending it to include fascinating new discoveries made during the previous decade. As with the first edition, it is written at an introductory level appropriate for high-school and undergraduate students, while also providing fresh, current insights that will appeal to professionals as well as general readers with an interest in planetary science. This is accomplished in a light and uniform style, including everyday metaphors and many spacecraft images.

This second edition is filled with vital new facts and information, and lavishly illustrated in color throughout. Hundreds of new images have been provided. Most of these illustrations have never appeared together in print before, and many of them have a beauty comparable to works of art.

An Internet site for use by the instructor, students or casual reader also supports this book. It contains all of the images in this second edition, together with their legends and overview bullets of their seminal content. This site also includes similar material for the author's books about the Sun, including the second edition of *Sun, Earth and Sky* and the second edition of *The Sun from Space*. The website address is http://ase.tufts.edu/cosmos/.

Striking examples of new images from contemporary planetary spacecraft include the *Chandrayaan-1* and *LCROSS* missions to the Moon, the *MESSENGER* spacecraft that is viewing the unseen half of Mercury, the *2001 Mars Odyssey*, *Spirit* and *Opportunity Exploration Rovers* and *Phoenix* lander on Mars, the *Cassini–Huygens* mission to Saturn and its moons Enceladus and Titan, the *Deep Impact* and *Stardust* encounters with comets, with *Stardust*'s sample return to Earth, and the *Hayabusa* encounter with the asteroid Itokawa.

The more effective illustrations from previous spacecraft have been retained, without an excessive increase in the length of the book, including those from the *Apollo* missions to the Earth's Moon, the *Viking 1* and *2* missions to Mars, the *Mars Global Surveyor*, the *Voyager 1* and *2* missions to the four giant planets, the *Galileo* mission to Jupiter, and several spacecraft encounters with asteroids and comets.

We have not forgotten our home planet Earth, which continues to provide the reference background for discussions of volcanoes, water, geology, atmospheres and magnetospheres. The second edition also includes an updated appraisal of the effects of global warming, with attempts to combat it, and investigations of space weather that can threaten astronauts and influences the performance, reliability and lifetime of interplanetary spacecraft, Earth-orbiting satellites, and terrestrial communications and power systems.

A new chapter, entitled 'Beyond Neptune', takes us to the outer precincts of the planetary realm, with the discovery of several worlds that orbit the Sun beyond the orbit of Neptune. At least three of these dwarf planets are either larger than Pluto or comparable to it in size, and along with Pluto they have also been designated Plutoids. Three companion moons are also now known to orbit Pluto, and its tenuous atmosphere and surface markings have been scrutinized in anticipation of the encounter of the *New Horizons* spacecraft with Pluto in 2015. In the meantime, the *Voyager 1* and *2* spacecraft have traveled far beyond the planets and measured the termination shock of the Sun's winds at about 100 times the distance between the Earth and the Sun.

An entirely new end-chapter describes the origin of our solar system and the exciting new discoveries of hundreds of planets orbiting nearby stars other than the Sun. These extrasolar planets, or exoplanets for short, include multiplanet systems and are mainly hot Jupiters, which revolve unexpectedly close to their star and have masses comparable to that of Jupiter. But atmospheres have been found on some of these new exoplanets, and astronomers are also finding super-Earths of just a few times the mass of the Earth. Astronomers are now finding Earth-size planets that reside in the warm habitable zone where liquid water can exist on the planet's surface and living things might be present.

The main body of this second edition does not simply consist of a few updates or cosmetic patch ups of the material in the first edition. It instead contains all of the relevant new discoveries, ideas and information resulting from recent planetary spacecraft. They include: water on the Moon; evidence for widespread volcanic activity on young Mercury; ubiquitous evidence for ancient water flows and vast amounts of frozen water now on Mars; jets of ice particles, water vapor and organic compounds from Saturn's active water moon Enceladus; organic dunes and rain and lakes of liquid methane and ethane on Saturn's moon Titan; close-up details of Saturn's rings and its enigmatic moons Hyperion and Phoebe, immersed within a newly discovered ring; asteroid Itokawa's loose collection of rubble; the return of comet dust, containing organic molecules, from comet Wild 2 to Earth; and the *Deep Impact* collision with comet Tempel 1, with forced ejections of water vapor, water ice and other substances.

In addition to bringing images and discoveries up to date, the second edition also adds more scientific substance with set-aside *focus boxes* that emphasize basic planetary physics at the algebra level, such as the luminosity and temperature of the Sun; conservation of angular momentum in the Earth–Moon system; escape velocity, temperature and retention of planetary atmospheres; the Roche limit, and gravitational collapse. These interesting boxed set-asides can be used in introductory university courses that include fundamental scientific topics, but they are not a crucial aspect of the main text and can be bypassed by students with non-scientific interests or by the general reader.

In short, this is a fine, stimulating collection of exciting spacecraft images and marvelous discoveries about the planets and moons, as well as exoplanets, and I hope the reader derives as much pleasure as I have from finding out about them.

I am grateful to three bright, alert Tufts students, Laura Costello, Jeffrey Gottlieb and Nathaniel Eckman, for reading draft chapters of this book, offering insightful comments and spotting necessary corrections. This volume was also substantially improved by the careful editing of Sue Glover. Special thanks are extended to Joe Bredekamp at NASA Headquarters who has actively encouraged the writing of this book and helped fund it through NASA Grant NNX07AU93G with NASA's Applied Information Systems Research Program.

Kenneth R. Lang
Tufts University

Preface to the first edition

The planets have been the subject of careful observations and myth for millennia and the subject of telescopic studies for centuries. Our remote ancestors looked into the night sky, and wondered why the celestial wanderers or planets moved across the stellar background. They saw the planets as powerful gods, whose Greek and Roman names are still in use today. Then progressively larger telescopes enabled the detection of faint moons and remote planets that cannot be discerned with the unaided eye, and resolved fine details that otherwise remain blurred.

Only in the past half century have we been able to send spacecraft to the planets and their moons, changing many of them from moving points of light to fascinating real worlds that are stranger and more diverse than we could have imagined. Humans have visited the Moon, and robot spacecraft have landed on Venus and Mars. We have sent vehicles to the very edge of the planetary system, capturing previously unseen details of the remote giant planets, dropping a probe into Jupiter's stormy atmosphere, and perceiving the distant satellites as unique objects whose complex and richly disparate surfaces rival those of the planets. Probes have also been sent to peer into the icy heart of two comets, and robotic eyes have scrutinized the battered and broken asteroids.

The Cambridge Guide to the Solar System is a complete modern guide, updating and extending the prize-winning *Wanderers in Space* – Prix du livre de l'Astronomie in 1994. This book, written by the author and Charles A. Whitney, was completed before the *Clementine* and *Lunar Prospector* spacecraft were sent to the Moon, the *Magellan* orbiter penetrated the veil of clouds on Venus, the *Mars Pathfinder* landed on the red planet with its mobile roving *Sojourner*, the *Mars Global Surveyor* obtained high-resolution images of the surface of Mars, the *NEAR-Shoemaker* spacecraft orbited the asteroid 433 Eros, the *Galileo* orbiter and probe visited Jupiter and its four large moons, *Deep Space 1* peered into the nucleus of Comet Borrelly, and Comet Shoemaker–Levy 9 collided with Jupiter. *The Guide* updates *Wanderers* to include the captivating results of all these missions, presenting more than a half century of extraordinary accomplishment.

The Cambridge Guide to the Solar System provides comprehensive accounts of the most recent discoveries, from basic material to detailed concepts. It is written in a concise, light and uniform style, without being unnecessarily weighed down with incomprehensible specialized materials or the variable writing of multiple authors. Metaphors, similes and analogies will be of immense help to the lay person and they add to the enjoyment of the material. Vignettes containing historical, literary and even artistic material make this book unusual and interesting, but at a modest level that enhances the scientific content of the book and does not interfere with it.

The book is at once an introductory text of stature and a thorough, serious and readable report for general readers, with much compact reference data. The language, style, ideas and profuse illustrations will attract the general reader as well as students and professionals. In addition, it is filled with vital facts and information for astronomers of all types and for anyone with a scientific interest in the planets and their satellites.

The many full-color images, photographs, and line drawings help make this information highly accessible.

Each chapter begins with a set of pithy, one-sentence statements that describe the most important or interesting things that will be described in that chapter. A summary diagram, placed at the end of each chapter, captures the essence of our knowledge of the subject.

Set-aside *focus boxes* enhance and amplify the discussion with interesting details, fundamental physics and important related topics. They will be read by the especially curious person or serious student, but do not interfere with the general flow of the text and can be bypassed by the general educated reader who wants to follow the main ideas. Equations are kept to a minimum and, when employed, are almost always placed within the set-aside *focus elements*.

Numerous tables provide fundamental physical data for the planets and large moons. Many graphs and line drawings complement the text by summarizing what spacecraft have found. Guides to other resources are appended to the book as an annotated list of books for further reading, all published after 1990, and a list of relevant Internet addresses.

The Cambridge Guide to the Solar System has been organized into four main parts. The first introduces the planets and their moons, with a brief historical perspective followed by a discussion of their common properties. These unifying features include craters, volcanoes, water, atmospheres and magnetic fields. The second part discusses the rocky worlds found in the inner solar system – the Earth with its Moon, Mercury, Venus and Mars. The third part presents the giant planets, their satellites and their rings – worlds of liquid, ice and gas. The last part discusses the smaller worlds, the comets and asteroids, as well as collisions of these bodies with Jupiter, the Sun and Earth.

Chapter 1 traces our evolving understanding of the planets and their satellites made possible by the construction of ever-bigger telescopes. They resulted in the discovery of new planets and satellites, and resolved details on many of them. Here we include, in chronological order, the discoveries of Jupiter's moons, Saturn's rings, Uranus, Neptune, the asteroids, the icy satellites of the giant planets, tiny Pluto with its oversized moon, and the small icy objects in the Kuiper belt at the edge of the planetary realm. Other fundamental discoveries have been woven into the fabric of this chapter, including the realization that planets are whirling endlessly about the Sun, refinements of the scale and size of the planetary realm, and the spectroscopic discovery of the main ingredients of both the Sun and the atmospheres of the planets.

Chapter 2 begins with a description of how spacecraft have fundamentally altered our perception of the solar system, providing detailed close-up images of previously unseen landscapes and detecting incredible new worlds with sensors that see beyond the range of human vision. These new vistas have also resulted in a growing aware-ness of the similarities of the major planets and some moons. In the rest of Chapter 2 and in Chapter 3, they are therefore interpreted as a whole, rather than as isolated objects, by presenting comparative aspects of common properties and similar processes. This provides a foundation for subsequent examination of individual objects in greater detail.

Impact craters are found on just about every body in the solar system from the Moon and Mercury to the icy satellites of the distant planets, but in different amounts that depend on their surface ages and with varying properties. Ancient impacts on Venus have, for example, been erased by outpourings of lava, and the debris from subsequent impacts has been shaped by the planet's thick atmosphere. Numerous volcanoes have also been found throughout the solar system, including fiery outbursts on the Earth, towering volcanic mountains on Mars, numerous volcanoes that have resurfaced Venus, currently active volcanoes that have turned Jupiter's satellite Io inside out, and eruptions of ice on Neptune's largest moon, Triton.

Liquid water, which is an essential ingredient of life, covers seventy-one percent of the Earth's surface. Catastrophic floods and deep rivers once carved deep channels on Mars, and spring-like flows have been detected in relatively recent times. Water ice is ubiquitous in the outer solar system, including the clouds of Jupiter, the rings of Saturn, and the surfaces of most satellites. There is even evidence for subsurface seas beneath the water-ice crusts of Jupiter's satellites Europa, Ganymede and Callisto, and liquid water might also reside beneath the frozen surface of Saturn's satellite Enceladus.

Chapter 3 describes the atmospheres and magnetic fields that form an invisible buffer zone between planetary surfaces and surrounding space. Venus has an atmosphere that has run out of control, smothering this nearly Earth-sized world under a thick blanket of carbon dioxide. Its greenhouse effect has turned Venus into a torrid world that is hot enough to melt lead and vaporize oceans. Mars now has an exceedingly thin, dry and cold atmosphere of carbon dioxide. The red planet breathes about one-third of its atmosphere in and out as the southern polar cap grows and shrinks with the seasons. Jupiter's powerful winds and violent storms have remained unchanged for centuries, and Neptune has an unexpectedly stormy atmosphere. Saturn's largest moon, Titan, has a substantial Earth-like atmosphere, which is mainly composed of nitrogen and has a surface pressure comparable to that of the Earth's atmosphere. Temporary, rarefied and misty atmospheres cloak the Moon, Mercury, Pluto, Triton, and Jupiter's four largest moons.

Magnetic fields protect most of the planets from energetic charged particles flowing in the Sun's ceaseless winds, but some electrons and protons manage to penetrate this barrier. Jupiter's magnetism is the strongest and largest of all the planets, as befits the giant, while the magnetic fields of Uranus and Neptune are tilted. Guided by magnetic fields, energetic electrons move down into the polar atmospheres of Earth, Jupiter and Saturn, producing colorful auroras there.

Our description of individual planets begins in Chapter 4, with our home planet Earth. Earthquakes have been used to look inside our world, determining its internal structure and locating a spinning, crystalline globe of solid iron at its center. At the surface, continents slide over the globe, colliding and coalescing with each other like floating islands, as ocean floors well up from inside the Earth.

A thin membrane of air protects life on this restless world, and that air is being dangerously modified by life itself. Synthetic chemicals have been destroying the thin layer of ozone that protects human beings from dangerous solar ultraviolet radiation, and wastes from industry and automobiles are warming the globe to dangerous levels. The world has become hotter in the last decade than it has been for a thousand years, and at least some of this recent rise in temperature is due to greater emissions of greenhouse gases by human activity. The politicized debate over global warming is also described in Chapter 4, as are the probable future consequences if we don't do something about it soon.

This fourth chapter also discusses how the Sun affects our planet, where solar light and heat permit life to flourish. The amount of the Sun's radiation that reaches the Earth varies over the 11-year solar cycle of magnetic activity, warming and cooling the planet. Further back in time, during the past one million years, our climate has been changed by the recurrent ice ages, which are caused by variations in the amount and distribution of sunlight reaching the Earth.

An eternal solar gale now buffets our magnetic domain and sometimes penetrates it. Forceful mass ejections can create powerful magnetic storms on Earth, and damage or destroy Earth-orbiting satellites. Energetic charged particles, hurled out during solar explosions, endanger astronauts and can also wipe out satellites that are so important to our technological society. Space-weather forecasters are now actively searching for methods to predict these threats from the Sun.

In Chapter 5 we continue on to the still, silent and lifeless Moon, a stepping stone to the planets. Most of the features that we now see on the Moon have been there for more than 3 billion years. Cosmic collisions have battered the lunar surface during the satellite's formative years, saturating much of its surface with impact craters, while lunar volcanism filled the largest basins to create the dark maria.

Twelve humans went to the Moon more than three decades ago, and brought back nearly half a ton of rocks. The rocks contain no water, have never been exposed to it, and show no signs of life. Yet, orbiting spacecraft have found evidence for water ice deposited by comets in permanently shaded regions at the lunar poles.

The fifth chapter also describes how the Moon generates tides in the Earth's oceans, and acts as a brake on the Earth's rotation, causing the length of day to steadily increase. The satellite also steadies our seasons by limiting the tilt of Earth's rotation axis. The story of the Moon's origin is given the latest and most plausible explanation: a glancing impact from a Mars-sized object knocked a ring of matter out of the young Earth; that ring soon condensed into our outsized, low-density Moon.

We discover in Chapter 6 that Mercury has an unchanging, cratered and cliff-torn surface like the Moon, but in a brighter glare from the nearby Sun. Although the planet looks like the Moon on the outside, it resembles the Earth on the inside. Relative to its size, Mercury has the biggest iron core of all terrestrial planets, and it also has a relatively strong magnetic field. Here we also mention tiny, unexplained motions of Mercury. As demonstrated by astronomers long ago, the planet does not appear precisely in its expected place. This discrepancy led Einstein to develop a new theory of gravity in which the Sun curves nearby space.

Chapter 7 discusses veiled Venus, the brightest planet in the sky. No human eye has ever gazed at its surface, which is forever hidden in a thick overcast of impenetrable clouds made of droplets of concentrated sulfuric acid. Radar beams from the orbiting *Magellan* spacecraft have penetrated the clouds and mapped out the surface of Venus in unprecedented detail, revealing rugged highlands, smoothed-out plains, volcanoes and sparse, pristine impact craters. Rivers of outpouring lava have resurfaced the entire surface of Venus, perhaps about 750 million years ago, and tens of thousands of volcanoes are now found on its surface. Venus exhibits every type of volcanic edifice known on Earth, and some that have never been seen before. Some of them could now be active. Unlike Earth, there is no evidence for colliding continents on Venus, its surface moves mostly up and down, rather than sideways. Vertical motions associated with upwelling hot spots have buckled, crumpled, deformed, fractured and stretched the surface of Venus.

Our voyage of discovery continues in Chapter 8 to the red planet Mars, long thought to be a possible haven for life. Catastrophic flash floods and deep ancient rivers once carved channels on its surface, and liquid water might have lapped the shores of long-vanished lakes and seas. But its water is now frozen into the ground and ice caps, and it cannot now rain on Mars. Its thin, cold atmosphere lacks an ozone layer that might have protected the surface from lethal ultraviolet rays from the Sun, and if any liquid water were now released on the red planet's surface it would soon evaporate or freeze. Yet underground liquid water may have been seeping out of the walls of canyons and craters on Mars in recent times.

Three spacecraft have landed on the surface of Mars, failing to detect any unambiguous evidence for life. Corrosive chemicals have destroyed all organic molecules in the Martian ground, which means that the surface now contains no cells, living, dormant or dead. A meteorite from Mars, named ALH 84001, exhibits signs that bacteria-like micro-organisms could have existed on the red planet billions of years ago, but most scientists now think that there is nothing in the meteorite that conclusively indicates whether life once existed on Mars or exists there now. The future search for life on Mars may include evidence of microbes that can survive in hostile environments, perhaps energized from the planet's hot interior.

Chapter 9 presents giant Jupiter, which is almost a star and radiates its own heat. Jupiter radiates nearly twice as much energy as it receives from the Sun, probably as heat left over from when the giant planet formed. Everything we see on Jupiter is a cloud, formed in the frigid outer layers of its atmosphere. The clouds are swept into parallel bands by the planet's rapid rotation and counter-flowing winds, with whirling storms that can exceed the Earth in size. The fierce winds run deep and are driven mainly from within by the planet's internal heat. The biggest storms and wind-blown bands have persisted for centuries, though the smaller eddies are engulfed by the bigger ones, deriving energy from them. The little storms pull their energy from hotter, lower depths. Jupiter has a non-spherical shape with a perceptible bulge around its equatorial middle, and this helps us determine what is inside the planet. It is almost entirely a vast global sea of liquid hydrogen, compressed into a fluid metal at great depths. And above it all, Jupiter has a faint, insubstantial ring system that is made of dust kicked off small nearby moons by interplanetary meteorites.

Chapter 9 additionally provides up-to-date accounts of the four large moons of Jupiter, known as the Galilean satellites. The incredible complexity and rich diversity of their surfaces, which rival those of the terrestrial planets, are only visible by close-up scrutiny from spacecraft. Although the *Voyager 1* and *2* spacecraft sped by with just a quick glimpse at them, it was time enough for their cameras to discover new worlds as fascinating as the planets themselves, including active volcanoes on Io, smooth ice plains on Europa, grooved terrain on Ganymede, and the crater-pocked surface of Callisto. Then the *Galileo* spacecraft returned for a longer look, gathering further data on the satellites' surfaces and using gravity and magnetic measurements to infer their internal constitution. Changing tidal forces from nearby massive Jupiter squeeze Io's rocky interior in and out, making it molten inside and producing the most volcanically active body in the solar system. Jupiter's magnetic field sweeps past the moon, picking up a ton of sulfur and oxygen ions every second and directing them into a doughnut-shaped torus around the planet. A vast current of 5 million amperes flows between the satellite Io and the poles of Jupiter and back again, producing auroral lights on both bodies. There are no mountains or valleys on the bright, smooth, ice-covered surface of Europa. The upwelling of dirty liquid water or soft ice has apparently filled long, deep fractures in the crust. Large blocks of ice float like rafts across Europa's surface, lubricated by warm, slushy material. A subsurface ocean of liquid water may therefore lie just beneath Europa's icy crust, perhaps even harboring alien life that thrives in the dark warmth. Ganymede has an intrinsic magnetic field. As far as we know, it is the only satellite known that now generates its own magnetism. Callisto is one of the oldest, most heavily cratered surfaces in the solar system. Both Callisto and Europa have a borrowed magnetic field, apparently generated by electrical currents in a subsurface ocean as Jupiter's powerful field sweeps by.

Our voyage of discovery continues in Chapter 10 with Saturn, second only to Jupiter in size. Like Jupiter, the ringed planet radiates almost twice as much energy as it receives from the Sun, but Saturn is not massive enough to have substantial heat left over from its formation. Its excess heat is generated by helium raining down inside the planet. It is Saturn's fabled rings that set the planet apart from the other wanderers. The astonishing rings consist of billions of small, frozen particles of water ice, each in its own orbit around Saturn like a tiny moon. They have been arranged into rings within rings by the gravitational influences of small nearby satellites that generate waves, sweep out gaps and confine the particles in the rings. Saturn's rings are thought to be relatively young, less than 100 million years old. They may have originated when a former moon strayed too close to the planet and was torn apart by its tidal forces.

Saturn's largest satellite, Titan, has a substantial atmosphere composed mainly of nitrogen molecules, also the principal ingredient of Earth's air. Clouds of methane, raining ethane, and flammable seas of ethane, methane and propane could exist beneath the impenetrable haze. We should find out what lies beneath the smog when the *Cassini* spacecraft arrives at Saturn, in July 2004, and parachutes the *Huygens* probe through Titan's atmosphere four months later. Six medium-sized moons revolve around Saturn, each covered with water ice. They are scarred with ancient impact craters, and some of them show signs of ice volcanoes and internal heat. A number of small irregularly shaped moons of Saturn have remarkable orbits. The co-orbital moons move in almost identical orbits, the Lagrangian moons share their orbit with a larger satellite, and the shepherd moons confine the edges of rings.

Uranus and Neptune are treated together in Chapter 11, because of their similar size, mass and composition. Unlike all the other planets, Uranus is tipped on its side and rotates with its spin axis in its orbital plane and in the opposite direction to that of most of the other planets. No detectable heat is emitted from deep inside Uranus, while Neptune emits almost three times the amount of energy it receives from the Sun. This internal heat drives Neptune's active atmosphere, which has fierce winds and short-lived storms as big as the Earth. Both planets are vast global oceans, consisting mainly of melted ice with no metallic hydrogen inside. The magnetic fields of both Uranus and Neptune are tilted from their rotation axes, and are probably generated by currents in their watery interiors. The ring systems of both planets are largely empty space, containing dark narrow rings with wide gaps. One of Neptune's thin rings is unexpectedly lumpy, with material concentrated in clumps by a nearby moon. The rings we now see around these planets will eventually be ground into dust and vanish from sight, but they can easily be replaced by debris blasted off small moons already embedded in them. The amazingly varied landscape on Miranda, the innermost mid-sized satellite of Uranus, indicates that the satellite may have been shattered by a catastrophic collision and reassembled, or else it was frozen into an embryonic stage of development. Neptune's satellite Triton revolves about the planet in the opposite direction to its spin. The glazed satellite has a very tenuous, nitrogen-rich atmosphere, bright polar caps of nitrogen and methane ice, frozen lakes flooded by past volcanoes of ice, and towering geysers that may now be erupting on its surface. Triton may have formed elsewhere in the solar system and was captured into obit around Neptune. Triton is headed for a future collision with Neptune as the result of tidal interaction with the planet.

Chapter 12 discusses the icy comets. They light up and become visible for just a few weeks or months when tossed near the Sun, whose heat vaporizes the comet's surface and it grows large enough to be seen. A million, million comets are hibernating in the deep freeze of outer space, and they have been out there ever since the formation of the solar system 4.6 billion years ago. We can detect some of them in the Kuiper belt reservoir at the edge of the planetary system, but billions of unseen comets reside in the remote Oort cloud nearly halfway to the nearest star. Two spacecraft have now passed close enough to image a comet nucleus, of Comet Halley and Comet Borrelly, showing that they are just gigantic, black chunks of water ice, other ices, dust and rock, about the size of New York City or Paris. When these comets come near the Sun, their icy nuclei release about a million tons of water and dust every day, from fissures in their dark crust. Some comets develop tails that flow away from the Sun, briefly attaining lengths as large as the distance between the Earth and the Sun, but other comets have no tail at all. Comets can have two kinds of tails: the long, straight, ion tails, that re-emit sunlight with a faint blue fluorescence, and a shorter, curved, dust tail that shines by reflecting yellow sunlight. They are blown away from the Sun by its winds and radiation, respectively. Meteor showers, commonly known as shooting stars, are produced when sand-sized or pebble-sized pieces of a comet burn up in the Earth's atmosphere, never reaching the ground. Any comet that has been seen will vanish from sight in less than a million years, either vaporizing into nothing or leaving a black, invisible rock behind. Some burned-out comets look like asteroids, and a few asteroids behave like comets, blurring the distinction between these two types of small solar-system bodies.

We continue in Chapter 13 with the rocky asteroids. There are billions of them in the main asteroid belt, located between the orbits of Mars and Jupiter, but they are so small and widely spaced that a spacecraft may safely travel though the belt. The combined mass of billions of asteroids is less than five percent of the Moon's mass. The Earth resides in a smaller swarm of asteroids, chaotically shuffled out of the main belt. Many of these near-Earth asteroids travel on orbits that intersect the Earth's orbit, with the possibility of an eventual devastating collision with our planet. The asteroids are the pulverized remnants of former, larger worlds that failed to coalesce into a single planet. The colors of sunlight reflected from asteroids indicate that they formed under different conditions prevailing at varying distance from the Sun. We could mine some of the nearby ones for minerals or water. An asteroid's gravity is too weak to hold on to an atmosphere or to pull most asteroids into a round shape. The close-up view obtained by passing spacecraft and radar images indicates that asteroids have been battered and broken apart during catastrophic collisions in years gone by. One spacecraft has circled the near-Earth asteroid 433 Eros for a year, examining its dusty, boulder-strewn landscape in great detail, obtaining an accurate mass for the asteroid, and showing that much of it is solid throughout. Other asteroids are rubble piles, the low-density, collected fragments of past collisions held together by gravity. Meteorites are rocks from space that survive their descent to the ground, and most of them are chips off asteroids. Organic matter found in meteorites predates the origin of life on Earth by a billion years; but the meteoritic hydrocarbons are not of biological origin.

The concluding Chapter 14 discusses colliding worlds, including pieces of a comet that hit Jupiter, comets that are on suicide missions to the Sun, and an asteroid that wiped out the dinosaurs when it hit the Earth 65 million years ago. The Earth is now immersed within a cosmic shooting gallery of potentially lethal, Earth-approaching asteroids and comets that could collide with our planet and end civilization as we know it. The lifetime risk that you will die as the result of an asteroid or comet striking the Earth is about the same as death from an airplane crash, but a lot more people will die with you during the cosmic impact. It could happen tomorrow or it might not occur for hundreds of thousands of years, but the risk is serious enough that astronomers are now taking a census of the threatening ones. With enough warning time, we could redirect its course.

The Cambridge Guide to the Solar System continues with an annotated list of books for further reading, all published after 1990, and a list of Internet addresses for the topics discussed.

The illustrator Sue Lee has combined artistic talent with a scientist's eye for detail in producing the fantastic line drawings and diagrams in this book. The text has been substantially improved by the careful attention of copy-editor Brian Watts.

This book was stimulated by the author's visit to the Jet Propulsion Laboratory, when the main results of the recent planetary missions were summarized by its director, Edward C. Stone, and the Project Scientists of many of them. Andrew P. Ingersoll, Torrence V. Johnson, Kenneth Nealson, R. Stephen Saunders, Donald K. Yeomans and Richard W. Zurek provided comprehensive scientific summaries matched only by the extraordinary accomplishments of the missions themselves. Planetary scientists with comprehensive knowledge have assured the accuracy, completeness and depth of individual chapters through critical review. I am grateful to my expert colleagues who have read portions of this book, and substantially improved it, either by thorough review or by expert commentary on some isolated sections. They include Reta Beebe, Doug Biesecker, Mark A. Bullock, Owen K. Gingerich, Torrence V. Johnson, Brian G. Marsden, Steven J. Ostro, Carl B. Pilcher, Roger A. Phillips, David Senske, Paul D. Spudis, David J. Stevenson and Donald K. Yeomans.

Kenneth R. Lang
Tufts University

Principal units

This book uses the International System of Units (Système International, SI) for most quantities, but with two exceptions. As is the custom with planetary scientists, we often use the kilometer unit of length and the bar unit of pressure. The familiar kilometer appears on most automobile speedometers. There are one thousand meters in a kilometer, and a mile is equivalent to 1.6 kilometers. One bar corresponds to the surface pressure of the Earth's air at sea level. For conversion to the SI pressure unit of pascals, $1 \, \text{bar} = 10^5$ pascals, or $1 \, \text{pascal} = 10^{-5}$ bar.

Some other common units are the millibar, equivalent to 0.001 bar, the nanometer (nm) with $1 \, \text{nm} = 10^{-9}$ meters, the micron or micrometer (μm) with $1 \, \mu\text{m} = 10^{-6}$ m, the ångstrom unit of wavelength, where $1 \, \text{ångstrom} = 1 \, \text{Å} = 10^{-10}$ meters, the nanotesla (nT) unit of magnetic flux density, where $1 \, \text{nT} = 10^{-9}$ tesla $= 10^{-5}$ gauss, and the ton measurement of mass, where $1 \, \text{ton} = 10^3$ kilograms $= 10^6$ grams.

The reader should also be warned that centimeter-gram-second (c.g.s.) units have been, and still are, widely employed in astronomy and astrophysics. The table provides unit abbreviations and conversions between units.

Quantity	SI units	Conversion to c.g.s. units
Length	meter (m)	100 centimeters (cm)
Mass	kilogram (kg)	1000 grams (g)
Time	second (s)	
Temperature	kelvin (K)	
Velocity	meter per second (m s^{-1})	100 centimeters per second (cm s^{-1})
Energy	joule (J)	10 000 000 erg
Power	watt (W) = joule per second (J s^{-1})	10 000 000 erg s^{-1} (= 10^7 erg s^{-1})
Magnetic flux density	tesla (T)	10 000 gauss (G) (= 10^4 G)
Force	newton (N) (= kg m s^{-2})	100 000 dyn (= 10^5 dyn)
Pressure	pascal (Pa) (= N m^{-2}) (= kg m^{-1} s^{-2})	10 dyn cm^{-2} (= 10^{-5} bar)

1   Evolving perspectives: a historical prologue

- The wandering planets move in a narrow track against the unchanging background stars, and some of these vagabonds can suddenly turn around, apparently moving in the opposite direction before continuing on their usual course.

- The ancient Greeks noticed that the Earth always casts a curved shadow on the Moon during a lunar eclipse, demonstrating that our planet is a sphere.

- For centuries, astronomers tried to describe the observed planetary motions using uniform, circular motions with the stationary Earth at the center and with the distant celestial sphere revolving about the Earth once a day.

- Around 145 AD, Claudius Ptolemy devised an intricate system of uniform motion around small and large circles to model the motions of the Sun, Moon and planets around a stationary Earth; his model was used to predict their location in the sky for more than a thousand years.

- The stars seem to be revolving around the Earth each night, but the Earth is instead spinning beneath the stars. This rotation also causes the Sun to move across the sky each day.

- Mikolaj Kopernik, better known as Nicolaus Copernicus, argued in 1543 that the Earth is just one of several planets that are whirling endlessly about the Sun, all moving in the same direction but at different distances from the Sun and with speeds that decrease with increasing distance.

- Almost four centuries ago, Johannes Kepler used accurate observations, obtained by Tycho Brahe, to conclude that the planets move in ellipses, or ovals, with the Sun at one focus, and to infer a precise mathematical relation between the mean orbital distance and period of each planet.

- More distant planets take longer to move once around the Sun and they move with slower speeds; their orbital periods are in proportion to the cubes of their distances.

- Astronomy is an instrument-driven science in which novel telescopes and new technology enable us to discover cosmic objects that are otherwise invisible and hitherto unknown.

- Many major astronomical discoveries have been unanticipated and serendipitous, made while new telescopes were used to study other, known cosmic objects; the earliest of these accidental discoveries include the four large moons of Jupiter, the planet Uranus, and the first known asteroid, Ceres, discovered respectively by Galileo Galilei in 1610, William Herschel in 1781, and Giuseppe Piazzi in 1801.

- The asteroid belt between the orbits of Mars and Jupiter contains more than 500 000 asteroids, but it is largely empty space and has a total mass that is much less than that of the Earth's Moon.

- Two kinds of telescopes, the refractor and the reflector, enable astronomers to detect faint objects that cannot be seen with the unaided eye, and to resolve fine details on luminous planets that otherwise remain blurred.

- Jupiter, Saturn and Uranus have a retinue of large satellites, and Neptune has only one really large moon that moves in the opposite direction to all the other large satellites. Mercury and Venus have no moons, the Earth has one satellite, our Moon, and Mars has two very small ones.

- Christiaan Huygens discovered Saturn's rings in 1659; they are completely detached from the planet and consist of innumerable tiny satellites each with an independent orbit about Saturn.

- In his *Principia*, published in 1686, Isaac Newton showed how the laws of motion and universal gravitation describe the movements of the planets and everything else in the Universe.

- The solar system is held together by the Sun's gravitational attraction, which keeps the planets in their orbits; they move at precisely the right speed required to just overcome the pull of solar gravity.

- The gravitational attraction between two objects increases in proportion to the product of their masses and in inverse proportion to the square of the distance between them.

- The planet Neptune was discovered in 1846, near the location predicted by mathematical calculations under the assumption that the gravitational pull of a large, unknown world, located far beyond Uranus, was causing the observed positions of Uranus to deviate from its predicted ones.

- Estimates for the mean Earth–Sun distance, known as the astronomical unit or AU, were gradually refined over the centuries, eventually setting the scale of the solar system at $1\,AU = 149.6$ million kilometers. At this distance, it takes 499 seconds for light to travel from the Sun to the Earth.

- The nearest star other than the Sun is located at a distance of 4.24 light-years; it is about 270 000 times further away from the Earth than the Sun.

- The Sun is the most massive and largest object in our solar system. The Sun's mass, which is 333 000 times the Earth's mass, can be inferred from Kepler's third law using the Earth's orbital period of one year and the Earth's mean distance from the Sun, the AU.

- The Sun's size, at 109 times the diameter of the Earth, can be inferred from the Sun's distance and angular extent.

- The temperature of the Sun's visible disk is 5780 kelvin; it can be determined from the Sun's total irradiance of the Earth, the Earth–Sun distance or the AU, and the radius of the Sun.

- The temperature at the center of the Sun is 15.6 million kelvin, estimated from the speed a proton must be moving to counteract the gravitational compression of the massive Sun.

- The composition of the Sun is encoded in absorption lines that appear in the visible spectrum of sunlight.

- The lightest element, hydrogen, is the most abundant element in the Sun, and the next most abundant solar element, helium, was first discovered in the Sun.

- The regular spacing of hydrogen's spectral lines can be explained by quantum theory, in which the angular momentum and energy of an orbiting electron are quantized, depending on an integer quantum number.

- The eight major planets can be divided into two groups: the four rocky, dense, terrestrial planets, Mercury, Venus, Earth and Mars, located relatively near the Sun, and the four giant, low-density planets, Jupiter, Saturn, Uranus and Neptune, that are further from the Sun.

- The temperature and density increase systematically with depth in the giant planets, owing to the greater compression by overlying material.

- As the result of differentiation in their originally molten interiors, the rocky terrestrial planets contain dense iron cores surrounded by less-dense silicate mantles.

- The terrestrial planets contain partially molten, liquid cores, but their internal temperatures cool as time goes on due to the depletion of radioactive elements and the emission of internal heat.

1.1 Moving points of light

The ancient wanderers

Our remote ancestors spent their nights under dark skies, becoming intimately familiar with the stars. They looked up on any moonless night, and watched thousands of stars embedded in the black dome of the night, ceaselessly moving from one edge of the Earth to overhead and back down to another edge, night after night without end.

The brightest stars received names, and patterns, now called constellations, were noticed among groups of them. These permanent stellar beacons are always there, firmly rooted in the dark night sky, and the constellations remain unchanged over the eons.

As ancient astronomers watched the stars, they focused attention on seven objects that did not move with the stars. These celestial vagabonds changed position on the sphere of background stars from hour to hour or night to night, and unlike the stars, they would appear in the night sky at different times from year to year. Ranked in order of greatest apparent brightness, they are the Sun, Moon, Venus, Jupiter, Saturn, Mercury and Mars, each with the Latinized name of a Greek god or goddess. Our ancestors called them

planetes, the ancient Greek word for "wanderers"; and the designation planet is still used for all but the Sun and Moon.

The Sun and Moon move with a rhythm, pattern and beat, marking out the time of our first clocks. The rising and setting Sun ticked off the days, the Moon's changing phase set the monthly cycle, and the seasons marked off the years.

The Sun does not rise at precisely the same point on the horizon each day. Instead, the location of sunrise drifts back and forth along the horizon in an annual cycle. Ancient astronomers used monuments to line up the limits of these excursions (Fig. 1.1). The length and height of the Sun's arc across the sky also change with a yearly rhythm. In the northern hemisphere, the Sun rises highest in the sky on the summer solstice, around June 21 each year, with its longest trajectory and the most daylight hours (Fig. 1.2).

Like the Sun, the Moon rises and sets at different points along the horizon, and reaches varying heights in the sky. Since the full Moon always lies nearly opposite to the Sun, the winter full Moon rises much higher in the sky than the summer full Moon.

Fig. 1.1 Stonehenge The ancient stone pillars of Stonehenge in southern England, shown in this photograph, frame the rising Sun. This monument was used to find midsummer and midwinter 4000 years ago – before the invention of writing and the calendar. The Sun rises at different points on the horizon during the year, reaching its most northerly rising on Midsummer Day (summer solstice on 21 June). After this, the rising point of the Sun moves south along the horizon until it reaches its most southerly rising on Midwinter Day (winter solstice on 22 December). An observer located at the center of the main circle of stones at Stonehenge watched midsummer sunrise over a marker stone located outside the circle; other stones within the circle framed midwinter sunrise and sunset. (Courtesy of Owen Gingerich.)

The Moon repeats its motion around the Earth on a monthly cycle, periodically changing its appearance (Fig. 1.3). Once each month, the Moon comes nearly in line with the Sun, vanishing into the bright daylight. On the next night the Moon has moved away from this position, and a thin lunar crescent is seen. The crescent thickens on successive nights, reaching the rotund magnificence of full Moon in two weeks. Then, in another two weeks, the Moon disappears into the glaring Sun, completing the cycle of the month and providing another natural measure of time.

Even the earliest sky-watchers must have noticed that the wanderers are confined to a narrow track around the sky. Babylonian astronomers noticed it thousands of years ago, identifying constellations that lay along its path. Twelve of these constellations subsequently became known as the zodiac, from the Greek word for "animal". The Sun's annual path, called the ecliptic, runs along the middle of this celestial highway, and the paths of all the other wanderers lie within it. Its narrowness is a sign that the planets move almost like marbles on a table because the planes of their orbits are closely aligned with each other.

It was obvious to astronomers from the earliest times that the wanderers do not move at uniform speeds or follow simple paths across the sky. Mars apparently moved in a backwards loop for weeks at a time, seemingly disrupting its uniform progress across the night sky. It gradually came to a stop in its eastward motion, moved backward toward the west, and then turned around again and resumed moving toward the east (Fig. 1.4). Jupiter and Saturn also displayed such a temporary backwards motion in the westward retrograde direction before continuing on in the eastward prograde direction.

But why did these planets behave in such an unusual and singular manner? The ancient Greeks first proposed logical explanations, based on geometry and uniform motion, but modern explanations differ in both the locations and motions of the planets.

Circles and spheres

The ancient Greeks used geometrical models to visualize the cosmos, incorporating the symmetric forms of the circle and sphere. Their aim was to describe the regularities that underlay the planetary motions against the unchanging stellar background, and to thereby predict the locations of the planets at later times. They wanted to provide a reliable guide to the future, which is still one of the main points of science.

In arguments used by the Pythagoreans, and subsequently recorded by Aristotle (384–322 BC), it was shown that the Earth is a sphere. During a lunar eclipse, when the Moon's motion carries it through the Earth's shadow, observers at different locations invariably saw a curved shadow on the Moon (Fig. 1.5). Only a spherical body can cast a round shadow in all orientations. The curved surface of the ocean was also inferred by watching a ship disappear over the horizon; first the hull and then the mast disappear from view.

According to Plato, writing around 380 BC, the simplest and purest sort of motion was circular, so circles ought to describe the visible paths of the moving planets.

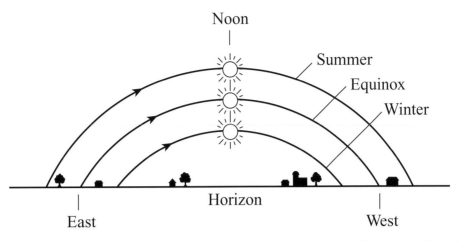

Fig. 1.2 The Sun's trajectory   The Sun's motion across the sky as seen from the northern hemisphere. The maximum height of the Sun in the sky, and the Sun's rising and setting points on the horizon, change with the seasons. In the summer, the Sun rises in the northeast, reaches its highest maximum height, and stays up longest. The Sun rises southeast and remains low in the winter when the days are shortest. The length of day and night are equal on the Vernal, or Spring, Equinox (March 20) and on the Autumnal Equinox (September 23) when the Sun rises exactly east and sets exactly west.

After all, a wheel moves so well because it is round, without rough, sharp edges to get in the way. The circle also has no beginning or end, seemingly appropriate for describing the endless motion of the heavenly wanderers. And the central Earth would be separated from the heavens, like a magician who draws a boundary circle around him to seal off the region in which magical powers are brought into play.

Following Plato's suggestion, astronomers spent centuries trying to discover those uniform, perfectly regular, circular motions that would "save the appearances" presented by the planets. They supposed that the Earth stood still, an immobile globe at the center of it all. The imaginary celestial sphere of fixed stars wheeled around the central, stationary Earth once every day, with uniform circular motion and perfect regularity, night after night and year after year. Such a celestial sphere would also explain why people located at different places on Earth invariably saw just half of all the stellar heavens, and why travelers to new and distant lands would see new stars as well as new people.

The Sun, Moon and planets were once supposed to be carried on concentric, transparent crystalline spheres, which revolved around the stationary Earth, but their hypothetical uniform and circular motions contradicted observations. The Earth-centered model did not explain, for example, why each planet moved with changing speed across the sky, not at an unchanging, uniform rate.

So the Egyptian astronomer Claudius Ptolemy (fl. 150 AD) shifted the Earth from the exact center of the Universe by just a small amount, and described the planetary appearances with an intricate system of circles moving on other circles, like the gears of some fantastic cosmic machine. A planet in uniform circular motion about a center offset from the Earth would appear to a terrestrial observer to be moving with varying speed, faster when it is closest to Earth and slower when further away. Combinations of uniform circular motion were additionally required to account for the looping, or retrograde, paths of the planets (Fig. 1.6). Each planet was supposed to move with constant speed on a small circle, or epicycle, while the center of the epicycle rotated uniformly on a large circle, or deferent. In this way Ptolemy, in his *Mathematical Compilations*, or *Almagest*, written about 145 AD, was able to predict the motions of every one of the seven wanderers, compounding them from circles upon circles. By selecting suitable radii and speeds of motion, Ptolemy reproduced the apparent motions of the planets with remarkable accuracy. He succeeded so well that his model was still being used to predict the locations of the planets in the sky more than a thousand years after his death.

The Earth moves

The ancient Indians of Asia had a different point of view, supposing that the Earth moves around the Sun, as did the Greek mathematician and astronomer Aristarchos, born on the island of Samos in 310 BC. Aristarchos moved the center of the Universe from the Earth to the Sun, and set the Earth in motion, supposing that the Earth and other planets travel in circular orbits around the stationary Sun. He further stated that the fixed stars do not move, and that their apparent daily motion is due to the Earth's rotation on its axis.

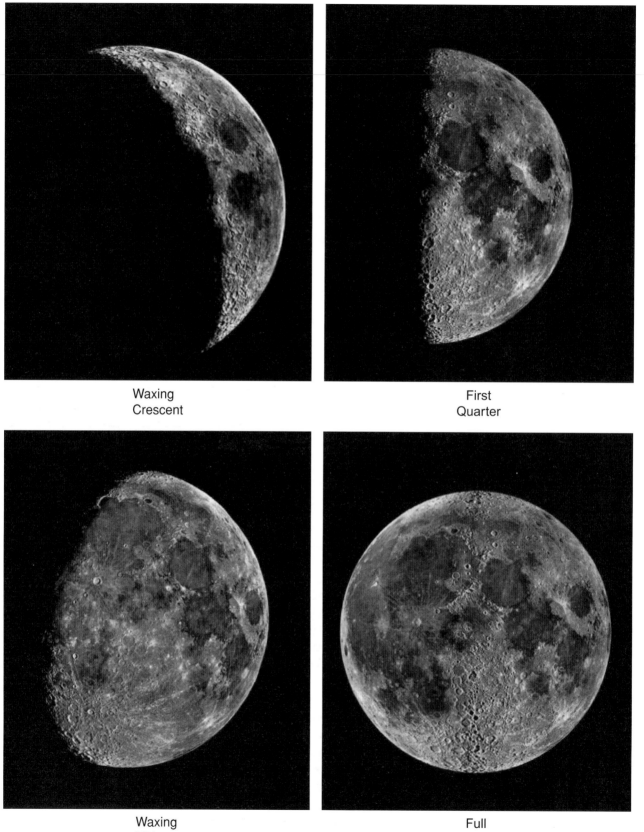

Waxing
Crescent

First
Quarter

Waxing
Gibbous

Full
Moon

Waning
Gibbous

Third
Quarter

Waning
Crescent

Fig. 1.3 The Moon's varying appearance   During the monthly cycle, the illuminated part of the Earth's Moon waxes (grows) from crescent to gibbous, and then after full Moon, it wanes (decreases) to a crescent again. The term crescent is applied to the Moon's shape when it appears less than half-lit; it is called gibbous when it is more than half-lit but not yet fully illuminated. The reason for the Moon's changing shape is described in Fig. 1.8. (Lick Observatory Photographs.)

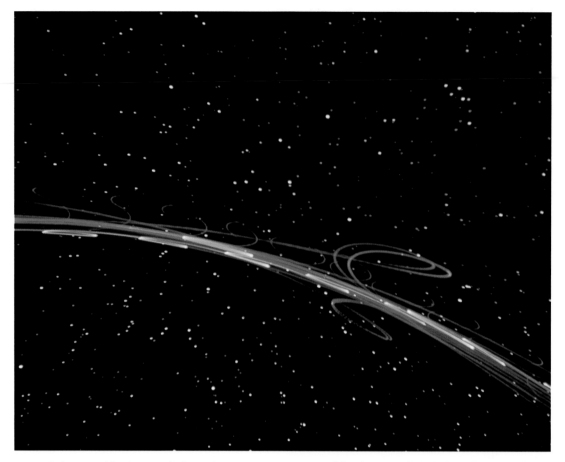

Fig. 1.4 Retrograde loops   This photograph shows the apparent movements of the planets against the background stars. Mars, Jupiter and Saturn appear to stop in their orbits, then reverse direction before continuing on – a phenomenon called retrograde motion by modern astronomers. Ancient and modern explanations for this temporary backward motion are illustrated in Figs. 1.6 and 1.9, respectively. (Courtesy of Erich Lessing/Magnum.)

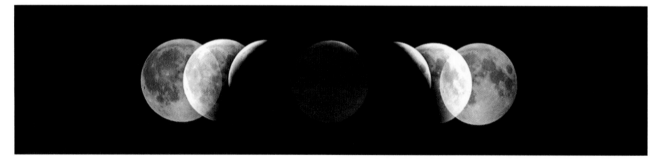

Fig. 1.5 Curved shadow of Earth   This multiple-exposure photograph of a total lunar eclipse reveals the curved shape of the Earth's shadow, regarded by ancient Greek astronomers as evidence that the Earth is a sphere. Only a spherical body will cast the same circular shadow on the Moon when viewed from different locations on Earth or during different lunar eclipses. This photograph was taken by Akira Fujii during the lunar eclipse of 30 December 1982.

As we now know, Aristarchos was right. The stars seem to be revolving about the Earth each night, but appearances can be deceiving. The Earth could instead be spinning beneath the stars. As the Earth rotates, the stars slide by, accounting for the wheeling night sky, which just seems to be revolving.

And the Sun might not be moving across the bright blue sky each day, for the Earth's rotation could produce this motion. Every point on the surface of a spinning Earth can be carried across the line of sight to an unmoving Sun, from sunrise to sunset, producing night and day (Fig. 1.7). Since the Earth rotates from west to east, the Sun appears

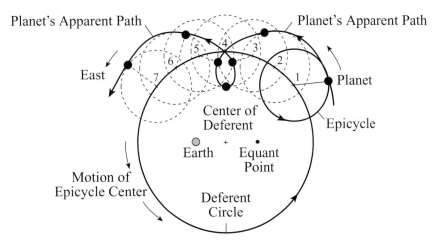

Fig. 1.6 Circles upon circles to explain retrograde loops   To explain the occasional retrograde loops in the apparent motions of Mars, Jupiter and Saturn, astronomers in ancient times imagined that each planet travels with uniform speed around a small circle, known as the epicycle. The epicycle's center moves uniformly on a larger circle, the deferent. A similar scheme was used by Ptolemy (fl. 150 AD) to explain the wayward motions of the planets in his *Almagest*. In the Ptolemaic system, the Earth was displaced from the center of the large circle, and each planet traveled with uniform motion with respect to another imaginary point, the equant, appearing to move with variable speed when viewed from the Earth.

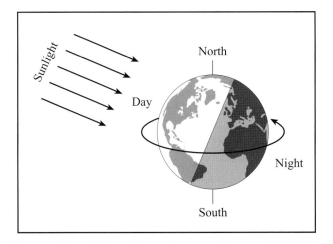

Fig. 1.7 Night and day   The Earth rotates with respect to the Sun once every 24 hours, causing the sequence of night and day. Each point on the Earth's surface moves in a circular track parallel to the equator, and each track spends a different time in the Sun depending on the season. This drawing depicts summer in the northern hemisphere and winter in the southern hemisphere. Because the northern part of the Earth's rotational axis is tipped toward the Sun, circular tracks in the northern hemisphere spend a longer time in the Sun than southern ones.

to rise in the east and set in the west. Such a perspective involves a certain amount of detachment – the ability to separate yourself from the ground and use your mind's eye to look down on the spherical, rotating Earth, like a spinning ball suspended in space.

The Moon's motion from horizon to horizon each night could also be neatly explained by the rotation of the Earth, and the Moon's monthly circuit against the background stars could be ascribed to its slower orbital motion around the Earth. This would also account for the Moon's varying appearance (Fig. 1.8). The Moon borrows its light from the Sun, and the Sun illuminates first one part of the Moon's face and then another as the Moon orbits the Earth. On any given night, all observers on Earth will see the same phase of the Moon as our planet's rotation brings it into view.

The concept of a moving Earth nevertheless seems to violate common sense. The ground certainly seems to be at rest beneath our feet, providing the terra firma on which we carry out our daily lives. As Aristotle noticed, an arrow shot vertically upward falls to the ground where the archer stands, suggesting that the ground has not moved while the arrow was in flight. Moreover, if the Earth is rotating, then its surface regions have to be moving at high speeds (Focus 1.1).

Yet the globe on which we live might not only spin on its axis; it could also be whirling ceaselessly around the Sun, completing one circuit each year as Aristarchos had supposed. But his proposals had little impact on his contemporaries. It took another eighteen centuries before the Polish cleric and astronomer Mikolaj Kopernik (1473–1543), better known as Nicolaus Copernicus, revived this heliocentric, or Sun-centered, model. By 1514 Copernicus was privately circulating a manuscript, the *Commentariolus*, or *Little Commentary*, in which the planets were placed in uniform motion about a central Sun. His longer, more influential book, *De Revolutionibus Orbium Coelestium Libri VI*, or *Six Books Concerning the Revolutions of the Celestial*

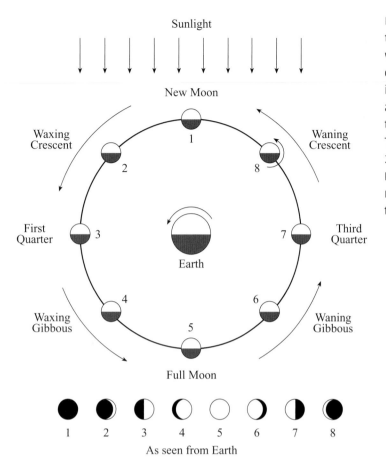

Sunlight

New Moon

Waxing
Crescent

Waning
Crescent

First
Quarter

Earth

Third
Quarter

Waxing
Gibbous

Waning
Gibbous

Full Moon

1   2   3   4   5   6   7   8

As seen from Earth

Fig. 1.8 Phases of the Moon   Light from the Sun illuminates one half of the Moon, while the other half is dark. As the Moon orbits the Earth, we see varying amounts of its illuminated surface. The phases seen by an observer on Earth (*bottom*) correspond to the numbered points along the lunar orbit. The period from new Moon to new Moon is 29.53 days, the length of the month. As the Earth completes its daily rotation, all night-time observers see the same phase of the Moon.

Focus 1.1   Location and rotation speed on the Earth

The length of the day and the rotation period is the same for every place on Earth, but the speed of rotation around its axis depends on the surface location. A grid of great circles on the spherical Earth defines this location. A great circle divides the sphere in half, and the name comes from the fact that no greater circles can be drawn on a sphere. A great circle halfway between the North and South Poles is called the equator, because it is equally distant between both poles. Circles of longitude are great circles that pass around the Earth from pole to pole perpendicular to the equator, with 0 degrees at the Prime Meridian that passes through the Royal Observatory in Greenwich, England. The latitude is the angle measured northward (positive) or southward (negative) along a circle of longitude from the equator to the point.

The surface speed of rotation is greatest at the equator and reduces to almost nothing at the poles. Using an equatorial radius of about 6378 kilometers, which is close to the value inferred long ago (by Eratosthenes about 200 BC), the Earth would have to be rotating at a velocity of about 460 meters per second to spin about its equatorial circumference once every 24 hours. To calculate this speed, just multiply the equatorial radius by 2π to get the equatorial circumference, and divide by 24 hours and 3600 seconds per hour. At higher latitudes, closer to the poles, the circumferential distance around the Earth, and perpendicular to a great circle of longitude, is less, so the speed is less. The speed diminishes to almost nothing at the geographic poles, which are pierced by the rotation axis.

Bodies, was published almost thirty years later, in 1543, the year of its author's death.

For Copernicus, the Sun was located at the heart of the planetary system and the center of the Universe. The only thing to orbit the Earth was the Moon, and the Earth was supposed to rotate on its axis to make the stars swing by. In this model, the Earth was just one of several planets revolving around the Sun, in the same direction but at different distances and with various speeds, always passing each other without ever intersecting. In order of increasing

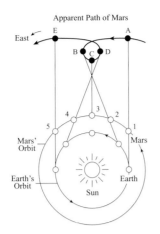

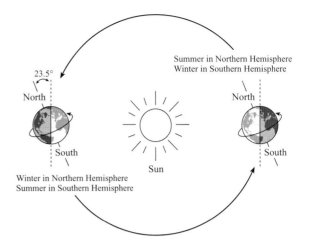

Fig. 1.9 Retrograde loops in a Copernican Universe   A Sun-centered model of the solar system explains the looping path of Mars in terms of the relative speeds of the Earth and Mars. The Earth travels around the Sun more rapidly than Mars does. As Earth overtakes and passes the slower moving planet (*points 2 to 4*), Mars appears to move backward (*points B to D*) for a few months.

Fig. 1.10 The seasons   As the Earth orbits the Sun, the Earth's rotational axis in a given hemisphere is tilted toward or away from the Sun. This variable tilt produces the seasons by changing the angle at which the Sun's rays strike different parts of the Earth's surface. The greatest sunward tilt occurs in the summer when the Sun's rays strike the surface most directly. In the winter, the relevant hemisphere is tilted away from the Sun and the Sun's rays obliquely strike the surface. When it is summer in the northern hemisphere, it is winter in the southern hemisphere and vice versa. (Notice that the radius of the Earth and Sun and the Earth's orbit are not drawn to scale.)

distance from the Sun, they are Mercury, Venus, Earth, Mars, Jupiter and Saturn. As Copernicus noticed, the further a planet is from the Sun, the longer it takes to complete a circuit.

There was no definite proof of this heliocentric hypothesis; but it did provide natural explanations for observed phenomena. Venus and Mercury, for example, are never to be seen far from the Sun. They rise and set with the Sun, unlike Mars, Jupiter and Saturn, which can be seen at any time of night. Since the orbits of Venus and Mercury lie inside that of Earth and closer to the Sun, these planets are only seen around dawn or dusk. In contrast, the orbits of Mars, Jupiter and Saturn lie outside that of the Earth, so they are visible throughout the night.

The Sun-centered view also provides a simple explanation of the retrograde motions that were so hard to reproduce using an Earth-centered, or geocentric, model. The jerky backwards motions were attributed to the uniform motion of the Earth and other planets at different speeds around the Sun. Planets moving at a slower speed than the Earth would sometimes appear to move ahead of Earth, and sometimes fall behind (Fig. 1.9).

Most of the time we see Mars, Jupiter and Saturn moving around the Sun in the same direction as the Earth, but during the relatively short time that the Earth overtakes one of these planets, that planet appears to be moving backward (Fig. 1.9). Moreover, one could confidently predict when a planet's apparent motion would come to a halt and turn around, and for how long it would seem to move backwards.

We now realize that the tilt of the Earth's rotational axis and the annual orbit of the Earth cause sunlight to fall

differently on our planet at different times of year, explaining the seasons (Fig. 1.10). As the Earth orbits the Sun, its rotational axis remains pointed in the same direction in space, toward the North Star Polaris, so the orientation of the rotation axis in space remains unchanged throughout the year. But the orientation of the axis toward the Sun changes over the course of each orbit when the northern and southern hemispheres are tilted toward or away from the Sun by up to 23.5 degrees. The greatest sunward tilt in a given hemisphere occurs in summer when the Sun is more nearly overhead and its rays strike the surface more directly. Winter occurs half an orbit later, when that hemisphere is at its greatest tilt away from the Sun. Notice that the seasons are caused by the change in tilt, toward or away from the Sun, and not by any noticeable change in distance from the Sun.

The semi-annual alteration in incident sunlight is less pronounced in the equatorial regions, where the seasonal weather changes are not as great as they are at higher latitudes. People living near the equator have rainy and dry seasons, with the rains coming when the Sun is higher in the sky.

Copernicus' goal was to provide a geometric model that could replicate the planetary motions, but transforming their center to the Sun did not by itself improve the predictions. Proof of his Sun-centered model required improved

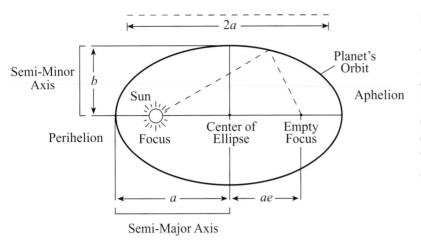

Fig. 1.11 Elliptical orbit   Each planet moves in an ellipse with the Sun at one focus. The length of a line drawn from the Sun, to a planet and then to the empty focus, denoted by the dashed line, is always 2*a*, or twice the semi-major axis, *a*. The eccentricity, or elongation, of the planetary ellipse has been greatly overdone in this figure; planetary orbits look much more like a circle.

observations and the introduction of non-circular motions. Yet Copernicus' book did become a symbol for a new perspective of the heavens, a view that was ultimately to unite the Earth and planets in the domain of terrestrial physics. He opened the way to the study of not only how the celestial bodies move, but to an investigation of the forces that propel them and the underlying laws that govern their motion.

The harmony of the world

In the hope of developing a more precise description of planetary motions, the Danish astronomer Tycho Brahe (1546–1601) built, with royal patronage, an observatory, Uraniborg, on the island of Hven, where he amassed a great number of observations that were more accurate and complete than any previous ones, including detailed records of the orbit of Mars. This was before the days of telescopes, and he used ingenious measuring instruments that resembled large gun sights with graduated circles. Johannes Kepler (1571–1630), Tycho's assistant and successor after his death, eventually interpreted these data and was able to determine precise mathematical laws from them.

Since circular motions could not describe Tycho's accurate observations, Kepler concluded that non-circular shapes were required. In 1605, after four years of computations, Kepler found that the observed planetary orbits could be described by ellipses, or ovals, with the Sun at one focus (Fig. 1.11; Focus 1.2). This ultimately became known as Kepler's first law of planetary motion. A planet also speeds up when it approaches the Sun, and slows down when it moves away from the Sun, and that accounts for a planet's varying speed when observed from Earth; this is described when the modern concept of conservation of angular momentum is applied to elliptical orbits (Focus 1.2).

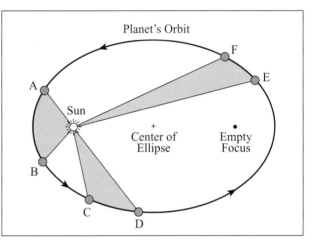

Fig. 1.12 Kepler's first and second laws   Kepler's first law states that the orbit of a planet about the Sun is an ellipse with the Sun at one focus. The other focus of the ellipse is empty. According to Kepler's second law, the line joining a planet to the Sun sweeps out equal areas in equal times. This is also known as the law of equal areas. It is represented by the equality of the three shaded areas ABS, CDS and EFS. It takes as long to travel from A to B as from C to D and from E to F. A planet moves most rapidly when it is nearest the Sun (at perihelion); a planet's slowest motion occurs when it is farthest from the Sun (at aphelion).

Kepler was able to describe a planet's changing speed along its orbit in a precise mathematical form that can be explained with the help of Fig. 1.12. Imagine a line drawn from the Sun to a planet. As the planet swings about its elliptical path, the line (which will increase and decrease in length) sweeps out a surface at a constant rate. This is Kepler's second law of planetary motion, also known as the law of equal areas. During the three equal time intervals shown in Fig. 1.12, the planet moves through different arcs because its orbital speed changes, but the areas swept out are equal.

Focus 1.2   Elliptical planetary orbits

According to Kepler's first law, the planets move in elliptical orbits with the Sun at one focus. The planet's closest point to the Sun, when the planet moves most rapidly, is called the perihelion; and its most distant point is the aphelion, where the planet moves most slowly. The distance between the perihelion and aphelion is the major axis of the orbital ellipse. Half that distance is called the semi-major axis, designated by the symbol a.

The semi-major axis of the Earth's elliptical orbit about the Sun is called the astronomical unit, abbreviated AU. It sets the scale of the solar system, and when combined with the Earth's year-long orbital period permitted the determination of the Sun's mass and the Earth's orbital velocity, but only after astronomers had found out how large an AU is.

The shape of an ellipse is determined by its eccentricity, e. If $e = 0$ its shape is a circle. The ellipse becomes more elongated and squashed as its eccentricity increases toward $e = 1.0$. The eccentricity of the planetary ellipse has been greatly exaggerated in Fig. 1.11, with an eccentricity of about $e = 0.5$.

At perihelion the distance between the planet and the Sun is $a(1 - e)$ and at aphelion that distance is $a(1 + e)$. With the exception of Mercury, all of the major planets have orbits that are nearly circular, with eccentricities of less than $e = 0.1$. This means that the Sun is very near the center of each orbital ellipse. For Mercury, $a = 0.387$ AU and $e = 0.206$, so its distance from the Sun is just 0.307 AU at perihelion and quite a lot greater at aphelion, located at 0.467 AU.

One of the great laws of physics, known as the conservation of angular momentum, explains why a planet keeps on whirling around the Sun, and why its speed is fastest at perihelion. For a planet of mass M, orbiting the Sun with speed or velocity V at a distance D:

$$\text{Angular momentum} = M \times V \times D$$

By the way, in physics velocity has both a magnitude and a direction, and speed is the magnitude of the velocity. In astronomy the velocity is often just given by its magnitude, the speed, so the orbital velocity is its speed along the orbit.

The conservation law says that as long as no outside force is acting on the planet, its angular momentum cannot change. So the planet just keeps on moving along without anything pushing or pulling it. The mass does not change, so when the distance from the Sun decreases, at perihelion, the velocity increases to compensate and keep the angular momentum unchanged; at aphelion the distance increases so the speed has to decrease.

Kepler labored another decade before publication of *Harmonice mundi*, or *Harmony of the World*, in 1619, where he claimed to have listened to, and described mathematically, the music of the heavenly spheres. Kepler investigated arithmetic patterns between the periods and sizes of the planetary orbits, discovering the harmonic relation that is now known as Kepler's third law. It states that the squares of the planetary periods are in proportion to the cubes of their average distances from the Sun.

These periods and distances are given in Table 1.1 with other mean orbital parameters of the major planets. Here the periods are given in units of the Earth's orbital period of one Earth year, and the distances from the Sun are specified in units of the Earth's mean distance from the Sun, the astronomical unit (AU). One Earth year is equivalent to 31.557 million (3.1557×10^7) seconds; a precise measurement for the length of 1 AU, which is equal to 149.598 million ($1.495\,98 \times 10^8$) kilometers, took centuries to determine. The orbital velocities are given in kilometers per second (km s^{-1}).

If P_P denotes the orbital period of a planet measured in Earth years, and a_P describes its semi-major axis (Focus 1.2) measured in AU, then Kepler's third law states that $P_P^2 = a_P^3$, where the subscript "P" denotes the planet under consideration. This expression is illustrated in Fig. 1.13 for the major planets and for the brighter moons of Jupiter. The mean orbital velocity of each planet is proportional to the ratio a_P/P_P, so the velocity varies inversely with the square root of the distance, or as $(a_P)^{-1/2}$.

In other words, the more distant planets have longer orbital periods and they move around the Sun with a slower speed. For example, Jupiter is 5.2 times as far away from the Sun as the Earth is, and it takes Jupiter 11.86 Earth years to travel once around the Sun. The Earth's mean orbital velocity is nearly 30 kilometers per second, while Jupiter's orbital velocity is about half that amount. Both planets are whirling around the Sun with awesome speed.

1.2 Telescopes reveal the hitherto unseen

Astronomers use telescopes to extend their vision to otherwise invisible cosmic objects. Without a telescope, for example, you can't see Neptune, or any of the numerous asteroids and planetary satellites – other than the Earth's

Table 1.1 Mean orbital parameters of the major planets[a]

Planet	Semi-major axis, a_P (AU)	Orbital period, P_P (years)	Eccentricity, e	Inclination to the ecliptic, i (degrees)	Orbital velocity (km s^{-1})
Mercury	0.387 099	0.2409	0.2056	7.00	47.890
Venus	0.723 332	0.6152	0.0068	3.39	35.030
Earth	1.000 000	1.0000	0.0167	0.01	29.790
Mars	1.523 688	1.8809	0.0933	1.85	24.130
Jupiter	5.202 834	11.8622	0.048	1.31	13.060
Saturn	9.538 762	29.4577	0.056	2.49	9.640
Uranus	19.191 391	84.0139	0.046	0.77	6.810
Neptune	30.061 069	164.793	0.010	1.77	5.430

[a] The dashed line divides the six planets known in Kepler's time from the two major outer planets discovered later.

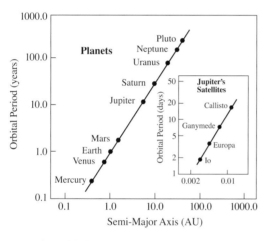

Fig. 1.13 Kepler's third law　The orbital periods of the planets are plotted against their semi-major axes, using a logarithmic scale. The straight line that connects the points has a slope of 3/2, thereby verifying Kepler's third law that states that the squares of the orbital periods increase with the cubes of the planetary distances. This type of relation applies to any set of bodies in elliptical orbits about a much larger mass, including Jupiter's four largest satellites shown in the inset.

Moon. When we step outside to look at the dark night sky, our eyes can detect only about 3000 of the 100 billion stars in the Milky Way, and billions of galaxies similarly remain invisible.

The historical record indicates that the known cosmos which can be observed at any given time is just a modest part of a much vaster one that remains to be found. Moreover, several of the more significant astronomical discoveries have been unanticipated and serendipitous, made when new telescopes were used to study known cosmic objects (Focus 1.3).

Focus 1.3　Serendipitous astronomy

The unanticipated discovery of hitherto unknown cosmic objects began with the four large satellites of Jupiter, which were found when Galileo was observing the nearby full Moon. Serendipity continued with the discovery of Uranus, found while preparing to measure the distances of stars; the first asteroid Ceres, discovered when determining accurate positions of the stars; the high velocities of galaxies in the expanding Universe, discovered while measuring the rotations of spiral nebulae thought to be newborn stars with nascent planetary systems; and the energetic radio Universe, found while measuring interference with terrestrial radio communications.

More recently, the Nobel Prize in Physics has been awarded for four serendipitous astronomical discoveries, including cosmic X-ray sources, discovered while studying solar X-rays reflected from the Moon; the three-degree cosmic microwave background radiation, found when measuring noise sources in a microwave detection system; the discovery of pulsars while observing the twinkling of cosmic radio sources caused by the Sun's winds; and the indirect detection of gravitational radiation when carrying out a search for new pulsars. Although outside the general solar-system theme of this book, these discoveries demonstrate the instrument-driven, accidental nature of astronomy, in which significant discoveries result from new technology and novel telescopes, enabling us to "see" the invisible and permitting us to look at the Universe in new ways.

The unanticipated discovery of Jupiter's four large moons

One of the most fascinating and lively books in astronomy, *Sidereus Nuncius* or *Starry Messenger*, was published in 1610. In it, Galileo Galilei (1564–1642) described how he turned the newly devised spyglass, which shows faraway things as though nearby, toward the heavens, bringing the sky down to Earth and the Earth into the sky.

After presenting the Venetian Doge with a spyglass, or telescope, as a valuable tool of war, Galileo turned a telescope of his own making toward our Moon on 25 August 1609, discovering craters, rugged mountains and valleys that you need a telescope to see. At least one cosmic object, the Moon, was no longer the polished, smooth and perfectly spherical body imagined by the ancients. Even the Sun was found to be spotty and impure under telescopic scrutiny by Galileo, Thomas Harriot (1560–1621), and other pioneering observers.

Although few of Galileo's many telescopes survive, he probably used one with an objective lens of about 0.05 meters (2 inches) in diameter to make his startling discoveries. The collecting area of such a lens is roughly fifty times that of the pupil of the unaided eye, which is about 0.007 meters across. The increase in light-collecting power of even this small telescope enabled Galileo to resolve objects that remain blurred to the unaided eye, detecting previously unseen craters on the Moon, viewing thousands of stars in the Milky Way that had never been seen before, and detecting the four large moons of Jupiter.

When directing his rudimentary telescope at the nearly full Moon on 7 January 1610, Galileo must have naturally moved his spyglass just a little to look at Jupiter, which was then located just above the Moon and was also the next brightest object in the sky. As reported in the *Starry Messenger*, Galileo found four companions lined up on each side of Jupiter, accompanying the planet and orbiting it at different distances (Fig. 1.14). The smaller their orbit, the faster their orbital speed and the shorter their orbital period.

Any object that orbits a planet is now called a satellite, and a natural satellite is also now called a moon. We designate the Earth's Moon, or our Moon, with a capital M to distinguish it from all the other planetary moons. The four large moons that Galileo discovered are now often called the "Galilean satellites" in his honor. Galileo named them the "Medicean stars" after his patron Cosmo II de' Medici, but they now go by the names of four of Jupiter's lovers. Io is the innermost of the four Galilean satellites, succeeded by Europa, Ganymede, and Callisto.

The discovery of these satellites was most likely an unexpected result of Jupiter's proximity to the Moon when

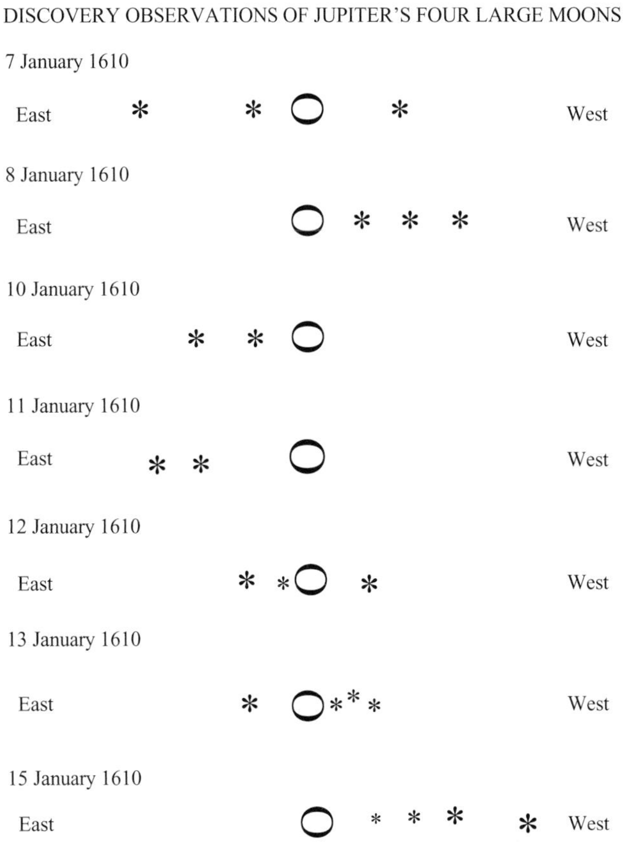

DISCOVERY OBSERVATIONS OF JUPITER'S FOUR LARGE MOONS

Fig. 1.14 Moons of Jupiter   Some of Galileo Galilei's (1564-1642) observations of the "Medicean stars", lined up on each side of Jupiter's disk and changing position while accompanying Jupiter. (Adapted from Galilei, Galileo: *Sidereus Nuncius*, or *The Sidereal Messenger*, 1610. Translation by Albert Helden with Introduction, Conclusion and Notes. The University of Chicago Press, Chicago, 1989.)

Galileo happened to be observing it with his spyglass. No one predicted the possible existence of moons orbiting any other object than the Earth, and Galileo's important discovery of more than one center of motion in the Universe contradicted the geocentric Ptolemaic system in which all astronomical objects move around the central Earth.

Galileo's *Dialogo Massimi Sistemi Del Mondo, Tolemaico e Copernicano*, or *Dialogue of the Two Great World Systems, Ptolemaic and Copernican*, published in 1632, demonstrated the advantages of the Sun-centered Copernican cosmology, and provided telescopic evidence in its favor, including his discovery that Venus exhibits phases like our Moon. If nearby Venus orbited the Earth inside the Sun's orbit, then it could never appear completely illuminated, but Venus could appear from Earth in all its phases if it orbited the Sun.

Galileo's adoption of Copernicus' theory, in which the Earth moves around the Sun, was nevertheless opposed by theologians of the time, since a strict interpretation of the Bible (Psalm 104) indicated that "God fixed the Earth on its foundation, so it will never be moved". After trial by Inquisition in 1633, the Roman Catholic Church forced Galileo to recant his support of the Copernican system as "abjured, cursed and detested". He was banished to confinement at his house in Arcetri, in the hills surrounding Firenze, where he spent his last years. Legend has it that as Galileo rose from kneeling before his inquisitors, he murmured, "*e pu, si muove*" – "even so, it does move", but he would hardly have been foolish enough to risk even greater punishment. Not until 1992, more than 350 years after his trial, did Pope John Paul II in effect apologize for the harshness of Galileo's sentence.

The serendipitous discovery of Uranus

The first planet to be discovered since the dawn of history was found accidentally, by a professional musician and self-taught amateur astronomer, William Herschel (1738–1822), who moved from Germany to England in 1750 and became the organist in the town of Bath. He was both an excellent observer and a skilled mirror-maker, constructing telescopes with then unsurpassed light-collecting ability. Since they could not build such a fine instrument and obtain similar results, other astronomers were initially skeptical of Herschel's observations, but he simply replied that they would have to learn to see.

In 1781 Herschel was using one of his unique telescopes, with a metal mirror of 0.15-meters (6.2-inches) in diameter and 2.1-meters (7-feet) focal length, to examine all the brighter stars, of less than 8th magnitude, for faint companions. (In the peculiar magnitude system adopted by early astronomers, the brighter stars have smaller magnitudes.) He was hoping to determine the parallax, or distance, of the bright star from its changing position relative to the faint, adjacent one.

On 13 March 1781, Herschel unexpectedly found an uncommonly bright object, of 6th magnitude, that had a well-defined disk, unlike stars, and moved slowly from one night to another against the background stars. It was suspected by Herschel to be a comet without a tail. After several months of observations by Herschel and others, the moving object was recognized as a new planet, named Uranus, which orbits the Sun at about twice Saturn's orbital distance. Herschel became world-famous almost overnight. He was eventually appointed the King's Astronomer with a pension, which permitted him to give up music as a career and devote full time to astronomy.

Herschel proposed that the new planet be named the "Georgian Planet" in honor of King George III, England's reigning monarch and a patron of the sciences. After some controversy, the new planet was instead named Uranus, after the Greek personification of the sky. One consequence of this naming was that a newly discovered, heavy element was designated uranium in honor of the discovery of a new world.

When he found Uranus, Herschel was apparently unaware of a numerical sequence that predicted its relative distance from the Sun. Known as the Titius–Bode law, after the last names of the first persons to state it, the sequence describes the regular spacing of the planets, suggesting that the next planet beyond Saturn would be located at 19.6 AU, or at about twice Saturn's distance (Focus 1.4). The so-called "law" also indicated a missing planet at 2.8 AU, in the gap between Mars and Jupiter. Encouraged by the discovery of Uranus, astronomers began a search

---

Focus 1.4   The Titius–Bode law

In the inner solar system, each planet's orbit is about 1.5 times the distance of its inward neighbor, and this ratio increases to roughly a factor of 2.0 in the outer solar system. This relative spacing of the planets is described by the Titius–Bode law, first noted in 1766 by Johann Daniel Titius (1729–1796), and brought to prominence by Johann Elert Bode (1749–1826) in a 1772 edition of his popular book on astronomy.

The law states that the relative distances of the planets from the Sun can be approximated by taking the sequence 0, 3, 6, 12, 24, ..., adding 4, and dividing by 10. Mathematically, the semi-major axis, a_n, of the nth planet, in order of increasing distance from the Sun, is given by:

$$a_1 = 0.4 \text{ for } n = 1$$
$$a_n = 0.1 \times [4 + 3 \times 2^{n-2}] \text{ for } n = 2, 3, \ldots, 9$$

where a_n is the relative distance compared with that of the Earth.

A comparison of the observed semi-major axes, a_P, of the planets with the distance predicted by this law is given in Table 1.2. The Titius–Bode law predates the discovery of Uranus at $n = 8$ by 15 years, the discovery of the first asteroid at $n = 5$ by 35 years, and the discovery of Neptune at $n = 9$ by 80 years. Although there is no well-accepted explanation for why this expression works so well, it probably has something to do with the dynamics, evolution or origin of the solar system.

Table 1.2 Comparison of measured planetary distances from the Sun with those predicted from the Titius–Bode law

Planet	n	Measured a_P (AU)	Predicted a_n (AU)
Mercury	1	0.387	$0.4 = (0 + 4)/10$
Venus	2	0.723	$0.7 = (3 + 4)/10$
Earth	3	1.000	$1.0 = (6 + 4)/10$
Mars	4	1.524	$1.6 = (12 + 4)/10$
Ceres (asteroid)	5	2.767	$2.8 = (24 + 4)/10$
Jupiter	6	5.203	$5.2 = (48 + 4)/10$
Saturn	7	9.537	$10.0 = (96 + 4)/10$
Uranus	8	19.19	$19.6 = (192 + 4)/10$
Neptune	9	30.07	$38.8 = (384 + 4)/10$

for the unknown planet that ought to be located at the predicted distance of 2.8 AU from the Sun. The first object to be found in this location was nevertheless discovered quite unexpectedly by the Sicilian monk and astronomer Giuseppe Piazzi (1746–1826) while he was preparing a catalog of accurate star positions.

The unexpected discovery of the first asteroid, Ceres

In the late 18th century, Giuseppe Piazzi (1746–1826) was compiling a catalog of the accurate positions of stars in the sky, using a finely calibrated telescope built by the celebrated instrument maker, Jesse Ramsden (1735–1800) of London, and installed in Piazzi's observatory at Palermo, Sicily (Fig. 1.15). On 1 January 1801, Piazzi unexpectedly discovered a new "star" of 8th magnitude that changed position from night to night. It was moving at a slow uniform rate against the background stars, and was thought to possibly be a comet without a nebulosity or tail.

Piazzi observed the new object for six weeks until it moved too close to the Sun to be observed, and when it returned to dark skies it could not be located. Hearing of the lost object, Carl Friedrich Gauss (1777–1855) developed a method of establishing an orbit from just a few observations, and used Piazzi's observations to predict a location. It was found a year after it had been first sighted, independently by Baron Franz Xaver von Zach (1754–1832) and Heinrich Wilhelm Olbers (1758–1840), close to Gauss's estimated position. A hitherto unknown, small solar-system body, named Ceres Ferdinandea by Piazzi, had been found, with an orbit between those of Mars and Jupiter. A few months later Olbers found another tiny

object, named Pallas, orbiting the Sun at about the same distance as Ceres, and William Herschel named the two objects "asteroid" because they could not be resolved into disks and appeared to be "star-like" points of light.

The asteroids remained unresolved because they are very small and relatively nearby, rather than very large and distant like the stars. Even the largest asteroid, Ceres, has a radius of 476 kilometers, which is less than a third of the radius of the Moon, less than a tenth the radius of the Earth, and less than one hundredth the radius of Jupiter.

No other asteroids were identified for 38 years, but the hunt for new ones became something of an astronomical sport in the last half of the 19th century. More than 300 asteroids had been discovered by 1891, and the pace of discovery subsequently increased by using long-exposure photographs of several hours to detect the motions of asteroids against the stars.

Each asteroid is given a number corresponding to its chronological place in the discovery list, and a name that is usually provided by the discoverer, but they do not receive official numbers until their orbits are reliably known. For instance, 433 Eros was the 433rd asteroid to be discovered with a reliable orbit. The list of known asteroids reached the 2000 mark in 1977, and there were 50 000 known in the early 21st century. Astronomers estimate that there may be as many as half a million (500 000) faint asteroids smaller than one kilometer across, many with orbits that have now been determined.

Yet, despite their vast numbers, the combined mass of all the asteroids is estimated to be no more than 10 percent of that of the Earth's Moon, and nowhere near the mass of a single large planet.

Most of the asteroids with well-determined orbits lie in a great asteroid belt between the orbits of Mars and Jupiter (Fig. 1.16), at distances from the Sun of 2.2 to 3.3 AU and with orbital periods of 3 to 6 Earth years. The asteroids are so little, and distributed across such a large range of distances, that the asteroid belt is largely empty space. This leaves plenty of room for spacecraft to pass through to the giant planets, undamaged by collision with any asteroid.

Refractors and reflectors

Galileo's use of the telescope to extend the human senses marked the beginning of a new age in astronomy – an age in which telescopes are used to view objects hitherto unseen and unknown, and to scrutinize known ones in greater detail. Telescopes extend our vision by collecting enough light to detect intrinsically faint and otherwise invisible sources or to resolve bright sources whose individual features are too near to each other to separate with the unaided eye. This era continues today, as we build new

Fig. 1.15 The Palermo circle This instrument, built by Jesse Ramsden (1735–1800) of London, was used by Giuseppe Piazzi (1746–1826) to obtain precise measurements of stellar positions, with an accuracy of a few seconds of arc after observing each star for at least two nights. In the process, Piazzi unexpectedly discovered the first asteroid, 1 Ceres, in 1801. The telescope has a 7.5-cm objective lens. The altitude scale (5 feet in diameter) was read with the aid of two diametrically opposed micrometer microscopes; the azimuth scale (3 feet in diameter) was also read by means of a micrometer microscope.

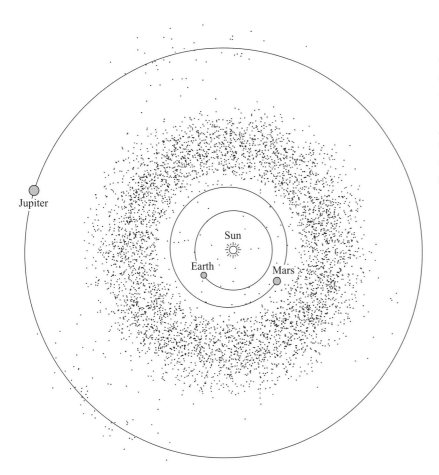

Fig. 1.16 Asteroid belt The exact locations of 5000 flying rocks, called asteroids or minor planets, whose orbits are accurately known. The vast majority of the asteroids orbit the Sun in the main belt located between the orbits of Mars and Jupiter. A few of them pass inside the orbit of Earth, while others move about 60 degrees ahead of and behind Jupiter in similar orbits. (Courtesy of Jeff Bytof, University of California at San Diego.)

telescopes on the ground and in space, to discover new worlds and to investigate familiar ones in different ways, and use spacecraft to carry instruments for close-up views of the planets and satellites, revealing unanticipated features that are far beyond the range of human vision with even the best telescope on Earth or in orbit around it.

There are two kinds of telescopes that are used to collect and focus radiation visible to the human eye. They are the refractor, used by Galileo, and the reflector, used by William Herschel. As the names suggest, the refractor uses a lens to gather, bend and focus light, employing the principle of refraction, while the reflector uses a curved mirror to collect, reflect and focus light (Fig. 1.17). Since the science of optics is used to describe the refraction or reflection of light rays, both kinds of telescopes are known as optical telescopes. They are used to carry out visible-light optical astronomy. Radio astronomers and X-ray astronomers use different kinds of telescopes that detect invisible radiation.

The first telescopes, such as the one Galileo used in 1610, were refractors using combinations of glass lenses to magnify and focus light. They were long, slender tubes with a convex objective lens at the far end that focused light on a second, smaller lens, termed the eyepiece, a few feet away. The eyepiece was used to magnify or enlarge the image before being observed, permitting more detail to be seen.

Galileo used an objective lens just one or two inches in diameter, and a smaller, concave eyepiece about 30 or 40 inches away, giving an erect image but a very narrow field of view. His telescopes had a magnification of about 20 and a field of view of about a quarter of a degree. Kepler introduced a convex eyepiece behind the focal plane, which widened the viewing angle and inverted the image. Many subsequent astronomers became accustomed to viewing this upside-down world.

The distance from the objective lens to the focal plane, called the focal length, determines the overall size of an image – the greater the focal length, the bigger the image. The diameter of the objective lens is called the aperture, and the focal ratio of the lens is the focal length divided by the aperture. The magnification, or power, of such a telescope is equal to the ratio of the focal lengths of the objective and the eyepiece.

Because the glass objective lens in a refractor could not be shaped into an accurate curve, these early telescopes yielded blurry images, distorted by spherical aberration. Moreover, because light is composed of different colors or wavelengths, the imperfect glass lens bent or refracted

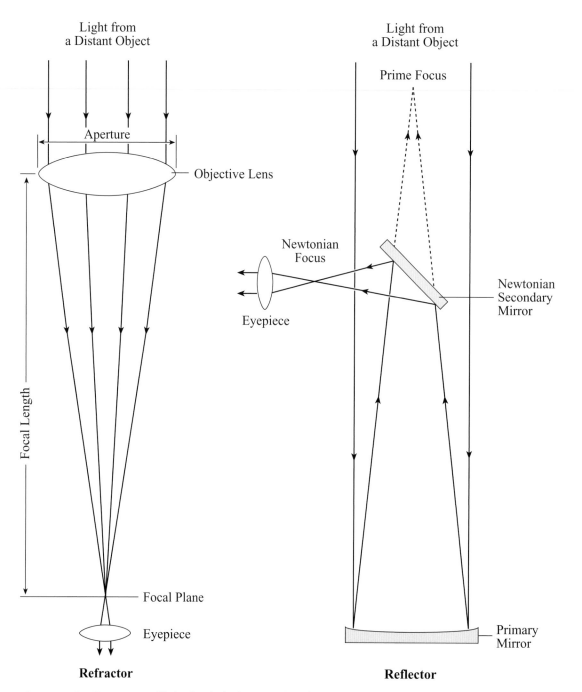

Fig. 1.17 Refractor and reflector   Two kinds of optical telescopes, the refractor and the reflector, are used to gather and concentrate light at visible wavelengths. Light waves that fall on the Earth from a distant object are parallel to one another, and are focused to a point by the lens or mirror of a telescope. The earliest telescopes were refractors (*left*). The curved surfaces of the convex objective lens bend the incoming parallel light rays by refraction, and bring them to a focus at the center of the focal plane, where the light rays meet and an image is created. A second, smaller lens, called the eyepiece, was used to magnify the image in the early refractors; later versions placed photographic or electronic detectors at the focal plane. The reflecting telescope (*right*) uses a large, concave, or parabolic, primary mirror to collect and focus light. A small, flat secondary mirror, inclined at an angle of 45 degrees to the telescope axis, reflects the light sideways, at a place now known as the Newtonian focus. Other light-deflecting mirror arrangements can be used to obtain any desired focal length, which varies with the curvature and position of small convex mirrors.

each wavelength through a slightly different angle, focusing light of different colors differently and resulting in chromatic aberration.

Such blurring was avoided in a reflecting telescope that uses mirrors instead of lenses. A curved primary mirror gathers the parallel rays of light entering the open end of a telescope and focuses them to a point (Fig. 1.17). Light does not pass though a mirror, as it does through a lens, and the mirror concentrates light of all colors to the same focus, producing a sharp image.

In 1668 Isaac Newton (1642–1727), born on Christmas day the year of Galileo's death, placed a second, flat mirror, angled at 45 degrees, just before the focal point of the primary mirror, to direct the light to the side where an eyepiece was located. This Newtonian focus remains popular for many amateur astronomers because of its elegant simplicity, but other light-deflecting mirror arrangements are used with the biggest optical telescopes to obtain any desired focal length.

The critical parameter of a telescope is the diameter of the light-gathering lens or mirror. The larger the diameter, the more light is gathered, the brighter will be the image, and the fainter the objects that can be seen or recorded. The amount of light that can be gathered is proportional to the area of the lens or mirror, and consequently to the square of its diameter, so a small change in diameter can make a large change in light-collecting area.

Bigger telescopes also provide better angular resolution, which is the ability to detect the separation between things that are close together. This ability to discriminate fine details is called the resolving power of a telescope, and it depends on the diameter of the light-gathering lens or mirror and the wavelength of observation (Focus 1.5). At a given wavelength, a bigger objective lens or primary mirror provides better angular resolution.

The angular resolution of ground-based optical telescopes operating at visible wavelengths is limited by turbulence in the Earth's atmosphere. Similar variations cause the stars to twinkle at night. This atmospheric limitation to angular resolution is called "seeing". The best seeing, of about 0.2 seconds of arc in unusual conditions, is found only at a few high-altitude sites in the world, and observatories are located at most of them. Better visible images with even finer detail can be obtained from the unique vantage point of outer space, using satellite-borne telescopes unencumbered by our atmosphere.

Satellites and rings

In order to get greater magnification, and at the same time to reduce distortion of images and the colored halo around them, or to reduce spherical and chromatic aberrations,

---

Focus 1.5   Angular resolution

The angular resolution, θ, of a telescope is determined by the diameter, D, of the objective lens or primary mirror, as well as the wavelength, λ, of observation. The mathematical expression is:

$$\text{Angular resolution } \theta = \frac{\lambda}{D} \text{ radians}$$

$$= 2.063 \times 10^5 \frac{\lambda}{D} \text{ seconds of arc}$$

where one radian is equivalent to 206 265 ($2.062\,65 \times 10^5$) seconds of arc. This equation tells us that a bigger lens or mirror provides finer angular resolution at a given wavelength. The resolving power of a telescope operating at the wavelengths that we detect with our eye is about $0.13/D$ seconds of arc if D is in meters.

Atmospheric effects limit the resolution of any telescope operating at visible wavelengths to about one second of arc, so you cannot improve the angular resolution by building an optical telescope bigger than about 0.13 meters in diameter. Nevertheless, a bigger telescope still gathers more light than a smaller one, permitting the detection of fainter sources. If a large telescope is placed in space, above our distorting atmosphere, greater angular resolution can also be achieved.

The equation applies at radio wavelengths where very big telescopes are required to achieve significant angular resolution. At a radio wavelength of 0.1 meters, an angular resolution of 1 second of arc requires a telescope with a diameter of 20 kilometers. The advantage of radio signals is that the atmosphere does not distort them, or limit the angular resolution. We can observe cosmic radio sources on a cloudy day, just as your home radio or cell phone work even when it rains or snows outside.

---

objective lenses of great focal length were necessary. As a result, early refractor telescopes became longer and longer, with an objective lens placed on a tower or tall pole and a separate eyepiece near the ground. This novel arrangement permitted the use of an objective lens with slight curvature and long focal length to help correct for aberration and bring the image into sharp focus.

Such "aerial" refractors enabled the discovery of new planetary satellites and the rings of Saturn. In 1655 the Dutch astronomer Christiaan Huygens (1629–1695), for example, discovered Titan, the first known and largest satellite of Saturn, named after Saturn's older brother. Huygen's telescope had a 0.05-meter (2-inch) objective lens with a focal length of 7 meters (23 feet), connected to an eyepiece by just a string for alignment.

Within a few decades the Italian astronomer Giovanni Domenico Cassini (1625–1712), working at the Paris Observatory, had used an aerial refractor to discover four more moons circling Saturn; they are named Iapetus, Rhea, Tethys and Dione. Like the Earth's Moon, Saturn's second-largest moon, Iapetus, always presents the same face to its planet. According to Greek mythology, Gaia (Earth) gave birth to Uranus (Heaven) without the aid of any male, and coupled with her son to conceive six male Titans, including Iapetus, and six female Titanesses, including Rhea and Tethys.

During the next three centuries, the discovery of planetary satellites progressed more or less in tandem with the development of increasingly large and more powerful reflecting telescopes, whose greater light-gathering capability permitted the detection of smaller satellites that reflected less sunlight. William Herschel, the discoverer of Uranus, used his 0.15-meter (6.2-inch) reflector to identify four new moons, two each of Uranus (Oberon and Titania in 1787) and Saturn (Mimas and Enceladus in 1789). It wasn't until 1851 that William Lassell (1799–1880) found two more Uranian satellites, Ariel and Umbriel, using a 0.61-meter (24-inch) reflector, and a fifth moon, Miranda, wasn't found until nearly a century later – in 1948 by Gerard P. Kuiper (1905–1973) at the McDonald Observatory in Texas using a 2.1-meter (82-inch) reflector.

Saturn's satellite Mimas has the name of one of the giants who fought against the gods in Greek mythology. The ringed planet's moon Enceladus is named for the giant who was crushed in a battle between the Olympian gods and the Titans. Earth that was piled on top of him became the island of Sicily. Two other large satellites of Saturn, discovered in the 19th century, are named after Hyperion, a Titan, and Phoebe, a Titaness.

The five large satellites of Uranus are named for characters in literature. Oberon and Titania are the king and queen of the fairies in Shakespeare's *A Midsummer Night's Dream*. Inside their orbits is Umbriel, a "dusky, melancholy sprite" in Alexander Pope's *Rape of a Lock*. Close to the planet is Ariel, described by Shakespeare as "an airy spirit" in *The Tempest*. Closer yet is Miranda, named for Prospero's daughter in *The Tempest*.

Lassell found Neptune's largest satellite in 1846, just a few weeks after the discovery of the planet. The satellite was named Triton – a sea god in Greek mythology, the son of Poseidon, the Greek equivalent of Neptune, the Roman god of the sea. A second, much smaller Neptunian satellite was not definitely known for more than a century; Kuiper located it on photographic plates in 1949. It was named Nereid for a sea nymph lured by Triton's conch-shell music in mythology.

Neptune's satellites differ from those of Jupiter, Saturn and Uranus. Each of these three giant planets has a group of large satellites that revolve in regularly spaced, circular orbits in the same direction as the rotation of the planet and close to the planet's equatorial plane, presumably because they share the rotation of the material from which the planet and its satellites formed. In contrast, Neptune's largest satellite, Triton, moves in the backward retrograde direction, opposite to the direction of the planet's rotation and the orbital direction of all planets and most satellites.

The large planetary satellites, with radii larger than 100 kilometers, or 10^5 meters, were all discovered by the mid 20th century, and most of them were known by the end of the 19th century (Table 1.3). Altogether 21 of them are known – one for Earth, four for Jupiter, nine for Saturn, five for Uranus, and two for Neptune.

In the meantime, back in the 17th century, Huygens turned his telescope toward Saturn itself, and explained its mysterious handle-like appendages. Galileo had noticed that the planet was not round, but had blurry objects on each side. When these objects disappeared two years later, Galileo wondered if Saturn "had devoured his own children". In 1656 Huygens, then only 27 years old, realized that the planet is surrounded by "a thin flat ring, nowhere touching it, and inclined to the ecliptic" (Fig. 1.18). Because the ring is tipped with respect to the plane of the Earth's orbit around the Sun, it changes its shape when viewed from Earth, slowly opening up and then turning edge-on as Saturn makes its 29.5-year orbit around the Sun. When the ring is opened up, it resembles handle-like appendages, but when it is viewed edge-on the ring virtually disappears.

Cassini also observed Saturn's ring, suggesting that it was composed of swarms of satellites too small to be resolved individually, circling the planet with different velocities. In 1675 he discovered a dark separation in the ring that is now known as the "Cassini Division".

In fact, there are three main rings of Saturn visible from the Earth, the outer A, central B and inner C rings (Table 1.4). The Cassini Division separates the A and B rings. The C ring is also known as the "crepe ring" since it is the most transparent of the three main rings.

But why don't Saturn's rings fall down onto the planet? The rings stay up because they are moving, supported by their motion against the downward pull of Saturn's gravity. That is the same reason our Moon stays apart from the Earth, yet always accompanying it. Except, as James Clerk Maxwell (1831–1879) showed in 1859, the wide, thin rings of Saturn are composed of a vast number of small particles, each pursuing its individual orbit in the

Table 1.3 Large planetary satellites

Name	Mean radius (km)[a]	Mass (10²⁰ kg)[a]	Mean mass density (kg m⁻³)[a]	Distance from planet (10³ km)	Period of revolution[b] (days)	Year of discovery
EARTH						
Moon	1738	734.8	3344	384.4	27.3217	
JUPITER						
Io	1822	843.2	3528	422	1.77	1610
Europa	1561	480.0	3013	671	3.55	1610
Ganymede	2631	1481.9	1942	1070	7.16	1610
Callisto	2410	1075.9	1834	1883	16.7	1610
SATURN						
Mimas	198	0.38	1150	186	0.94	1789
Enceladus	252	1.08	1608	238	1.37	1789
Tethys	533	6.18	973	295	1.89	1684
Dione	562	10.96	1476	377	2.74	1684
Rhea	764	23.07	1233	527	4.52	1672
Titan	2576	1345.5	1880	1222	15.9	1655
Hyperion	135	0.06	542	1481	21.3	1848
Iapetus	736	18.06	1083	3561	79.3	1671
Phoebe	107	0.08	1634	12 952	550 R	1898
URANUS						
Miranda	236	0.66	1214	130	1.41	1948
Ariel	579	12.6	1592	191	2.52	1851
Umbriel	585	12.2	1479	266	4.14	1851
Titania	789	34.2	1662	436	8.71	1787
Oberon	761	28.8	1559	583	13.5	1787
NEPTUNE						
Triton	1353	214.2	2064	354	5.88 R	1846
Nereid	170	0.3	1500	5510	360	1949

[a] The radii are given in units of kilometers (km), the mass is in kilograms (kg), and the mean mass density in kilograms per cubic meter (kg m⁻³). By way of comparison, the equatorial radius of the planet Mercury is 2440 kilometers, so Ganymede and Titan are both bigger than Mercury.

[b] The letter R following the period denotes a satellite revolving about its planet in the retrograde direction, opposite to that of the planet's rotation and orbital motion about the Sun.

plane of the planet's equator. The innumerable particles that make up Saturn's rings act as tiny satellites that move in accordance with Kepler's third law, with the inner parts moving at a faster speed than the outer ones.

1.3 What holds the solar system together?

Gravity and motion – a delicate balance

The great English scientist Isaac Newton (1643–1727) showed that the same unchanging physical laws apply to the terrestrial and the celestial. Motions everywhere, whether up above in the heavens or down below on the ground, are described by the same concepts, and all material objects in the cosmos are subject to universal gravitation, with its unlimited capacity to act on matter. So everything in the cosmos moves in predictable and verifiable ways, described by Newton's laws of motion and gravitation. The basic ideas are that a moving body will continue to move in a straight line unless acted upon by an outside force, and that every object in the Universe attracts every other object as the result of universal gravitation.

According to tradition, Newton was sitting under an apple tree when an apple fell next to him on the grass.

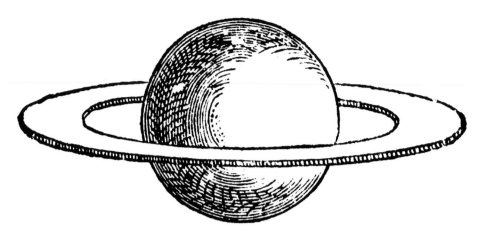

Fig. 1.18 Saturn's ring   In 1659 Christiaan Huygens (1629–1695) published this drawing of Saturn and its ring in his monograph *Systemia Saturnium*. Huygens recognized that a detached ring would explain the planet's ever-changing appearance, and announced his discovery in the form of an anagram, a succession of scrambled letters. The drawing shown here was accompanied by the deciphered anagram "Saturn is girdled by a thin flat ring, nowhere touching it, and inclined to the ecliptic."

Table 1.4 The main rings of Saturn

Name	Distance from planet center (Saturn radii[a])	Orbital period[b] (hours)	Width (km)
A ring	2.025 to 2.267	12.1 to 14.2	14 670
Cassini division	1.949 to 2.025	11.4 to 12.1	4 585
B ring	1.525 to 1.949	7.9 to 11.4	25 580
C ring	1.235 to 1.525	5.8 to 7.9	17 490

[a] The equatorial radius of Saturn at the one-bar pressure level is 60 268 kilometers, or 6.0268×10^7 meters, nearly ten Earth radii.
[b] Saturn's rotation period is 10.6562 hours, so the inner B ring and the entire C ring move around the planet at a faster rate than the planet rotates.

This reminded him that the power of gravity, whose pull influences the motion of falling bodies, seems undiminished even at the top of the highest mountains. He therefore argued that the Earth's gravitational force extends to our Moon, and showed that this force can pull the Moon into an orbit. It was as if the Earth's Moon is perpetually falling toward the planet while always keeping the same mean distance from it. The Sun's gravity similarly deflects the moving planets into their curved paths, so they forever revolve around the Sun.

So Newton discovered the cosmic reach of gravity, which keeps our feet on the ground. Gravity has pinned us there, so we rotate with the spinning Earth and stay on it. The air and oceans are similarly held close to the planet by its gravitational pull.

The English genius was a self-isolated intellect, a bit obsessed, famously distracted, and frequently depressed. Newton didn't like interacting with people. He declined most invitations, avoided personal contact, never traveled abroad, and, they say, died a virgin at the age of eighty-five. He was also a rebel against authority, and spent much of his life immersed in experiments in alchemy and theological or mystical speculations, hoping to understand the origin of the elements and the eternal mysteries of health and mortality, examining mystic clues left by God.

It was his friend, the English astronomer Edmond Halley (1656–1742), who persuaded the secretive Newton to write his greatest work, the *Philosophiae naturalis principia mathematica*, or the *Mathematical Principles of Natural Philosophy*, commonly known as the *Principia*. It was presented to the Royal Society of London in 1686, which withdrew from publishing it owing to insufficient funds, so Halley, a wealthy man, paid for the publication the following year.

As Newton wrote in the *Principia*, "the Copernican system of the planets stands revealed as a vast machine working under mechanical laws here understood and explained for the first time". These were his laws of motion and the law of universal gravitation, achievements that resulted in Sir Isaac becoming the first person in England to be knighted, in 1705, for his scientific work.

The enormous reach of gravity can be traced to two causes. In the first place, gravitational force decreases relatively slowly with distance, and this gives gravitation a much greater range than other natural forces, such as those that hold the nuclei of atoms together. In the second place, gravitation has no positive and negative charge as electricity does, or opposite polarities as magnets do. This means that there is no gravitational repulsion between masses.

In contrast, the repulsive and attractive forces among like and unlike electrical charges in an atom cancel each other, shielding it from the electrical forces of any other atom.

The gravitational force is mutual, so any two objects attract each other, and every atom in the Universe feels the gravitational attraction of every other atom. Their attraction is proportional to the product of their masses and inversely proportional to the square of the distance between them. Their mass possesses inertia, the tendency to resist any change in its motion. Mass is incidentally an intrinsic aspect of an object, different from its weight which alters with distance from the main source of gravity. An astronaut weighs less when leaving the Earth, but retains the same mass.

Expressed mathematically, any mass M_1 produces a gravitational force $F_{gravity}$ on another mass M_2, given by the expression:

$$\text{Gravitational force} = F_{gravity} = \frac{G M_1 M_2}{D^2}$$

where G is the universal gravitational constant, $G = 6.6726 \times 10^{-11} \, m^3 \, kg^{-1} \, s^{-2}$, and D is the distance between the centers of the two masses. This is sometimes called an inverse square law, since the force of gravity is inversely proportional to the square of the distance or separation.

Newton used observations of planetary motions to determine just how the force decreases with increasing distance. Contrary to popular belief, Newton did not use his theory to show that planets move in elliptical orbits, but instead used Kepler's third law, connecting a planet's orbital period and distance, to show that the force of gravity must fall off as the inverse square of the distance.

The concept of universal gravitation, and Newton's expression for the gravitational force, can be used to derive Kepler's third law in the form:

$$P_P^2 = \frac{4\pi^2}{G} \frac{a_P^3}{(M_P + M_\odot)} = 5.9 \times 10^{11} \frac{a_P^3}{M_\odot} \, s^2$$

where a_P is the semi-major axis of the planet's orbital ellipse in meters, P_P is the orbital period in seconds, and M_P and $M_\odot$ respectively denote the mass of the planet and the mass of the Sun in kilograms.

Within the solar system, the dominant mass is that of the Sun, which far surpasses the mass of any other object there. That is why we call it a solar system, governed by the central Sun. The sum $(M_P + M_\odot)$ is therefore, to the first approximation, a constant equal to the Sun's mass, $M_\odot$, regardless of the planet under consideration.

So it is the Sun's gravitational attraction that keeps the planets in their orbits, and holds the solar system together. But why doesn't the immense solar gravity pull the entire planetary system into the Sun? Motion holds the planets up, opposing the relentless pull of the Sun's gravity and keeping the planets from falling into the Sun.

The reason that the planets do not plunge into the Sun is that each planet is also moving in a direction perpendicular to an imaginary line connecting it to the Sun, at exactly the speed required to overcome the Sun's gravitational pull, keeping the planet in perpetual motion. This orbital speed depends only on the Sun's mass and the planet's distance, but it is independent of the planet's mass.

If a planet moved any faster, it would leave the solar system, and if it moved any slower it would be pulled into the Sun. This delicate balance between motion and gravity also explains why the Moon revolves around the Earth, and why Saturn's ring particles remain separate from the planet.

You might say that motion seems to define existence, for there is nothing in the Universe that is completely at rest. Everything that exists, from atoms to planets and stars to galaxies, moves through space, all surely going somewhere. It is this motion that shapes the Universe, giving it form, structure and texture. When you stop moving it is all over, or as Bob Dylan (1941–) sang: "better start swimming, or you'll sink like a stone".

Neptune's discovery, triumph of Newtonian gravitational theory

Neptune's discovery was no accident, in contrast to those of Uranus and the first asteroid. It was a direct consequence of precise mathematical calculations of Uranus' motion. Uranus had been detected by professional astronomers and mistaken for a star on no less than 22 occasions during the century that preceded the realization that it was a planet. These additional observations could be combined with the post-discovery ones to determine Uranus' trajectory and calculate its future position. Before long it was found that the planet was wandering from its predicted path.

A large unknown world, located far beyond Uranus, was evidently producing a gravitational tug on Uranus, causing it to deviate from the expected location. Two astronomer–mathematicians, John Couch Adams (1819–1892) in England and Urbain Jean Joseph Le Verrier (1811–1877) in France, independently located the planet by a mathematical analysis of the wanderings of Uranus.

Adams, a recent graduate from Cambridge University, finished his work first, deriving a precise position of the planet in mid-1845. He left a summary of his results with the then Astronomer Royal, George Biddell Airy (1801–1892), who did not feel compelled to look for the unknown world.

Le Verrier finished his best calculations about a year later, and, unlike Adams, published his results. Both scientists had assumed that the undiscovered planet occupied the next place in the sequence of the Titius–Bode law, and they arrived at nearly identical locations for it.

When Le Verrier's memoir reached Airy, he persuaded James Challis (1803–1882), professor of astronomy at Cambridge University, to make a search for the undiscovered planet. For a variety of reasons, Challis began the investigation slowly, and Le Verrier had in the meantime sent his results to the Berlin Observatory where Johann Gottfried Galle (1812–1910) and his student Heinrich Louis d'Arrest (1822–1875) found the planet. They identified it on the first night of their search, on 23 September 1846, using a 0.23-meter (9-inch) refractor; it was located within a degree of both Adams' and Le Verrier's predicted positions. Only later did Challis realize that he had previously observed the planet twice when beginning his own search.

The discovery of the new planet, named Neptune after the Roman god of the sea, was acclaimed as the ultimate triumph of Newtonian science. It resulted from mathematical calculations, based on Newton's theories, of the effects of an unknown planet whose gravity was pulling Uranus from its predicted place. If proof were needed, this achievement certified the validity of gravitational theory.

Neptune is located at a mean distance of about 30 times the distance between the Earth and the Sun. The remote planet takes about 165 Earth years to travel once around the Sun, so it will not make a full orbit since its discovery until 2011.

1.4   Physical properties of the Sun

What are the distances to the Sun and other nearby stars?

How far away is the Sun from the Earth, and how fast is the Earth moving through space? Kepler's model of planetary motion only provided a scale model for the relative distances of the planets from the Sun, and for a long time no one knew exactly how big the solar system was. Our planet's true distance from the Sun remained unknown for centuries. And since the distance was not reliably known, the velocity of the planet around the Sun could not be determined. The Earth's mean orbital speed is equal to the circumference of the orbit divided by one year, the time for the Earth to complete one trip around the Sun.

The crucial unit of distance for the planets is the mean Earth–Sun distance, known as the astronomical unit (AU). It can be determined by first estimating the distance between Earth and a nearby planet, and using the measurement together with geometry and Kepler's third law to infer the Earth–Sun distance.

The distance to a planet can be estimated by measuring the angular separation of the planet when observed simultaneously from two widely separated locations. This angle is known as parallax, from the Greek *parallaxis*, for the "value of an angle". If both the parallax and the separation between the two observers are known, then the distance of the planet can be determined by triangulation. It is based on the geometric fact that if you know the length of one side of a triangle and the angles of the two corners, all the other dimensions can be calculated.

The parallax technique of estimating a planet's distance is similar to the way your eyes infer how far away things are. To see the effect, hold a finger up in front of your nose, and look at your finger with one eye open and the other closed, and then with the open eye closed and the closed one open. Any background object near to one side of your finger seems to move to the other side, making a parallax shift. When this is repeated with your finger held farther away, the angular shift is smaller. In other words, the more distant an object, the smaller the parallax shift, and vice versa.

Giovanni Domenico Cassini (1625–1712), an Italian astronomer and the first director of the Paris Observatory, obtained an early triangulation of Mars in 1672, combining his observations from Paris with those taken from Cayenne, French Guiana. The planet was then in opposition, at its closest approach to Earth. From the two sets of observations, made 7200 kilometers apart, it was possible to estimate the distance to Mars and to infer a value of 139 million (1.39×10^8) kilometers for the astronomical unit.

Astronomers in the 18th and 19th centuries attempted to improve the measurement accuracy of the Sun's distance during the rare occasions when Venus crossed the face of the Sun, in 1761, 1769, 1874 and 1882. The method also involved comparison of observations from widely separated locations to determine the distance by triangulation. Subsequent determinations of the distance to the nearby minor planet 433 Eros during its closest approaches to the Earth resulted in an estimated 150 million kilometers for the astronomical unit.

Significant improvements in the precision of planetary distances came in the late 1960s by bouncing pulsed radio waves off Venus and timing the echo (Fig. 1.19). The round-trip travel time – about 276 seconds when Venus is closest to the Earth – was measured using accurate atomic clocks, and a precise distance to Venus was then obtained by multiplying half the round-trip time by the speed of light (Focus 1.6). The distance of Venus from the Sun is equal to one half of the difference between the Earth and Venus

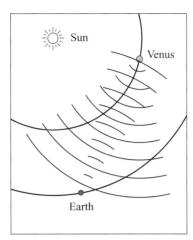

Fig. 1.19 Radar-ranging to Venus Accurate distances to the nearby planets have been determined by sending radio pulses from Earth to the planet, and timing their return several minutes later. The figure shows the emission of a pulse toward Venus; when it bounces from Venus the radiation spreads over the sky and we receive only a small fraction of the original signal, delayed by the round-trip travel time. If T is the round-trip time and c is the speed of light, the total distance traveled is cT and the distance to Venus is cT/2. For Venus, the round-trip time is 4.6 minutes when the planet is nearest Earth and increases to 28.7 minutes when it is furthest away from us.

when it is closest and furthest away from us, on the other side of the Sun. The value of the astronomical unit inferred from the radar determination of the distance of Venus is 149 597 870 kilometers, with an accuracy of about 1 kilometer, or for the accuracy required in most astronomical calculations 1 AU = 149.6 million kilometers.

Once the Venus–Sun distance is known, we can infer the distance of any other planet from the Sun using Kepler's third law, which relates the orbital periods and orbital distances of the planets. The average distances of the planets from the Sun are: Mercury 0.39 AU, Venus 0.72 AU, Earth 1.00 AU, Mars 1.52 AU, Jupiter 5.2 AU, Saturn 9.54 AU, Uranus 19.19 AU and Neptune 30.07 AU.

Nowadays the accuracy of the mean Earth–Sun distance is fixed by the exact value for the speed of light. The time τ_{AU} for light to travel across 1 AU is given as a primary astronomical constant:

$$\text{Earth–Sun light travel time} = \tau_{AU}$$
$$= 499.004\,782 \text{ seconds}$$

with a derived value for the mean Earth–Sun distance of:

$$1\,\text{AU} = c\tau_{AU} = 1.495\,978\,70 \times 10^8 \text{ kilometers}$$

where the speed of light c is 299 792.458 kilometers per second. By way of comparison, one light-year is equal to 63 240 AU.

Once you have an accurate value for the Sun's distance, the Earth's mean orbital velocity can be determined by assuming, to a first approximation, a circular orbit and dividing the Earth's orbital circumference by the Earth's orbital period of $P_E = 1$ year $= 3.1557 \times 10^7$ seconds, or:

$$\text{Earth's mean orbital velocity}$$
$$= \frac{2\pi \times (1\,\text{AU})}{P_E} = 29.8 \text{ kilometers per second}$$

which is equivalent to about 170 000 kilometers per hour.

And what about the distances to the other nearby stars? Even the closest stars, other than the Sun, are too far away for us to detect a shift in position from any two points on Earth. To triangulate the distances of these nearest stars, astronomers needed a wider baseline, measuring their parallax from opposite sides of the Earth's annual orbit, or from a separation of twice the astronomical unit. The measurement involves careful scrutiny of two stars that appear close together in the sky, a bright one that is relatively nearby and the other fainter one that is much further away (Fig. 1.20).

The annual parallax of the nearest, brighter star can then be determined by measuring its angular separation from the fainter, distant one for a year or more. During the course of the year, the nearby star will seem to sway to and fro, in a sort of cosmic minuet that mirrors the Earth's orbital motion. Measurements separated by six months, from opposite sides of the Earth's orbit, can reveal a shift in the position of a nearby star with respect to the more distant ones. Half of this angular displacement, known as the annual parallax π, is the ratio of the AU and the star's distance D, or to be precise $\sin \pi = (1\,\text{AU})/D$ radians. The nearer the star, the larger the annual-parallax sways.

The first star whose distance was reliably determined in this way was 61 Cygni. As reported by Friedrich Wilhelm Bessel (1784–1836) in 1838, this star lies at an estimated distance of 10.4 light-years, corresponding to an annual parallax of just 0.314 seconds of arc. The star 61 Cygni had been christened the "Flying Star" in 1792 by Piazzi who discovered its unusually large angular motion across the sky; if all stars moved at the same velocity perpendicular to the line of sight, such a large angular motion would indicate a nearby distance. Modern measurements provide a parallax of 0.287 18 seconds of arc for 61 Cygni, yielding a distance of 11.36 light-years or about 718 000 times further away than the Sun.

Within a century of Bessel's result, the annual parallax of about 2 000 nearby stars had been determined using long exposures on photographic plates, and the number tripled in succeeding decades. The closest star, known as Proxima Centauri, is 4.22 light-years away. And many of the brightest stars are hundreds of light-years away, so you

Focus 1.6   Light, the fastest thing around

It was once thought that light moves instantaneously through space, but we now know that it travels at a very fast, but finite, speed, which was first inferred from observations of Jupiter's moon Io in the 18th century. The King of France had directed Giovanni Domenico Cassini (1625–1712), the Director of the Paris Observatory, to use such observations to improve knowledge of terrestrial longitude and maps of France. While working at the observatory the Danish astronomer Ole Roemer (1644–1710) and Cassini noticed a varying time between eclipses of the satellite by the planet. Although Io's orbital period was about 42 hours, the duration of the orbit seemed to grow shorter when Jupiter was closer to Earth, and larger when it would move away, with a total time difference of about 22 minutes. Both Cassini and Roemer concluded that it was not the orbit itself that varied, but the time it took Jupiter's light to cross the Earth's orbit and reach our planet. Neither astronomer gave a value for the speed of light, which would have been equal to the diameter of the Earth's orbit divided by the time difference, or a velocity of $c = 2$ AU / 22 minutes or about 227 000 kilometers per second, where 1 AU is the mean distance between the Earth and the Sun.

More refined laboratory measurements during succeeding centuries indicated that light is always moving at a constant speed with the precise velocity of $c = 299\,792.458$ kilometers per second. Light emitted by any star will move through empty space for all time, never stopping or slowing down and never coming to rest. Moreover, nothing outruns light; it's the fastest thing around.

The unvarying speed of light was first demonstrated in 1887 by the American physicist Albert A. Michelson (1852–1931), assisted by his friend the chemist Edward W. Morley (1838–1923), when they attempted to precisely measure how the speed of light depends on the Earth's motion through a hypothetical, space-filling medium, the ether, in which light waves were supposed to propagate and vibrate.

As the Earth moves through the stationary medium, an ether wind should blow past the Earth in the direction of its motion, and the speed of light would vary, like a swimmer moving downstream or struggling upstream. But Michelson and Morley found that there was no detectable difference in the speed of light measured in the direction of the Earth's motion or at right angles to it. So the experiment meant that there was no light-carrying ether. It also implied that the velocity of light is constant, exactly the same in all directions and at all seasons, and independent of the motion of the observer.

The speed of light, c, enters into Albert Einstein's (1879–1955) *Special Theory of Relativity* through the factor

$$\gamma = \frac{1}{\sqrt{\left[1 - \left(\frac{v}{c}\right)^2\right]}}$$

for an object moving at a velocity v. We normally regard time as absolute and immutable, with nothing disturbing its relentless, steady tick. But for Einstein, time was relative and variable. In rapid travel, the rate at which time flows decreases, so clocks run slower by the factor γ. Lengths are diminished at high speed, shrinking in the direction of motion by the amount γ. At very high velocities, all is relative, even mass, which increases with the speed by the same infamous γ factor. If any material body reached the velocity of light, it might shrink into an insignificant nothing, perhaps even disappearing, while time might be stretched to infinity. That doesn't happen because an object's mass increases without bound when it moves as fast as light, and there is nothing that can propel it so fast.

can walk outside at night and see stars whose light was emitted before your parents were born.

Distant stars with parallaxes smaller than 0.05 seconds of arc cannot be measured with Earth-based telescopes because of atmospheric distortion that limits their angular resolution. However, instruments aboard the *HIPPARCOS* satellite, which orbited the Earth above its atmosphere in the 1990s, pinpointed the positions of more than 100 000 stars with an astonishing precision of 0.001 seconds of arc, determining the parallax and distance of many of them out to a few hundred light-years.

How big, massive and hot is the Sun?

Once an accurate value for the mean distance between the Earth and the Sun is known, we can use it with the orbital period of the Earth to infer the mass of the Sun, $M_\odot$, from Newton's formulation of Kepler's third law:

$$\text{Sun's mass} = M_\odot = 5.9165 \times 10^{11} \times \frac{(1\,\text{AU})^3}{P_E^2}$$

$$= 1.989 \times 10^{30}\ \text{kilograms}$$

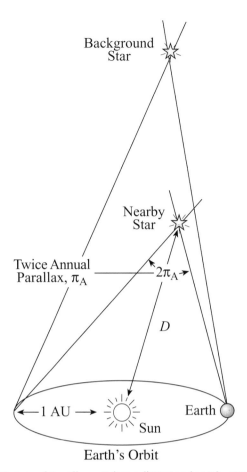

Fig. 1.20 Annual parallax When a distant and nearby star are observed at six-month intervals, on opposite sides of the Earth's orbit around the Sun, astronomers measure the angular displacement between the two stars. It is twice the annual parallax, designated by π_A, which can be used to determine the distance D of the nearby star. The German astronomer Friedrich Wilhelm Bessel (1784–1846) announced the first reliable measurement of the annual parallax of a star in 1838.

using the Sun–Earth distance of $1\,\mathrm{AU} = 1.495\,978\,7 \times 10^{11}$ meters, and the Earth's orbital period of $P_E = 1$ year $= 3.1557 \times 10^7$ seconds.

The linear radius of the Sun, $R_\odot$, can be determined from its angular diameter, θ, using:

$$R_\odot = \theta \times (1\,\mathrm{AU})/2 = 6.955 \times 10^8 \text{ meters}$$

where $\theta = 31.97$ minutes of arc $= 0.0093$ radians, and a full circle subtends 2π radians and 360 degrees.

The mass density of the Sun, $\rho_\odot$, is obtained by dividing this mass by the Sun's volume, $4\pi R_\odot{}^3/3$, where the Sun's radius $R_\odot = 6.955 \times 10^8$ meters. It is $\rho_\odot = 1409$ kilograms per cubic meter, only about one-fourth of the mass density of the Earth, which is 5520 kilograms per cubic meter.

The entire Sun is nothing but a big luminous ball of gas, very hot and concentrated at the center and cooler and more tenuous further out. From the Sun's size and luminous output, we can infer the temperature of its visible disk, 5780 kelvin (Focus 1.7).

The material deep down inside the Sun must become hotter and more densely concentrated to support the overlying weight and to keep the star from collapsing. Calculations show that the temperature reaches 15.6 million (1.56×10^7) kelvin at the center of the Sun (also see Focus 1.7). The center is also extremely compacted with a density of 151 300 kilograms per cubic meter. All of the essential physical properties of the Sun are given in Table 1.5.

What is the Sun made out of?

The ingredients of the Sun can be determined when the intensity of sunlight is spread out into its visible wavelengths or different colors. Such a display is called a spectrum, and the study of spectra is known as spectroscopy.

Each chemical element or compound produces a unique set, or pattern, at certain specific wavelengths in the spectrum, and only at those wavelengths. They resemble a barcode or a fingerprint that can be used to identify the element or compound.

A hot, glowing body like the Sun emits radiation at all wavelengths, with a continuous spectrum. If this radiation passes through a cool, tenuous gas, such as the outer layers of the Sun's visible atmosphere, part of the radiation is absorbed at discrete wavelengths. These spectral features are called absorption lines because they look like a line in the spectrum. If the gas is heated to incandescence, it will emit radiation at the same specific wavelengths, and the spectral features are called emission lines. The patterns of either the absorption or emission lines tell us the atoms or molecules that are present in the gas.

The technique of astronomical spectroscopy was first developed using the bright light of the Sun. When its spectrum is examined carefully, with fine wavelength resolution, numerous fine, dark absorption lines are seen crossing the rainbow-like display (Fig. 1.21). The separate colors of sunlight are somewhat blurred together when coarser spectral resolution is used, and the dark places are no longer found superimposed on its spectrum.

These dark gaps, or absorption lines, were first noticed in 1802 by the English chemist William Hyde Wollaston (1766–1828), and investigated in far greater detail by the Bavarian telescope-maker Joseph Fraunhofer (1787–1826). By 1815, Fraunhofer had catalogued the wavelengths of hundreds of them, assigning Roman letters A, B, C, . . . to the darkest and most prominent of them, starting from

Focus 1.7 Taking the Sun's temperature

Satellites have been used to accurately measure the Sun's total irradiance just outside the Earth's atmosphere, establishing the value of the solar constant:

$$f_\odot = 1361 \text{ joule per second per square meter}$$

where one joule per second is equivalent to one watt. The solar constant is defined as the total amount of radiant solar energy per unit time per unit area reaching the top of the Earth's atmosphere at the Earth's mean distance from the Sun. We can use it to determine the Sun's absolute luminosity, $L_\odot$, from:

$$L_\odot = 4\pi f_\odot (1\,\text{AU})^2 = 3.854 \times 10^{26} \text{ joule per second}$$

where the mean distance between the Earth and the Sun is $1\,\text{AU} = 1.496 \times 10^{11}$ meters.

The effective temperature, $T_{e\odot}$, of the visible solar disk, called the photosphere, can be determined using the Stefan–Boltzmann law:

$$L_\odot = 4\pi\sigma\, R^2 T_{e\odot}^4$$

where the Stefan–Boltzmann constant is $\sigma = 5.670 \times 10^{-8}$ J m^{-2} K^{-1} s^{-1}, and the Sun's radius is $R_\odot = 6.955 \times 10^8$ meters. Solving for the temperature:

$$T_{e\odot} = [L_\odot/(4\pi R_\odot^2)]^{1/4} = 5780 \text{ kelvin}$$

Incidentally, the Stefan–Boltzmann law applies to other stars, indicating that at a given temperature giant stars, with greater radii than most stars, have greater luminosity.

The temperature, $T_{c\odot}$, at the center of the Sun can be estimated by assuming that a proton must be hot enough and move fast enough to counteract the gravitational compression it experiences from all the rest of the star. That is:

Thermal energy

$$= \frac{3}{2} k\, T_{c\odot} = \frac{G m_P M_\odot}{R_\odot} = \text{Gravitational potential energy}$$

where Boltzmann's constant k is $1.380\,66 \times 10^{-23}$ joule per kelvin, the gravitational constant G is 6.6726×10^{-11} m^3 kg^{-1} s^{-2}, the Sun's mass $M_\odot$ is 1.989×10^{30} kilograms, and the mass of the proton is $m_P = 1.6726 \times 10^{-27}$ kilograms. Solving for the central temperature we obtain:

$$T_{c\odot} = \frac{2 G m_P\, M_\odot}{3k\, R_\odot} = 1.56 \times 10^7 \text{ kelvin}$$

So the temperature at the center of the Sun is 15.6 million kelvin.

Table 1.5 Physical parameters of the Sun[a]

Mass, $M_\odot$	1.989×10^{30} kilograms (332 946 Earth masses)
Radius, $R_\odot$	6.955×10^8 meters (109 Earth radii)
Volume	1.412×10^{27} m^3 (1.3 million Earths)
Density (center)	151 300 kg m^{-3}
(mean), $\rho_\odot$	1409 kg m^{-3}
Pressure (center)	2.334×10^{11} bars
(photosphere)	0.0001 bar
Temperature (center) $T_{c\odot}$	15.6 million kelvin
(photosphere) $T_{e\odot}$	5 780 kelvin
(corona)	2 million to 3 million kelvin
Luminosity, $L_\odot$	3.854×10^{26} J s^{-1}
Solar constant, $f_\odot$	1361 J s^{-1} m^{-2} = 1361 W m^{-2}
Mean distance, AU	$1.495\,9787 \times 10^{11}$ m = 1.0 AU
Age	4.55 billion years
Principal chemical constituents (by number of atoms)	
Hydrogen	92.1 percent
Helium	7.8 percent
All others	0.1 percent

[a] Mass density is given in kilograms per cubic meter, or kg m^{-3}; the density of water is 1000 kg m^{-3}. The unit of pressure is bars, where 1.013 bars is the pressure of the Earth's atmosphere at sea level. The unit of luminosity is joule per second; power is often expressed in watts, where 1.0 watt = 1.0 joule per second.

Fig. 1.21 Visible solar spectrum A spectrograph has spread out the visible portion of the Sun's radiation into its spectral components, displaying radiation intensity as a function of wavelength. When we pass from long wavelengths to shortest ones (*left to right* and *top to bottom*), the spectrum ranges from red through orange, yellow, green, blue and violet. Dark gaps in the spectrum, called Fraunhofer absorption lines, represent absorption by atoms or ions in the Sun. The wavelengths of these absorption lines can be used to identify the elements in the Sun, and the relative darkness of the lines helps establish the relative abundance of these elements. (Courtesy of NSO/NOAO.)

the long-wavelength, red end of the visible solar spectrum and progressing to its short-wavelength side.

An explanation for the Sun's absorption lines was provided in the mid 19th century in a laboratory in Heidelberg, Germany, where the chemist Robert Bunsen (1811–1899), inventor of the Bunsen burner, and his physicist colleague Gustav Kirchhoff (1824–1887) unlocked the chemical secrets of the Universe. When they vaporized an individual element in a flame, the hot vapor produced a distinctive pattern of bright emission lines whose unique wavelengths coincided with the wavelengths of some of the dark absorption lines in the Sun's spectrum, identifying that element as an ingredient of the solar gas. Solar lines designated by Fraunhofer by D and E are respectively ascribed to sodium and iron, while both the H and K lines are due to calcium.

The lightest element, hydrogen, was also identified in the solar spectrum, in 1862 by the Swedish physicist Anders Jonas Ångström (1814–1874); it accounts for Fraunhofer's C and D lines. Subsequent investigations of the great strength of these lines indicated that hydrogen is the most abundant element in the visible solar gases.

Since the Sun was most likely chemically homogeneous, a high hydrogen abundance was implied for the entire star. This accounts for the Sun's low mass density. We now know that hydrogen accounts for 92.1 percent of the number of atoms in the Sun, and that hydrogen is the most abundant element in most stars, in interstellar space, and in the entire Universe.

Helium, the second-most abundant element in the Sun, is so rare on Earth that it was first discovered in the Sun. A previously unknown emission line was initially noticed during a solar eclipse on 18 August 1868, at a wavelength of 587.56 nanometers near the two yellow sodium lines. Since this feature had no known Earthly counterpart, it was thought that a new chemical element had been discovered, which Norman Lockyer (1836–1920) named "helium" after the Greek Sun god Helios. Helium was not found on Earth until 1895, when William Ramsay (1852–1919) discovered it as a gaseous emission from a mineral called clevite. Helium accounts for 7.8 percent of the number of atoms in the Sun, and all the heavier elements in the Sun amount to only 0.1 percent (Table 1.6).

Today, helium is used on Earth in a variety of ways, including the inflation of party balloons and in its liquid state to keep sensitive electronic equipment cold. Though plentiful in the Sun, helium is almost non-existent on the Earth. It is so terrestrially rare that we are in danger of running out of helium during this century.

A more complete understanding of spectral lines followed the discovery that atoms are mostly empty space, just as the room you are sitting in is mainly empty. A tiny, heavy, positively charged nucleus lies at the heart of an atom, surrounded by a cloud of relatively minute, negatively charged electrons that occupy most of an atom's space and govern its chemical behavior. The charge of a proton is exactly equal to that of an electron, so the complete atom, in which the number of

Table 1.6 The ten most abundant elements in the Sun

Atomic number[a]	Element	Symbol	Number of atoms (silicon = 10.0)	Date of discovery
1	Hydrogen	H	279 000	1766
2	Helium	He	2 720	1868[b]
6	Carbon	C	101	(ancient)
7	Nitrogen	N	31.3	1772
8	Oxygen	O	23.8	1774
10	Neon	Ne	34.4	1898
12	Magnesium	Mg	10.7	1755
14	Silicon	Si	10.0	1823
16	Sulfur	S	5.15	(ancient)
26	Iron	Fe	9.00	(ancient)

[a] The atomic number is equal to the number of protons in the nucleus of an atom.
[b] Helium was discovered on the Sun in 1868, but it was not found on Earth until 1895.

electrons equals the number of protons, is electrically neutral.

Hydrogen is the simplest atom, consisting of a single electron circling around a single proton. The nucleus of a more complex atom includes neutrons, which were proposed in order to keep the nuclear protons from repelling each other out of the nucleus. The nucleus of helium, for example, contains two neutrons and two protons, and has two electrons in orbit to balance the charge of the two protons. Ernest Rutherford (1871–1939) established the existence of a small, positively charged nucleus at the center of the atom in 1911; James Chadwick (1891–1974) discovered the neutron in 1932.

The sole electron in a hydrogen atom revolves about the central proton according to very specific rules that were proposed to explain observations of regularities in the Sun's hydrogen lines; adjacent lines of the hydrogen atom systematically crowd together and become stronger at shorter wavelengths.

The Swiss mathematics teacher Johann Balmer (1825–1898) published an equation that describes the regular spacing of the wavelengths of the four lines of hydrogen detected in the spectrum of visible sunlight, and they are still known as Balmer lines. The strongest one, with a red color, is also called the hydrogen alpha line.

In the early 20th century, the Danish physicist Niels Bohr (1885–1962) explained Balmer's equation by an atomic model, now known as the Bohr atom (Fig. 1.22), in which the electron in a hydrogen atom revolves about the proton in specific orbits with definite quantized values of energy, which are characterized by a quantum number that takes on integer values of $n = 1, 2, 3, 4, \ldots, \infty$ (infinity).

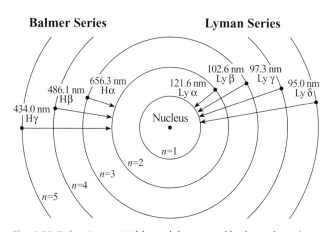

Fig. 1.22 Bohr atom   In this model, proposed in the early 20th century by the Danish physicist Niels Bohr (1885-1962), a hydrogen atom's one electron revolves around the hydrogen nucleus, a single proton, in well-defined orbits described by the integer $n = 1, 2, 3, 4, 5, \ldots$ An electron absorbs or emits radiation when it makes a transition between these allowed orbits. The electron can jump upward, to orbits with larger n, by absorption of a photon of exactly the right energy, equal to the energy difference between the orbits; the electron can jump down to lower orbits, of smaller n, with emission of radiation of that same energy and wavelength. Transitions that begin or end on the $n = 2$ orbit define the Balmer series that is observed at visible wavelengths. They are designated by Hα, Hβ, Hγ, ... The Lyman series, with transitions from the first orbit at $n = 1$, is detected at ultraviolet wavelengths. The orbits are not drawn to scale for the size of the radius increases with the square of the integer n.

Focus 1.8   Satellite motions and planetary mass

Kepler's third law also applies to the motion of a natural satellite, which can be used to determine the planet mass M_P from the expression:

$$M_P = \frac{4\pi^2}{G} \frac{a_S^3}{P_S^2} = 5.9 \times 10^{11} \frac{a_S^3}{P_S^2} \text{ kilograms}$$

for a satellite orbiting the planet at a distance a_S in meters and with period P_S in seconds, with $\pi = 3.14159$ and the Newtonian constant of gravitation $G = 6.673 \times 10^{-11} \text{ m}^3 \text{ kg}^{-1} \text{ s}^{-2}$. The Earth's Moon, for example, orbits the Earth at a mean distance of $a_M = 384400$ kilometers $= 3.844 \times 10^8$ meters with a period of $P_M = 27.322$ days $= 2.36 \times 10^6$ seconds, giving a mass $M_E = 60 \times 10^{23}$ kilograms for the Earth.

As another example, we can obtain Jupiter's mass from observations of any one of its four large satellites. The satellite Io's period is $P_S = 1.77$ days $= 152928$ seconds and Io's distance from Jupiter is $a_S = 4.22 \times 10^6$ meters, yielding a mass $M_J = 4\pi^2 a_S^3 / (G P_S^2) = 1.9 \times 10^{27}$ kilograms, or about one-thousandth the mass of the Sun.

This expression can also be used to determine the orbital distance of an artificial geosynchronous satellite that is launched to always hover above the same location on the Earth, using:

$$a_S^3 = \frac{G M_E P_S^2}{4\pi^2} = 1.7 \times 10^{-12} M_E P_S^2 \text{ meters}$$

This yields $a_E = 4.2 \times 10^7$ meters $= 42000$ kilometers from the Earth's center for an orbital period of one day, or $P_S = 86400$ seconds and an Earth mass of $M_E = 6 \times 10^{24}$ kilograms. Since the Earth's mean radius is 6400 kilometers, the geosynchronous satellite orbits at a distance of about 35600 kilometers above the ground.

The energy of the nth orbit is proportional to the inverse square of n, or to $1/n^2$.

The electron only emits or absorbs radiation when jumping between these allowed orbits, each jump being associated with a specific energy equal to the difference in orbital energies, and to a single wavelength, like one pure note. If an electron jumps from a low-energy orbit to a high-energy one, it absorbs radiation at this wavelength; radiation is emitted at exactly the same wavelength when the electron jumps the opposite way. Bohr was awarded the 1912 Nobel Prize in Physics for his investigation of the structure of atoms and the radiation emanating from them.

1.5 Terrestrial and giant planets

The eight major planets have been divided into two groups, the terrestrial and the giant planets, which differ in size, composition and distance from the Sun. The four planets close to the Sun, Mercury, Venus, Earth and Mars, are known as terrestrial planets because they are similar to the Earth. These inner planets are rocky and relatively compact and dense. In contrast, the four giant planets, Jupiter, Saturn, Uranus and Neptune, which reside in the outer parts of the planetary system, are big, gaseous and have relatively low mean mass densities. Unlike the inner terrestrial planets, rings and numerous satellites also encircle each of the outer giant planets.

Once the distance of a planet is known, its radius can be determined from its angular extent. For instance, Jupiter has an angular diameter of $\theta = 46.86$ seconds of arc when it is closest to Earth, or at a distance of $D = 4.2\,\text{AU}$ from us. That corresponds to a radius of $R = 7.14 \times 10^7$ meters. You can do the arithmetic yourself using $R = \theta D/2$ with $1\,\text{AU} = 1.496 \times 10^{11}$ meters and converting the angle to radians with 1 radian $= 2.06265 \times 10^5$ seconds of arc. The radius of Jupiter is 11.2 times the Earth's radius, and its volume is more than 1000 times that of the Earth.

The distance to a planet can be combined with the angular separation of one of its satellites to determine the orbital distance of that satellite from its planet, which can then be combined with the satellite's orbital period to establish the planet's mass. The motion of a satellite is governed by the mass of its planet, all in accordance with the inverse square law of gravity and Kepler's third law (Focus 1.8). Newton used it centuries ago to infer the masses of the Earth, Jupiter and Saturn from their satellite motions, obtaining values comparable to modern ones.

The masses of the other giant planets can be inferred from their satellites in the same way. The technique can also be used to obtain the Earth's mass, from the motion of its Moon. If a planet has no satellite, like Mercury and Venus, its mass can be obtained from detailed observations of its gravitational effects on spacecraft that pass or orbit near it.

The mean, or average, mass density of planet can be computed by dividing its mass, in kilograms, by its volume in cubic meters. The volume of a planet of radius R is $4\pi R^3/3$, so the mean mass density of a planet of mass M is $3M/(4\pi R^3)$. For instance, the mean mass density of Jupiter is $1330\,\text{kg m}^{-3}$ (kilograms per cubic meter). That is comparable to the Sun's mean mass density of $1409\,\text{kg m}^{-3}$. The mean mass density of the major planets, which is also known as their bulk density, is given in Table 1.7 together with their angular size at closest approach, radius, and mass. Here they are divided into two groups: the four

Table 1.7 Angular diameter, radius, mass and bulk density of the major planets

Terrestrial planets	Mercury	Venus	Earth	Mars
Angular diameter (seconds of arc)[a]	10.9	61.0		17.88
Equatorial radius, R_P (km)	2439.7	6051.8	6378.14	3396.19
Equatorial radius, R_P ($R_E = 1.0$)	0.382	0.949	1.000	0.533
Mass, M_P (kg)	3.3010×10^{23}	4.8673×10^{24}	5.9722×10^{24}	6.4169×10^{23}
Mass, M_P ($M_E = 1.0$)	0.0553	0.815	1.000 00	0.107
Bulk density (kg m^{-3})	5429	5243	5513	3934

Giant planets	Jupiter	Saturn	Uranus	Neptune
Angular diameter (seconds of arc)[a]	46.86	19.52	3.60	2.12
Equatorial radius, R_P (km)	71 492	60 268	25 559	24 764
Equatorial radius, R_P ($R_E = 1.0$)	11.19	9.46	3.98	3.81
Mass, M_P (kg)	1.8981×10^{27}	5.6832×10^{26}	8.6881×10^{25}	1.0241×10^{26}
Mass, M_P ($M_E = 1.0$)	317.89	95.18	14.54	17.13
Bulk density (kg m^{-3})	1326	687	1270	1638

[a] The largest angular diameter seen from the Earth when the planet is at its closest approach to Earth.

Table 1.8 Distribution of mass in the solar system

Mass of the Sun	$M_\odot = 1.989 \times 10^{30}$ kilograms
Mass of Jupiter	$M_J = 1.899 \times 10^{27}$ kilograms
Mass of the Earth	$M_E = 5.974 \times 10^{24}$ kilograms
Total mass of the planets	2.668×10^{27} kilograms $= 446.6 \, M_E$
Total mass of the satellites	6.2×10^{23} kilograms $= 0.104 \, M_E$
Total mass of the Kuiper-belt objects	3.0×10^{23} kilograms $= 0.05 \, M_E$
Total mass of the asteroids	1.8×10^{21} kilograms $= 0.0003 \, M_E$
Total mass of the planetary system	2.669×10^{27} kilograms $= 446.7 \, M_E = 0.001 \, 34 \, M_\odot$

rocky, dense and relatively small terrestrial planets, and the four giant, massive, low-density worlds.

When the mass of the Sun and planets are determined, we find that the Sun doesn't just lie at the heart of our solar system; it dominates it. Some 99.866 percent of all the matter between the Sun and halfway to the nearest star is contained in the Sun (Table 1.8). All of the objects that orbit the Sun – the planets and their satellites, the comets and the asteroids – add up to just 0.134 percent of the mass in our solar system. As far as the Sun is concerned, the planets are insignificant specks, left over from its formation and held captive by its massive gravity.

1.6 What is inside the major planets?

Density is a measure of compactness of an object, and all planets have greater density in their compact centers. The temperature in the interiors of the giant planets also increases with depth. To understand this, imagine a hundred mattresses stacked into a pile. The mattresses at the bottom must support those above so they will be squeezed thin. Those at the top have little weight to carry, and they retain their original thickness. The material at the center of a gaseous giant planet is similarly squeezed into a smaller volume by the overlying material, so the central regions become hotter and more densely concentrated. All of the giant planets are hot inside, with temperatures that increase with depth, and three of them still radiate more heat from their cloud tops than absorbed from the Sun.

The rocky terrestrial planets were all so hot in their formative stages, beginning about 4.6 billion years ago, that their interior rock and metal melted and gravity separated them by density. The denser material sank toward the center, while less dense rocky material remained closer to the surface. This process is called differentiation, since the internal layers are composed of different material. The Moon, for example, contains a small, partially molten liquid core, while the Earth, Mars and Mercury have larger ones hidden deep within their surfaces.

The highest-density material, consisting primarily of metals such as iron, resides in the central core of the terrestrial planets. Rocky material of moderate density is found in a thick mantle that surrounds the core. The mantle consists mostly of rocky silicate material containing silicon, oxygen and other elements. A terrestrial planet's thin outer crust contains the lowest density rocks such as granite and volcanic basalt.

Although the terrestrial planets and rocky satellites started off with hot interiors, as the result of their origin by colliding objects and the heat associated with radioactive decay, they have cooled from the outside in as time went on. After all, we now walk across solid rock rather than the originally molten crust.

When all the radioactive elements inside a rocky planet or moon are depleted, or decay into non-radioactive elements, there will be no more sources of internal heat, and these objects should become cooler inside while retaining a high internal pressure. As the remaining heat escapes from the surface, more heat will flow upward to replace it, until their interior is no hotter than its surface.

Since the total amount of internal heat depends on the volume, the time required to lose its heat depends on the ratio of the surface area $4\pi R^2$ to its volume $4\pi R^3/3$, so the rate of heat loss scales as $3/R$ for a planet of radius R. This means that larger planets or satellites with a greater radius will cool more slowly than smaller ones, provided they started off with the same internal temperature. Moreover, bigger objects most likely started off with more internal heat and higher temperatures. These two factors explain why the Moon's interior is so much cooler than the Earth's interior.

The terrestrial planets all now have solid surfaces, and on these surfaces is preserved a long record of their evolution. They have been shaped and modified by geological processes that include impact craters, produced when meteoroids similar to today's asteroids or comets struck them, and the volcanic eruption or flow of molten rock, from the planet's interior onto its surface. They are discussed in the next chapter.

2 The new close-up view from space

- Close examination of the planets and moons began when spacecraft flew past them, providing an initial reconnaissance. Orbiting spacecraft that mapped out the global terrain of the Earth's Moon, Venus and Mars, as well as the realms of Jupiter and Saturn, followed this. Probes have been parachuted down into the atmospheres of Jupiter, Titan and Venus, and landers and rovers have been sent to the surfaces of the Moon and Mars.

- The space-age investigation of the solar system began in a cold-war competition between the Soviet Union, which launched the first artificial satellite, and the United States, which won the race to the Moon.

- The *Voyager 1* and *2* flyby spacecraft transformed our understanding of the four giant planets, Jupiter, Saturn, Uranus and Neptune, and revealed fascinating, unexpected aspects of their moons and rings.

- The *Giotto* spacecraft was the first to provide a close-up view of a comet, showing that its nucleus is a black, city-sized chunk of water ice and dust that emits sunward jets of water when passing near the Sun.

- Orbiting spacecraft have greatly increased the time for study of the planets and moons, revealing ancient water flow on Mars, vast outpourings of lava on Venus, Jupiter's volcanic moon Io, and an ice-covered ocean on its satellite Europa, and Saturn's marvelous rings, water-spewing satellite Enceladus, and haze-shrouded moon Titan.

- Three rovers have explored the surface of Mars and provided evidence for water flow across its surface roughly 4.0 billion years ago.

- The *Huygens Probe* and radar from the orbiting *Cassini* spacecraft have discovered rain, rivers and lakes of liquid methane on Saturn's moon Titan.

- The planets and moons gathered together as the result of the collisions of smaller bodies beginning about 4.6 billion years ago.

- Every solid planet or satellite contains impact craters, but in different amounts that depend on the ages of their surfaces.

- Impact craters on the Moon, Mercury, and Jupiter's icy moon Callisto all record an ancient, intense rain of meteorites, which occurred about 4.0 billion years ago, and a continued cosmic bombardment at lower rates since then.

- Ancient craters on the relatively young surfaces of Earth, Venus and Io have been erased by geologic and volcanic activity.

- The round craters on the Moon were formed by the explosive impact of large meteorites that came from interplanetary space, releasing enormous energy, melting rock and excavating circular craters with raised rims on impact.

- Massive impacts of exceptionally big objects gouged out large impact basins on the Earth's Moon and the planet Mercury. Concentric, ring-like rims as tall as mountains can surround the impact basins, and the basins have been subsequently filled with lava.

- Giant impacts in the early history of the solar system may account for the origin of the Earth's Moon, the removal of Mercury's low-density mantle, the backwards rotation direction of Venus, and the crustal dichotomy between the low-lying northern plains and southern highlands of Mars.

- Upon impact with the surface of Mars, ground water ice can be melted, lubricating the ejected material that flows like mud.

- The material ejected from craters on Venus has been shaped by the planet's hot, thick atmosphere into asymmetric, lobate forms. Small impacting projectiles have been burnt up in the thick atmosphere, so there are no small craters on Venus.

- Internal heat can be produced by radioactive decay of rocks inside a terrestrial planet, or within a satellite as the result of varying gravitational interaction with its planet. The giant planets still retain the heat of their formation.

- Molten rock, or magma, that is localized in underground chambers of a planet can rise to the surface and cause two types of basaltic volcanism – tall shield volcanoes and smooth volcanic flows known as plains.

- Earth has unique underwater volcanoes found in mid-ocean ridges that supply a spreading sea-floor, as well as chains of hot-spot volcanoes, such as the Hawaiian Islands, and volcanoes arising from the downward plunge of moving plates.

- Upwelling of internal magma is cracking part of Africa open, in a great rift valley.

- Extensive lava flows filled large impact basins on the Moon, creating the dark lunar maria between 3.9 and 3.2 billion years ago.

- Ancient, smooth volcanic flows on Mercury have obliterated small craters, filled the interiors of large impact basins, and spread out between large craters, producing about 40 percent of the planet's surface. Most of this volcanic activity occurred after the heavy bombardment about 4.0 billion years ago, but before some craters were formed on the smooth plains.

- Extensive volcanic activity on Venus resurfaced the planet about 750 million years ago.

- Mars has the tallest volcanoes in the solar system, and most of its northern hemisphere is covered with volcanic flows of lava.

- The volcanoes on Jupiter's satellite Io have turned the satellite inside out; it is heated inside by the tidal flexing action of nearby massive Jupiter.

- Liquid water flows out of cracks in the icy surface of Jupiter's moon Europa, and erupts as jets of water ice and water vapor from Saturn's moon Enceladus.

- Volcanoes of ice may have created some of the features now frozen into the bright smooth surface of Neptune's largest moon, Triton; dark geyser-like plumes have been observed in the process of eruption on the satellite.

- Seventy-one percent of the Earth's surface is covered with liquid water, and our bodies are largely composed of water.

- Dark, permanently shadowed regions inside craters in the Moon's polar regions could contain water. The *Clementine, Lunar Prospector, Chandrayaan-1* and *LCROSS* spacecraft have provided evidence of very small amounts of water on the Moon.

- Strong radar echoes from the highly reflective polar regions of Mercury suggest that thick deposits of water ice reside in the permanently shadowed interiors of craters near the planet's poles.

- Although Venus is now dried out, it may have once contained a small ocean.

- Small amounts of water vapor are found in the atmosphere of Mars, together with clouds and fogs of water ice.

- Vast amounts of frozen water now exist in the polar caps of Mars and beneath the surface of the polar, mid-latitude, and equatorial regions of the red planet.

- It cannot now rain on Mars, and liquid water cannot now exist for any length of time on the planet's surface.

- Catastrophic floods and deep rivers once carved channels on Mars, and an ancient ocean may have once covered the planet's northern lowlands.

- Saturn's rings consist of billions of particles of water ice.

- Jupiter's satellite Europa is covered with bright, smooth water ice, which has cracked due to the contorting tidal effects of Jupiter's strong gravity. The warmth generated by tidal heating may have been sufficient to form an ocean of liquid water below Europa's icy covering.

- Magnetic measurements provide indirect evidence for an ocean of salty, liquid water below the icy crust of Jupiter's satellite Europa.

- Jupiter's satellite Ganymede also probably contains an ocean of liquid water under its ice-covered surface.

- Saturn's satellite Enceladus has a frozen covering of water ice, and the moon emits icy jets, feeding the E ring that encircles Saturn.

2.1 Flybys, orbiters, probes and landers

We live at an incredible time, when all of the major planets and most of their satellites have been viewed close up with the inquisitive eyes of robotic spacecraft, revealing awesome, unanticipated features that cannot be seen in any other way. No two of these fascinating new worlds are exactly the same. Most of them have been investigated many times, with increasingly sophisticated instruments aboard many different spacecraft. At the same time, ground-based telescopes and telescopes aboard Earth-orbiting satellites have provided other new insights about the planets and their moons.

This captivating voyage of discovery began close to home, in 1957, when the Soviet Union launched the first artificial satellite, the beeping *Sputnik*, stunning the American public and initiating the space race with the United States. The subsequent cold-war competition included the Russian *Luna 3* that swung once around the far side of the Earth's Moon, which had never been seen before, and

Fig. 2.1 Lunar rover   The battery-powered lunar rovers, used in the last three *Apollo* missions, could carry two astronauts and all their equipment for kilometers across the lunar surface. The astronauts deployed instruments and returned rocks to the *Lunar Module*, which carried them back to their orbiting *Command Module* while the rover remained on the Moon. Because there is no substantial atmosphere, water or weather on the Moon, both the rover and the footprints in the lunar soil may last for millions of years. By that time micrometeorites will have pitted the rover and erased the footprints. (Courtesy of NASA.)

culminated on 20 July 1969 when the *Lunar Module Eagle* carried two *Apollo 11* astronauts to the shores of the Moon's Sea of Tranquility.

Altogether twelve astronauts roamed the surface of the Earth's Moon, during the four-year *Apollo* program, dissipating the Soviet Union's lead in space, tarnishing the image of Soviet competence, and teaching the American people that even the frontier of space can be conquered with resolve and willpower, especially in a democratic nation that stresses individual freedom.

The exploration of our Moon served as a stepping-stone to the planets, and established a blueprint for subsequent planetary missions. For both our Moon and the planets, the initial reconnaissance was provided by spacecraft that flew by them, obtaining just a brief glimpse. In the lunar case, three *Ranger* spacecraft were also sent crashing into the Moon's surface, transmitting high-definition pictures on the way down. This was followed by a more-detailed exploration with orbiting spacecraft, such as *Lunar Orbiter 1, 2, 3, 4* and *5*, that can circle a moon or planet many times and map out its global terrain. Like the explorers of new territories on Earth, the orbiters were sent to reconnoiter, to get the lay of the land, and to disclose possible dangers awaiting future visits. The next step involves landers, like *Luna 9* and *13*, and *Lunar Surveyor 1, 3, 5, 6* and *7*,

that explore the surfaces and probes that plunge into the atmospheres.

So far, humans have only visited the Moon. An estimated half billion people watched the televised first visit, on 20 July 1969, when Neil Armstrong (1930–) moved cautiously down the lander's ladder and stood firmly on the fine-grained lunar surface. An ancient dream had come true – man had set foot on another world.

In all, there have been six manned landings on the Moon, beginning with *Apollo 11* in July 1969 and ending with *Apollo 17* in December 1972. The actual landings were performed by the bug-like *Lunar Module* that separated from the main spacecraft while in orbit around the Moon, and returned to it. At first the astronauts traveled on foot, staying near to the *Lunar Module*, but they subsequently moved to more-remote locations in roving vehicles (Fig. 2.1). Altogether, 382 kilograms of rocks were brought back from the Moon for analysis in the terrestrial laboratory, determining the Moon's age, chemical composition, history and probable origin.

Robotic missions then took over, first exploring Earth's nearest planetary neighbors, Venus and Mars, and eventually ranging to the distant giant planets and beyond. The US *Mariner 2* spacecraft had already flown past Venus in 1962, and on the way detected a perpetual wind of

Table 2.1 Important flyby missions in the solar system

Spacecraft[a]	Launch date	Encounter date	Object	Discovery
Luna 3[b]	4 Oct. 1959	7 Oct. 1959	Moon	Photographed backside of Moon
Mariner 2	26 Aug. 1962	14 Dec. 1962	Venus	First successful planetary flyby Measured solar wind
Mariner 4	28 Nov. 1964	14 July 1965	Mars	Craters on Mars
Pioneer 10	3 Mar. 1972	3 Dec. 1973	Jupiter	Passage through asteroid belt
Mariner 10	3 Nov. 1973	29 Mar. 1974	Mercury	Heavily cratered surface
Voyager 1	5 Sept. 1977	5 Mar. 1979	Jupiter	Ring, volcanoes on satellite Io
Voyager 1		12 Nov. 1980	Saturn	Titan's dense atmosphere
Voyager 1		16 Dec. 2004		Termination shock of solar wind at 94 AU
Voyager 2	20 Aug. 1977	9 July 1979	Jupiter	
Voyager 2		25 Aug. 1981	Saturn	
Voyager 2		24 Jan. 1986	Uranus	Rings, magnetic field
Voyager 2		24 Aug. 1989	Neptune	Excess heat, winds, rings, magnetic field
Voyager 2		30 Aug. 2007		Termination shock of solar wind at 84 AU
Giotto[c]	2 July 1985	14 Mar. 1986	Comet Halley	Flyby, image of nucleus
Galileo	18 Oct. 1989	29 Oct. 1991	Asteroid 951 Gaspra	Flyby, image
Galileo		28 Aug. 1993	Asteroid 243 Ida	Ida and its moon Dactyl
NEAR[d]	17 Feb. 1996	27 June 1997	Asteroid 253 Mathilde	Flyby
Deep Space	24 Oct. 1998	28 July 1999	Asteroid 9969 Braille	Flyby
Deep Space	24 Oct. 1998	22 Sept. 2001	Comet Borrelly	Flyby; image of nucleus
Stardust	7 Feb. 1999	2 Jan. 2004	Comet Wild 2	Encounter; sample returned to Earth on 15 Jan. 2006
Hayabusa[e]	9 May 2003	30 Sept. 2005	Asteroid 25143 Itokawa	Encounter, rubble-pile asteroid, landing 19 Nov. 2005
MESSENGER	3 Aug. 2004	14 Jan. 2008	Mercury	Flybys on 14 Jan. 2008, 6 Oct. 2008, and 29 Sep. 2009; Mercury orbit insertion, 18 Mar. 2011
Dawn	27 Sept. 2007	Sept. 2011	Asteroid 4 Vesta	Flyby
Dawn		Feb. 2015	Dwarf-planet 1 Ceres	Flyby
New Horizons		July 2015	Pluto	Flyby

[a] Unless otherwise noted, these spacecraft are missions of the United States National Aeronautics and Space Administration (NASA).
[b] Spacecraft launched by the former Soviet Union.
[c] *Giotto* was a mission of the European Space Agency (ESA).
[d] The acronym *NEAR* stands for *Near Earth Asteroid Rendezvous*.
[e] *Hayabusa*, meaning "falcon" in Japanese, is a mission of the Japan Aerospace Exploration Agency (JAXA).

Fig. 2.2 First visit to Mercury   A photomosaic of Mercury's southern hemisphere produced from images acquired by *Mariner 10* during its first encounter with the planet in March 1974. Mercury has a heavily cratered surface that resembles the lunar highlands. Bright rayed craters are also present on Mercury, as they are on the Moon. (Courtesy of NASA/JPL.)

charged particles in interplanetary space, emanating from the Sun. In 1970, an unmanned entry probe, the *Venera 7* spacecraft, was sent from the USSR into the thick, carbon-dioxide atmosphere of Venus, measuring the temperature and pressure all the way down to the surface. These data showed that the surface of Venus is hot enough to melt lead and that the atmosphere is ninety times as heavy as our air. This was the first craft to touch down successfully on another planet's surface, and it was followed by other *Veneras*, which had just enough time to send back pictures of the volcanic surface before being wiped out by the intense heat and pressure.

The first spacecraft to be launched on lengthy journeys beyond the Earth's Moon were flyby missions – the *Mariners*, *Pioneers* and *Voyagers* – that passed near the planets and their satellites to give us new vistas, unavailable from the ground, making important discoveries in the process (Table 2.1). *Mariner 10*, for example, obtained the first spacecraft photographs of the atmosphere of Venus in 1974, and traveled on to reveal the heavily cratered surface of Mercury (Fig. 2.2).

In 1972–74, the *Pioneer 10* and *11* missions to Jupiter showed that spacecraft could pass safely through the

asteroid belt, blazing a trail for the extraordinarily successful *Voyager 1* and *2* flyby missions to far-distant worlds (Figs. 2.3, 2.4). Their "Grand Tour" included Jupiter (1979; Fig. 2.5), and Saturn (1980, 1981; Fig. 2.6). *Voyager 2* went on to Uranus (1986) and Neptune (1989), and both *Voyager 1* and *2* encountered the termination shock of the solar wind (2004, 2007).

Voyager 1 and *2* vastly improved our understanding of the atmospheres of the giant planets, and discovered unexpected rings, moons and magnetic fields. They also transformed the satellites of the giant planets into unique and distinctive places with diverse surfaces and in some cases atmospheres or magnetic fields.

Comets are so tiny and so far away that you cannot detect them until they come near the Sun, and their centers are then hidden within the brilliant glare of fluorescing gases and reflected sunlight. As a result, no one had ever seen the bare surface of a comet's nucleus until 1986, when the *Giotto* spacecraft peered into the core of comet Halley. It found the nucleus to be a black, oblong chunk of ice and dust, roughly the size of Paris or Manhattan (Fig. 2.7). At the moment of encounter, the comet was spewing out about 25 tons of water every second, propelled

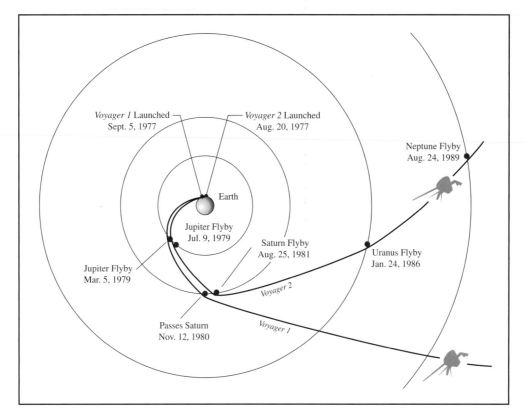

Fig. 2.3 Grand tour of *Voyager 1* and *2*   The flights of the two *Voyager* spacecraft through the solar system. Both spacecraft were launched in 1977 and flew past Jupiter in 1979, transmitting remarkable details of the giant planet's weather and the surfaces of its four largest satellites. *Voyager* 1 and 2 used the gravity of Jupiter to accelerate them on toward Saturn, providing close-up images of its rings and satellites in 1980–81. Saturn provided another gravity assist to propel *Voyager* 2 on to Uranus, in 1986, and then on to Neptune in 1989. *Voyager* 1 was targeted differently at Saturn, sacrificing its grand tour for close views of the satellite Titan. Both spacecraft have now crossed the termination shock of the solar wind, marking its outer boundary.

Fig. 2.4 New perspectives of the giant planets   This montage of the giant planets was prepared from images taken by the *Voyager 2* spacecraft. They include banded Jupiter, ringed Saturn, and Uranus and Neptune with their blue clouds. (Courtesy of NASA/JPL.)

Fig. 2.5 Giant Jupiter's clouds   Jupiter's clouded world with its alternating structure of light zones and dark belts. The two innermost Galilean satellites are also visible. Bright orange Io is seen just above the cloud tops, and icy-white Europa lies to the right. (Courtesy of NASA/JPL.)

into sunward jets by the vaporizing ice. So comets provide evidence that large quantities of water ice can be found in the outer solar system.

Flyby missions are still being used. The *MErcury Surface, Space ENvironment, GEochemistry and Ranging (MESSENGER)* spacecraft, for example, investigated roughly half of the previously unmapped surface of Mercury in 2008 and 2009, settling down into Mercury orbit in March 2011. *Dawn* is on its way for a rendezvous with the asteroids Vesta and Ceres, in 2011 and 2015, respectively. *New Horizons* is expected to fly by Pluto in 2015.

Orbiting spacecraft followed the initial planetary explorations using flybys. The orbiters greatly increased the time available for detailed study, often for years at a time (Table 2.2). They revealed many features that previous flyby missions had missed, and forever changed our view of the planets and their satellites.

The *Mariner 9* mission to Mars first demonstrated the extraordinary promise of planetary orbiters in 1971–72. The three previous flyby missions, *Mariner 4, 6* and *7*, had discovered the ancient cratered terrain on Mars, but missed all of the younger geologic features. The orbiting

Fig. 2.6 Ringed Saturn with icy moons   Saturn's yellow-brown clouds are swept into bands by the planet's rapid rotation. Two of its white moons (*left*), Tethys (*above*) and Dione (*below*), are covered with water ice. The shadows of Saturn's main rings and Tethys are cast onto the cloud tops. The outer A ring is separated from the central B ring by the dark Cassini Division, which is 3500 kilometers wide. This gap is so tenuous that the edge of Saturn can be seen through it. The faintest of Saturn's main rings, the inner C ring or crepe ring, is barely visible against the planet. This image was obtained from the *Voyager 1* spacecraft on 3 November 1980. (Courtesy of NASA/JPL.)

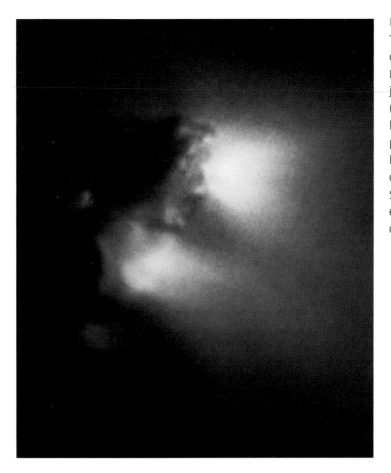

Fig. 2.7 The black heart of comet Halley
The coal-black nucleus of comet Halley is a dirty ball of ice, about the size of Paris or Manhattan. It is silhouetted against bright jets of water and dust that stream sunward (*right*) from at least three places that have been warmed by the Sun's radiation. In this projection, the nucleus measures 14.9 kilometers by 8.2 kilometers. This is a composite of images taken by the European Space Agency's *Giotto* spacecraft near its encounter with the nucleus of comet Halley on 14 March 1986. (Courtesy of ESA.)

Mariner 9 had sufficient time to completely explore the planet, revealing for the first time Mars' great volcanoes, the vast canyon system Valles Marineris, and evidence of ancient stream-beds and water erosion.

The *Viking 1* and *2* orbiters amplified and enhanced this new perspective of Mars, beginning in 1976 (Fig. 2.8). Close-up, high-resolution views of the surface of Mars were next obtained from the *Mars Global Surveyor* at the end of the 20th century and the beginning of the 21st century. The images showed much finer detail than those obtained with the *Viking* orbiters. Beginning in April 2001, the *2001 Mars Odyssey* mapped the amount and distribution of the chemical elements and minerals on the Martian surface, and obtained evidence for subsurface water ice. The mission was named after the movie *2001: A Space Odyssey*. In August 2005 the *Mars Reconnaissance Orbiter* began obtaining new information about the surface, subsurface and atmosphere of Mars, to characterize the planet's climate and geology, to determine if life arose there, and to prepare for eventual exploration of Mars by humans.

The *Pioneer Venus* spacecraft began orbiting Venus in 1978, measuring strong winds in the planet's upper atmosphere and using radar to penetrate the thick atmosphere and map much of the planet's surface with low resolution. Five probes were also launched from *Pioneer Venus*, measuring the properties of the planet's atmosphere as they descended through it.

Five years later, the *Venera 15* and *16* orbiters also used radar to map the surface of Venus, but it wasn't until the 1990s that the *Magellan* orbiter used radar to map the entire planet with a clarity and resolution not available for much of Earth. Since Venus is perpetually shrouded in opaque, sulfurous clouds, this was the only way to establish its global surface terrain. *Magellan's* radar images have revealed an unearthly world that was resurfaced about 750 million years ago by rivers of outpouring lava, and disclosed numerous volcanoes that now pepper its surface (Fig. 2.9).

The *Galileo* orbiter–probe spacecraft, launched in October 1989, was so massive that no existing rocket had the power to launch it directly to Jupiter, its primary target. Instead, the spacecraft was placed on a looping trajectory that took it past Venus once and Earth twice (Fig. 2.10). The gravity of these planets was used to accelerate and propel the spacecraft in slingshot fashion toward its eventual rendezvous with the giant planet, somewhat like a pitcher winding up to throw a high-velocity strike. While

Table 2.2 Important orbital missions in the solar system

Spacecraft[a]	Launch date	Encounter date	Object	Discovery
Lunar Orbiter 1	10 Aug. 1966	14 Aug. 1966	Moon	Global photographs of lunar surface
Mariner 9	30 May 1971	13 Nov. 1971	Mars	Global image, volcanoes, canyons, outflow channels
Viking 1	20 Aug. 1975	19 June 1976	Mars	Orbiter and lander, surface photographs, life search
Viking 2	9 Sept. 1975	7 Aug. 1976	Mars	Orbiter and lander, surface photographs, life search
Pioneer Venus	20 May 1978	4 Dec. 1978	Venus	Orbiter and multi-probe, global radar images
Magellan	4 May 1989	10 Aug. 1990	Venus	Orbiter, radar maps of surface, volcanic resurfacing
Clementine	25 Jan. 1994	21 Feb. 1994	Moon	Mapped global surface composition and topography of Moon; possible evidence for water ice at lunar poles
Galileo	18 Oct. 1989	7 Dec. 1995	Jupiter	Orbiter and probe, atmosphere and four largest satellites; *Galileo* impacted Jupiter on 21 Sept. 2003
NEAR[g] Shoemaker	17 Feb. 1996	14 Feb. 2000	Asteroid	First orbiter of an asteroid, 433 Eros, determining its composition, size, shape, and mass
Mars Global Surveyor	7 Nov. 1996	12 Sept. 1997	Mars	Orbiter, laser altimeter, magnetometer, high-resolution images, mapping began on 4 April 1999, water flow and volcanic activity, ancient magnetism
Lunar Prospector	7 Jan. 1998	15 Jan. 1998	Moon	Maps global elemental abundance, magnetic field and gravity, detection of lunar core, evidence for substantial water ice at lunar poles
2001 Mars Odyssey	7 Apr. 2001	24 Oct. 2001	Mars	Maps the amount and distribution of chemical elements and minerals on the Martian surface, and provides evidence for substantial subsurface water ice; contact lost 14 Nov. 2006
Cassini[c]	15 Oct. 1997	1 July 2004	Saturn	Atmosphere, rings, magnetic environment, methane rain and lakes of methane and ethane on moon Titan, active water jets on moon Enceladus
Mars Reconnissance Orbiter	12 Aug. 2005	1 July 2006	Mars	Information about surface, subsurface and atmosphere of Mars to determine whether life ever arose on Mars, to characterize the climate and geology of Mars, and to prepare for human exploration of Mars
Venus Express[d]	09 Nov. 2005	11 April 2006	Venus	Measurements of atmosphere, clouds, polar vortex

Table 2.2 (*cont.*)

Spacecraft	Launch date	Encounter date	Object	Discovery
Kaguya[e] (Selene)	14 Sept. 2007	3 Oct. 2007	Moon	Lunar topography maps, gravity map of far side of Moon, south pole region, Shackleton crater
Chang'e-1[f]	24 Oct. 2007	5 Nov. 2007	Moon	Composition and images of surface
Chandrayaan-1[g]	22 Oct. 2008	8 Nov. 2008	Moon	Small amounts of water on Moon
Lunar Reconnaissance Orbiter and *LCROSS*	18 June 2009	18 Sept. 2009	Moon	High-resolution mapping of lunar surface in preparation for manned landing; *LCROSS* impacts lunar surface providing evidence for water on Moon

[a] Unless otherwise noted these spacecraft are missions of the United States National Aeronautics and Space Administration (NASA), except the *Cassini* mission, which is a joint ESA and NASA venture.
[b] The acronym *NEAR* stands for Near Earth Asteroid Rendezvous.
[c] The *Cassini–Huygens* mission is a joint ESA and NASA venture, where ESA denotes the European Space Agency.
[d] *Venus Express* is a mission of the European Space Agency.
[e] *Kaguya* is a Japanese mission.
[f] *Chang'e-1* is a mission of the People's Republic of China.
[g] *Chandrayaan-1* is a mission of India's national space agency.

Fig. 2.8 Mosaic of Mars   This computer-generated mosaic of *Viking 1* and *2* orbiter images of Mars shows three volcanoes as dark spots to the west (*left*), while the bottom center of the scene shows the entire Valles Marineris canyon system, from Noctis Labyrinthus (*left*) to the chaotic terrain (*right*). Outflow channels are found in the north (*top*), and a variety of clouds and hazes are also visible, especially near the planet's edge. (Courtesy of NASA/USGS.)

Fig. 2.9 Venus unveiled with radar A cloud-penetrating radar system on board the *Magellan* spacecraft has mapped the global landforms and features on Venus with a resolution of 120 meters, more completely than any planet including Earth. This hemisphere, centered at 180 degrees east longitude, shows the bright, planet-wide, equatorial highlands that contain towering volcanoes, long lava flows and deep faults and fractures. They run from lower left to upper right through Aphrodite Terra (*left of center*), a continent-sized highland, and the bright highland Atla Regio (*just right of center*) to Beta Regio (*far right and north*). Dark areas correspond to terrain that is smooth on the scale of the radar wavelength (0.13 meters); bright areas are rough. The orange tint, based on color images taken by the *Venera 13* and *14* landers, simulates the color of sunlight at ground level after being filtered through the planet's thick atmosphere and clouds. (Courtesy of NASA/JPL.)

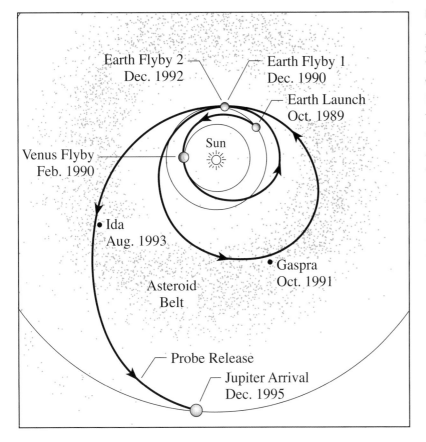

Fig. 2.10 Galileo's long flight to Jupiter After launch in October 1989, the *Galileo* spacecraft used the gravity of the Earth and Venus to accelerate it on to its encounter with Jupiter, six years after launch. In its long, indirect flight path, *Galileo* was able to fly past two asteroids at close range, 951 Gaspra in October 1991 and 243 Ida in August 1993 (see Fig. 2.11). The spacecraft entered into orbit around Jupiter in December 1995, when its descent probe, which had been released five months earlier, dove into the giant planet's atmosphere. The main spacecraft continued to orbit the planet and examine its satellites until 21 September 2003.

Fig. 2.11 Asteroids close up   The Jupiter-bound *Galileo* spacecraft took images of the asteroids 951 Gaspra (*left*) and 243 Ida (*right*) on 29 October 1991 and 28 August 1993, respectively. Both objects have irregular, elongated shapes, suggesting numerous past collisions. Gaspra has dimensions of 19 × 12 × 11 kilometers, and it contains a striking abundance of small craters. Ida is almost three times as long as Gaspra, and a small moon, named Dactyl, accompanies Ida, though the tiny satellite is not visible in this image. Ida is a member of the Koronis family of asteroids, presumed fragments left from the breakup of a larger precursor asteroid in a catastrophic collision. (Courtesy of NASA/JPL.)

the roundabout route took six years, in comparison to the direct 21-month flights of *Pioneer 10* and *11*, it also took *Galileo* on close encounters with two asteroids along the way (Fig. 2.11).

Galileo carried an entry probe that penetrated Jupiter's kaleidoscopic clouds, obtaining the first direct, or in-situ, sampling of a giant planet's atmosphere. The main orbiting spacecraft looped around Jupiter for more than five years, until 2003, obtaining high-resolution images and analysis of the planet's stormy weather, its sparse ring system, and the four large moons. It has provided new insights to these four Galilean satellites (Fig. 2.12), including: volcanic activity on Io; compelling evidence for a global ocean beneath the ice crust of Europa; and the discovery of a global dipolar magnetic field generated within Ganymede, the first such field to be found on a moon.

Seventeen countries and several space agencies, including NASA and ESA, contributed to the *Cassini–Huygens* mission to Saturn and its moons. After a seven-year journey, the spacecraft reached Saturn in July 2004, beginning unparalleled investigations of the planet's atmosphere, rings, satellites and magnetic environment (Fig. 2.13).

Landers and probes are the third category of spacecraft used to investigate the moons and planets (Table 2.3). They have landed on the Moon and Mars, and impacted comet Tempel 1 and the Moon's south polar region, while entry probes have parachuted into the atmospheres of Venus, Jupiter and Saturn's satellite Titan.

The robotic exploration of Mars began with the two *Viking* landers, each a one-ton laboratory that safely landed

on the planet's surface in 1976. They obtained beautiful panoramas of the Martian surface and measured the properties of the thin, freezing atmosphere. The *Viking* landers were also sent to search for extant life on Mars, but the results were inconclusive.

On 4 July 1997 the *Mars Pathfinder* landed near the mouth of a canyon system carved by massive floods more than 3.5 billion years ago (Fig. 2.14). Unlike the *Viking* landers, which were shackled to one location, *Pathfinder* contained the small mobile *Sojourner Rover*, which could roam across the surrounding terrain. It was about the size of a small microwave oven and equipped with six-wheel drive. *Sojourner* explored about 250 square meters of the Martian surface, measuring the chemical makeup of the rocks and surface. The distribution of nearby rocks, dust and pebbles were consistent with the downstream deposit of flowing water from an outflow channel long ago (Fig. 2.15).

Sojourner paved the way for the two *Mars Exploration Rovers*, named *Spirit* and *Opportunity*, which landed on opposite sides of Mars in 2003 and continued to explore the surface for more than six years. Each the size of a small dune buggy, they contained cameras and instruments that demonstrated the existence of past flowing water at both locations.

The *Cassini* spacecraft carried the *Huygens Probe* that descended into the hazy, dense atmosphere of Saturn's moon Titan, landing on its surface on 14 January 2005. *Huygens* determined the properties of Titan's Earth-like atmosphere and its mysterious surface below, providing evidence for methane rain and rivers of liquid methane

Fig. 2.12 Giant red spot and Galilean satellites   The edge of Jupiter with its Great Red Spot and on the same scale the planet's four largest moons, known as the Galilean satellites. From top to bottom the moons are Io, Europa, Ganymede and Callisto. Winds blow counter-clockwise around the Great Red Spot, which has been observed for more than 300 years and is larger than one Earth diameter. Europa is about the size of Earth's Moon, and Ganymede, the largest moon in the solar system, is bigger than the planet Mercury. (Courtesy of NASA/JPL.)

on Titan's surface (Fig. 2.16). Radar from *Cassini* revealed lakes of liquid methane and ethane.

In the early 21st century, we can reflect in amazement at the incredible new worlds that have been discovered by the flybys, orbiters, landers and probes. Future spacecraft are now poised to continue the exploration in greater detail, focusing on issues such as the search for life outside the Earth, and the origin and discovery of planetary systems around the Sun and other stars. Scientists will, for example, ultimately return samples of the surface of Mars for study in our Earth-bound laboratories, to examine them for fossil or recent evidence of life.

Fig. 2.13 In Saturn's shadow This marvelous panoramic view of Saturn and its rings was created from 165 images taken by a wide-angle camera aboard the *Cassini* spacecraft on 15 September 2006, as it drifted into the darkness of Saturn's shadow and permitted observations of the planet's tiny ring particles. The narrowly confined G-ring resides just outside the bright, inner main rings. The outer, wide E-ring encircles the entire system; icy plumes feed the E-ring from Saturn's satellite Enceladus. The exaggerated color contrast in this mosaic view can be used to infer processes that are sorting the ring particles according to their size. (Courtesy of NASA/JPL/SSI.)

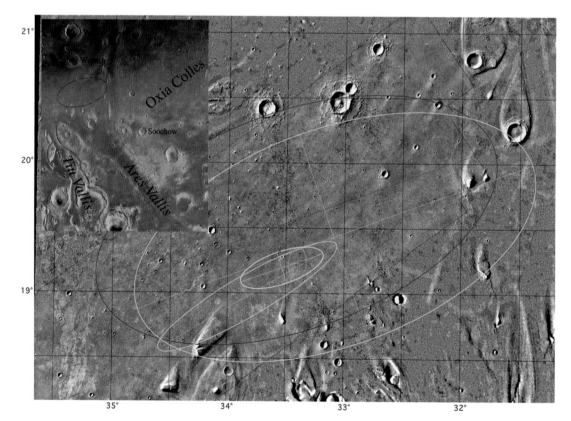

Fig. 2.14 *Mars Pathfinder* lands on an ancient flood plain About 3.7 billion years ago, great floods rushed out of the outflow channel, Ares Vallis, and emptied into the Chryse Planitia, or Plains of Gold, region of Mars (*color inset*). The flowing water carved out streamlined islands around craters (*top right*). This area was chosen as the *Mars Pathfinder* landing site for three reasons: it seemed safe, with no steep slopes or rough surfaces; it had a low elevation, which provided enough air density above the surface for a parachute to work; and it appeared to offer a variety of rock types deposited by the floods. The ellipses mark the area targeted for landing of *Mars Pathfinder*, as refined several times during the final approach to Mars. An X within the smallest ellipse marks the location of the lander at 19.33 degrees north and 33.55 degrees west. The site is about 850 kilometers southeast of the location of *Viking 1* lander. (Courtesy of NASA/JPL.)

Fig. 2.15 Flood debris on Mars   A panorama of the surface of Mars taken from the *Mars Pathfinder* spacecraft soon after its landing on 4 July 1997. The scene is littered with boulders and rocks, the debris of catastrophic floods early in the planet's history. The flood residue is found between a few meters away from the lander to the two modest hills, the Twin Peaks, which are about 30 meters tall and located at a distance of about one kilometer. Between and partially covering the rocks is rust-red, iron-oxide dust, the result of chemical weathering of exposed rock surfaces here and elsewhere. (Courtesy of NASA/JPL.)

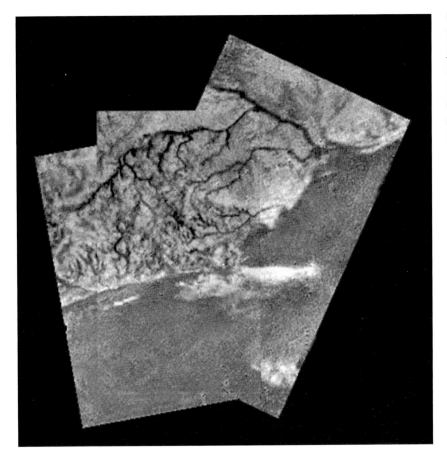

Fig. 2.16 Methane flows on Titan   This image was taken form the *Huygens Probe* just before it landed on the surface of Saturn's large satellite Titan on 14 January 2005. It shows flows down a high ridge into a major river channel from different sources. The feature has been attributed to liquid methane fed by the fall of methane rain. (Courtesy of NASA/JPL/ESA/U. Arizona.)

Table 2.3 Landers and probes

Spacecraft[a]	Launch date	Encounter date	Object	Discovery
Luna 9[b]	31 Jan. 1966	3 Feb. 1966	Moon	Soft landing, Oceanus Procellarum
Surveyor 1	30 May 1966	2 June 1966	Moon	Soft landing near Flamsteed
Apollo 11	16 July 1969	20 July 1969	Moon	First humans on Moon, Mare Tranquillitatis, sample return
Venera 7[b]	17 Aug. 1970	15 Dec. 1970	Venus	High surface temperature and pressure
Venera 9[b]	8 June 1975	22 Oct. 1975	Venus	Surface photograph, volcanic rocks
Viking 1	20 Aug. 1975	20 July 1976	Mars	Chryse Planitia, negative life search, monitor environment for more than one Martian year
Viking 2	9 Sept. 1975	3 Sept. 1976	Mars	Utopia Planitia, negative live search, monitor environment for more than one Martian year
Galileo	18 Oct. 1989	7 Dec. 1995	Jupiter	Atmosphere winds, ingredients
Mars Pathfinder	4 Dec. 1996	4 July 1997	Mars	Chryse Planitia, surface rover *Sojourner* examines dust, pebbles and rocks near mouth of outflow channel
Cassini–Huygens[c]	15 Oct. 1997	14 Jan. 2005	Titan	Huygens probe, methane rain, vast lakes of liquid methane
Spirit	10 June 2003	3 Jan. 2004	Mars	Gusev crater, rover, past water flow
Opportunity	7 July 2003	14 Jan. 2004	Mars	Meridiani Planum, rover, past water flow
Deep Impact	12 Jan. 2005	4 Jul. 2005	Comet	Water ice on comet Tempel 1
Phoenix	3 Aug. 2007	25 May 2008	Mars	Water ice in Martian subsurface at north polar plains of Mars
LCROSS[d]	18 June 2009	9 Oct. 2009	Moon	Evidence for water ice by impact in permanently shadowed lunar crater, south polar region
Rosetta	2 Mar. 2004	Nov. 2014	Comet	Landing on Comet 67/Churyumov-Gerasimenko. The spacecraft passed by two asteroids, 2867 Steins in Sept. 2008 and 21 Lutetia in July 2010.

[a] Unless otherwise noted the spacecraft are missions of the United States National Aeronautics and Space Administration, abbreviated NASA.

[b] Spacecraft launched by the USSR.

[c] The *Cassini–Huygens* mission is a joint ESA and NASA venture, where ESA denotes the European Space Agency.

[d] The acronym LCROSS designates *Lunar CRater Observatory and Sensing Satellite.*

The space-age investigations of the moons and planets have shown us that each moon or planet is unique, the result of different combinations of physical, chemical and dynamical processes that have formed and shaped it. Yet there are fascinating similarities between the major planets and some moons, despite the individual differences that make each of them stand apart. They all exhibit common properties and similar processes, such as impact craters, volcanoes, water and atmospheres, reminding us of the basic elements in ancient Greek philosophy – earth, fire, water and air.

2.2 Impact craters

Ubiquitous craters, ancient records of formation

Impacts have played an important role in the early history of the planets and their subsequent evolution. The collisions of small bodies to create larger bodies resulted in the formation of the planets and their satellites, which gradually cleaned out interplanetary space. Even about 4.0 billion years ago, the impact rate was still sufficient to

Fig. 2.17 Craters on Jupiter's moon Callisto The ancient surface of Callisto shows one of the highest densities of impact craters in the solar system. The satellite's icy surface is as rigid as steel, permitting it to record the bright scars of a heavy bombardment by meteorites roughly 4.0 billion years ago. As shown in this image, taken from the *Galileo* spacecraft in May 2001, Callisto's surface seems uniformly cratered, but it is not uniform in color or brightness. Scientists believe the brighter areas are mainly water ice, while the darker areas are highly eroded, ice-poor material. (Courtesy of NASA/JPL/DLR.)

produce large impact basins with diameters measured in hundreds to thousands of kilometers. Impacts by smaller bodies have always been more frequent, since there are more small objects in space than large ones, and they also contributed to the creation of the ancient, heavily cratered terrain on the moons and planets.

The similarity of the highland crusts of the Moon, Mars and Mercury, despite their differing masses and locations, indicates that impacting objects were spread throughout the inner solar system during its early days. The ubiquitous craters found on the icy satellites of Jupiter, such as Callisto (Fig. 2.17), suggests that they were also subject to a heavy bombardment in their early formative stages. All the planets and satellites most likely experienced the most intense hail of impacting projectiles at about the same time as the Earth's Moon, for scientists think that the entire solar system, with its Sun, planets and their satellites, gathered together beginning 4.6 billion years ago.

Using impact craters to estimate surface age

Radioactive dating of the rocks that *Apollo* astronauts have brought back from the Earth's Moon has permitted a reconstruction of the rate of impact during most of the Moon's history. Some of these rocks are the oldest ones ever found, and they show that the Moon accumulated by the aggregation of rocky projectiles about 4.6 billion years ago. When the Moon was very young, its outer layers were probably molten, but the crust cooled and became solid. The battered lunar surface that we see today remains a museum of impact scars that were created back then. It records an intense, heavy bombardment of leftover formation material that created the large impact basins and most of the lunar craters roughly 4.0 billion years ago.

When the measured ages of lunar rocks are combined with crater counts, scientists obtain a record of the cratering rate on the Moon and its variation with time (Fig. 2.18). The earliest pace of bombardment declined very rapidly for the first 800 million years following the start of planetary formation 4.6 billion years ago. The late heavy bombardment, which ended about 3.9 billion years ago on the Moon, is thought to account for the most heavily cratered regions on Mercury and Mars, as well as the lunar highlands, all saturated with craters upon craters.

Gradually the hail of impacting meteorites decreased, as most of the interplanetary meteoroids were swept up and pulled in, and the rate of cratering slowed during

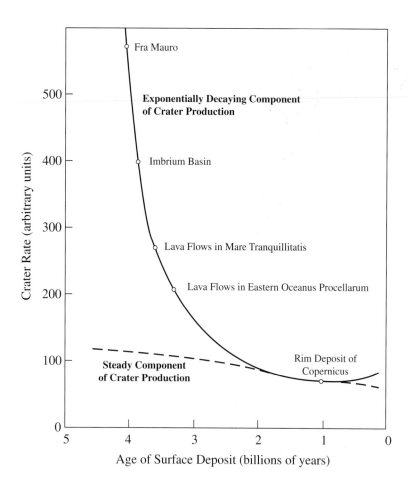

Fra Mauro

**Exponentially Decaying Component
of Crater Production**

Imbrium Basin

Lava Flows in Mare Tranquillitatis

Lava Flows in Eastern Oceanus Procellarum

Rim Deposit of
Copernicus

**Steady Component
of Crater Production**

Crater Rate (arbitrary units)

Age of Surface Deposit (billions of years)

Fig. 2.18 Varying crater rate on the Earth's Moon   The rate of forming craters on the lunar surface is plotted against time. The circles denote the crater rate and rock ages at various *Apollo* landing sites. The crater rate was very high during an intense bombardment that occurred 3.9 billion years ago. The rate dropped rapidly during the subsequent billion years, giving way to the lower steady rate of crater production that has persisted for the last 3 billion years. With such a curve, we can obtain approximate surface ages just by counting the number of craters in different parts of the Moon. Estimates of the ages of planetary surfaces can similarly be obtained from the density of craters on them.

the subsequent billion years. A much lower, steady rate of crater production has persisted for the last 3.0 billion years, so relatively young craters are hard to find on the Moon. White rays that splash across the lunar surface distinguish young craters. The rays of older craters are darkened by eons of continued impact by small meteorites.

Similar declining rates of impact crater formation are thought to have occurred in the final stages of planetary and satellite formation throughout the solar system. Detailed counts of the numbers of craters at different part of the ancient surfaces of these bodies can be used to estimate their ages to within a few million years, particularly in locations on Mars or Mercury that date back to the tail end of the heavy bombardment.

In contrast, the surfaces of the Earth, Venus and Jupiter's satellite Io have relatively few craters, for they have been erased by subsequent geological or volcanic activity. Erosion by wind and water, the deposition of sediments, the collisions of continents, and the internal churning of its rocks have erased any record of an ancient intense bombardment of the Earth. Massive volcanic outpourings covered the surface of Venus with lava about 750 million years ago, and Io is so volcanically active that its surface is continually renewed and contains no impact craters whatsoever.

How to make a crater

Comparatively recent craters still exhibit the details of the impact that created them (Fig. 2.19); older craters have been worn away by small particles that continuously bombard the moons and planets.

There are several lines of evidence that the lunar craters were formed by the explosive impact of interplanetary projectiles:

(i) The amount of material piled on a crater's raised rim is nearly equal to the material excavated from the interior, so if the rim was pushed back into the crater its depressed floor would rise to the level of the neighboring surface.

(ii) Nearly all of the Moon's craters are round. The explosive force of a large impacting object, or meteorite, will produce round craters despite the fact that the projectiles that produced them must have arrived in a variety of directions — some nearly vertically, others at a glancing angle.

(iii) The rocks returned from the heavily cratered regions on the Moon consist of fragments of pre-existing rocks that have been welded together by the enormous pressures of impact.

Fig. 2.19 Lunar crater Timocharis
Astronauts on board the *Apollo 15* mission took this image of the medium-sized crater Timocharis, about 34 kilometers across, in August 1971. The deposits and ejected material have been thrown outward in the radial direction by the meteorite impact that created the primary crater with its circular rim. Smaller secondary craters are located outside this material (*lower left*). (Courtesy of NASA.)

So, solid, rocky objects, named meteoroids, which came from interplanetary space and hit the Moon, must have created the lunar craters, as well as those found on the terrestrial planets and some other moons. Rocky meteoroids were strewn throughout the solar system in its youth, orbiting the Sun for millions and even billions of years until they happened to collide with a moon or planet. This leftover residue from planet formation was more abundant in the early history of the solar system, for it was gradually swept up through collisions with larger bodies.

When the meteoroids strike the surface of a planet or satellite they are called *meteorites*. Meteorites of all sizes have hit the Moon, and its crust records the impact of more small meteorites than large ones.

Although the projectile vaporizes on impact, the explosion excavates material and hurls the pulverized debris outward, creating a raised circular rim 10 to 20 times as wide as the impacting meteorite (Fig. 2.20). Many of the ejected rocks were large enough to create their own craters in turn; they are known as secondary craters. The depth of the crater hole is about one-tenth of its diameter, a relationship that holds for simple craters on the Moon, Mars, Mercury, Earth and Callisto.

The interplanetary bodies, the meteoroids, travel at rapid velocities of tens of kilometers per second, and carry a high kinetic energy. When they impact the surface of a planet or satellite, about half of this kinetic energy is transferred to the target when it stops the moving object. Its energy of motion is suddenly transformed to shocks which engulf and vaporize the impacting object and heat the surface. The high temperature and pressure melt material at the point of impact, and push it downward, while the shocks excavate the crater cavity and throw up a rim of pulverized and melted rock around it. This material is carried out in all directions from the point of impact, creating circular craters and radial ejecta.

The rim is tossed out almost nonchalantly, something like flicking a particle off the end of a whip, but the excavated material is still many times as massive as the impacting projectile and about 10 000 times the volume. This is essentially due to the energy released by the impacting object, which can be equivalent to the explosion of tens of thousands of hydrogen bombs. Violent rebound of the crater floor, from the greater energy and shock of larger meteorites, gives rise to a central peak or peaks. In addition, many of the larger craters have terraced walls caused by rim material slumping in toward the crater center.

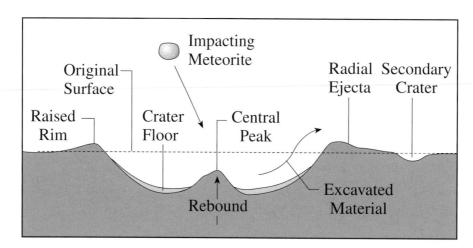

Fig. 2.20 Cross-sectional anatomy of a crater   An impacting meteorite excavates a circular crater that is about 20 times the diameter of the meteorite. The depth of the crater is roughly one-tenth its diameter, and the crater floor is depressed below the surrounding terrain. The explosion gouges out a circular hole, depositing material around its rim and ejecting debris outward in the radial direction. The surface rebounds from the impacting force of a large meteorite, creating a central peak in the floor of the biggest craters.

Giant impacts

The most distinctive features on the Moon are the circular craters that closely pepper its surface, and the largest lunar craters are the impact basins. A typical one is the Imbrium Basin, with a diameter of 1500 kilometers. Its outline can be seen with the unaided eye, forming an "eye socket" of the face of the "Man on the Moon". Its outer rim is defined by prominent mountain ranges, such as the Apennine Mountains (Fig. 2.21). Such basins were created early in the Moon's history, and they were subsequently flooded and nearly filled with dark molten lava from the interior.

Ejecta from the Imbrium Basin gouged out radial ridges and valleys that went a quarter of the way around the Moon, scattering a thick blanket of debris over most of the near side of the Moon. The energy of impact was so great that the floor of the crater rebounded, surging up and down and creating multiple, concentric, ring-like rims as the lunar surface vibrated like the head of a drum.

As on the Moon, there are numerous small bowl-shaped craters on Mercury, smaller than 100 meters in diameter, and large impact craters up to a thousand kilometers across. Both worlds also contain a few young craters with bright rays as well as many older craters without rays, and both the Moon and Mercury have no atmosphere or weather to erode their surface.

Mercury's surface also contains multi-ringed impact basins, such as the Caloris Basin (Fig. 2.22). It has been named *Caloris*, the Latin name for "heat", because it is located at a place on Mercury that faces the Sun when the planet is at the point in its orbit that is closest to the Sun. The rim of mountains that marks the outer boundary of the Caloris Basin is about 1550 kilometers in diameter.

The cataclysmic impact that created the Caloris Basin occurred an estimated 3.85 billion years ago when a meteorite roughly 150 kilometers across hit Mercury, like a cosmic bomb with an energy of a trillion 1-megaton hydrogen bombs. The violent explosion reverberated through the young planet, sending strong seismic waves along the surface and through the deep interior (Fig. 2.23). These waves converged to a focus on the side of Mercury opposite to the Caloris Basin, producing a huge region of cracks, faults, hills and valleys.

The Earth's Moon probably formed from the remains of a giant impact with our planet. A Mars-sized object apparently struck the young Earth, melting the surface material at the point of impact and sending debris into orbit that eventually congealed to become the Moon.

A similar massive impact may have sent Venus spinning in the opposite direction to that of the rotation and orbital motion of all the other major planets. It has also been proposed that a giant impact during the early stages in the history of Mars gouged out most of the Martian crust in the planet's northern hemisphere, resulting in a global crustal dichotomy between the low-lying north and the southern highlands. An original low-density mantle of young Mercury might have been similarly blasted off, leaving its dense iron core behind. But all of these ideas are speculations, hypotheses that have not been fully confirmed.

Unique craters on Mars and Venus

The shape of virtually all impact craters is a circular depression with an upraised rim. The details differ according to

Fig. 2.21 The Moon's Apennine mountains   The radial structure and steep inner slopes of these mountains (*lower right*) mark a section of the outer rim of the Imbrium Basin. The huge excavation was subsequently filled with lava to form the smooth Mare Imbrium and partially submerge the inner ring of mountains (*upper left*). The smaller circular craters include Timocharis, which is also shown in Fig. 2.19 and is about 34 kilometers in diameter (*center left*) and the largest round structure Archimedes (*upper center*) with a diameter of 83 kilometers. (Photo courtesy of UCO/Lick Observatory.)

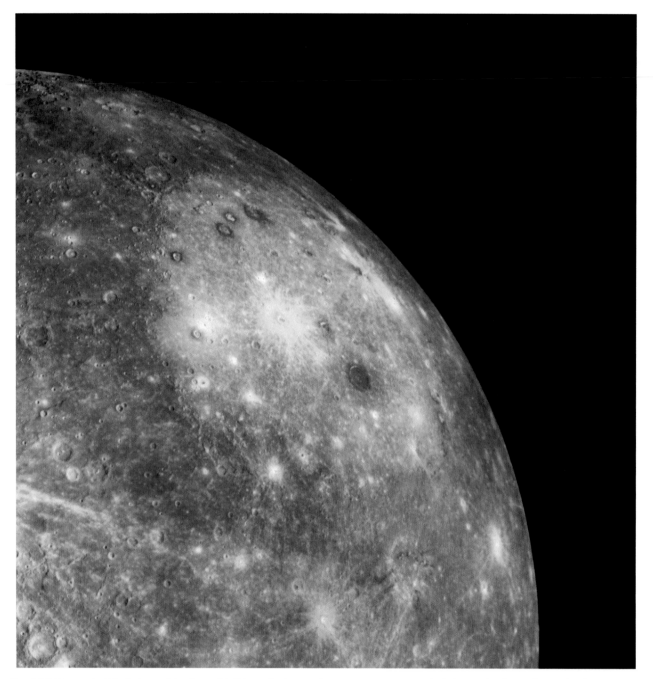

Fig. 2.22 Mercury's Caloris impact basin   This false-color image shows Mercury's great Caloris impact basin as a large, circular orange feature in the center of the image. It was acquired on 14 January 2008 from the *MESSENGER* spacecraft. The smaller, bright orange spots just inside the rim of Caloris basin are thought to mark the location of volcanic features. The color variations in the surrounding plains indicate Mercury's variable surface composition. (Courtesy of NASA/JHUAPL/ASU/CIW/*Science*/AAAS.)

the varying size of the impacting object, from small, simple bowl-shaped holes to larger complex craters with central peaks and internally terraced rims and the largest multi-ring impact basins. But given these variations, the craters that have been excavated from the Moon and terrestrial planets are quite similar.

In contrast, there are noticeable variations in the patterns of ejected material. The round, fresh craters on the Moon are surrounded by secondary craters and bright rays thrown out in ballistic trajectories and undisturbed by any atmosphere. Many craters on Mars display mud-like ejecta, most likely associated with impact melting of subsurface ice. Asymmetric ejecta that surround craters on Venus are attributed to the oblique impacts of meteoroids interacting with the planet's thick atmosphere.

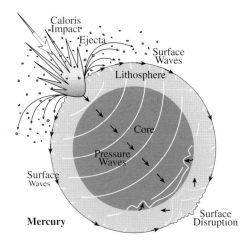

Fig. 2.23 Explosive impact on Mercury   When an exceptionally large meteorite hit Mercury an estimated 3.85 billion years ago, it sent intense waves around the planet and through its core. They came to a focus on the opposite side of Mercury, disrupting the surface and producing hilly and lineated terrain there. The Caloris Basin was excavated at the impact site (also see Fig. 2.22), and it now exhibits concentric waves that froze in place after the impact.

Upon heating of ground ice by impact on Mars, liquid water is most likely incorporated into the ejecta, lubricating the material that flows along the ground after ejection (Fig. 2.24). As a result, some craters on Mars resemble those produced by impacts into mud.

The largest impact craters on Mars are also shallower than their lunar counterparts, with more subdued rims and flatter floors, and there are fewer smaller craters on Mars than there are on the Moon. These differences might be explained by enhanced erosion that modified the worn, old-looking craters and wiped out many of the existing small craters during the planet's early history, when the majority of craters were still forming.

Following impact, large objects left craters on Venus that at first sight resemble those on the Moon, with central peaks, flat floors and distinct circular rims. But the dense atmosphere on Venus affected both the incoming projectile and its ejected debris, creating features that are unlike any other craters in the solar system. The bright apron of debris that surrounds large craters on Venus often has a lobate, petal-like appearance with an unexpected asymmetry (Fig. 2.25). Material that was ejected from the crater became entrained in the hot, thick atmosphere, transforming it into a turbulent, fluid-like substance. The material flowed and spread out from the crater, creating patterns that resemble flowers or butterflies, rather than hurtling away from it to great distances.

Moreover, when the impact on Venus was oblique, the atmospheric wake of the incoming object prevented the ejecta from scattering back in the direction from which

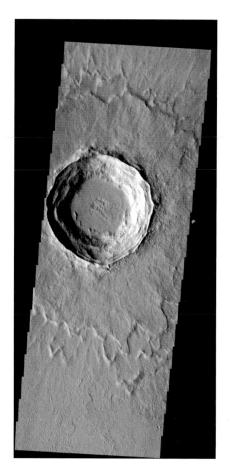

Fig. 2.24 Material ejected from impacts on Mars   Some Martian craters are surrounded by discrete lobes of fresh-appearing flows, each surrounded by a low ridge or rampart and sometimes layered. The unique pattern of ejected material can be attributed to melting of water ice by the heat of impact. An instrument aboard the *2001 Mars Odyssey* orbiter took this image. (Courtesy NASA/JPL/ASU.)

the impacting meteoroid came, so ejecta are missing in this region. Small incoming projectiles never made it to the ground, for they were burned up in the thick atmosphere. There are consequently no very small craters on Venus.

Large impact craters on Venus are relatively scarce when compared with the closely spaced, overlapping lunar craters. At one time Venus was probably as heavily cratered as the Earth's Moon, but the relatively small number and wide spacing of the craters now on Venus indicate that the surface we now see is much younger. When the Moon's cratering rate is scaled to Venus, the relative paucity of craters on Venus's surface indicates an average surface age of about 750 million years, but the planet originated about 4.6 billion years ago. The relatively few craters we now see are due to meteorite impact since rivers of outpouring lava resurfaced the entire planet about 750 million years ago.

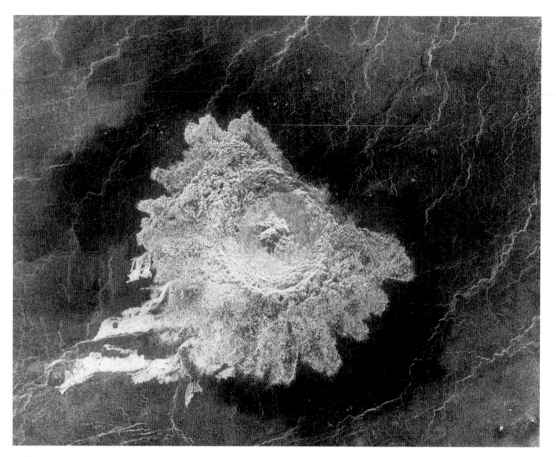

Fig. 2.25 Aurelia impact crater on Venus   The unusual crater shapes on Venus are illustrated in this *Magellan* radar image of the Aurelia crater. Like the large impact craters on the Moon, it contains a circular rim, terraced walls and a central peak. But unlike lunar craters, flows emanate from the crater, and a sector of the flow is missing, apparently due to an oblique impact from the upper-right. Interaction with the dense, thick atmosphere on Venus caused the ejected debris to act like a fluid, producing the lacy, rounded lobes. Crater Aurelia, which is 32 kilometers in diameter, has been named in honor of the mother of Julius Caesar; apparently, Aurelia is also the name of Arnold Schwarzenegger's mother. (Courtesy of NASA/JPL.)

2.3 Volcanism

Trapped heat

Volcanoes, another common aspect of the solar system, are driven by internal heat. For a large rocky planet, internal heat is continuously generated by the slow decay of radioactive material. Because of their large size and rocky composition, both Earth and Venus have internal heat powered by radioactive decay. Satellites can be heated by varying gravitational interaction with their planet; Io and Europa are both heated inside through tidal flexing by Jupiter's immense gravitational forces. Heat was also provided when the planets and satellites originated, as the result of high-speed collisions between smaller bodies, creating an ocean of melted rock on the newly formed surfaces of the Earth's Moon and the terrestrial planets.

Volcanic activity has transformed about three-quarters of the surface of the Earth and Venus, extensive parts of the surfaces of Mercury, Mars and the Earth's own Moon, and all of the surface of Jupiter's satellite Io. The volcanism occurs when heat produced in the interior of a planet or satellite rises to the surface, either as tall volcanoes or flatter, surface flows. Internal heat also generates liquid water within the large ice-covered satellites of the giant planets; the water works its way out to erupt as volcanoes of ice.

The underground molten rock, known as magma, is trapped inside a terrestrial planet and the pent-up heat wants to rise up and get out. The liquid magma is swollen by the heat, becoming lower in density than solid rock, and tends to rise through cooler, higher-density material. The surrounding solid rock can also squeeze the molten rock within it, helping to drive the magma upward under pressure. Gases locked within the molten rock expand as it rises, providing another thrust to the upward rise.

The interior of a satellite or planet is not full of molten rock. Within the Earth, for example, the magma is bottled up in chambers that are surrounded by solid rock, and the partially molten material is localized in pockets below the solid mantle and crust. Eventually some of the hot magma rises beneath the surface, pushes up, and sometimes melts and punches a hole in the crust, like a welder's torch. The lava then erupts as a volcano in a relatively small area of the surface. In contrast to the outer parts of a planet's interior, only the internal core is fully molten, but it lies so far down that the core material never makes its way to the surface.

Shield volcanoes and volcanic plains

The two most common types of volcanism are known as shield volcanoes and volcanic plains, and they are both found on Earth, Mars, Venus and Io.

The lava emitted in shield volcanoes solidifies relatively quickly and builds up tall features with gentle slopes after hundreds and even thousands of eruptions. Collapsed depressions called calderas are found at or near the summits of many shield volcanoes, arising when the source of magma drains back into its underground chamber, awaiting the next eruption.

Smooth volcanic plains are created when flowing lava spreads out along the surface, instead of rising far upward, and flattens out before solidifying.

Both the shield volcanoes and lava plains are made of a mixture of minerals called basalt, which arise from metal-rich and silica- and volatile-poor magmas.

Earth's unique volcanoes

Long lines of volcanoes are found under the Earth's oceans, forming volcanic ridges near the centers of the water-filled basins. The underwater volcanoes are fed by magma rising from partially melted mantle rock at the tops of internal convection cells.

Basalt lava has been flowing out from both sides of the mid-ocean ridges for the past 300 million years, producing spreading sea-floors and helping to propel drifting continents across the globe. They are related to the theory of plate tectonics discussed in greater detail in Chapter 4.

Terrestrial volcanoes also rise up from "hot spots", giving rise to volcanism that is not related to the mid-ocean ridges. The Hawaiian Islands are an example. The islands form a chain of shield volcanoes created as a moving plate crosses an underlying hot spot. Mauna Kea and Mauna Loa, on the big island of Hawaii, together form a mountain of basaltic lava that is much broader than it is tall and has gentle slopes; it is more than 120 kilometers across at its base and rises 9 kilometers above the ocean floor. Mauna Loa is still erupting and growing, with repeated surges

of lava that flow down its flanks. The smaller Hawaiian Islands mark out a string of extinct volcanoes (Fig. 2.26).

When a moving plate bends downward into the Earth's interior, and is at least partly re-melted, it produces another type of unique terrestrial volcano whose magma is rich in silicas and volatiles like water and carbon dioxide. These volcanoes often erupt explosively and they are found in a ring of fire that encircles the Pacific Ocean.

Cone-shaped terrestrial volcanoes with steep slopes are formed when the lava is propelled out by hot gas. Examples are Vesuvius in Italy and Mount Fuji in Japan. The eruptions that formed these steep-walled mountains often expelled large clouds of volcanic ash.

Terrestrial volcanoes can have direct, unanticipated effects on our lives. In April 2010, for example, a volcano in Iceland, which had been dormant since 1821, suddenly burst through its glacial covering and filled the skies with volcanic ash. Fearing that the metallic particles would damage jet engines, and perhaps cause the airplanes to plunge to the sea or ground, European air traffic was stopped for days. The recent Iceland eruption was nevertheless relatively modest as far as terrestrial volcanoes go. In June 1991 Mount Pinatubo spewed out enough material to shade and cool much of the planet, altering its climate. About a century earlier, on 27 August 1883, the volcano-island of Krakatoa exploded, producing an immense tsunami that killed tens of thousands of people. Vesuvius, the volcano that buried the Roman city of Pompeii in 79 AD, is a disaster waiting to happen. The Italian government has emergency evacuation plans for more than half a million people that now live in the densely populated area near the volcano.

The upwelling of pent-up heat and magma also forms rift valleys on Earth, with steep sides, sunken floors, and copious outpouring of lava. An example is the Great Rift Valley in Africa (Fig. 2.27), a long forking gash that crosses 4500 kilometers of the continent. It extends from Mozambique in the south to Ethiopia in the north, branching out through the Red Sea in one direction and diverging through the Gulf of Aden in another.

Lava flows on the Earth's Moon

Lunar rocks, which were brought to Earth by the *Apollo* astronauts in the 1970s, are of two main types. There are relatively bright rocks from the heavily cratered highlands of the Moon, and dark rocks from the lunar maria. The highlands were cooled long ago from a formerly melted magma ocean, and were extensively modified by the heavy bombardment that formed the craters upon craters between 4.3 and 3.8 billion years ago. The largest craters and impact basins were subsequently flooded episodically

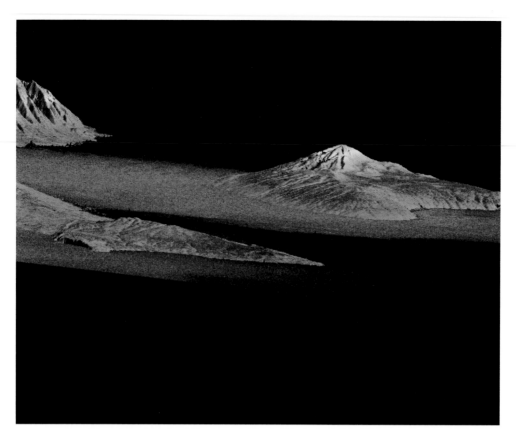

Fig. 2.26 Hawaiian Islands   Molokai (*left*), Lanai (*right*), and the northwest tip of Maui (*upper left*) are shown in this radar image obtained from the *Space Shuttle*. These islands are now extinct volcanoes, destined to eventually erode away into sunken islands known as seamounts. (Courtesy of NASA/JPL/NIMA.)

Fig. 2.27 Nyiragongo volcano flow in Africa   The continent of Africa is being split apart by the pent-up pressure of hot, rising magma in numerous underlying hot spots along the Great Rift Valley. Volcanic outpourings like Nyiragongo fill the valley with lava as the rift slowly widens. (Courtesy of Bruce Coleman.)

Fig. 2.28 Lava flows in a lunar maria   Volcanism on the Earth's Moon is seen frozen into place on the Sea of Serenity in this *Apollo 17* image taken in December 1972. Craters (*bottom*) are superposed on the lava, but the lunar maria contain relatively few craters when compared with the lunar highlands. The maria formed a secondary crust on the Moon when lava filled the giant impact basins over a period of several hundred million years ending around 3.2 billion years ago. The fluid spread rapidly, creating thin extensive sheets rather than piling up to form volcanoes. (Courtesy of NASA.)

by extensive lava flows, forming the dark maria between 3.9 and 3.2 billion years ago. They have been filled with exceptionally fluid basaltic lava, forming relatively crater-free surfaces (Fig. 2.28).

Substantial amounts of volatile elements like water and carbon dioxide have not been detected in the rocks returned from the Moon, presumably being vaporized away during its origin by a giant impact of a Mars-sized body with the young Earth. Comets may have subsequently deposited water on the cooled lunar surface, perhaps accounting for the small amounts of water that have been detected by remote sensing of the lunar surface and in recent examination of the lunar rocks.

Volcanic activity on Mercury

In the heavily cratered terrain on Mercury, there is a paucity of craters of decreasing size when compared to those of the lunar highlands. The smaller craters were probably obliterated by vast volcanic flows that occurred during the period of late heavy bombardment about 4.0 billion years ago. *Mariner 10* scientists designated this terrain as intercrater plains, since they are found between large craters and have relatively smooth surfaces.

Images from the *Mariner 10* and *MESSENGER* spacecraft have shown that large craters and impact basins on Mercury have an internal smoothness (Fig. 2.29). They resemble the maria on Earth's Moon, and are similarly attributed to ancient volcanic flow that is younger than the basins they occupy. These smooth plains cover approximately 40 percent of the surface of Mercury, and are evidence for the volcanic origin of a large part of the planet's crust, after the heavy bombardment that excavated the older craters on the planet but before the smaller, younger craters that are superposed on the smooth plains.

Some source regions of volcanism on Mercury have also been discovered in images taken during *MESSENGER*'s flybys. They show bright areas surrounding irregular-shaped depressions that have been identified as volcanic vents.

Extensive volcanism on Venus

Tens of thousands of shield volcanoes have been identified on the face of Venus, by their round shapes and gentle slopes. They range in size from major, Hawaii-sized edifices that are several hundred kilometers across (Fig. 2.30) to more numerous, smaller domes that pop up everywhere on the surface. These shield volcanoes have probably been built up from runny basaltic lava that spreads out over large distances, with the ease of spilt olive oil. The relatively low number of superimposed impact craters indicates that the extensive lava flows resurfaced the planet about 750 million years ago.

A smaller number of volcanic flows on Venus appear to be built from lava that is as stiff and thick as batter. In places, the sluggish lava has oozed onto the hot, flat surface of Venus, forming volcanic domes as round and flat as pancakes (Fig. 2.31). Each one has a dark feature almost precisely at the center, suggesting a vent from which the pasty lava flowed, like pancake batter on a hot griddle. Some of them even have little craters or pits on them that resemble bubbles that have burst in the batter. So, depending on the internal conditions when the magma formed in Venus, the resulting lava has the consistency and viscosity of either motor oil or toothpaste, and this helps determine the size and shape of the resulting volcanic formations.

Shields and plains on Mars

When *Mariner 9* neared Mars in 1971, the planet was engulfed in a dust storm. The eyes of the spacecraft – its cameras – could only peer at a disappointing, featureless ball, but as the dust storms began to settle four dark, round spots – the Tharsis volcanoes – poked out of the gloom. Even the thick blanket of dust could not cover these towering volcanic mountains. Thus, although Mars has just half the radius of the Earth, the red planet is still large

Fig. 2.29 Volcanic activity on Mercury   This double-ringed impact basin on Mercury has a smooth inner floor attributed to lava flows that partially flooded the basin some time after impact. The basin is approximately 60 kilometers in diameter. This image was acquired from the *MESSENGER* spacecraft on 29 September 2009. (Courtesy of NASA/JHUAPL/CIW.)

enough to retain significant amounts of internal heat and to sustain long periods of volcanic activity.

The large volcanoes on Mars have the gentle slopes and rounded profiles of shield volcanoes on Earth, but the volcanoes on Mars stand higher. A striking example is Olympus Mons (Fig. 2.32). Highly fluid lava has flowed out of Olympus Mons for 100 million years or longer. The caldera at its peak was formed after the most recent volcanic episode ceased and the roof of the emptied magma chamber collapsed.

Olympus Mons is very much larger than any volcano on Earth. The major volcanic edifice is about 600 kilometers across at its base, stands more than 27 kilometers above the mean surface level of Mars, and is rimmed by a cliff that is 6 kilometers high in some places. For comparison, the diameter of the base of Hawaii's Mauna Loa is just one fifth that of Olympus Mons, and the height of the Hawaiian volcano is a third the height of the Martian one.

The impressive size of the volcanoes on Mars is attributed to the planet's thick outer shell, which does not move sideways and remains fixed over the internal sources of magma. This gives the volcanoes on Mars a long time to grow, sometimes for billions of years. In contrast, the Earth's thinner crust is broken into pieces and moves over an internal magma chamber, limiting the growth of individual terrestrial volcanoes and producing chains of smaller ones, such as the Hawaiian Islands.

Images of the Martian surface suggest that volcanic activity might have persisted from the planet's youth into relatively recent times. Like the Moon and Mercury, the red planet bears the scars of a steady rain of meteorites, and the relative ages of volcanoes and lava flows can be determined from the density of impact craters on them. While the most recent lava flows on Olympus Mons may be only a few million years old, and the average lava age is about 30 million years, lava could have been flowing out of this volcano for a long time before that, as long as 3 billion years ago.

The bulk of the volcanism on Mars is found in the flat, low-lying plains that cover most of the northern hemisphere and about 40 percent of the planet's surface. These lava flows are covered by wind-blown dust, as well as rock debris sent down into them by ancient episodes of water

Fig. 2.30 Maat Mons, a volcano on Venus   This three-dimensional perspective of Maat Mons on Venus was obtained from radar data taken from the *Magellan* spacecraft in October 1991. The volcano is 8 kilometers high, the second highest peak on the planet. Fresh, dark lava extends for hundreds of kilometers in the foreground, perhaps flowing from a relatively recent eruption. Maat Mons is a giant shield volcano similar in size and shape to the big island of Hawaii. *Maat* is the name of the ancient Egyptian goddess of truth and justice, and *Mons* is the Latin term for "mountain". The orange tint simulates the color of sunlight at ground level after filtering by the dense, thick atmosphere of Venus. (Courtesy of NASA/JPL.)

flow from the southern highlands. The low impact-crater densities on some volcanic flows in the Martian plains suggest relatively young ages, as low as 100 million years or less, but lava most likely also flowed across the surface of Mars billions of years ago.

Volcanoes on Jupiter's moon Io

There is one place in the solar system that is now more volcanically active than any other place; it is Jupiter's innermost satellite Io. It is so hot inside that you can see it melting before your eyes. Io is now spewing out 100 times more lava than all the volcanoes on the Earth. This is a totally unexpected discovery, made by the inquisitive camera eyes of *Voyager 1* in 1979, and confirmed by the *Galileo* spacecraft in 1999–2004 (Figs. 2.33, 2.34). Volcanoes are literally turning the satellite inside out, so parts of Io's surface are younger than your backyard. Because of the satellite's low

gravity and lack of substantial atmosphere, the volcanic plumes spread out in graceful fountain-like trajectories, depositing circular rings of material up to 1400 kilometers in diameter.

What drives Io's continuous volcanism? The satellite is too small to now retain internal heat created during its formative years or to be significantly heated by radioactive rocks. The heat released during the satellite's formation and subsequent heating of its interior should have been lost to space long ago. After all, Io is about the same size as the Earth's Moon, which shows no signs of volcanism other than that associated with its earliest history between 3 and 4 billion years ago. Io's internal heat is instead generated by tides that massive Jupiter raises in the solid body of the satellite.

The gravitational force of the nearby giant planet decreases with increasing distance, so Jupiter pulls hardest on the side of Io facing it, and least on the opposite

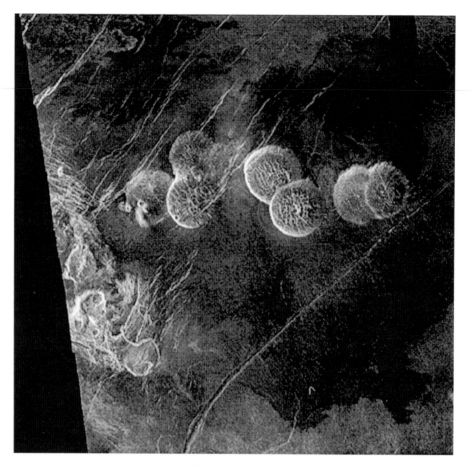

Fig. 2.31 Pancake domes on Venus   These seven volcanic domes were discovered in radar images of Venus taken from the *Magellan* orbiter. They all have round shapes that are about 25 kilometers across, and steep sides that are less than 750 meters high. Their central vents may be lined up along a crack in the surface. These domes are attributed to very thick, stiff and sluggish lava flows, rather than the fluid and runny type. Eruptions of the pasty, viscous lava, coming from a central vent on a relatively level surface, would form the circular, flattened shapes that resemble giant pancakes. Since there is little or no erosion by wind or water on Venus, newer pancakes look much the same as the ones on which they are superimposed. (Courtesy of NASA/JPL.)

side; the center of Io is pulled with an intermediate force. These differences in the gravitational attraction of Jupiter on opposite sides of Io produce two tidal bulges in the solid rock inside the satellite – one facing Jupiter and one facing away. The giant planet thus effectively squeezes Io into the shape of an egg.

If Io remained in a circular orbit with the same face toward Jupiter, its tidal bulges would not change in height and no heat would be generated; but its orbit is not perfectly circular. Io's orbit has a forced eccentricity due to the combined gravitational interaction of Io with Jupiter and the other satellites of the giant planet. When the elongated orbit carries Io closest to Jupiter, the shape of Io is distorted more than when the satellite is further away. The resultant variation in the tides flex Io's surface, bending it in and out by as much as 100 meters during each orbit. Friction associated with this tidal flexing heats Io inside, melting its rocks and producing volcanoes at its surface.

Instruments on *Galileo* measured the temperatures of the volcanoes, showing that the lava is at 1700 to 2000 kelvin, up to twice the temperature of volcanoes on Earth. The high-temperature eruptions emit gaseous sulfur and sulfur dioxide; the bright surface flows are attributed to sulfur and the white surface deposits to sulfur dioxide. The very high temperatures apparently rule out liquid sulfur as a dominant volcanic fluid, and they have certainly driven off any water that might have been on Io.

Volcanoes of ice

Some of the large icy satellites of the giant planets exhibit fountains, geysers and smooth surface flows of water ice. This is a sort of ice volcanism, in which the generation of liquid water inside the satellite mimics the partial melting of rocks within the terrestrial planets and Io. The melted ice works its way out from inside the

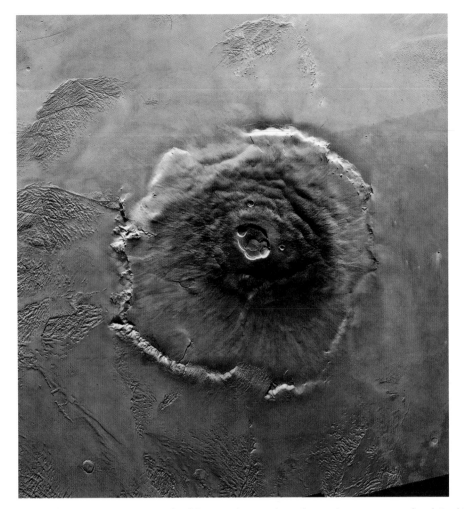

Fig. 2.32 Olympus Mons, a volcano on Mars   A mosaic of the towering Martian volcano Olympus Mons, using data obtained from the *Viking* 1 orbiter in the late 1970s. It is the largest known volcano in the solar system, rising about 27 kilometers and spreading over 600 kilometers at its base. Counts of impact craters suggest that the lava flows on the gentle slopes of this volcano are relatively young, averaging only about 30 million years old. The summit caldera, or central depression, is a composite of as many as seven roughly circular depressions that formed by recurrent collapse when magma was withdrawn from within the volcano. The caldera is almost 3 kilometers deep and up to 70 kilometers across. The volcano is surrounded by a well-defined scarp, or cliff, that is up to 6 kilometers high. Many of the plains surrounding the volcano are covered by terrain containing ridges and grooves; it is called an *aureole*, the Latin term for "circle of light". *Mons* is the Latin term for "mountain". Mount Olympus, the highest mountain in Greece, is the home of the gods in Greek mythology. (Courtesy of NASA/JPL/USGS.)

satellite and either erupts from it or flows across the surface.

Jupiter's satellite Europa, for example, is almost perfectly smooth and exceptionally bright, with no mountains and valleys in sight (Fig. 2.35). Very few impact craters are present on its face, indicating that the smooth surface was formed relatively recently, geologically speaking. Some process must be keeping it young on timescales of a few hundred million years or less. Liquid water or slush apparently oozes out within cracks in the ice, resurfacing the globe (Fig. 2.36). Some cracks in the icy moon are as long as the distance from Los Angeles to New York, and when you look at them you might see water rising.

Saturn's moon Enceladus has a bright, smooth, icy surface that also contains cracks and grooves (Fig. 2.37), suggesting the release of water from below the surface. This would be consistent with the satellite's low mean mass density of just 1240 kilograms per cubic meter, suggesting that it is just a big ball of water ice.

The satellite is embedded within a ring of water ice, dubbed the E ring, which is derived from the satellite. Instruments aboard the *Cassini* spacecraft have revealed ice jets erupting from fractures in Enceladus' surface (Fig. 2.38). Geysers of water, carbon dioxide and organic molecules spray far out from the moon at high speeds. Some of the fountains and plumes of water ice erupt at

Fig. 2.33 Volcanic activity on Io   Massive eruptions continuously disfigure the surface of Jupiter's satellite Io, the most volcanically active body in the solar system. As shown in this color-enhanced image, taken from the *Galileo* spacecraft on 19 September 1997, Io's surface is continuously being covered by lava flowing from its volcanoes, erasing any impact craters. A bright red ring surrounds the volcano Pele, marking the site of sulfur compounds deposited by its volcanic plumes. A dark circular area, about 400 kilometers in diameter, intersects the upper-right part of the red ring and surrounds another volcanic center named Pillan Patera. Deposits of sulfur dioxide frost appear white and gray in this image, while other sulfurous materials probably cause the yellow and brown shades. Pele is the Hawaiian goddess of the volcano, and Pillan Patera is named for the Araucanian thunder, fire and volcano god. (Courtesy of NASA/JPL/U. Arizona.)

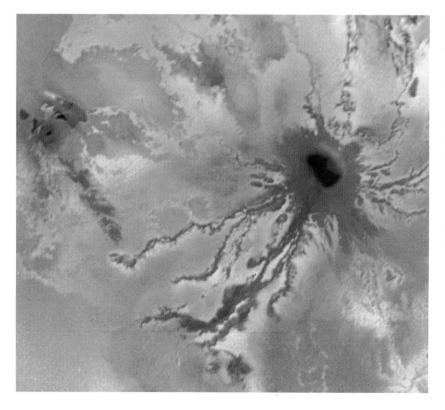

Fig. 2.34 Lava flows on Jupiter's satellite Io   Numerous volcano calderas and lava flows were discovered on Jupiter's innermost large moon, Io, in 1979 using an instrument aboard the *Voyager 1* spacecraft. This *Voyager 1* image of Ra Patera, a large shield volcano, shows flows up to 300 kilometers long emanating from a dark volcanic vent. The diffuse reddish and orange colorations are probably surface deposits of sulfur compounds. The bright whitish patches probably consist of freshly deposited sulfur dioxide frost. Ra is an Egyptian Sun god. (Courtesy of NASA/JPL.)

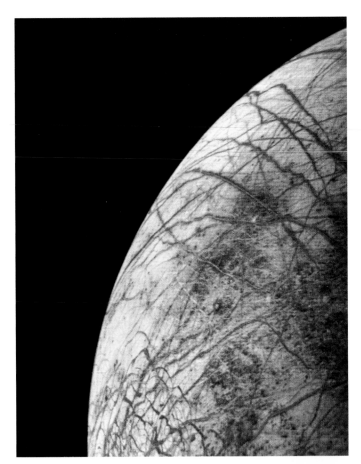

Fig. 2.35 Jupiter's satellite Europa Dark streaks mark Europa's smooth surface, forming a spidery, veined network in this *Voyager 2* image taken on 9 July 1979. In contrast to Jupiter's satellite Callisto (see Fig. 2.17), Europa has very few impact craters; the absence of craters suggests that the ice crust is relatively young. Internal stresses have apparently fractured Europa's icy mantle, producing intersecting cracks that extend 2000 kilometers but reach depths of less than 100 meters. The fractures may have been filled by liquid water gushing out from a global ocean in the satellite's interior, warmed by tidal heating. (Courtesy of NASA/JPL.)

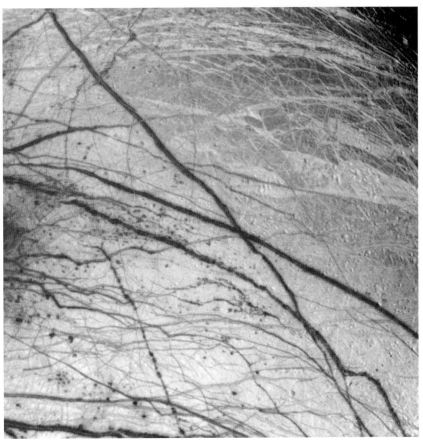

Fig. 2.36 Water oozes out from Jupiter's moon Europa A composite, color-enhanced image of the Minos Linea region of Jupiter's moon Europa, taken on 28 June 1996 by imaging cameras on *Galileo*. The icy plains, shown here in bluish hues, reflect different amounts of light, probably as the result of differences in the sizes of the ice grains. The long red cracks in the ice could mark the sites of liquid water oozing out from the warm interior of Europa. The area covered in this image is about 1.26 kilometers across. In Greek mythology, Minos is the son of Zeus and the king of Crete, who kept a monster named Minotaur in a labyrinth. *Linea* is a "dark or bright elongate marking". (Courtesy of NASA/JPL/U. Arizona.)

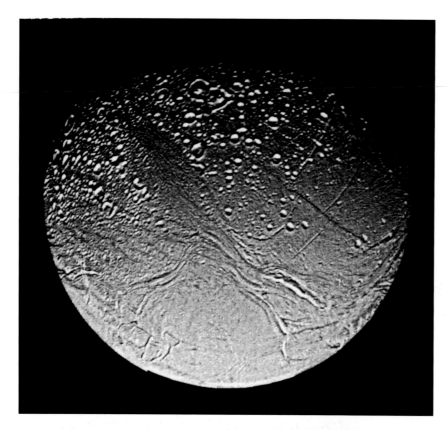

Fig. 2.37 Saturn's satellite Enceladus
The bright, smooth surface of Enceladus, shown in this *Voyager 2* image obtained on 25 August 1981, reflects almost 100 percent of the incident sunlight, making it one of the most reflective objects in the solar system. When viewed up close, part of its surface is scarred with impact craters. Other parts of the surface contain cracks and grooves, suggesting that internal stresses may have discharged water that froze into smooth ice. (Courtesy of NASA/JPL.)

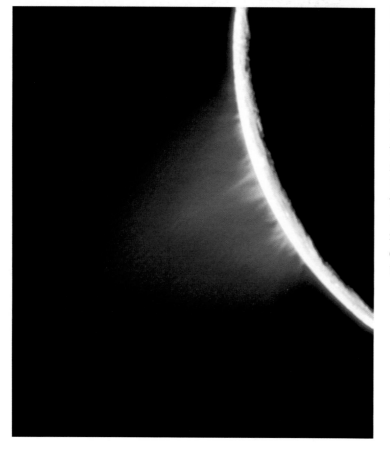

Fig. 2.38 Ice plumes on Saturn's moon Enceladus   Enormous jets and fountains of ice are erupting on Saturn's moon Enceladus, feeding the planet's E-ring. This image, taken from the *Cassini* spacecraft on 27 November 2005, exhibits several geyser-like jets, which vent and spurt plumes of ice particles, water vapor and trace amounts of organic compounds. Eight source locations were identified in this image, all on the prominent tiger stripe features, or sulci, in the moon's south polar region. These features were under close scrutiny from *Cassini* for years after their discovery in 2005. (Courtesy of NASA/JPL/SSI.)

fast enough speeds to escape Enceladus and feed the E ring with ice particles.

Enceladus is caught in a gravitational tug-of-war between Saturn and its satellite, Dione, whose orbital period is about twice that of Enceladus. Dione's repeated gravitational tug produces Enceladus' eccentric orbit, and causes Saturn's recurrent tidal flexing of Enceladus, which warms the moon's interior and apparently maintains a liquid ocean beneath its ice.

Neptune's largest satellite, Triton, is the coldest moon ever recorded, with a temperature of just 38 kelvin, approaching absolute zero where all motion stops. It is so cold because Neptune is so far away from the Sun, therefore receiving little sunlight, and also because Triton reflects more of the incident sunlight than most satellites – only Enceladus and Europa are comparable. Yet the frozen moon is a dynamic world, set in motion and molded by volcanic eruptions.

Triton's surface has a smooth, youthful appearance, with no large impact craters and few small ones. Global resurfacing by volcanoes of ice might have wiped out pre-existing craters on Triton, perhaps about a billion years ago when tidal flexing may have heated the satellite's insides. The deep internal heat may have turned the ice into liquid that rose to the surface, like a squeezed slush cone, filling the vast frozen basins that are now found there. These frozen lakes of ice look like inactive volcanic calderas; complete with smooth filled centers, successive terraced flows and vents.

Numerous dark plumes and streaks, found in the midst of the bright southern cap of Triton, suggest a different kind of volcanic activity, propelled by relatively recent eruptions of nitrogen gas (Fig. 2.39). Nitrogen boils at very low temperatures, at just 77 kelvin on Earth, and when it boils it expands, producing enormous pressures that can shoot gas and other material high into Triton's thin nitrogen atmosphere. Thus, geyser-like eruptions may have lofted the dark material outward from beneath the surface. The prevailing winds would then carry it across the satellite, depositing it on the ice as dark streaks.

Four active plumes were observed during the *Voyager 2* encounter with Triton. They rose in narrow, straight columns to an altitude of 8 kilometers, where dark clouds of material were left suspended and carried downwind horizontally for over 100 kilometers, like smoke wafted away from the top of a chimney. Most of the dark streaks are probably remnants of such plumes.

Since the active plumes occur where the Sun is overhead, it is possible that sunlight produces the weak subterranean heat required to make the nitrogen boil and break through the overlying layer of ice. The sunlight would pass through the translucent ice and become absorbed by

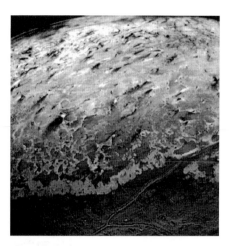

Fig. 2.39 Triton's dark plumes and streaks   This image of the south polar terrain on Triton reveals about 50 elongated dark plumes, or "wind streaks", on the moon's highly reflective surface. The plumes originate at very dark spots, generally several kilometers across, probably marking vents where nitrogen gas was driven outward in geyser-like eruptions from beneath the surface. Winds in Triton's thin nitrogen atmosphere may have carried the dark erupted material along, depositing it in the elongated streaks. This image was taken on 25 August 1989 from *Voyager 2*. (Courtesy of NASA/JPL.)

darker material encased beneath. The overlying nitrogen ice would trap the solar heat, for it is opaque to infrared heat radiation, producing a solid-state greenhouse effect. Nitrogen gas, pressurized by the subsurface heat, might then explosively blast off the iced-over vents or lids, launching volcanic plumes of gaseous nitrogen and ice-entrained darker material into the atmosphere, just as the water in an overheated car radiator is explosively released when the radiator cap is removed. The nitrogen geysers also resemble water-driven geysers on Earth, like Yellowstone's Old Faithful, but water boils at a much higher temperature of 373 kelvin.

This discussion of ice volcanism on Europa, Enceladus and Triton brings us to water, another common property of many planets and satellites.

2.4 Water

Earth, the water planet

From outside, our home planet Earth looks like a tiny, fragile oasis in space, a glistening blue and turquoise ball of water, flecked with delicate white clouds and capped with glaciers of ice (Fig. 2.40). Seventy-one percent of the Earth's surface is now covered with water. The oceans contain so much water that if the Earth were perfectly smooth the oceans would cover the entire globe to a depth

Fig. 2.40 Earth, the water planet   Almost three-quarters of the Earth's surface is covered by water, as suggested by this view of the North Pacific Ocean. Earth is the only planet in the solar system where substantial amounts of water exist in all three possible forms – gas (water vapor), liquid and solid (water ice). Here white clouds of water ice swirl near Alaska. The predominantly white ground area, consisting of snow and ice, is the Kamchatka Peninsula of Siberia. Japan appears near the horizon. From this orientation in space, we also see both the day and night sides of our home planet. (Courtesy of NASA.)

of 2.8 kilometers. They contain about one billion trillion (10^{21}) kilograms of water, and provide the home for about half of all species on Earth, including the one-celled plants called phytoplankton, which supply roughly half of the oxygen to our air.

The Earth's oceans absorb heat in hot months, and release it in cold months. Global ocean currents such as the Gulf Stream circulate heat and cold around the world. The oceans also absorb carbon dioxide from the atmosphere, removing roughly half of the heat-trapping gas that is added to our air.

Water is a marvelous substance. As a liquid, it will dissolve almost anything to some extent, and it can hold and release very large quantities of heat. When liquid water freezes, it expands and becomes less dense, in contrast to most substances. As a result, ice floats on the surface of lakes and oceans, so they freeze from the top down.

Water is crucial to life here on Earth, for the chemical reactions that sustain life must take place in water. We ourselves are largely water. Just about anywhere there is water here on Earth, there is some sort of life – even deep inside the Earth. It follows that if liquid water was found on another planet, or on one of its satellites, that place might also be hospitable to life.

Water is made of the two most abundant, chemically reactive elements in the Universe, hydrogen and oxygen, and so ought to be very common. Yet the Earth is the only place in the solar system where substantial quantities of water exist in all three possible states – as a gas (water vapor), liquid, and solid (ice). Ours is the only planet whose surface temperature matches the temperature of liquid water, between water's freezing and boiling temperatures of 273 and 373 kelvin, respectively. The narrow range of distances from the Sun, or other star, at which temperatures from stellar radiation allow liquid water to exist is known as the habitable zone; but the greenhouse warming of a planet's atmosphere can raise the temperature above that due to starlight alone.

When we look at our nearest neighbors, we see that Venus is too hot and Mars is too cold for significant amounts of liquid water to exist on their surfaces. Any water on Venus would now be in the form of steam, and water on Mars is now mainly locked beneath the surface in the form of ice. The terrestrial planets nevertheless had different surface temperatures in the past and these temperatures will change in the future.

It was too hot for liquid water to exist on any of the rocky inner planets in their very early history. Because of the energy released by the colliding rocks that merged to form these planets, they probably began with molten surfaces, too hot for any water to accumulate on them. The heat of impacts would have vaporized any water, and sterilized the young planet. Only when the initial bombardment slowed, and the growing planet cooled, could liquid surface water collect into lakes or oceans, which might have sustained life.

This water might have been liberated from the planetary interiors by volcanoes or carried to the planets by icy comets or water-rich asteroids. The water vapor released from the hot internal materials would cool and condense, falling as rain. Other water may have arrived by the direct impact of comets or asteroids. On Earth, the water fed rivers that flowed across the planet's surface, dissolving salty minerals that settled into the early oceans.

Some astronomers argue that the Earth's oceans were supplied from outside the planet after it first came together. The water was supposed to be carried to our planet by small bodies of rock or ice, similar to today's asteroids or comets, coming in from the cold outer parts of the solar system. Some geologists reason that the ancient oceans were instead steamed out of the Earth's interior by erupting volcanoes, supplying water vapor to the primitive atmosphere. No one knows for sure whether external impact or internal volcanism resulted in most of our water. But since the Earth formed by the accumulation of colliding objects, the water expelled by the volcanoes also had to be originally supplied by cosmic impact.

Water on the Moon

The rocks returned from the Earth's Moon by astronauts of the *Apollo* missions are dryer than a terrestrial desert. They resemble an Earth rock that has had all the water boiled out of it. For this reason, most scientists have assumed that there is no water on the Moon, and that there never was any.

There has nevertheless been speculation since the 1960s that water ice might be found in the polar regions of the Moon, which the astronauts did not visit. This possibility hinges on the fact that the Sun never rises more than a scant 2 degrees above the horizon as seen from the lunar poles. This means that the deepest polar craters have regions inside them that remain permanently in shadow, eternally dark and cold with temperatures that never exceed about 100 kelvin. Consequently, any water that chanced to enter the craters would be frozen solid, remaining permanently frozen, never melting or vaporizing and escaping to space.

If there were water on the Moon, it would probably have been deposited by impacting comets, which are comprised largely of water ice, or by the impact of water-rich asteroids. They have both been bombarding the moons and planets for billions of years, ever since their formation. Liquid water and water vapor, formed by the heat of impact on the Moon, would normally be evaporated into space, but over the eons of impact some of the water could have collected in the shadowed craters near the lunar poles. If the water settled down to the crater floors, it would freeze in their "cold trap". Or perhaps hydrogen from the Sun's winds could combine with oxygen atoms in the Moon's dust and rock, forming water molecules that might enter the cold traps.

For more than a decade, successive observations have either strengthened evidence or dampened hopes for water on the Moon. Early evidence for the water began in 1994 when the US Department of Defense and NASA launched the *Clementine* spacecraft into polar orbit around the Moon, obtaining the first global perspective of its surface composition and topography. Bright radar echoes obtained from an instrument on *Clementine* suggested deposits of water ice in a deep, cold, shadowed crater near the lunar south pole. Indirect supporting evidence was obtained from the *Lunar Prospector* spacecraft, launched in 1998. Its neutron and gamma ray spectrometers discovered enhanced signals from hydrogen atoms that were being struck by energetic cosmic rays. If the hydrogen, designated H, was taken as evidence for water, or H_2O, with O for oxygen, then there seemed to be plenty of water ice near both lunar poles.

Astronomers hoped that more direct evidence of water would be obtained by crashing *Lunar Prospector* into a crater near the Moon's south pole, at the end of the spacecraft's lifetime. The impact might release icy rock and dust, with indications of water vapor. The *Hubble Space Telescope* and numerous ground-based telescopes were focused on the impact, on 31 July 1999, but no debris or spectral signatures of the impact were detected. There were several explanations, including the possibility that the spacecraft missed its target. It is also possible that many of the hydrogen nuclei that have been detected were delivered to the Moon by the Sun's winds, and that they are not a component of water.

Hopes for water on the Moon were further dampened in 2004–06 when the world's most powerful radar transmitter, at the Arecibo Observatory in Puerto Rico, was beamed into the permanently shadowed crater floors at the Moon's polar regions. The absence of strong radar echoes indicated that there are no thick ice deposits at the lunar poles, but thin ones with relatively low concentrations of water were possible.

Then, in September 2009, a trio of satellites obtained evidence for water molecules on the Moon. A spectral mapping instrument carried aboard India's *Chandrayaan-1* spacecraft detected infrared absorption patterns attributed to water molecules and hydoxyl, or OH. Spectrometers aboard the *Cassini* and *Deep Impact* spacecraft contributed to confirmation of this finding.

Within a month, the *Lunar CRater Observation and Sensing Satellite* (*LCROSS*) sent two spacecraft in to bomb the Moon. Although the craft successfully struck their target, in the Cabeus crater near the lunar south pole, no billowing clouds of dust and ice were detected by telescopes trained on the impact site and the televised event was quite a disappointment. Careful scrutiny of the *LCROSS* camera images nevertheless revealed plumes of material ejected from the bottom of the crater on impact, and the satellite's spectrometers showed the spectral signatures of water vapor.

We therefore no longer consider the Earth's Moon to be a completely dry and desolate place. But there aren't any lakes or even puddles of water on the Moon, and there isn't even enough water to drink. The lunar water molecules are bound to other molecules in rock and dust, and exist in only trace amounts.

So why is so much attention being given to the search for water on the Moon? Significant reservoirs of water could reduce the payloads needed to maintain a future human outpost on the Moon. The lunar water could be purified to drink, or it could be chemically split into hydrogen, to burn as a rocket propellant, and oxygen to breathe. This would make it easier to support a colony on the Moon, or to build a fueling station on it for interplanetary spacecraft.

Radar evidence for water ice near Mercury's poles

Since Mercury is so close to the Sun, with boiling temperatures on its sunlit side, it has long been assumed that the planet retains no water on its surface. But a strong tidal lock with the Sun makes Mercury's rotation axis point straight up, without any tilt, so its equator points directly at the Sun at all times, and its polar regions are always in shadow. This means that crater interiors near the poles are never exposed to direct sunlight. The permanently shadowed spots may have remained colder than 120 kelvin for eons, permitting substantial quantities of water ice to accumulate at the frigid crater floors near the poles.

When Mercury's poles tip toward the Earth, while never deviating from the north–south direction and remaining hidden from the Sun, astronomers have beamed radio signals at them and examined the echoes. These are return signals from pulses of radio radiation sent from both the Arecibo and the linked Goldstone – Very Large Array radar facilities. Unusually strong radar echoes coming from the polar regions show prominent, radar-reflective material that is plausibly attributed to water ice. They suggest that the planet's north and south polar regions may contain substantial deposits of water ice, at least a couple of meters thick. The similar radar-scattering properties of Mercury polar craters and those of Jupiter's ice-covered satellites supports the suggestion that the craters contain water ice.

Unlike the weak radar echoes obtained from the Moon, the powerful signals returned to Earth from Mercury's highly reflective polar regions suggest relatively thick sheets of water ice at the planet's poles. They appear to be concentrated only in the young, fresh craters seen in *Mariner 10* images, rather than the ancient, degraded ones, which do not exhibit strong radar reflectivity. This is most likely because the older low-rimmed, shallow craters do not contain permanently shadowed floors that act as cold traps for the water.

A former ocean on Venus

Today the surface and atmosphere of Venus are exceptionally dry, which is what you would expect for a planet whose surface temperature is now a scorching 735 kelvin. The planet may nevertheless once have had liquid oceans, perhaps 4 billion years ago, until a runaway greenhouse effect boiled it all away.

Models of planet formation predict, for example, that the Earth and Venus were once endowed with roughly equal amounts of water. When the strong greenhouse effect of Venus's thick atmosphere raised the planet's surface temperature, most of its water evaporated and was lost to space, while the Earth, with a much thinner atmosphere and weaker greenhouse effect, remained cool enough to keep its oceans. And if the very small quantities of water vapor now found in the atmosphere of Venus are a remnant of an ancient reservoir, then Venus has lost the equivalent of a very large lake or a small ocean.

Evidence that Venus once had an ocean is found in an excess of deuterium now in its atmosphere. Deuterium is an atom chemically identical to hydrogen but heavier and therefore more likely to be retained in the atmosphere. On Earth, it is found in heavy water, which comprises only about 0.016 percent (0.000 16) of the oceans. The natural explanation of the atmospheric deuterium on Venus is that the planet once had vast quantities of normal water, containing light hydrogen, and heavy water, containing deuterium. When these liquids were subsequently boiled away by the intense heat, the lighter hydrogen easily escaped from the planet, but some of the heavier deuterium remained behind as a residue. The amount of remaining deuterium suggests that Venus once had enough liquid water to uniformly cover the planet's surface with a global lake at least 4 meters deep, or just 0.12 percent of a full terrestrial ocean.

Water on Mars

Mars orbits the Sun at the outer edge of the solar system's habitable zone, where water might be either liquid or frozen solid. As expected from its thinner atmosphere and greater distance from the Sun, Mars is a much colder planet than the Earth, with surface temperatures that are usually below the freezing point of water. Instruments aboard orbiting spacecraft have shown that large quantities of water ice exist in the polar caps of Mars and as subsurface ice in many other locations on the planet. Spectroscopic instruments aboard the *2001 Mars Odyssey* spacecraft have, for example, found evidence of subsurface water ice on Mars in large regions surrounding both of the planet's polar regions, with lesser amounts of subsurface ice at mid-latitudes and equatorial regions. The concentration of ice in the upper meter of the ground in the polar plains is surprisingly high – one-fifth to one-third by weight and more than 50 percent water ice by volume. So if you heated one full bucket of this polar material it would be more than half a bucket of water.

Ninety-five percent of the Martian atmosphere is gaseous carbon dioxide; and it contains only small quantities of water ice crystals, in the form of clouds and haze. Such small amounts of water vapor indicate that the Martian atmosphere is drier than the driest of the Earth's deserts. If the water vapor were collected and condensed, it would amount to no more than a good-sized lake. Yet, despite the small amount of water vapor, the cold, thin atmosphere is close to saturation. It is about as wet as it can be. Consequently, the formation of clouds and fogs of water ice is a common feature of Martian weather. Such clouds or fogs have been observed along the flanks of volcanoes, above the polar caps, and in low-lying areas such as canyon floors.

Fig. 2.41 Streamlined islands on Mars   An image of the Mangala Vallis region on Mars, taken with an instrument aboard the *2001 Mars Odyssey* orbiter. The scoured floors and teardrop-shaped islands were probably created by powerful, ancient flows of liquid water. The flowing water ran from the heavily cratered southern highlands to the northern lowland plains. The name Mangala is the word for Mars in Sanskrit. (Courtesy of NASA/JPL/ASU.)

Liquid water cannot now exist for any length of time on the surface of Mars. It would immediately begin to boil, evaporate and freeze — all at the same time. Because of the low pressure and freezing temperatures of the thin Martian atmosphere, any liquid water would quickly vaporize or freeze into ice.

Nevertheless, Mars almost certainly contained liquid rivers, lakes and possibly oceans in the distant past, roughly 4 billion years ago. Huge, dry river beds and flood channels, imaged by the *Mariner 9*, *Viking 1* and *2*, *Mars Global Surveyor* and *Mars Reconnaissance Orbiter* spacecraft, provide unmistakable signs of former torrents of flowing water that cascaded across the surface of Mars (Figs. 2.41, 2.42). The flow channels that have been carved and etched into the surface of Mars are immense by terrestrial standards, as much as 100 kilometers wide and 2000 kilometers in length. The amount of water required

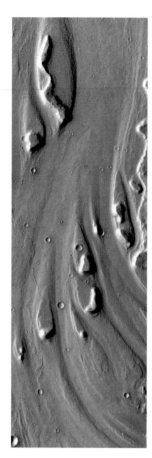

Fig. 2.42 Ancient water flow in Mars' Ares Vallis   As shown in this image, obtained from an instrument aboard the *2001 Mars Odyssey* orbiter, large quantities of water were diverted around pre-existing craters in Ares Vallis. As the water made its way downstream, the interference with the flow was reduced, and the water flow reformed at the narrow ends of the islands. The orientation of the islands therefore indicates the direction of flow, with the narrow end of the island pointing downstream. In this case, the flow is from the lower right to upper left. Ares Vallis is an outflow channel opening on to the Chryse Planitia, landing site of *Mars Pathfinder* (see Figs. 2.14 and 2.15). Ares is the Greek god of war, and possibly connected with the Roman god Mars. (Courtesy of NASA/JPL/ASU.)

to gouge out these river-like outflow channels is enormous, requiring catastrophic floods containing million of tons, or billions of kilograms of liquid water.

The outflow channels must have drained and emptied into the vast low-lying plains in the northern hemisphere of Mars. The *Mars Exploration Rover, Opportunity*, has found sedimentary layers and round, mineral "blueberries" that must have formed when water flowed across the northern plains about 3.7 billion years ago. Instruments aboard orbiting spacecraft have also identified clay minerals and chloride salt deposits formed at diverse watery environments in the distant past on Mars.

As everyone knows, water flows downhill, which is toward the low-lying northern hemisphere of Mars (Fig. 2.43). Meandering river-like valleys found in the southern highlands of Mars apparently flow down from higher elevations (Fig. 2.44). This suggests that the floodwaters that cut the outflow channels also drained northward and pooled in the vast northern lowlands at the ends of the channels. When topographical maps are combined with measurements of the red planet's gravity, buried subsurface canyons are found located beneath the sediment, emanating from the visible outflow channels. The transport of water therefore continued far into some parts of the northern plains. It has even been argued that the northern lowlands were once the sites of an ancient ocean, covering up to one-third of the surface area of the planet and up to 1.6 kilometers deep.

Most geologists agree that vast amounts of water flowed across the Martian landscape long ago, perhaps 3 to 4 billion years in the past. In one interpretation, the red planet was warmer and wetter in its early history, and was therefore much different from the cold, arid Mars we see today. A thicker, warmer atmosphere would have permitted flowing water that generated the outflow channels and filled low-lying areas with pools and lakes of liquid water. Another explanation involves frequent comet impacts in the early history of the planet.

Water ice in the outer solar system

Dense rocky substances dominate the four terrestrial planets (Mercury, Venus, Earth and Mars) and the Earth's Moon, which are nearest the Sun, while the lighter gaseous and icy substances dominate the outer giant planets (Jupiter, Saturn, Uranus and Neptune) and their moons. These compositional differences appear to result from the fact that the terrestrial planets formed close to the hot, bright, young Sun, and they suggest that water ice might be common in the colder, outer parts of the planetary system. That is, the oxygen, O, and hydrogen, H, in the frigid outer precincts of the young planetary system could have combined to make water ice, or frozen H_2O.

In the inner regions of the solar nebula, the higher temperatures would vaporize water ice so it would not condense, leaving only high-density, rocky substances to coalesce and merge together to form the terrestrial planets. Further out, in the colder realm of the giant planets, large amounts of water ice could survive and not evaporate. The asteroid belt, that forms the great divide between the orbits of Mars and Jupiter, is thus known as the water line of the primeval solar system; it is the first place that a 1-kilometer chunk of ice could form.

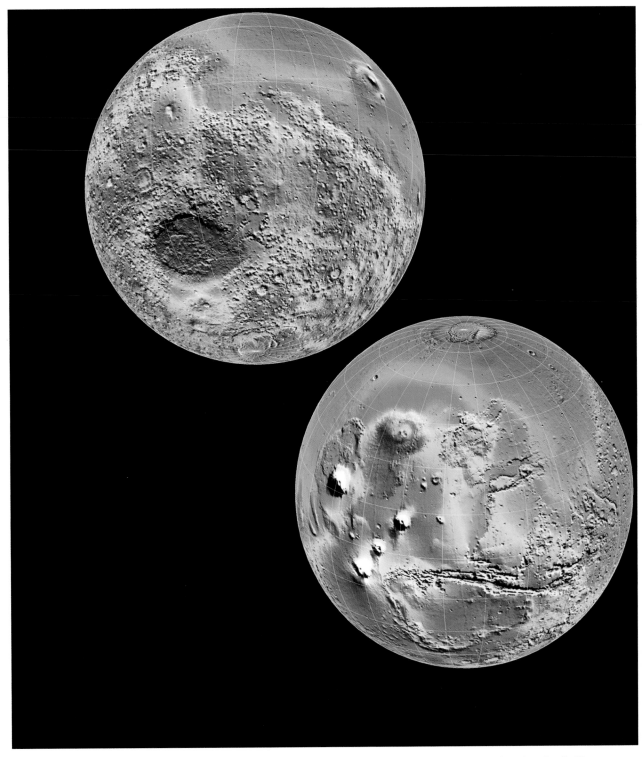

Fig. 2.43 Highs and lows of Mars' topography   Global topographic maps of two sides of the planet Mars, based on detailed laser altimeter measurements from the *Mars Global Surveyor* spacecraft. In these images, color represents height above (or below) the mean planetary radius, ranging from dark blue (−8 kilometers) through green, yellow and red (+4 kilometers) to white (over 8 kilometers in altitude). Flowing water would run downhill, collecting in the low-lying blue regions including the Hellas impact basin (*upper image*), about 2800 kilometers across, the Valles Marineris (*lower image*), shown as a horizontal gash beside the Tharsis volcanoes (*pink*), and the extensive lowland plains in the north (*top of both images*). (Courtesy of NASA/JPL/ GSFC.)

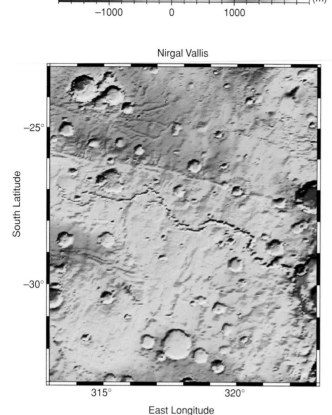

Nirgal Vallis

Fig. 2.44 Nirgal Vallis on Mars   This valley meanders over 500 kilometers across the heavily cratered southern highlands on Mars, with few tributaries. The valley network is shown in topographic relief with a vertical accuracy of approximately 1 meter provided by a laser altimeter on the *Mars Global Surveyor*. The direction of water flow would be downhill, to the lower right. Nirgal Vallis is located at 28.4 degrees south and 42.0 degrees west longitude (*east longitude shown here*). Nirgal is the Babylonian word for Mars. (Courtesy of NASA/JPL/GSFC.)

Infrared spectroscopy from the ground and space have shown that the rings of Saturn are composed of pure water ice, with no trace of dirt, dust or rock. And the smooth, cold, highly reflective surfaces of Jupiter's moon Europa and Saturn's satellite Enceladus are now known to consist of water ice, encasing great balls of liquid water. After all, ice freezes from the top down, so you might expect Europa and Enceladus to be cold and desolate on the outside, with a warm ocean on the inside.

Images from the *Galileo* spacecraft indicate that Europa almost certainly had a liquid ocean at one time, and they have considerably strengthened the evidence for an ocean of liquid water existing just beneath its icy surface at the present time. The *Galileo* images show evidence for near-surface melting and movements of large blocks of ice, the first icebergs found off planet Earth. In some cases, large ice rafts the size of cities have broken off and drifted apart, sliding away from each other, with edges that fit like the pieces of a jigsaw puzzle. Warm ice or even liquid water must lubricate the moving ice from below.

There is indirect magnetic evidence of a hidden ocean beneath the smooth, icy surface of Europa. Magnetic measurements from *Galileo* indicate that the moon generates internal electrical currents as it sweeps through Jupiter's powerful magnetic field, and these electrical currents generate a temporary magnetic field that briefly alters Jupiter's field near the satellite. For electrical currents to flow in Europa, some part of the satellite must conduct electricity. Ice is not a good conductor, but salty water is. A saltwater ocean about 10 kilometers below the surface can produce the measured changes of Jupiter's magnetic field as it sweeps by in different orientations to the satellite.

What keeps Europa's subsurface ocean from freezing solid? Europa is probably kept warm inside by the gravitational tugging and flexing it experiences when moving toward and away from massive Jupiter in the course of the satellite's elliptical orbital motion. A similar effect melts the inside of rocky Io, producing its ubiquitous volcanoes. The orbit of Europa is pulled slightly out of round by the gravitational action of Io, which is closer to Jupiter, and Ganymede, the next satellite out. Jupiter's varying gravitational pull as Europa moves along its eccentric orbit produces tides of different size, causing the satellite to stretch and distort, heating its interior and keeping the water liquid beneath its icy crust. These internal tides will also cause Europa's overlying ice shell to flex, producing cracks that open and close as Jupiter squeezes the moon in and out.

Jupiter's satellite Ganymede, the largest moon in the solar system, probably contains substantial amounts of water inside, which would help account for its low mean mass density of just 1940 kilograms per cubic meter. Its surface has large dark plates separated by lighter regions, and impact craters that are surrounded by bright material (Fig. 2.45). The dark regions are believed to be part of the original crust of Ganymede, which probably cracked and spread apart. The lighter regions are most likely water ice that has moved in, replacing about half of the ancient, dark surface. The brilliant white material that surrounds some craters is probably clean water ice that splashed out from inside the satellite.

Like Europa, Saturn's moon Enceladus is also an ice world being squeezed inside as it moves along an eccentric orbit. The tidal flexing caused by the varying gravitational pull of nearby massive Saturn apparently heats Enceladus enough to maintain a liquid ocean beneath its

Fig. 2.45 Jupiter's satellite Ganymede Large dark blocks are frozen within the icy surface of Ganymede. They are believed to be part of the original crust of the satellite, resembling frozen-over continents floating on a background of translucent ice. The brilliant white material that surrounds some craters is probably clean water ice or bright snow that was splashed out from inside the satellite. The enhanced color of this *Galileo* image of Ganymede, taken on 29 March 1998, also reveals the two predominant terrain features on the moon: bright grooved terrain and older, dark furrowed areas. The violet hues at the poles may be the result of small particles of frost. (Courtesy of NASA/JPL/DLR.)

ice. As we have previously mentioned, geysers of water ice, water vapor and other substances spray out of fractures in its surface.

Substantial amounts of water ice reside well beyond all the major planets. This is where comets hibernate for billions of years, in the distant Kuiper belt and even more remote Oort cloud (Chapter 14). When one of these comets is deflected into the inner solar system and comes close to the Sun, it can expel about 25 000 kilograms of water every second, propelled by water vapor generated when the Sun's heat melts the water ice. A typical comet contains one million billion (10^{15}) kilograms of water.

3 Atmospheres, magnetospheres and the solar wind

- An atmosphere is a gaseous layer of molecules, with smaller amounts of atoms and ions, which surrounds a planet or natural satellite, held near them by their gravity.

- The present-day atmospheres of the terrestrial planets are thought to be secondary, having originated after planetary formation about 4.6 billion years ago. The atmospheres may have been released from volcanoes or acquired during collisions of comets and asteroids.

- The giant planets retain their primeval atmospheres, created when these planets formed, capturing significant amounts of hydrogen and helium gas.

- An atmosphere is characterized by the pressure and temperature of the molecules in it.

- Pressure increases with the temperature and density of the gas.

- An atmosphere allows the warmth of sunlight in but prevents the escape of infrared heat radiation from the planet's surface; this global warming by heat-trapping gases in an atmosphere is now known as the greenhouse effect.

- The ability of a planet or satellite to retain an atmosphere depends on both the temperature of that atmosphere and the gravitational pull of the planet or satellite. If the gas is hot, the molecules move about with a greater velocity and are more likely to escape the gravitational pull of the planet. A planet with a larger mass is more likely to retain an atmosphere.

- Only the massive giant planets, like Jupiter and Saturn, have a high enough escape velocity that they can retain all atoms and molecules, including the lightest element, hydrogen.

- Not all molecules move at the same average speed. Some of them move faster than the average speed and others move slower, with speeds described by the Maxwellian distribution.

- The lighter, high-velocity molecules can slowly leak out or evaporate from the top of an atmosphere where collisions no longer dominate the velocity distribution; this process is called Jeans escape or thermal evaporation.

- A thin membrane of air protects, ventilates and incubates us.

- Winds move air from hot to cold regions, in an attempt to equalize the temperature differences.

- The surface of Venus now lies under a hot and heavy atmosphere of carbon dioxide; its greenhouse effect has raised the surface temperature on Venus to a torrid 735 kelvin, hot enough to melt lead and zinc.

- High-velocity winds on Venus whip its highest clouds around the planet 60 times faster than the planet rotates.

- Mars now has an exceedingly thin, dry and cold atmosphere of carbon dioxide, with less than one-hundredth the surface pressure of the Earth's atmosphere, but global winds can stir up enough dust to completely cover Mars.

- The red planet breathes about one-third of its atmosphere in and out as its southern polar cap grows and shrinks with the Martian seasons.

- Mercury and the Earth's Moon are surrounded by a tenuous, varying mist of atoms that is continuously escaping into the surrounding space and being replenished from below. It is essentially a layer of exit, or escape, known as an exosphere.

- The atmospheres of Earth, Venus and Mars have evolved to their present states as the result of their varying distances from the Sun, a runaway greenhouse effect on Venus, and the development of life on Earth.

- The Sun generated so little heat more than 2 billion years ago that the Earth's oceans should have been frozen solid. This faint-young-Sun paradox could be resolved if a thick carbon-dioxide atmosphere warmed the planet back then or if the young Sun was more magnetically active or more massive than it is now.

- Like the Sun, the most abundant element in the giant planets is the lightest element, hydrogen, and the next most abundant element is helium; but Uranus and Neptune have lesser amounts of these two gases and relatively greater amounts of the heavier hydrogen compounds like methane, ammonia and water.

- Jupiter is all atmosphere or liquid, with no solid surface to rub against or continents to disturb the flow. Its Great Red Spot has existed for more than 300 years; the location and speed of powerful winds and some violent storms on the giant planet have remained unchanged for at least one century.

- Helium rain has been falling toward the center of Saturn for the past 2 billion years, significantly depleting the amount of helium in the planet's upper atmosphere.

- Saturn's largest moon, Titan, has a substantial Earth-like atmosphere, which is mainly composed of nitrogen and has a surface pressure comparable to that of the Earth's atmosphere. An opaque smog-like haze hides the surface of Titan from view.

- A thin film of oxygen envelops Jupiter's large moon Europa, while a tenuous mist of sulfur dioxide surrounds Jupiter's moon Io.

- An energy-laden, electrically charged solar wind blows out from the Sun in all directions and never stops, carrying with it a magnetic field rooted in the star.

- The radial, supersonic outflow of the solar wind creates a huge bubble of electrons, protons and magnetic fields, with the Sun at the center and the planets inside, called the heliosphere.

- The Earth has a dipolar magnetic field, amplified and sustained by dynamo action in its liquid core.

- Ancient magnetic rocks indicate that Earth's magnetic poles keep switching places every 10 thousand to 10 million years, and that our planet's magnetic field may now be heading for a flip.

- The terrestrial magnetic field deflects the solar wind, hollowing out a cavity called the magnetosphere. The magnetosphere of any planet is the volume of space from which the main thrust of the solar wind is excluded.

- Energetic electrons and protons have penetrated the Earth's magnetic defense. Some of them are confined within two doughnut-shaped radiation belts that encircle the Earth's equator but do not touch it.

- The magnetosphere contains particles from the Sun that arrive via the solar wind and penetrate the Earth's magnetic defense through a temporary opening in it, which is produced by magnetic reconnection when the solar magnetic field and the Earth's magnetic field point in opposite directions at the place where they touch.

- A plasmasphere is found in the inner part of the Earth's magnetosphere. This is located just outside the upper terrestrial atmosphere, which is called the ionosphere. The plasmasphere contains oxygen ions, protons and electrons derived from the ionosphere.

- Cosmic rays produce neutrons in the Earth's atmosphere, and a small fraction of these neutrons move out into the Earth's inner radiation belt before they disintegrate, producing electrons and protons in places they could not otherwise have reached.

- Of the eight major planets, six are known to generate detectable, global magnetic fields; only Mars and Venus do not now have such a dipolar magnetic field.

- Jupiter's magnetosphere is the largest enduring structure in the solar system, more than 10 times larger than the Sun. It was discovered when Earth-based radio telescopes unexpectedly detected the synchrotron radio emission of high-speed electrons trapped in the giant planet's immense magnetic field.

- The magnetic fields of Uranus and Neptune are offset by large amounts from their centers, and tilted by enormous angles from their rotation axes.

- The Earth's aurora is a spectacular multi-colored light-show.

- When viewed from space, the aurora forms an oval centered on the magnetic poles of the Earth; similar aurora ovals have been detected in ultraviolet light at both the north and south magnetic poles of Jupiter and Saturn.

3.1 Fundamentals

What is an atmosphere?

An atmosphere is a gaseous layer of molecules, with smaller amounts of atoms and ions, which surrounds a planet or natural satellite, held near them by their gravity. Since gas has a natural tendency to expand into space, only bodies that have a sufficiently strong gravitational pull can retain atmospheres. So the ability of a planet or satellite to retain an atmosphere depends on its mass; but it also depends on the mass of the gas particles as well as the gas temperature, determined by both the planet's distance from the Sun and the atmosphere's greenhouse effect.

Where do atmospheres come from?

The present-day atmospheres of the terrestrial planets are thought to be secondary, having come from other sources after planetary formation about 4.6 billion years ago. They can acquire atmospheric gas by releasing vapors from within the planet or by capturing volatile materials from comets or asteroids when they strike the planet.

An important source of the atmospheres of the terrestrial planets is the volcanic release of gases trapped inside their hot interiors. Volcanoes can supply water vapor H_2O, carbon dioxide CO_2, nitrogen N_2, and sulfur-bearing gases H_2S and SO_2, where H denotes a hydrogen atom, O an oxygen atom, C a carbon atom, N a nitrogen atom, and S a sulfur atom.

Another source of the terrestrial atmospheres is comets and asteroids. Because these objects formed further from the Sun's heat, they could retain water and carbon dioxide. Early in the history of the solar system, many more comets and asteroids were in orbits that intersected the orbits of Mars, Earth and Venus. The collisions would have released ices and gases, supplying these planets with the volatile substances needed to form their early atmospheres and oceans.

Initially these atmospheres were probably dominated by carbon dioxide and water vapor, and some of these gases could condense to become surface liquids or ices. The Earth's oceans and polar caps probably originated from water vapor that condensed as rain or snow. The polar caps of Mars contain frozen carbon dioxide, and vast amounts of water ice are frozen into its surface. Water vapor or carbon dioxide gas can be returned to these atmospheres by evaporation of surface liquids or sublimation of surface ice into gas.

In contrast to the terrestrial planets, the giant planets retain their primeval atmospheres, created when these planets coalesced in the cold outer precincts of the planetary system. They were large and massive enough to capture significant amounts of hydrogen and helium gas, and far enough from the Sun to be unaffected by its heat and winds.

Pressure and temperature

An atmosphere is characterized by the pressure and temperature of the molecules, which vary as a function of height within the atmosphere. Collisions between molecules in an atmosphere create the pressure, and collisions occur more frequently when a gas is hotter or when the gas has a greater density. The pressure, P, of a gas with molecules of number density, n, and temperature, T, is given by the perfect gas law $P = nkT$, where Boltzmann's constant $k = 1.38 \times 10^{-23}$ joule per kelvin, and the number density $n = \rho/m$ for a gas of mean mass density ρ and a mean molecular mass m. So pressure increases with the temperature and density of the gas.

For comparison purposes, the pressure and temperature are specified at the surfaces of the terrestrial planets and at the cloud tops of the giant planets. The atmospheric pressure is usually measured in a unit called the bar, as in barometer, where one bar is roughly equal to the Earth's atmospheric pressure at sea level. A thick atmosphere has a relatively large surface pressure, while a thin atmosphere produces comparatively little surface pressure.

Solar radiation warms a planet's atmosphere, and as we would expect, the heat is greatest for objects that are closest to the Sun. That is because the intensity of sunlight falls off as the inverse square of distance from the Sun.

We can make an initial estimate for the temperature of the surface of a terrestrial planet, or the cloud tops of a giant planet, by assuming that the surface or cloud tops are not noticeably warmed by heat rising from the planet's interior and that there is no atmosphere above them. The planet is then heated solely by the Sun's radiation, and we can calculate the planet's effective temperature, T_{eff}, from the relation $T_{eff} = 279/\sqrt{D}$ kelvin, where D is the planet's distance from the Sun in AU (Focus 3.1).

The effective temperatures of the planets are compared to the mean observed surface or cloud-top temperatures in Table 3.1. The surface of Venus is much hotter than expected, and the surface of the Earth is somewhat hotter, both a consequence of the greenhouse effect. The giant planets are also hotter than expected, owing to internal heat left over from their formation or to helium raining down inside them.

Global warming by the greenhouse effect

The surface temperature of a terrestrial planet can increase when its atmosphere traps heat near the surface, warming it to a higher temperature than would be achieved by the Sun's radiation in the absence of an atmosphere. Incoming sunlight is partly reflected by clouds, but the rest passes through the atmosphere to warm the planet's surface. Much of the surface heat is re-radiated in the form of long infrared waves that are absorbed by atmospheric molecules such as carbon dioxide or water vapor. Some of the trapped heat is re-radiated downward to warm the planet's surface and the air immediately above it. The atmosphere thus acts as a one-way filter, allowing the warmth of sunlight in, and holding it close to the planet's surface and elevating the temperature there.

The idea that this atmospheric blanket might warm the Earth was suggested in 1827 by the French mathematician Jean-Baptiste Fourier (1768–1830) and developed by the Irish scientist John Tyndall (1820–1893) in the 1860s. Fourier wondered how the Sun's heat could be retained to keep the Earth hot, concluding that sunlight passes through the atmosphere, which also prevents the escape of heat from the planet's surface.

Global warming by heat-trapping gases in the air is now known as the greenhouse effect, but this is a misnomer. The air inside a garden greenhouse is heated because it is enclosed, preventing the circulation of air currents that would carry away heat and cool the interior. Nevertheless, the term is now so common that we continue to use it to designate the process by which an atmosphere traps heat near a planet's surface.

Focus 3.1   How hot is a planet?

The radiant energy per unit time that a planet of radius R_P receives from the Sun is:

Solar energy received

$$= \pi \, R_P^2 \, f = \pi \, R_P^2 \frac{\sigma \, R_\odot^2 \, T_\odot^4}{D_P^2} \text{ joule per second}$$

where R_P is the radius of the planet, f is the total amount of radiant solar energy per unit time per unit area reaching the top of the planet's atmosphere, D_P is the planet's distance from the Sun, the Stefan–Boltzmann constant σ is $5.670 \times 10^{-8} \, \text{J m}^{-2} \, \text{K}^{-4} \, \text{s}^{-1}$, the solar radius $R_\odot$ is 6.96×10^6 meters, and the effective temperature of the visible solar disk is $T_\odot = 5800$ kelvin. The Stefan–Boltzmann law gives the amount of radiation energy lost per unit time from a planet of effective temperature, T_{eff}:

Energy lost $= 4\pi\sigma \, R_P^2 T_{eff}^4$ joule per second

Assuming thermal equilibrium between energy lost and received, and collecting terms, we obtain:

$$T_{eff}^4 = \frac{R_\odot^2 \, T_\odot^4}{4 \, D_P^2} \approx 1.36 \times 10^{32} \left(\frac{1}{D_P^2} \right)$$

$$\approx 6.089 \times 10^7 \left(\frac{1 \, \text{AU}}{D_P} \right)^2$$

or

$$T_{eff} \approx 279 \left(\frac{1 \, \text{AU}}{D_P} \right)^{1/2} \text{ kelvin}$$

where $1 \, \text{AU} = 1.496 \times 10^{11}$ meters is the distance of the Earth from the Sun. Notice that the effective temperature is independent of the planet's radius, and that the effective temperature for planets around other stars depends upon the square root of the star's radius and the star's

disk temperature, as well as the planet's distance from the star.

This expression assumes that all of the sunlight falling on the planet is absorbed, but some of it is always reflected. The extent to which a planet or satellite reflects light from the Sun is specified by its albedo, A, the percentage of reflected light. The visual albedo measures the fraction of incoming visible sunlight that is reflected directly into space, on a scale of 0.0 to 1.0. Rocky bodies like the Earth's Moon or Mercury absorb a lot of incident sunlight, while clouds or icy surfaces reflect it. Thus, the Moon and Mercury have a visual albedo of 0.12, while cloud-covered Venus has an albedo of 0.65, helping to make it the brightest planet in the solar system when it is visible from Earth.

Taking the albedo, A, into account, we have:

$$T_{eff} = 279(1 - A)^{1/4} \left(\frac{1 \, \text{AU}}{D_P} \right)^{1/2} \text{ kelvin}$$

There are two kinds of albedo: the Bond albedo, which measures the total proportion of electromagnetic energy reflected, and the visual geometric albedo that refers only to electromagnetic radiation in the visible spectrum. The geometric albedo of an astronomical body is the ratio of its actual brightness to that of an idealized flat, fully and isotropically reflecting disk with the same cross-sectional area. The Bond albedos for Mercury, Venus, Earth and Mars are 0.119, 0.75, 0.29 and 0.16, respectively, while their visual geometric albedos are 0.106, 0.65, 0.367 and 0.150. When the formula is applied to the Earth we obtain $T_{eff} \approx 256$ kelvin using the Bond albedo and $T_{eff} \approx 249$ kelvin using the visual geometric albedo. The Bond albedo for the Earth's Moon is 0.123, so its effective temperature would be higher, at about 270 kelvin.

Tyndall built an instrument to measure the heat-trapping properties of various gases, examining the transmission of infrared heat radiation through them. He found that the main constituents of our atmosphere – oxygen (O_2) 21 percent, and nitrogen (N_2) 77 percent – were transparent to both visible and infrared radiation. These diatomic, or two-atom, molecules are incapable of absorbing any noticeable amounts of infrared heat radiation. He also found that water vapor and carbon dioxide, which are minor ingredients of the Earth's air, absorb significant heat. As Tyndall realized, these gases are transparent to sunlight, which warms the ground, but partially opaque to the infrared heat rays, which are trapped near the surface and warm our globe. Water vapor (H_2O) and carbon dioxide (CO_2)

molecules consist of three atoms and are more flexible and free to move in more ways than diatomic molecules, so they absorb the heat radiation.

Once the temperature of a planet has been established, by direct observation or from calculations of solar heating, primeval heat, and the greenhouse effect, we can determine the likely constituents of its atmosphere.

Losing an atmosphere

The escape of gases from a planetary atmosphere plays as big a role in determining its composition as the supply of gases does. One of the most important loss mechanisms is thermal escape, in which the gas gets too hot to hold

Table 3.1 Distances, visual albedos, effective temperatures, and mean temperatures of the planets[a]

Planet	Average distance, D_P (AU)	Visual geometric albedo, A	Effective temperature, T_{eff} (kelvin)	Mean temperature[b] (kelvin)
Mercury	0.387	0.106	436	440
Venus	0.723	0.65	252	730
Earth	1.000	0.367	249	281
Mars	1.524	0.150	217	210
Jupiter	5.203	0.52	102	165
Saturn	9.537	0.47	77.1	134
Uranus	19.19	0.51	53.3	76
Neptune	30.07	0.41	44.6	73

[a] Distances and mean temperatures are from http://www.jpl.nasa.gov/solar_system/planets. Effective temperatures are calculated from the visual geometric albedos, which are from http://ssd.jpl.nasa.gov.
[b] There are the mean surface temperatures for the terrestrial planets and the mean cloud-top temperatures for the giant planets.

on to. Thermal escape provides a straightforward explanation for why the large planets have atmospheres containing hydrogen and hydrogen compounds, the middle-sized planets have atmospheres containing oxygen compounds, and small objects, such as the Earth's Moon and Mercury, have no appreciable atmosphere at all.

George Johnstone Stoney (1826–1911) set forth the simple explanation for these differences in 1898. It partly depends upon the mass and gravitational pull of the object, and it also depends on the atmospheric temperature and the mass of the gas atom or molecule.

A molecule will overcome the gravitational pull of a planet or satellite if the molecule's velocity exceeds the object's escape velocity, which increases with the object's mass (Focus 3.2). Small bodies with low mass, such as our Moon, have a very small escape velocity and insufficient gravitational pull to retain any substantial atmosphere. Middle-sized planets, like the Earth, Venus and Mars, have moderate escape velocities and enough gravity to hold on to heavier, slower-moving molecules, but they are small enough and warm enough for hydrogen and helium to escape. Only the massive giant planets, like Jupiter and Saturn, have a high enough escape velocity that they can retain all molecules, including the lightest one, hydrogen.

Temperature also plays a role, for it helps determine if the molecules can move fast enough to escape an object's gravity. A planet will only retain molecules that are moving at velocities less than the planet's escape velocity, and a

molecule's velocity increases with temperature (Focus 3.2). Hotter molecules dart about at faster speeds, and colder molecules move with slower speeds. Since the outer atmospheric temperature falls off with increasing distance from the Sun, molecules tend to have lower velocities out in the realm of the giant planets. At a given temperature, a molecule's velocity increases with decreasing molecular mass (Focus 3.2), so lighter molecules move at faster speeds and are more likely to escape a given planet or satellite than heavier ones.

When the thermal velocity exceeds the escape velocity for a given type of molecule, all of those molecules will promptly flow out into space, and if this happens for every type of molecule, an airless body is left behind – like Mercury, the Earth's Moon, and the four large satellites of Jupiter. For the Earth, Mars and Venus, the thermal velocity of all molecules is smaller than the escape velocity, but the lightest gases can still slowly move out into space from the top of their atmospheres.

To understand why this might happen, note that the molecules in a gas can gain or lose speed by collisions with each other, so not all molecules move at the same average speed, or at the thermal velocity. Some of them move faster and others move slower, with a velocity probability distribution published in 1866 by the Scottish physicist James Clerk Maxwell (1831–1879). His equation, known as the Maxwellian distribution, gives the fraction of gas molecules moving at a specified velocity at any given temperature. In 1871 Ludwig Boltzmann (1844–1906) derived

Focus 3.2   Thermal escape of an atmosphere

The ability of a planet or satellite to retain an atmosphere depends on both the temperature of that atmosphere and the gravitational pull of the planet or satellite. If the gas is hot, the molecules move about with a greater speed and are more likely to escape the gravitational pull of the planet. This is one of the reasons that Mercury, the closest planet to the Sun and therefore the hottest, has no atmosphere. The other reason is that Mercury has a relatively small size and mass, as far as planets go, and thus has a comparatively low gravitational pull. On the other hand, a planet with a larger mass is more likely to retain an atmosphere, which helps explain why massive Jupiter retains the lightest element, hydrogen. Jupiter is also relatively far away from the Sun's heat, so molecules in Jupiter's atmosphere move at a relatively slow speed.

An atom, ion or molecule moves about because it is hot. Its kinetic temperature, T, is used to define its thermal velocity, $V_{thermal}$, given by equating the thermal energy to the kinetic energy of motion:

$$\text{Thermal energy} = \frac{3}{2}kT = \frac{1}{2}mV_{thermal}^2 = \text{Kinetic energy}$$

or solving for the thermal velocity:

$$V_{thermal} = \left[\frac{3kT}{m}\right]^{1/2}$$

where Boltzmann's constant is $k = 1.380\,66 \times 10^{-23}$ joule per kelvin, and the particle's mass is denoted by m. We see right away that, at a given temperature, lighter particles move at faster speeds. Colder particles of a given mass travel at slower speed. Anything will cease to move when it reaches absolute zero on the kelvin scale of temperature.

When the kinetic energy of motion of a particle of mass m moving at velocity V is just equal to the gravitational potential energy exerted on it by a larger mass, M, we have the relation:

$$\text{Kinetic energy} = \frac{mV^2}{2} = \frac{GmM}{D}$$
$$= \text{Gravitational potential energy}$$

where the Newtonian gravitational potential is $G = 6.6726 \times 10^{-11}\,\text{m}^3\,\text{kg}^{-1}\,\text{s}^{-2}$, and D is the distance between the centers of the two masses. When we solve for the velocity, we obtain:

$$V_{escape} = \left[\frac{2GM}{D}\right]^{1/2}$$

where the subscript "escape" has been added to show that the small mass must be moving faster than V_{escape} to leave the larger mass, M. This expression is independent of the value of the smaller mass, m. The escape velocities at the surfaces or cloud tops of the planets range between 4 and 60 kilometers per second.

the distribution function independently, as a result of his kinetic theory of gases. Although the function looks symmetric, it cuts off at low velocities and is enhanced in a high-velocity tail (Fig. 3.1). This means that at any instant, a tiny fraction of the molecules are moving fast enough to escape even when the average thermal velocity is less than the escape velocity.

The lighter, high-velocity molecules can slowly leak out or evaporate from the top of the atmosphere where collisions no longer dominate the velocity distribution. At lower altitudes, collisions confine the particles, but above a certain altitude known as the exobase, the atmosphere is so tenuous that gas particles hardly ever collide. Nothing stops an atom or molecule with sufficient velocity from flying away from the exobase into space.

The method of escape from the exobase is known as Jeans escape, after the British scientist James Jeans (1877–1946) who introduced it in 1916. It is also known as thermal evaporation since it is analogous to the slow evaporation of water from the ocean. On average, the molecules in the ocean water do not have enough thermal energy to escape

from the liquid, but some of them acquire enough near the surface.

At and below the exobase in the atmosphere, collisions between particles drive the velocity distribution into a Maxwellian distribution, while above the exobase collisions are essentially absent and particles that have velocities greater than the escape velocity may leave the planet. The upward moving atoms in the high velocity tail of the Maxwell distribution can exit the planet, hence the name exobase.

As the lightest element, hydrogen is the one that most easily overcomes the gravity of a terrestrial planet, but first it must reach the exobase. On Earth, the exobase is located about 500 kilometers above the surface, and calculations indicate that about a billion billion billion (10^{27}) hydrogen atoms are still being lost from the Earth's exobase every second. This value is confirmed by satellite ultraviolet observations of hydrogen escaping from the Earth's upper atmosphere.

Notice that lighter particles are lost by thermal evaporation at a much faster rate than heavier ones. Even over

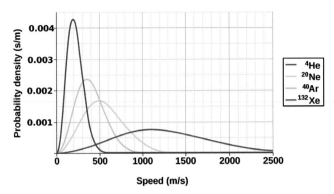

Maxwell-Boltzmann Molecular Speed Distribution for Noble Gases

Fig. 3.1 Maxwell-Boltzmann speed distribution   The probability distribution of speed density (*vertical axis*) at different speeds, v (*horizontal axis*) for a temperature of $T = 218.85$ kelvin and four noble gases, helium (*blue*), neon (*yellow*), argon (*turquoise*) and xenon (*purple*). The peak of each curve occurs at the most probable speed, $v_P = (2kT/m)^{1/2}$, where Boltzmann's constant is $k = 1.38 \times 10^{-23}$ joule per kelvin, T is the temperature, and m is the element's mass. This distribution was first derived by the Scottish scientist James Clerk Maxwell (1831–1879) in 1866. As indicated in this plot, the peak shifts to higher speeds at lower mass provided the temperature is unchanged, because less massive elements tend to move at faster speeds. A similar change occurs at higher temperatures for a given mass. Ninety-nine percent of all molecules have speeds greater than the average molecular speed, given by $v_{avg} = (3kT/m)^{1/2}$, for every curve has a high-velocity tail which is most evident in the purple line. The equation for the distribution f(v) is: $f(v) = 4\pi \ [m/(2\pi kT)]^{3/2} \ v^2 \ \exp[-mv^2/(2kT)]$.

the Earth's lifetime of 4.6 billion years, the total mass of all the hydrogen atoms lost by thermal evaporation is 2×10^{17} kilograms, and the amount lost by heavier molecules would be much less. By way of comparison the total mass of the Earth's atmosphere is about 5×10^{18} kilograms.

3.2 Atmospheres of the terrestrial planets

Earth's unique atmosphere

Our atmosphere forms an indispensable interface with nearby space, but it is often invisible. After all, you look right through the air in your room. Our atmosphere usually goes unseen on a warm, dry, windless day. Yet the slow drift of floating clouds or the sight of birds and airplanes supported by their motion proves that there is something substantial surrounding us. We can sense the touch of the wind on a stormy day, and on cold days we feel the air against our skin.

We find a further clue in the rise of smoke above a candle or a group of hawks circling above a warm meadow. Hot air rises around the flame of the candle, and the flowing air replenishes the supply of oxygen required to keep the candle burning. The hawks are getting free rides in the rising currents of hot air above ground.

When astronauts look down at the Earth at sunrise or sunset, they detect the thin atmosphere that warms and protects us, and permits us to breathe (Fig. 3.2). It is only 10 kilometers from the ground to the top of the sky, or no further than you might run in an hour. Everything beyond that thin layer of air is the black void of space. And everything below it is what it takes to sustain life.

If we were to weigh the air in a one-liter container we would find it tips the scales at slightly more than one gram. This is about one-thousandth the weight of the same amount of water. Determining its constituents is easy — just place the appropriate instrument in the air and see what is there. The major constituents of dry air on Earth are nitrogen molecules (77 percent), oxygen molecules (21 percent) that we breathe, and argon atoms (0.93 percent). Carbon dioxide is a miniscule 0.035 percent. There is almost no hydrogen in our air, and most of the hydrogen on Earth is found in water. The water vapor in wet air is variable in amount, usually no more than 1 percent.

Plants and animals respectively supply almost all of the oxygen and carbon dioxide molecules in our atmosphere, and they are continually being recycled in the photosynthesis and respiration processes. Animals breathe oxygen, and when they exhale they release carbon dioxide and water vapor. Green plants on the land and one-celled plants in the ocean water absorb carbon dioxide and water, use them in the photosynthesis of nourishment and then release oxygen into the atmosphere. This symbiotic relationship is one of the most remarkable features of life on Earth; you might call it the breath of life.

If plants did not continuously replenish the oxygen in our air, animals and humanity would exhaust the available supply in a mere 300 years. All the water on the Earth is split by photosynthesis and reconstituted by respiration every 2 million years or so. For millions of years, our ancestors have breathed the same oxygen and drank the same water, binding them temporarily in their bodies and then releasing them again to the atmosphere.

The Earth is the only place in the solar system where we can stand naked and survive. The air brings oxygen to our lungs and refreshes our bloodstream; sunlight and the "natural" greenhouse effect provide just enough heat to prevent our fluids from freezing or boiling. There is a very different situation on the other terrestrial planets, where there are no plants to supply the oxygen and the temperatures are either boiling hot or freezing cold.

Fig. 3.2 A thin colored line   The brilliant red of the setting Sun illuminates the thin atmosphere that warms and protects us. Without this atmospheric membrane we could not breathe and water would freeze. It is dust in the dense lower atmosphere that scatters red sunlight; molecules that are higher up in the air scatter blue sunlight, coloring the sky blue. This image was photographed on 3 June 2007 by an Expedition 15 crewmember on the *International Space Station*. (Courtesy of NASA.)

Earth's climate and weather

Because the Earth is a sphere, there is an unequal distribution of the Sun's heat at the terrestrial surface, producing the climate differences that starkly distinguish one part of the globe from another. The equator, for example, receives more sunlight than the poles, so the equatorial regions, known as the tropics, are the hottest places in the world and the poles are the coldest. The extra warmth at tropical latitudes evaporates seawater and causes the air to expand and rise, carrying freshwater moisture with it. As it rises, the tropical air cools and forms clouds of water ice. These clouds are moved over great distances by winds before condensing again to liquid water and falling to Earth as rain. When arriving on land, the water refreshes lakes and streams, and most of it eventually finds its way back to the sea.

The first to suggest a continuous, global circulation of the atmosphere was the English astronomer Edmond Halley (1646–1742), best known today for the comet that bears his name. Halley reasoned that the high-temperature air in equatorial regions would circulate toward the colder poles, and that the colder air from the north would move away from the poles to replace the warm tropical air. This movement of air in response to unequal temperatures is known as the wind, and the winds are blowing in an attempt to equalize global temperature differences. Since an increase in temperature produces higher pressure, the winds are also attempting to balance pressure differences, made unequal by different amounts of solar heating in various places.

Air currents tend to circulate from the tropics toward the poles and back, in the north–south direction, but they are deflected in the east–west direction by the Earth's rotation. Along the equator the near-surface air currents converge to form the trade winds, that blow mainly from east to west, almost every day of the year (Fig. 3.3). At mid-latitudes, the near-surface winds blow largely from the west and so are called westerlies. The high-altitude jet streams also blow eastward, at speeds of up to 40 meters per second, in a sinuous path that resembles the meandering of a river (Fig. 3.3).

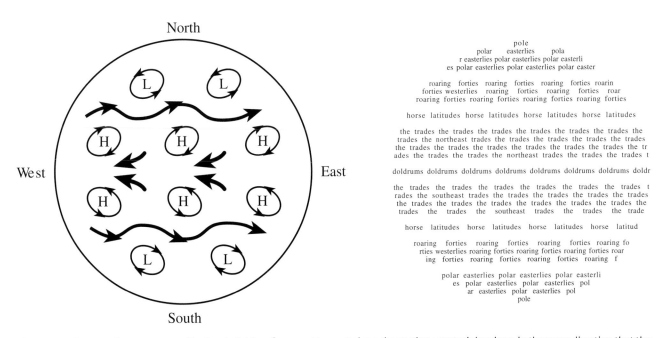

North

West

East

South

Fig. 3.3 Earth's weather patterns   Trade winds blow from east to west along the Earth's equatorial regions, in the same direction that the planet rotates. At higher latitudes, there are high-altitude jet streams that move at speeds of up to 40 meters per second in the opposite direction to the trade winds. In the northern hemisphere there are high-pressure cyclones, denoted by H, and low-pressure anti-cyclones, L, rotating in the clockwise and counter-clockwise directions, respectively; in the southern hemisphere the cyclones and anti-cyclones rotate in the opposite direction. The prevailing winds are given in Annie Dillard's poem, "the windy planet".

The atmosphere's attempt to overcome the Sun's unequal heating of the Earth is never ending. The equator is always hotter than the poles, and the Earth is always spinning. The temperatures and pressures are never balanced, and the winds always blow.

Moreover, the winds do not blow in a straight line. The Earth's rotation creates vast swirling eddies in the air, just like those in a river or stream. One type of eddy is known as a *cyclone*, from the Greek word for "wheel". It is a vast whirling mass of wind and precipitation.

A cyclone in the northern hemisphere is a low-pressure cell of air rotating counter-clockwise. An anticyclone is a high-pressure cell rotating clockwise in the northern hemisphere (Fig. 3.3). Between the trade winds and the westerlies, the pattern is primarily a series of high-pressure cyclones rotating clockwise in the north and counter-clockwise in the south. In the polar regions, the weather pattern is mainly a series of low-pressure cyclones rotating counter-clockwise in the north and clockwise in the south (Fig. 3.3).

As the cyclones pinwheel across the globe, they produce most of our stormy weather. Thunderstorms, blizzards and tornadoes are examples. They tend to move from west to east across the North American continent in an almost steady parade. Cyclones also occasionally develop into hurricanes (Atlantic Ocean and Caribbean Sea) and typhoons (Pacific Ocean and China Sea).

Venus' atmosphere

In many ways, Venus is Earth's twin sister, with almost the same weight and waistline. Her mass is 81 percent that of Earth, and her radius 95 percent, so the feel of gravity at the planet's surface is similar to that on Earth. The two planets also have nearly equivalent mean mass densities, of 5244 and 5520 kilograms per cubic meter respectively, indicating that their bulk composition must be nearly the same. Venus is also just a little closer to the Sun than the Earth, orbiting our star at a distance of 0.723 AU compared with the Earth's 1.000 AU. All of these similarities gave rise to the idea that the atmosphere of Venus might resemble the Earth's air, but with a more temperate climate due to its closer distance from the Sun.

About half a century ago, many astronomers believed that the surface of Venus was warm and wet, perhaps with steamy swamps and jungles and even living creatures. Then, in the late 1950s, Earthbound radio astronomers looked beneath the clouds and discovered the extremely hot and inhospitable surface of Venus. Cloudy atmospheres are transparent to the radio waves, so the radio radiation from Venus comes directly from its surface, and can be used to take its temperature. Beneath her gleaming clouds, the planet is an inferno with a temperature hot enough to melt lead or zinc.

Fig. 3.4 Raging winds on Venus   A high-velocity wind whips the upper layer of Venus's cloud deck around the planet's equator once every four Earth days, moving at speeds of up to 100 meters per second. These photographs taken at ultraviolet wavelengths on consecutive days by the *Pioneer Venus* spacecraft in 1979 illustrate this. The Y-shaped clouds move towards the west (*left*). A zonal east-to-west circulation dominates the winds of Venus. The low atmosphere and the planet's surface also rotate westward, but with the much slower period of 243 Earth days. (Courtesy of NASA/Larry Travis.)

Our knowledge of this torrid world was further enhanced when spacecraft directly measured the atmosphere. An entry probe parachuted from the Russian *Venera 7* spacecraft in 1970 transmitted measurements of the temperature and pressure all the way down to the bottom of the atmosphere, where the temperature reaches a sizzling 735 kelvin. Down there the thick, heavy atmosphere produces a pressure of 92 bars – that is, 92 times the air pressure at sea level on Earth. So the atmosphere on Venus is very hot and extremely heavy.

The principal constituent of the thick atmosphere on Venus is carbon dioxide. It was first identified in the upper atmosphere of Venus by infrared absorption lines in the planet's spectrum, obtained in 1932 using modest ground-based telescopes when infrared-sensitive photographic plates had been developed, and confirmed during the *Mariner 2* flyby in 1962. The fact that carbon dioxide is the major atmospheric gas on Venus was firmly established in 1967 and 1969, when three Russian space probes, *Veneras 4, 5* and *6*, descended by parachute into the atmosphere, and obtained direct measurements of its principal constituents, showing that it consists of 96 percent carbon dioxide.

Strong winds are blowing the highest clouds around Venus at speeds of up to 100 meters per second, racing around the planet's equator in the east–west direction once every four Earth days (Fig. 3.4). Curiously enough, Venus' surface rotates in the same westward direction but with a much longer period of 243 Earth days. So the winds blow the outer atmosphere around the planet much more rapidly than the planet spins. In comparison, terrestrial jet streams move at up to half the speed of the high-flying

clouds on Venus, but they are limited to narrow zones high in the Earth's atmosphere.

Near the surface of Venus, at the bottom of its massive atmosphere, the rapid winds have disappeared, and the atmosphere has become as sluggish and turgid as water at the bottom of Earth's oceans. Most of the atmosphere beneath the high-flying clouds probably rotates synchronously with the surface, just as most of the Earth's atmosphere does.

Mars' atmosphere

Mars, fourth planet from the Sun, was also long thought to resemble the Earth. The length of the day on Mars is only 37 minutes longer than our own, the rotational axes of both Earth and Mars are now tilted by about the same amount, and the Martian year is 687 Earth days, or almost two Earth years. Both planets have four seasons – autumn, winter, spring and summer – although the Martian seasons are nearly twice as long. Mars has white polar caps that wax and wane with the seasons. Their alternate growth and recession meant that gases were being extracted from, and released into, an atmosphere. White clouds are also found on Mars, resembling those on Earth, and clouds are not possible without an atmosphere.

Thus, Mars has an atmosphere, with clouds, polar caps and seasons, and these Earth-like qualities led astronomers to speculate that its atmosphere resembles Earth's. The red planet is nevertheless twice as far away from the Sun as the Earth, so it will be warmed less by solar radiation and ought to be colder — if there is no pronounced greenhouse effect on Mars.

Table 3.2 Percentage composition, surface pressures and surface temperatures of the atmospheres of Venus, Mars and Earth

	Venus	Mars	Earth
Carbon dioxide, CO_2 (%)	96.5	95	0.035
Nitrogen, N_2 (%)	3.5	2.7	77
Argon, Ar (%)	0.007	1.6	0.93
Water vapor, H_2O (%)	0.010	0.03 (variable)	1 (variable)
Oxygen, O_2 (%)	0.003	0.13	21
Surface pressure (bars)	92	0.007 to 0.010	1.0 (at sea level)
Surface temperature (kelvin)	735	183 to 268	288 to 293

The composition and extent of the Martian atmosphere wasn't understood until the Space Age. When the *Mariner 4* spacecraft passed behind Mars in 1964, its radio signal was sent through the Martian atmosphere, and a surface pressure of about 0.005 bars, or two-hundredths that of Earth, was inferred from the altered signal. So the atmosphere on Mars is exceedingly thin, with a surface pressure comparable to the pressure high in the Earth's rarefied stratosphere. In addition, by combining the spacecraft measurements of surface pressure on Mars with ground-based spectra of the planet, astronomers quickly concluded that carbon dioxide is the main ingredient of the Martian atmosphere. The exact chemical composition of the atmosphere on Mars was determined by direct measurements in 1976 when the *Viking 1* and *Viking 2* landers arrived at the surface.

The atmospheres of both Mars and Venus have the same ingredients as our air, but the proportions are different (Table 3.2). The principal atmospheric ingredient of both planets is carbon dioxide, at 96 and 95 percent respectively. Carbon dioxide accounts for only 0.035 percent of our air.

There are small, variable amounts of water vapor in the Martian atmosphere, and over the eons it has helped oxidize the planet. The water vapor has been broken down by ultraviolet sunlight into hydrogen and oxygen atoms. The light hydrogen leaked off into space and the heavier, surplus oxygen stayed to oxidize the surface rocks, turning them red. To account for its color, Mars must have lost an ocean of water equivalent to a global layer meters to tens of meters deep.

The atmospheres of Earth, Venus and Mars each have one or more gases that can saturate. That is, the

atmosphere fills up with as much of the vapor as it can hold, and then that substance condenses. In the Earth's atmosphere, water vapor condenses to form billowing white clouds of water ice. The same thing happens on Mars, where clouds of water ice are found near volcanoes. Carbon dioxide also condenses out of the thin, cold Martian atmosphere into clouds and onto the surface, to form carbon-dioxide ice, also known as dry ice. On Venus, it is sulfuric acid that condenses to form the thick, unbroken layer of yellow clouds that always enshroud the planet.

The atmospheric pressure on Mars increases and decreases as its largest polar cap, the southern one, grows and shrinks, producing a seasonal change in atmospheric pressure by about 30 percent. It is as if the planet was a giant lung that slowly breathes in and exhales the same gas, carbon dioxide. When the surface temperature drops during the southern winter, the atmospheric carbon dioxide condenses and freezes to enlarge the polar cap, resulting in a drop in atmospheric pressure. In the southern summer, the ice sublimates (goes directly from solid to vapor, without becoming liquid) back into the atmosphere, increasing the atmospheric pressure.

Powerful seasonal winds are driven by temperature differences between the northern and southern hemispheres of Mars. Warm air rises in the summer hemisphere and descends in the winter one, where carbon dioxide is condensing to make the seasonal polar cap grow. The strong Martian winds also strip away light-colored dust in some areas and deposit it in others, accounting for the seasonal growth and decay of large dark areas seen from Earth.

Martian winds stir up small, localized dust storms, and occasionally coalesce to produce a globe-encircling dust storm. When substantial amounts of dust have been tossed aloft, sunlight is absorbed in the atmosphere rather than at the surface, and the storm can sustain itself by converting the Sun's energy into wind energy. The entire planet then becomes wrapped in an opaque yellow veil.

Exospheres of Mercury and the Earth's Moon

A tenuous, variable shroud of hydrogen, helium, sodium, and potassium atoms envelops both Mercury and the Earth's Moon, but always with exceedingly low number densities of less than 0.1 million million (10^{11}) atoms per cubic meter. By way of comparison, a cubic meter of Earth's atmosphere at sea level contains about 25 million billion billion (2.5×10^{25}) molecules.

The gas enveloping Mercury and our Moon is so rarified that the surface pressure is about a trillionth (10^{-12}) of the pressure at the Earth's surface. The atoms are so far apart that they almost never hit and connect with each other. They interact primarily with the surface, rather than constantly ricocheting off each other as the molecules in our air do. Because collisions are rare, each constituent has unique sources and loss mechanisms.

Such a rarefied gaseous envelope is technically known as an exosphere, and not an atmosphere, for the atoms are not permanently bound by gravity but instead escape, or exit, into surrounding space on timescales of hours to days. The neutral, or un-ionized, atoms released from Mercury are, for example, accelerated by solar radiation pressure to form an extended tail of atoms pointing away from the Sun. So the atoms in the exosphere must be continuously replenished, and they are mainly liberated from the surface by the action of sunlight, solar wind particles or micrometeorites.

Small amounts of hydrogen and helium were detected when *Mariner 10* flew past Mercury in 1973–74. In 1985–86, Earth-based telescopic observations detected sodium and potassium in Mercury's exosphere, with abundance variations from hours to years. Sodium and potassium are also found in the Moon's exosphere. Subsequent Earth-based observations of Mercury revealed the presence of calcium, and demonstrated the presence of an extended, anti-sunward tail of sodium atoms.

Observations with an instrument aboard the *MESSENGER* spacecraft, during its flybys of Mercury in 2008 and 2009, revealed the presence of magnesium, calcium and sodium atoms in the anti-sunward tail of Mercury's exosphere, with differing spatial distributions. These atoms must be coming from the planet's surface, but the different sources, as well as the detailed transfer and loss processes, remain unknown.

Evolution of the terrestrial planetary atmospheres

Nothing in the cosmos is fixed and unchanging, and nothing escapes the ravages of time. The atmospheres of the Earth, Venus and Mars are no exception, for they have been slowly altered with the addition and removal of gases as time goes on. In fact, their atmospheres probably weren't even there when the planets formed. They had too little mass to attract and hold on to the abundant hydrogen gas that was around when they accumulated. The building blocks from which the terrestrial planets formed were primarily rocky and metallic objects, and these planets may have been initially too hot to retain substantial amounts of water vapor or carbon dioxide in their early atmospheres.

The atmospheres of Earth and Venus probably began with similar compositions about 4 billion years ago, but their subsequent histories have been quite different. The Earth's atmosphere is now depleted of carbon dioxide and has excessive oxygen, while Venus has no oceans.

A massive carbon-dioxide atmosphere is responsible for the high surface temperature of Venus through the greenhouse effect. On Earth, the surface temperature is raised by about 30 kelvin by this effect, resulting in the mild climate we enjoy today. But the greenhouse effect has raised the temperature of Venus's surface by hundreds of kelvin.

The atmosphere greenhouse effect raised the temperature of young Venus and boiled away any oceans that might have condensed. The increased water vapor blocked more heat, raising the surface temperature in a runaway greenhouse effect. It turned Venus into the torrid world we see today, with a surface temperature of 735 kelvin. Because Earth is slightly further from the Sun than Venus, with a slightly lower initial temperature, the water on Earth remained liquid and it kept its oceans.

Life has been an important influence on the evolution of the Earth's atmosphere. Over the past 4 billion years, living things have caused a decrease in atmospheric carbon dioxide, while also providing an increase in atmospheric oxygen. The carbon dioxide was absorbed in ocean water where tiny marine creatures extracted it to manufacture carbonate shells. When these creatures died, their shells sank, producing carbonate sediments and rocks on the ocean floor. The result was a gradual depletion of carbon dioxide from the atmosphere.

Plant life growing in the Earth's early oceans gradually supplied oxygen to the atmosphere. Substantial amounts of oxygen began to appear in our air about 2 billion years ago, making it breathable by animals and eventually humans.

Thus, the carbon dioxide on Earth probably moved from the atmosphere to the oceans and into the rocks. Some of the carbon dioxide is returned to the air when the spreading sea floor plunges into a deep ocean trench, producing volcanoes. But there is still as much carbon dioxide remaining at the bottom of the ocean and in rocks on Earth as there is in the atmosphere of Venus, enough to exert an atmospheric pressure of 70 bars if released into our air.

The balance is a delicate one. Because Earth is slightly further away from the Sun than Venus, the Earth evolved into a living world capable of sustaining a remarkable diversity of life. If the Earth were placed in Venus's orbit, its atmosphere would get hotter and thus capable of holding more water evaporating from the ocean. The additional water vapor would trap more heat from the Sun, raising the temperature further and evaporating more water. The runaway greenhouse would eventually raise the temperature to values as high as those on Venus today. All of Earth's water would be boiled away, and the Earth would turn into a dried-out, lifeless place like Venus.

Since Mars is further away from the Sun than any other terrestrial planet, it is warmed least by the Sun. The red planet has about half the size and a tenth of the mass of Earth, so its lower gravity would be less likely to hold on to a substantial atmosphere. We might therefore expect a thin, dry and cold atmosphere similar to the one we see today on Mars. Yet, in the distant past, the Martian atmosphere might have been warmer, denser and wetter than it is today, permitting torrents of water to flow across its surface. This would account for ancient water flow that occurred on the surface of Mars 3 to 4 billion years ago.

If Mars once had a thicker atmosphere, it could have slowly evaporated into space under the combined effects of energetic sunlight and the planet's weak gravitational field; ultraviolet light from the Sun breaks up the atmospheric molecules into lighter, more energetic atoms that can escape the relatively weak gravity. More recent climatic change on Mars, during the past few million years, has been induced by the changing tilt of its rotation axis, causing atmospheric water and carbon dioxide to move into polar ice caps and back into the atmosphere again.

The Sun is also changing as time goes on, growing slowly in luminous intensity with age, a steady, inexorable brightening that is a consequence of the nuclear reactions that make it shine. As the Sun burns hydrogen into helium in its energy-generating core, the increasing amounts of helium require a rise in temperature to sustain the nuclear burning, and hence an increase in the rate of the nuclear reactions and a slow brightening of the Sun. You couldn't detect the change over all of human history, but it has profound implications over cosmic periods of time.

Stellar evolution calculations indicate that when the Sun began to shine, about 4.6 billion years ago, it was 30 percent dimmer than it is today. Assuming an unchanging atmosphere on Earth, with the same composition, greenhouse effect, and reflecting properties as today, the decreased solar luminosity would have caused the planet's global surface temperature to drop below the freezing point of water at all times earlier than 2 billion years ago. The Earth's oceans would have been frozen solid, there would be no liquid water, and the entire planet would have been locked into a global ice age something like Mars seems to be in now.

Yet sedimentary terrestrial rocks, which must have been deposited in liquid water, date from 3.8 billion years ago. There is fossil evidence in those rocks for living things at about that time. Thus for billions of years the Earth's surface temperature was not very different from today, and conditions have remained hospitable for life on Earth throughout most of the planet's history.

The discrepancy between the Earth's warm climatic record and an initially dimmer Sun has come to be known as the faint-young-Sun paradox. It can be resolved if the Earth's primitive atmosphere contained about a thousand times more carbon dioxide than it does now. Greater amounts of carbon dioxide would enable the early atmosphere to trap greater amounts of heat near the Earth's surface, warming it by an enhanced greenhouse effect. That would keep the oceans from freezing.

Over time the Sun grew brighter and hotter. The Earth could only maintain a temperate climate by turning down its greenhouse effect as the Sun turned up the heat. Our planet's rocks, oceans, and life itself may have together removed carbon dioxide from the atmosphere. Thus, the increase over time in brightness of the Sun's radiation might have been compensated by a steadily decreasing greenhouse effect, so that the surface temperature of the planet has remained about the same for the past 4 billion years. This convenient thermostat might soon be disrupted as human civilization dumps more and more carbon dioxide into the atmosphere by burning fossil fuels like coal and oil.

A different remedy of the faint-young-Sun paradox is that the Sun was more active in its youth, explosively releasing greater amounts of magnetic energy, while still radiating faint visible sunlight. Because of its faster rotation, the young Sun might have had stronger magnetic fields with enhanced extreme-ultraviolet and X-ray radiation and a greater output of high-energy particles. Or the Sun might have begun shining as a brighter, more massive star, and wasn't faint after all, subsequently losing much of its mass in strong solar winds. As with humans, the Sun's energetic youth might have evolved into a calmer old age, with moderate magnetic activity and weaker winds while shining brighter in visible sunlight.

3.3 Atmospheres of the giant planets

Composition and temperature of giant planet atmospheres

The atmospheres of the giant planets – Jupiter, Saturn, Uranus and Neptune – are very unlike those of the Earth, Mars and Venus. The elemental composition of the giant planets resembles that of the Sun, and they have been

Table 3.3 Percentage composition of the Sun and the outer atmospheres of the giant planets[a]

Molecule or atom	Sun	Jupiter	Saturn	Uranus	Neptune
Hydrogen, H_2 (molecule)	84	86.4	97	83	79
Helium, He (atom)	16	13.6	3	15	18
Water, H_2O (molecule)	0.15	(0.1)	–	–	–
Methane, CH_4 (molecule)	0.07	0.21	0.2	2	3
Ammonia, NH_3 (molecule)	0.02	0.07	0.03	–	–

[a] The percentage abundance by number of molecules for the Sun, cooled to planetary temperatures so that the elements combine to form the compounds listed, and for the outer atmospheres of the giant planets below the clouds. Dashes indicate unobserved compounds. Courtesy of Andrew P. Ingersoll.

warmed by both sunlight and internal heat retained since their formation. Their main ingredient is hydrogen, the lightest element and the most abundant element in the Sun and most stars. This helps explain the low bulk mass densities of the giant planets, between 710 and 1670 kilograms per cubic meter, which are comparable to that of the Sun at 1409 kilograms per cubic meter. Like the Sun, the next most abundant element in the giant planets is helium. The overwhelmingly abundant hydrogen (H) would also combine with the abundant carbon (C), nitrogen (N) and oxygen (O), in the low-temperature environment far from the Sun, to form stable molecules of methane (CH_4), ammonia (NH_3) and water vapor (H_2O).

The composition of the Sun, cooled to planetary temperatures, and that of the outer atmospheres of the giant planets are given in Table 3.3. Jupiter has very nearly the same composition as the Sun, made up mainly of the light gases hydrogen and helium. Saturn has about the same composition, with a bit less helium, but Uranus and Neptune have lesser amounts of hydrogen and relatively greater amounts of the heavier hydrogen compounds like methane.

Definite spectroscopic proof that molecular hydrogen is the most abundant element in Jupiter's and Saturn's upper atmosphere did not occur until the 1960s and late 1970s, when high-dispersion infrared spectroscopy, from both the ground and space, showed several weak absorption features due to molecular hydrogen (Fig. 3.5).

Helium, the second most abundant element in the Sun, has no detectable spectral features to make its presence known in the cold outer atmospheres of the giant planets. Helium is chemically inert and does not combine with other atoms to make molecules. The presence and amounts of helium atoms have nevertheless been inferred from the hydrogen infrared spectral features.

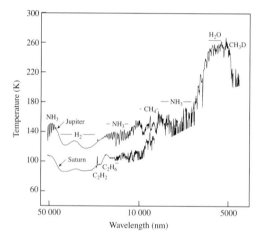

Fig. 3.5 Molecules in the atmospheres of Jupiter and Saturn The infrared radiation from the thin, cold upper atmospheres of Jupiter and Saturn exhibit numerous features that have no counterpart in the spectrum of sunlight. Strong features are seen in Jupiter for molecular hydrogen H_2, ammonia NH_3, methane CH_4, and water vapor H_2O. Saturn's outer atmosphere is also abundant in hydrogen and methane, but the ammonia features are missing and those of acetylene C_2H_2 and ethane C_2H_6 are enhanced. These spectra were taken with instruments aboard *Voyager 1* and *2* during their Jupiter and Saturn flybys in 1979 and 1980, respectively. (Courtesy of Rudolf A. Hanel.)

Collisions between helium atoms and hydrogen molecules alter the latter's ability to absorb infrared light, an effect that the *Pioneer 10* and *11* and *Voyager 1* and *2* instruments could detect. When the infrared measurements were combined with changes in these spacecraft's radio signals, observed when they passed behind the planets, the helium abundance could be determined with an uncertainty of only a few percent.

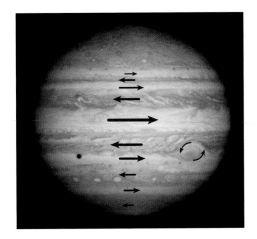

Fig. 3.6 Jupiter's counter-flowing winds   The rapid rotation of Jupiter has pulled its winds into bands that flow east to west and west to east, shown in this image taken from the *Cassini* spacecraft on 7 December 2001. The windswept clouds move in alternating light-colored, high-pressure zones and dark-colored, low-pressure belts. The arrows point in the direction of wind flow, and their length corresponds to the wind velocity, which can reach 180 meters per second in the equatorial regions (see Fig. 3.7). The Great Red Spot swirls in the counter-clockwise direction (*curved arrows*), like a high-pressure anticyclone in the Earth's southern hemisphere, but it has lasted for more than 300 years, much longer than terrestrial storms. Jupiter's moon Europa casts a shadow on the planet. (Courtesy of NASA/JPL.)

The helium abundance for Uranus and Neptune is consistent with that expected from a solar composition, but helium has been significantly depleted from Saturn's upper atmosphere and somewhat reduced in Jupiter's. Theoretical calculations indicate that helium rain has been falling toward the center of Saturn for the past 2 billion years, generating heat and producing lower amounts of helium in its outer atmosphere. Helium rain must be similarly settling toward Jupiter's core, but in lesser amounts.

Raging winds on the giant planets

Astronomers have been using telescopes, both small and large, to scrutinize weather patterns on Jupiter for more than a century. They observe clouds of various ices, such as ammonia, that are formed in the cold outer atmosphere. The clouds have been pulled into counter-flowing winds, moving in opposite eastward or westward directions and remaining confined to specific latitudes. These windswept clouds move in alternating light-colored bands called zones and dark ones known as belts (Fig. 3.6). Since Jupiter is all atmosphere and liquid, with no solid surface to rub against or continents to disturb the flow, its winds are free to rage unabated in response to the planet's rotation, with large-scale configurations that have remained unchanged for as long as they have been observed.

The weather patterns on Uranus and Neptune, where clouds of methane ice are observed, more nearly resemble those on Earth, which has low-latitude trade winds that blow westward and a meandering eastward current, the jet stream, in each hemisphere. The Earth has the weakest winds in the solar system; its fastest jet streams move at speeds of about 40 meters per second. In contrast, Jupiter's winds move at constant speeds of up to 180 meters per second, and Uranus's fastest winds are just a little faster (Fig. 3.7). The high clouds on Venus and Mars also move at faster speeds than those on Earth, both with speeds of

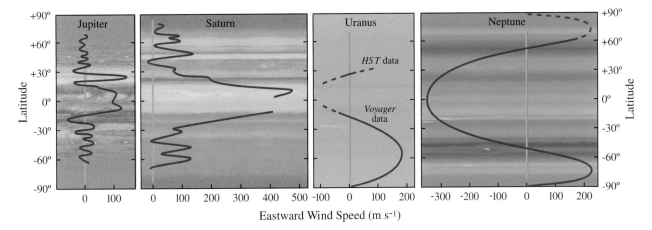

Fig. 3.7 Winds on the giant planets   Variation of wind speed and direction as a function of latitude. Since the giant planets have no solid surfaces, the winds are measured relative to the internal rotation speeds; the rotation period is determined from observations of the planet's periodic radio emission. Positive velocities correspond to winds blowing in the same direction but faster than the internal rotation; negative velocities are winds moving more slowly than the rotation. The winds are faster on Saturn than any other planet. (Courtesy of Andrew P. Ingersoll.)

up to 100 meters per second. The winds on Neptune and Saturn move at speeds of up to 400 and 450 meters per second, respectively, ten times the fastest winds on Earth. The speed of the Earth's winds are measured with respect to the rapidly rotating surface beneath them; for the giant planets, which have no solid surface, the wind speed is measured with respect to an internal rotation rate inferred by using radio emission from the magnetic fields that are generated within their cores.

Despite the slow motion of the Earth's winds, the solar energy available to drive them is greater than on any planet. Because it is relatively near the Sun, the Earth intercepts more intense sunlight to power its winds than the more distant planets do. Although Venus is nearer the Sun than the Earth is, the bright clouds on Venus reflect most of the incident sunlight. The nearest planet, Mercury, is too hot to retain a substantial atmosphere.

The circulation of the winds on the giant planets is powered by solar energy, as on Earth, plus internal energy left over from their formation. Even though Jupiter, Saturn and Neptune radiate about 1.67, 1.79 and 2.7 times more energy, respectively, than they absorb from the Sun, the total power per unit area, from both sunlight and internal heat, is much less than that on Earth. The amount of power available to drive winds near the Earth's surface is 25 times that at the cloud tops of Jupiter and 400 times that in Neptune's atmosphere.

Small-scale turbulence in any planet's atmosphere dissipates the energy that is available to drive large-scale winds, and there is more dissipation for the planets that are nearer the Sun. Thus, both the global energy available to drive the winds and the amount of that energy that is dissipated decrease with increasing distance from the Sun, but at different ratios so there is more wind-producing power at the larger distances. The energy-dissipating turbulence is greatest at Earth, where the winds are weak, and least in Neptune's low-dissipation atmosphere, where the winds are stronger. Jupiter lies in between these two extremes.

Stormy weather on the giant planets

Giant anticyclones create continent-sized ovals on Jupiter that roll like ball-bearings between the oppositely directed east–west winds. The large ovals revolve between the jet streams in the anticyclonic direction – clockwise in the northern hemisphere and counter-clockwise in the southern hemisphere. The smaller eddies are soon torn apart by the counter-flowing winds, lasting only about a day or two, but the larger ones can persist for decades or centuries. Jupiter's Great Red Spot, about three times the

diameter of Earth in size, has survived for more than 300 years.

How can the violent storms on Jupiter persist for months, decades and even centuries? Some external source must be feeding energy into the spinning vortex, either from the sides or from below. Perhaps each whirling spot draws energy from the shearing motion of the clouds blowing in opposite directions on its sides. Both the biggest long-lived storms and the powerful banded winds on Jupiter may be energized by smaller eddies that merge into them. It is as if the larger features devour the smaller ones, consuming them to help maintain their flow and replenish their energy.

The little, short-lived storms may receive their energy from deep within the hot interior of the planet. In response to heating, gases in lower levels of the atmosphere will expand and thereby become less dense than the gas in the overlying layers. The heated material, due to its low density, rises, just as a hot-air balloon does, cooling at the cloud tops and sinking again. Similar wheeling convective motions occur in a kettle of boiling water. They produce towering thunderclouds here on Earth; lightning is also created in the thunderstorms on both the Earth and Jupiter.

Neptune also has a stormy atmosphere with raging winds, but the big storms are not as long-lasting. The largest observed storm system on Neptune, detected when *Voyager 2* sped past the planet on 24 August 1989, is as broad as the Earth. It is called the Great Dark Spot because it resembles the Great Red Spot of Jupiter. Both storms are found in the planetary tropics, at about one-quarter of the way from the equator to the pole; both rotate counter-clockwise; and both are about the same size relative to their planet. As *Voyager 2* watched the Great Dark Spot it contracted in and stretched out with a regular rhythm, like the mouth of a feeding fish, and drifted slowly toward the planet's equator. Yet, when the *Hubble Space Telescope* was turned toward Neptune in 1995, no trace of the Great Dark Spot could be found. Perhaps it simply ran out of food to supply its energy, or perhaps it moved into strong equatorial winds that could not support it.

3.4 Titan, a satellite with a substantial atmosphere

Saturn's largest moon, Titan, is the only satellite with a substantial atmosphere. Detailed investigations with instruments aboard *Voyager 1* in 1980 showed that the dominant gas surrounding Titan is molecular nitrogen N_2, at 82 to 99 percent, similar to Earth (77 percent). In fact, the satellite is enveloped by about 10 times more nitrogen than we are, yielding a surface pressure 1.5 times greater than

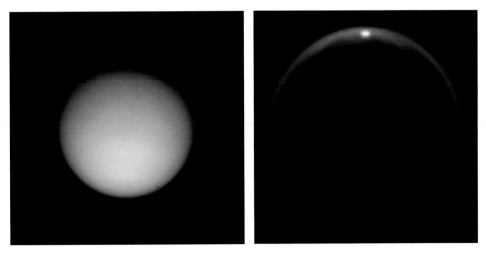

Fig. 3.8 Titan's dense, smoggy atmosphere   The surface of Saturn's moon Titan is hidden from view by a hazy layer of smog, giving it a fuzzy, tennis-ball appearance in a *Cassini* natural-color view (*left*) taken on 15 February 2005. When illuminated from behind, the dense atmosphere forms a crescent. By observing radiation at infrared wavelengths, on 8 July 2009, a *Cassini* instrument was able to penetrate through the moon's atmosphere and catch the glint of sunlight reflected off a huge lake in the northern hemisphere of Titan (*right*), which is probably filled with liquid methane and ethane. The lake is named Kraken Mare. [Courtesy of NASA/JPL/SSI (*left*) and NASA/JPL/U. Arizona/DLR (*right*).]

the sea-level pressure of Earth's atmosphere. The surface temperature on Titan is 94 kelvin, as expected for a body so far from the Sun.

The spectrometers on *Voyager 1* showed that the next most abundant gas enveloping Titan is methane CH_4, with an abundance between 1 and 6 percent. Gerard Kuiper (1905–1973) had discovered signs of methane in Titan's spectrum as early as 1944.

Methane molecules rise up to high levels in Titan's atmosphere, where they are broken apart by ultraviolet sunlight and electrons coming from Saturn's magnetic environment. These molecular fragments recombine to form heavier hydrocarbon molecules such as ethane C_2H_6, and familiar gases like acetylene C_2H_2, propane C_2H_8, and hydrogen cyanide HCN.

It doesn't rain water on Titan, but it does rain methane, in large drops that fall like snow. Given the known atmospheric composition and temperatures, scientists speculated that thin clouds of methane ice crystals may form in the lower atmosphere, and that methane, ethane and propane could rain all the way down to the surface, forming seas, lakes and ponds.

Images from the *Voyager* spacecraft in 1981 and the *Cassini* spacecraft in 2005 showed that an opaque haze completely enshrouds the satellite, hiding any lakes or seas from direct view (Fig. 3.8). The smog is unimaginably worse than that over any city on Earth. Compared with any such urban smog, there are relatively few smog particles per unit volume of Titan's atmosphere, but the

haze extends to an altitude of about 200 kilometers. This makes the smog thick enough to completely hide Titan's surface from view.

The orange smog must result from ultraviolet sunlight that breaks simple molecules like methane apart and chemical reactions that create more complex substances from these fragments. The exact composition of this photochemical smog remains unknown, but its mere existence suggests that heavy compounds may be falling through the atmosphere and sinking into the hypothetical seas to form an organic sludge on their floors.

As described in greater detail in Section 10.7, radar instruments aboard the *Cassini* spacecraft have been used to map out large lakes of liquid methane and ethane on Titan, and the *Huygens Probe* has been parachuted to the moon's surface, detecting river-like channels suggesting flows of liquid methane.

All of the large satellites in the solar system except Titan have tenuous atmospheres, or no atmosphere at all.

Tenuous exospheres of Europa and Io

The larger moons of the giant planets have sizes that are comparable to the Earth's Moon or Mercury, and these satellites are similarly cloaked in a thin film of gas. Their exospheres are also temporary features, and must be continuously re-supplied from the moon's surfaces.

A tenuous veil of oxygen molecules has been found around Europa, a large ice-covered moon of Jupiter. The

Focus 3.3   Why does Titan have a dense atmosphere?

Why does Saturn's largest satellite, Titan, have such a substantial atmosphere when it is only slightly bigger than the planet Mercury and almost as big as Jupiter's largest satellite, Ganymede, which have exceedingly tenuous atmospheres? The ability of a planet or satellite to retain an atmosphere is determined primarily by the body's mass and temperature, but because Titan, Ganymede and Mercury have nearly the same mass, differences in temperature should account for their atmospheric differences. Mercury is so hot that even the heaviest molecules move fast enough to escape the planet's gravity, while the temperature on Titan is so low that only the lighter molecules, like hydrogen, can escape.

But Ganymede is now sufficiently cold and massive to retain a Titan-like atmosphere. The difference between Titan and Ganymede is more likely a consequence of the temperature at the time of their birth. Titan was born in the remote cooler regions of the solar system, and nearby Saturn never became as hot as Jupiter did during its birth. The low temperatures permitted ammonia, methane and water ice to form on Titan's surface when it was born, and these ices probably sublimated to form a primeval atmosphere of ammonia and methane, while water remained locked into its surface as ice. On Ganymede, water was probably the only ice that formed in the slightly warmer climate. If there was no ammonia or methane ice on Ganymede, and if the temperatures never became high enough for its water ice to sublimate, then Ganymede would be left without any substantial atmosphere.

surface pressure of this atmosphere is barely one-hundred-billionth (10^{-7}) that of Earth's. Scientists believe that the atmospheric oxygen (O_2) on Europa is either created when energetic charged particles bombard water ice (H_2O) on the moon's surface, or ejected from it by volcanoes of ice. Exposure to sunlight and meteor impacts could also create some of the gas. The relatively lightweight hydrogen, H, escapes into space, leaving the heavier oxygen molecules to accumulate to form an atmosphere.

Jupiter's innermost large satellite Io has a varying, thin atmosphere of sulfur dioxide, SO_2, ejected from the moon's active volcanoes. The largest of Jupiter's moons, Ganymede, also has an exceedingly tenuous atmosphere. So we wonder why Saturn's moon Titan is the only large moon to have a dense atmosphere (Focus 3.3).

3.5 The planets are inside the expanding Sun

The space just outside the Earth is not empty. It is filled with pieces of the Sun. Our star is expanding in all directions, filling interplanetary space with electrically charged particles that are forever blowing from the Sun. This solar wind moves past the planets, carrying the Sun's rarefied atmosphere out to the space between the stars. So we are actually living in the outer part of the Sun.

The sharp outer edge of the visible solar disk is illusory. An invisible, rarefied atmosphere expands away from the Sun and extends all the way to the Earth and beyond. This expanding solar atmosphere is so tenuous that we can look right through it to the Sun's bright disk, just as we see through the Earth's clear air.

The outer atmosphere of the Sun, known as the corona, becomes visible to the unaided eye for only a few minutes when the Sun's light is blocked, or eclipsed, by the Moon. During such a total solar eclipse, the corona is seen at the limb, or apparent edge of the Sun, against the blackened sky as a faint halo of white light, or all the visible colors combined (Fig. 3.9).

The amazing thing about the corona is that it is incredibly hot, with a temperature of a few million kelvin. It is so intensely hot that its abundant hydrogen is torn into its component parts. Each hydrogen atom consists of one electron moving about one central, nuclear proton. The solar atmosphere therefore consists mainly of electrons and protons, with smaller amounts of heavier ions created from the less abundant elements in the Sun.

The corona is so hot that it can't stay still. The sizzling heat cannot be entirely constrained by either the Sun's inward gravitational pull or its magnetic forces. An overflow corona is therefore forever expanding in all directions, filling the solar system with a great eternal wind known as the solar wind (Focus 3.4). Interplanetary space probes have been making in-situ (Latin for "in place") measurements of the solar wind for decades, both within the space near the Earth and further out in the Earth's orbital plane.

Unlike any wind on Earth, the solar wind is a tenuous mixture of charged particles and magnetic fields streaming outward in all directions from the Sun at speeds of hundreds of kilometers per second. The seemingly eternal wind carries a magnetic field with it, with one end anchored in the Sun. This interplanetary magnetic field has a spiral shape due to the combined effects of the radial solar wind flow and the Sun's rotation.

Although the Sun is continuously blowing itself away, the outflow can continue for billions of years without significantly reducing the Sun's mass. Every second, the solar wind blows away a million tons, or a billion

Fig. 3.9 Eclipse corona streamers   The million-degree solar atmosphere, known as the corona, is seen around the black disk of the Earth's Moon, photographed in white light, or all the colors combined, from atop Mauna Kea, Hawaii, during the solar eclipse on 11 July 1991. The electrically charged gas is concentrated in numerous fine rays as well as larger helmet streamers. (Courtesy of the HAO/NCAR.)

Focus 3.4   Discovery of the solar wind

The existence of the solar wind was suggested from observations of comet tails about half a century ago. When a comet is tossed into the inner solar system, the dirty ice on its surface is vaporized, sometimes forming two kinds of tails that point generally away from the Sun rather than toward it (Fig. 3.10). One is a curved dust tail, pushed away from the Sun by the pressure of sunlight. The other is a straight ion tail that is affected by the solar wind.

The German astronomer Ludwig Biermann (1907–1986) noticed, in the 1950s, that the ions in a comet's tail move with velocities many times higher than could be caused by the weak pressure of sunlight, and proposed that a continuous flow of electrically charged particles pours out of the Sun at all times and in all directions, accelerating the ions to high speeds and pushing them away from the Sun in straight ion tails.

In 1958, Eugene N. Parker (1927–) of the University of Chicago showed how such a relentless flow might work, dubbing it the solar wind. It would naturally result from the expansion of the Sun's million-degree atmosphere, the corona. He also demonstrated how a magnetic field would be pulled into interplanetary space from the rotating Sun, attaining a spiral shape.

The first direct measurements of the solar wind's corpuscular, or particle, content were made by a group of Soviet scientists led by Konstantin I. Gringauz (1918–1993), using four ion traps aboard the *Lunik 2* spacecraft launched to the Moon on 12 September 1959. In the following year, Gringauz reported that the maximum current in all four ion traps corresponded to a solar wind flux of 2 million million (2×10^{12}) ions (presumably protons) per square meter per second. This is in rough accord with all subsequent measurements.

All reasonable doubt concerning the existence of the solar wind was removed by measurements made on board NASA's *Mariner 2*, launched on 27 August 1962. Marcia Neugebauer (1932–) and Conway W. Snyder of the Jet Propulsion Laboratory used more than 100 days of *Mariner 2* data, obtained as the spacecraft traveled to Venus, to show that charged particles are continuously emanating from the Sun, for at least as long as instruments on *Mariner 2* observed them. It also unexpectedly indicated that the solar wind has a slow and a fast component. The slow one moves at a speed of 300 to 400 kilometers per second; the fast one travels at twice that speed.

The solar wind flux determined by Neugebauer and Snyder was in good agreement with the values measured with the ion traps on *Lunik 2*. The average wind ion number density was shown to be 5 million (5×10^6) protons per cubic meter near the distance of the Earth from the Sun. We now know that such a low density close to the Earth's orbit is a natural consequence of the wind's expansion into an ever-greater volume, but that variable wind components can gust with higher densities.

(10^9) kilograms. That sounds like a lot of mass loss, but it is four times less than the amount consumed every second during the thermonuclear reactions that make the Sun shine. To supply the Sun's present luminosity, hydrogen must be converted into helium within the Sun's energy-generating core, with a mass loss of about 4 million tons every second. It is carried away by the Sun's radiation, whose energy vastly exceeds that of the solar wind.

The perpetual solar wind brushes past the planets and engulfs them, carrying the Sun's corona out into interstellar space. As the corona disperses, gases welling up from below to feed the wind must replace it. Exactly where this material comes from is an important subject of contemporary space research, but the main interest for planetary research is how the solar wind affects the planets.

The reason that space looks empty is that these subatomic pieces of the Sun are very small and moving incredibly fast, and there really are not very many of them when compared even to our transparent atmosphere. By the time it reaches the Earth's orbit, the solar wind is diluted to about 5 million electrons and 5 million protons per cubic meter, a very rarefied gas (Table 3.4). By way of comparison, there are 25 million billion billion (2.5×10^{25}) molecules in every cubic meter of our air at sea level. Still, at a mean speed of about 400 kilometers per second, the flux of solar wind particles is far greater than anything else out there in space. Between one million million and ten million million (10^{12} to 10^{13}) particles in the solar wind cross every square meter of space each second (Table 3.4).

The radial, supersonic outflow creates a huge bubble of electrons, protons and magnetic fields, with the Sun at the center and the planets inside, called the heliosphere (Fig. 3.11), from *Helios*, the Greek word for the Sun. Within the heliosphere, conditions are regulated by the Sun. Its domain extends out to about 100 AU, or about 100 times the mean distance between the Earth and Sun. Out there, the solar wind has become so weakened by expansion that it can no longer repel interstellar forces (Section 15.3).

Table 3.4 Mean values of solar-wind parameters at the Earth's orbit[a]

Parameter	Mean Value
Particle density, N	$N \approx 10$ million particles per cubic meter (5 million electrons and 5 million protons)
Velocity, V	$V \approx 400$ kilometers per second and $V \approx 800$ kilometers per second
Flux, F	$F \approx 10^{12}$ to 10^{13} particles per square meter per second
Temperature, T	$T \approx 120\,000$ kelvin (protons) to $140\,000$ kelvin (electrons)
Particle thermal energy, kT	$kT \approx 2 \times 10^{-18}$ joule ≈ 12 eV
Proton kinetic energy	$0.5\, m_p V^2 \approx 10^{-16}$ joule ≈ 1000 eV $= 1$ keV
Particle thermal energy density	$NkT \approx 10^{-11}$ joule m^{-3}
Proton kinetic energy density	$0.25\, N\, m_p\, V^2 \approx 10^{-9}$ joule m^{-3}
Magnetic field strength, H	$H \approx 6 \times 10^{-9}$ tesla $= 6$ nanotesla $= 6 \times 10^{-5}$ gauss

[a] These solar-wind parameters are at the mean distance of the Earth from the Sun, or at one astronomical unit, 1 AU, where 1 AU $= 1.496 \times 10^{11}$ meters. Boltzmann's constant $k = 1.38 \times 10^{-23}$ joule per kelvin relates temperature and thermal energy. The proton mass m_p is 1.67×10^{-27} kilograms.

Fig. 3.10 Comet tails   Telescopic photograph of Comet Mrkos taken in August 1957, showing the straight, well-defined ion tail and the more diffuse, slightly curved dust tail. Both comet tails point away from the Sun. The electrified solar wind deflects the charged ions and accelerates them to high velocities, creating the relatively straight ion tails. The radiation pressure of sunlight suffices to blow away the un-ionized comet dust particles, forming a broad arc that can resemble a scimitar. (Courtesy of Lick Observatory.)

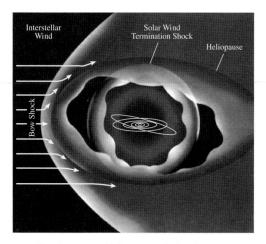

Fig. 3.11 Heliosphere With its solar wind going out in all directions, the Sun blows a huge bubble in space called the heliosphere. The heliopause is the name for the boundary between the heliosphere and the interstellar gas outside the solar system. Interstellar winds mold the heliosphere into a non-spherical shape, creating a bow shock where they first encounter it. The orbits of the planets are shown near the center of the drawing.

3.6 Magnetized planets and magnetospheres

Earth's magnetic dipole

In 1600, William Gilbert (1544–1603), physician to Queen Elizabeth I of England, authored a treatise in Latin with the grand title *De magnete, magneticisque corporibus, et de magno magnee tellure*, translated into English as *Concerning Magnetism, Magnetic Bodies, and the Great Magnet Earth*. In this work, which is still available in its English version, Gilbert showed that the Earth is itself a great magnet, which explains the orientation of compass needles. It is as if there were a colossal bar magnet at the center of the Earth (Fig. 3.12).

At the equator, the two ends of a compass needle point north or south, toward the Earth's magnetic poles. At each magnetic pole, the needle would stand upright, pointing into or out of the ground. And in between, at intermediate latitudes, the compass needles point north or south with a downward dip of one end, but not vertically as at a pole.

Since the geographic poles are located near the magnetic ones, a compass needle is aligned in the north–south direction. We usually put an arrow on the north end of the needle, and an arrowed compass therefore points north. Since the Earth's rotation axis is inclined 11.7 degrees with respect to its magnetic axis, a compass needle does not point exactly toward the geographic north pole, but within 11.7 degrees of it. Currently, the north magnetic

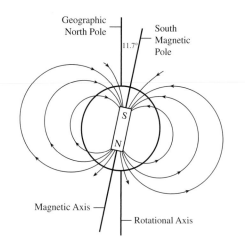

Fig. 3.12 Earth's magnetic dipole The Earth's magnetic field looks like that which would be produced by a bar magnet at the center of the Earth, with the north magnetic pole corresponding to the south geographic pole and vice versa. The Earth's magnetic dipole originates in swirling currents of molten iron deep in the Earth's liquid outer core, and extends more than 10 Earth radii, or 63.7 thousand kilometers, out into space on the side facing the Sun, and all the way to the Moon's orbit at 384 000 kilometers on the opposite side. Magnetic field lines loop out of the north magnetic pole and into the south magnetic pole. The lines are close together near the magnetic poles where the magnetic force is strong, and spread out where it is relatively weak. The magnetic axis is tilted at an angle of 11.7 degrees with respect to the Earth's rotational axis. This dipolar (two poles) configuration applies near the surface of the Earth, but further out the magnetic field is distorted by the solar wind.

pole is located in northern Canada, while the south magnetic pole is off the coast of Antarctica.

We can describe the Earth's magnetism by invisible magnetic field lines, which orient compass needles. These lines of magnetic force emerge out of the north magnetic pole, loop through nearby space and re-enter at the south magnetic pole (Fig. 3.12). According to this convention, the south magnetic pole corresponds to the north geographic pole and vice versa. The lines are close together near the magnetic poles where the magnetic force is strong, and spread out above Earth's equator where the magnetism is weaker. You cannot see the magnetic field lines, but compass needles point along them, and other instruments can be used to measure their strength.

The magnetic field strength at the Earth's magnetic equator is 0.000 030 5 tesla, or 3.05×10^{-5} tesla. Measurements of the surface magnetic fields of Earth show stronger fields near the poles where the magnetic field lines congregate, at roughly twice the strength of the field at the equator. The magnetic strength in both regions is several times weaker than a toy magnet, but the comparison is somewhat misleading since the Earth is a very big magnet.

Magnetism pervades the entire volume of the Earth. This global quality is expressed quantitatively by the magnetic dipole moment, equal to the product of the equatorial magnetic field strength and the cube of the planet's equatorial radius. For the Earth we have an equatorial radius of 6378 kilometers and a magnetic dipole moment of 7.91×10^{15} tesla meters cubed. In comparison, the magnetic dipole moment of a typical laboratory electromagnet is more than 100 billion billion, or 10^{20}, times weaker.

How is magnetism generated within the Earth's hot, molten interior? Heat cooks magnetism out of a permanent magnet, and liquefaction melts it away. But the compass needles are not guided by a permanent magnet. Instead, the inside of the Earth is an electromagnet, generating magnetism by changing electric currents. Heat-driven circulation and the Earth's rotation combine to produce electrically conducting streams of molten iron, which generate the terrestrial magnetic fields by dynamo action.

Electric currents give rise to magnetic fields, and moving magnets generate electric currents. These two effects, in the churning liquid outer core of the Earth, can amplify and sustain the small magnetic field that the planet captured from its surroundings when it formed. As opposing streams of molten iron, carrying tiny magnetic fields, sweep past one another, each induces currents in the other. This creates more magnetism, which induces more currents, and so on. In this way, the dynamo in the Earth's liquid outer core generates the magnetic fields detected at the terrestrial surface, taking energy from both the internal heat and the rotation energy of the planet. And because the currents deep down inside the Earth are always varying, the Earth's magnetism is a dynamic, changing thing.

Ancient magnetic rocks, found on the flanks of volcanoes, indicate that the Earth's magnetic field has not always been the same as it is today. When the molten volcanic lava flows to the surface and hardens into rock, its internal magnetism lines up with the Earth's magnetic field and freezes into position. These magnetic fossils record the direction and intensity of the terrestrial magnetic field when and where the lava solidified.

An inspection of magnetic fossils of differing ages from all parts of the world resulted in an amazing discovery. The great magnet of the Earth has flipped, or reversed its direction, many times in the past. During each flip, the north magnetic pole becomes the south one, and vice versa. The deep electric currents readjust; always remaining nearly aligned with the rotation axis, but with a swap in the magnetic poles.

An examination of volcanoes on land indicates that the Earth's magnetic field has reversed itself at least nine times over the past 3.6 million years. And the ordered succession of magnetized rocks on the spreading ocean floors records 100 and more full reversals of the direction of the magnetic poles during the past 200 million years. Tens of thousands of years separate some of the magnetic field reversals, while tens of million of years separate others.

So the terrestrial magnetic field is inevitably headed for a magnetic flip, but we don't know exactly when. The arrows on compass needles will then point south instead of north, reversing their direction. Animal species and satellites that depend on magnetic fields for guidance will lose their orientation, and will have to adapt to the changing field. The navigation systems of migrating birds and monarch butterflies, for example, depend in part on internal compasses that sense directions from the Earth's magnetic field. Honeybees, some wasps, some fish, sea turtles and even a species of mole rat take bearings magnetically.

The Earth's magnetic fields also influence the space near the Earth. They extend away from the Earth, decreasing in strength as the inverse cube of the distance. Yet they remain strong enough to shield the Earth from the full force of the Sun's charged winds.

Earth's protective magnetosphere

Fortunately for life on Earth, the terrestrial magnetic field deflects the Sun's winds away from the Earth, like a rock in a stream or a windshield that deflects air around a car. It hollows out a protective cavity in the solar wind called the magnetosphere. The magnetosphere of the Earth, or any other planet, is an enveloping bubble of magnetism. It is that region surrounding the planet in which its magnetic field dominates the behavior of electrically charged particles such as electrons, protons and other ions. It diverts most of the solar wind around our planet at a distance far above the atmosphere, thereby protecting humans on the ground from possibly lethal solar particles.

The dipolar (two-pole) magnetic configuration applies near the surface of the Earth, but further out the magnetic field is distorted by the Sun's perpetual wind. Although it is exceedingly tenuous, far less substantial than a terrestrial breeze or even a whisper, the solar wind is powerful enough to mold the outer edges of the Earth's magnetosphere into a changing asymmetric shape, like a tear drop falling toward the Sun (Fig. 3.13).

The hot, high-speed, magnetized solar wind confronts the Earth's magnetic field close to home, usually at a distance from the Earth's center of about 10 times the Earth's radius on the dayside that faces the Sun. Here the solar

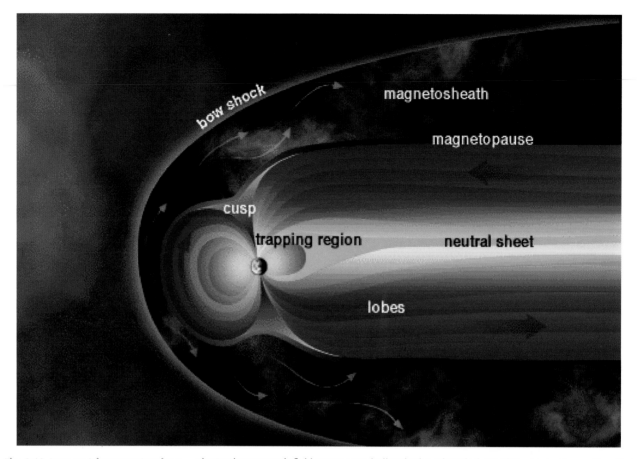

Fig. 3.13 Asymmetric magnetosphere   The Earth's magnetic field carves out a hollow in the solar wind, creating a protective cavity called the magnetosphere (*blue*). It is sculpted into an asymmetric shape by the solar wind, with a bow shock that forms at about 10 Earth radii on the sunlit, dayside facing the Sun (*left*). The location of the bow shock is highly variable since it is pushed in and out by the gusty solar wind. The magnetopause marks the outer boundary of the magnetosphere, at the place where the solar wind takes control of the motions of charged particles. The solar wind is deflected around the Earth, pulling the terrestrial magnetic field into a long magnetotail on the nightside (*right*). The red regions in the inner magnetosphere contain the plasmasphere, the ring current and the outer Van Allen belt, where electrons, protons and other ions are trapped in closed paths. (Courtesy of ESA.)

wind pushes the Earth's magnetism in, compressing its outer magnetic boundary and forming a shock wave. It is called a bow shock because it is shaped like waves that pile up ahead of the bow of a moving ship. The bow shock results because the solar wind is supersonic, moving faster than sound waves and other waves that might propagate through the wind. The motion of the solar wind around the magnetosphere has therefore been compared to the flow of air around a supersonic aircraft.

After forming the bow shock, the solar wind encounters and flows around the magnetopause, the boundary between the solar wind and the magnetosphere. The magnetic field carried in the solar wind merges with that of the planet, and stretches it out into a long magnetotail on the nightside of Earth. The magnetic field points roughly toward the Earth in the northern half of the tail and away in the southern. The field strength drops to nearly zero at the center of the tail where the opposite

magnetic orientations lie next to each other and currents can flow.

Thus, the Earth's magnetosphere is not precisely spherical. It has a bow shock facing the Sun and a long magnetotail in the opposite direction. The term "magnetosphere" therefore does not refer to form or shape, but instead implies a sphere of influence.

Trapped particles

The Earth's protective magnetic cocoon is not perfect. Energetic charged particles flowing from the Sun can penetrate the magnetic defense and become trapped within the magnetosphere. This was realized in 1958 when James A. Van Allen (1914–2006) and his students used instruments aboard the *Explorer 1* and *3* satellites to unexpectedly discover a large flux of high-energy electrons and protons that girdle the Earth far above the atmosphere, moving within

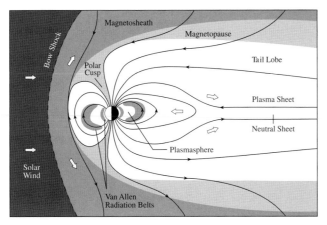

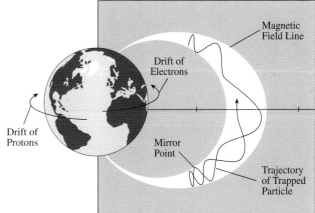

Fig. 3.14 Earth's magnetosphere   The Earth, its auroras, atmosphere and ionosphere, and the two Van Allen radiation belts all lie within the Earth's magnetosphere. Similar magnetic cavities are found around other magnetized planets. Electrons and protons in the solar wind are deflected at the bow shock (*left*), and flow along the magnetopause into the magnetic tail (*right*). Electrified particles can be injected back toward the Earth and Sun within the plasma sheet (*center*).

Fig. 3.15 Magnetic trap   Charged particles can be trapped by Earth's magnetic field. They bounce back and forth between polar mirror points in either hemisphere at intervals of seconds to minutes, or they also drift around the planet on timescales of about an hour. As shown by the Norwegian scientist Carl Størmer (1874-1957) in 1907, with the trajectories shown here, the motion is turned around by the stronger magnetic fields near the Earth's magnetic poles. Because of their positive and negative charge, the protons and electrons drift in opposite directions.

two belts that encircle the Earth's magnetic equator but do not touch it. They resemble a gigantic, invisible, torus-shaped doughnut. This was the first major discovery of the Space Age.

These regions are sometimes called the inner and outer Van Allen radiation belts. Van Allen used the term "radiation belt" because the charged particles were then known as corpuscular radiation; the nomenclature does not imply either electromagnetic radiation or radioactivity. The radiation belts lie within the inner magnetosphere at distances of 1.5 and 4.5 Earth radii from the center of the Earth, creating a veritable shooting gallery of high-speed electrons and protons in nearby space (Fig. 3.14).

In 1907, about half a century before the discovery of the radiation belts, the Norwegian geophysicist Carl Størmer (1874–1957) showed how electrons and protons can be almost permanently confined and suspended in space by the Earth's dipolar magnetic field. An energetic charged particle moves around the magnetic fields in a spiral path toward one magnetic pole. Its trajectory becomes more tightly coiled in the stronger magnetic fields close to a magnetic pole, where the intense polar fields act like a magnetic mirror, turning the particle around so it moves back toward the other pole.

Thus, the electrons and protons bounce back and forth between the north and south magnetic poles (Fig. 3.15). It takes about one minute for an energetic electron to make one trip between the two polar mirror points. The spiraling electrons also drift eastward, completing one trip around the Earth in about an hour. There is a similar drift for

protons, but in the westward direction. The bouncing can continue indefinitely for particles trapped in the Earth's radiation belts, until the particles collide with each other or some external force distorts the magnetic fields.

The problem at the time Størmer developed his theory was that there was no mechanism known to allow electrically charged particles into the dipolar magnetic field. After all, if electrons and protons cannot leave the magnetic cage, how could they get into it in the first place? Nevertheless, instruments aboard spacecraft have shown that energetic charged particles have entered the trap. They include particles from the Sun that arrive via the solar wind and penetrate the Earth's magnetic defense through a temporary opening in it.

The solar wind carries the Sun's magnetic field with it, and the solar magnetism is draped around the magnetosphere when encountering it. The solar magnetic field can open up the Earth's magnetic field when the two fields are pointing in opposite directions where they touch. When this happens, the two fields become linked, just as the opposite poles of two toy magnets stick together. The merging process, known as magnetic reconnection, can create an opening in the Earth's magnetic field, forming a portal through which the solar particles can flow.

The solar wind is then plugged into the Earth's electrical socket, and our planet becomes wired to the Sun along magnetic fields that can stretch all the way back to the solar corona. Tons of high-energy particles may then flow

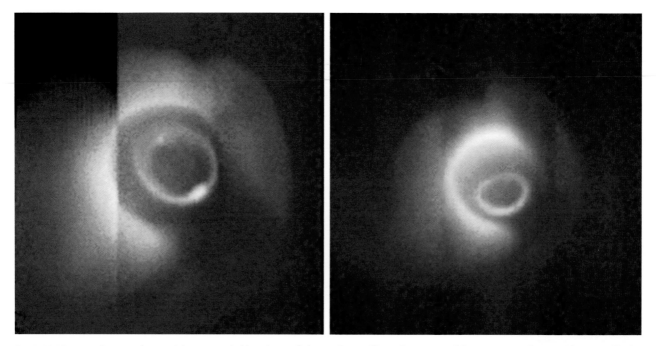

Fig. 3.16 Plasmasphere   The Earth is surrounded by a layer of plasma, located in an inner part of the magnetosphere and just outside the ionosphere, which is at the top of the Earth's atmosphere. This plasmasphere is created by energetic sunlight ionizing molecules and atoms in the Earth's upper atmosphere, and it is contained by the Earth's magnetic field. Plasma is an electrically neutral collection of electrons, protons and other ions. These images were obtained on 24 May 2000 in ultraviolet light from the *IMAGE* spacecraft. The Earth is at the center of the images, and the Sun is to the upper left. The view is toward Earth's north pole. The emission of the plasmasphere is brightest on the side pointing toward the Sun. The oval shape of the Earth's northern aurora is detected inside the arc of the plasmasphere. (Courtesy of NASA/EUV *IMAGE* science team.)

into the magnetosphere along this magnetic highway and through the opening before it closes again. The magnetic reconnection can occur during either a frontal assault near the bow shock and magnetic poles, or from the rear in the immense magnetotail.

In June 2007, for example, coordinated measurements from five identical *THEMIS* satellites, placed in carefully coordinated orbits, discovered a giant breach in the Earth's magnetic field wider than the Earth itself. The opening was explained by brief magnetic reconnection on the dayside of the Earth, facing the Sun. The magnetic portal formed over the Earth's equator, and then rolled over the planet's magnetically open polar regions.

The immense magnetic tail provides another location for breaking into the Earth's magnetic domain. When the solar and terrestrial magnetic fields touch each other, the magnetotail can be punctured, providing a back-door entry that funnels energy and particles into the magnetosphere. The magnetotail snaps like a rubber band that has been stretched too far. The snap catapults the outer part of the tail away from the Earth and propels the inner part back toward it. Once inside the magnetic trap, the charged particles can be additionally accelerated to higher energies.

A plasmasphere is found in the inner part of the Earth's magnetosphere (Fig. 3.16). It is located just outside the upper terrestrial atmosphere, called the ionosphere. The upper reaches of our planet's atmosphere are exposed to ultraviolet light from the Sun, which ionizes the atmosphere's atoms and molecules. The charged particles that result from the ionization are electrons, protons and other ions. The electrons have the lowest mass and highest velocities, and some of them move fast enough to overcome the Earth's gravitational pull. The growing number of escaped, negatively charged electrons electrically attracts the positively charged protons and other ions, pulling them out. The Earth's magnetic field then traps oxygen ions, protons and electrons derived from the ionosphere, creating the plasmasphere that rotates with the Earth's magnetic field and extends outward to include the radiation belts. Farther out, the magnetosphere is dominated by its interaction with the solar wind.

Energetic charged particles coming from interstellar space, known as cosmic rays, may also play a role in feeding the radiation belts, supplying the inner one with its high-energy protons. When cosmic rays bombard the Earth's atmosphere, which lies below the radiation belts, they collide with atoms in the air and eject neutrons from

Table 3.5 Planetary magnetic fields[a]

Planet	Magnetic dipole moment (Earth = 1)	Magnetic field at the equator, B_0 (Earth = 1)	Tilt of magnetic axis center (degrees)	Offset from planet R_{MP}	Bow shock stance, (R_P)	Planet equatorial radius, R_P (km)
Mercury	0.0007	0.0033	+14	$0.05\,R_M$	$1.5\,R_M$	$R_M = 2439$
Earth	1	0.305	+11.7	$0.07\,R_E$	$10\,R_E$	$R_E = 6378$
Jupiter	20,000	4.28	−9.6	$0.14\,R_J$	$42\,R_J$	$R_J = 71\,492$
Saturn	600	0.22	<1.0	$0.04\,R_S$	$19\,R_S$	$R_S = 60\,268$
Uranus	50	0.23	−58.6	$0.3\,R_U$	$25\,R_U$	$R_U = 25\,559$
Neptune	25	0.14	−47.0	$0.55\,R_N$	$24\,R_N$	$R_N = 24\,764$

[a] The magnetic field strengths are given at the surface of Mercury and the Earth and at the cloud tops for the giant planets. Venus and Mars have no detected global, dipolar magnetic field, with respective upper limits of 2×10^{-9} and 10^{-8} tesla. Here the magnetic dipole moment, $D_P = B_0 R_P{}^3$, is given in units of the Earth's magnetic dipole moment of $7.91 \times 10^{15}\,\mathrm{T\,m^3}$. The tilt is the angle between the magnetic axis and the rotation axis. Here we use the SI unit for magnetic field strength, the tesla (T). The c.g.s. unit of magnetic field strength, the gauss (G), can be computed from $1\,\mathrm{T}\,10^4\,\mathrm{G}$. The nanotesla (nT) is also used, with $1\,\mathrm{nT} = 10^{-9}\,\mathrm{T}$, and the nanotesla has historically also been called the gamma. A dipole moment of $1\,\mathrm{T\,m^3}$ equals $10^{10}\,\mathrm{G\,cm^3}$, where m and cm respectively denote meter and centimeter. The equivalent unit of $1\,\mathrm{G\,cm^3} = 10^{-3}\,\mathrm{A\,m^2}$ is also used, where the current is in units of amperes (A). The equatorial radius of the planets is given in kilometers (km).

the atomic nuclei. These neutrons travel in all directions, unimpeded by magnetic fields since they have no electrical charge.

Once it is liberated from an atomic nucleus, a neutron cannot stand being left alone. A free neutron lasts only about 10 minutes on average before it decays into an electron and proton. A small fraction of the neutrons produced in our atmosphere by cosmic rays move out into the inner radiation belt before they disintegrate, producing electrons and protons in places they could not otherwise have reached. These electrically charged particles are immediately snared by the magnetic fields and remain stored within them, accumulating in substantial numbers over time.

Planets with magnetospheres

Magnetic fields are ubiquitous in the solar system. Earth, Mercury and all the giant planets have strong magnetic fields generated within the planet, and Jupiter and Saturn have extensive magnetospheres. A magnetic field has been found on Mars, but the field's patchy nature suggests that it is not the result of an active internal dynamo. Instead the magnetic field on Mars is probably a remnant of former times, frozen into expanses of solidifying lava. Venus is the only major planet to have no detectable magnetic field. A dipolar magnetic field has even been found on at least one satellite, Jupiter's Ganymede.

Of the eight major planets, six are known to generate detectable, global magnetic fields. As with the Earth,

the magnetic fields near these planets can be described by a magnetic dipole, and the best characterization of each planet's intrinsic magnetism is the magnetic dipole moment. The dipole moment divided by the cube of the planet's radius yields the average strength of the magnetic field along the magnetic equator. Jupiter's magnetic moment is 20 000 times that of Earth. Saturn's magnetic moment is 600 times larger than the Earth's, but still about 30 times weaker than Jupiter's. The magnetic dipole moments, equatorial magnetic field strengths and planetary radii are given in Table 3.5 for the six planets with known dipolar magnetic fields.

Because the characteristic timescale for the decay of a magnetic field in a planetary interior is much less than the age of the planet, its global magnetic field must now be amplified and rejuvenated by an internal dynamo. Such a magnetic dynamo exists within a large, fluid, electrically conducting region.

A planetary magnetosphere is the volume in space from which the solar wind is deflected. The extent of a planet's magnetosphere depends upon both the strength of its magnetic field and the intensity of the solar wind at the planet's distance. The strong magnetic fields generated in Jupiter and Saturn, for example, as well as the weak solar-wind pressures present in the outer solar system, permit them to hollow out a larger cavity in the solar wind, with magnetospheres that are much larger than the Earth's.

The magnetospheres of Jupiter and Saturn are dominated by planetary rotation; satellites are a major source of

Focus 3.5   Planetary magnetospheres

Six planets are known to have magnetospheres. The size of the magnetosphere, on the day side facing the Sun, is determined by the distance, R_{MP}, along the planet–Sun line at which the pressure of the planetary magnetic field balances the dynamic ram pressure of the solar wind. The magnetic pressure at the surface of the planet is given by $B_o^2 / (2\mu_0)$, where B_o is the equatorial magnetic field strength, and $\mu_0 = 4\pi \times 10^{-7}$ is the permeability of free space. Since the dipole's magnetic field strength falls off as the cube of the distance from the planet, the magnetic pressure decreases as the sixth power of that distance. This means that the standoff point where the two pressures are equal occurs when:

$$\text{Magnetic pressure} = \frac{R_P^6 B_o^2}{2\mu_0 R_{MP}^6} = m_p N V^2$$
$$= \text{Wind ram pressure}$$

where the planet's radius is R_P, the proton mass is $m_p = 1.67 \times 10^{-27}$ kilograms, N is the number density of the protons in the solar wind at the planet's distance from the Sun, and V is the solar wind velocity at that distance. Solving for R_{MP} we have:

$$R_{MP} = \left(\frac{B_o^2}{2\mu_0 m_p N V^2} \right)^{1/6} R_P$$

At the Earth's distance from the Sun, the number density of the solar wind is about $N = 5$ million protons per cubic meter and the wind velocity is about $V = 400$ kilometers per second. The equatorial surface magnetic field strength of the Earth is $B_o = 3.05 \times 10^{-5}$ tesla. With these numbers our equation gives $R_{ME} = 10\ R_E$, so the bow shock of the Earth is out at about 10 times the Earth's radius. The bow shock distance can be reduced to half this value when a powerful coronal mass ejection from the Sun hits the Earth, producing extra ram pressure. Moreover, unusual drops in the wind's pressure have very occasionally inflated the leading edge of the Earth's magnetosphere five or six times farther out in space, until it engulfed the Moon.

The values of R_{MP} for the other planets can be inferred by noting that the solar wind number density N falls off with the inverse cube of the distance of the planet from the Sun, while the solar wind velocity remains relatively constant. Values of R_{MP} were given in Table 3.5 for the six planets with detected dipolar magnetic fields.

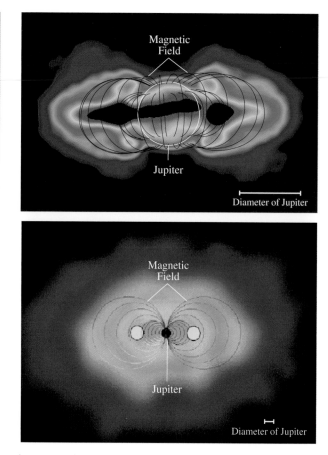

Fig. 3.17 Jupiter's magnetosphere   High-speed electrons that are trapped in Jupiter's magnetosphere emit steady radio radiation (*top*) by the synchrotron process; it is detected by ground-based radio telescopes such as the Very Large Array. An instrument aboard the *Cassini* spacecraft measured energetic atoms (*bottom*) created when fast-moving ions within the magnetosphere picked up electrons to become neutral atoms. The two open circles denote Io's orbital position on each side of the planet; the central black disk denotes Jupiter. [Courtesy of Imke de Pater, U. C. Berkeley (*top*) and NASA/JPL/JHUAPL (*bottom*).]

their charged particles, and internal forces shape those particles into an equatorial disk. Unlike Jupiter and Saturn, the magnetospheres of Uranus and Neptune are largely empty, and their magnetic fields are unexpectedly tilted and offset from the centers of these planets.

On the side facing the Sun, each of the planetary magnetic fields is compressed by the solar wind, forming a bow shock. It is located at the place where the pressure of the planet's magnetic field just equals the pressure of the solar wind (Focus 3.5). Such standoff distances for the six planets with detected magnetic fields were given in Table 3.5, but the varying solar wind pressure can alter this distance by a factor of 2.

The general form of Jupiter's magnetosphere resembles that of the Earth (Fig. 3.17), but its dimensions are at

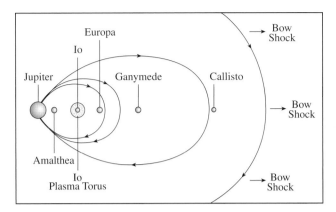

Fig. 3.18 Satellites within Jupiter's magnetic field This cross-section shows that the four Galilean satellites, Io, Europa, Ganymede and Callisto, are all embedded within Jupiter's magnetosphere. Small satellites orbit Jupiter within the orbit of Io, the innermost Galilean satellite. The outermost Galilean satellite, Callisto, orbits Jupiter near its bow shock. All of these satellites are being continuously bombarded with energetic charged particles that are trapped within Jupiter's magnetosphere. The distance from Jupiter to Callisto is 1.88 million (1.88×10^6) kilometers, while the radius of the Sun is 0.696 million kilometers, so Jupiter's bow shock is bigger than the Sun.

least 1200 times greater. It is larger than the Sun in size, with a bow shock at more than 3 million kilometers or at least 42 times the planet's radius. *Pioneer 10* and *Voyager 1* first encountered Jupiter's bow shock at 95 and 86 planetary radii, but the shock moved in and out due to the variable solar wind. Jupiter's largest satellites are all embedded within its magnetosphere and interact with it (Fig. 3.18).

Jupiter's enormous magnetic tail, driven outward by the solar wind, is almost a billion kilometers long. It spans the distance between the orbits of Jupiter and Saturn, which is as great as the distance from the Sun to Jupiter itself. In contrast, the Earth's magnetic tail barely flicks across our Moon's path, less than a half-million kilometers from the Earth. Thus, Jupiter's magnetosphere is the largest enduring structure in the solar system, although it is occasionally and temporarily exceeded in size by comet tails.

Jupiter's magnetic field was first recognized in 1954–55 by Earth-based observations of the planet's intense radio emission, and then directly measured by visiting spacecraft. High-speed electrons trapped within the planet's magnetic field generate the radio signals (Focus 3.6). The synchrotron radio emission is beamed in a direction nearly parallel to the magnetic equator, and it is therefore swept past an Earth-based observer as the planet rotates and brings the magnetic equator in and out of alignment with the line of sight. Periodic variations in the strength of the

Focus 3.6   Radio broadcasts from Jupiter

The discovery of intense radio emission from Jupiter is one of the many examples of the accidental discovery of an unexpected phenomenon while looking for something else. In 1954–55 Bernard F. Burke (1928–) and Kenneth L. Franklin (1923–2007) were using ground-based radio telescopes to observe the meter-wavelength emission of the Crab Nebula, a famous remnant of a stellar explosion in 1054 AD. They planned to study the changes in the Crab's radio signal, at a wavelength of 13.6 meters, as it passed behind the Sun, thereby determining properties of the outer solar atmosphere.

Their observations were hampered by radio bursts that resembled terrestrial interference, but the alert scientists noticed that they only appeared when the radio telescope was pointing in a certain direction in the sky. This meant that the radio bursts had an extraterrestrial origin. They were at first assumed to be coming from the Sun, an intense source of variable radio radiation, but the calculated position in the sky coincided with Jupiter.

Soon thereafter, a steady Jovian radio signal was found at shorter wavelengths of several centimeters, and scientists interpreted this emission as synchrotron radiation emitted by electrons trapped in the Jovian magnetic field and moving at relativistic speeds approaching the velocity of light. The electrons spiral about the magnetic field, emitting the radio radiation, named after the synchrotron particle accelerator on Earth where similar radiation was first observed visually.

The new theory for Jupiter's radio signals was confirmed by the observation of two characteristic signatures of synchrotron radiation. The radio broadcasts were weaker at shorter wavelengths, and stronger at the longer ones, unlike the thermal emission of a hot gas that is most intense at shorter wavelengths. The non-thermal radio radiation was also polarized, with a preferred orientation or direction, which ought to coincide with that of the magnetic fields. In addition, radio interferometer measurements indicated that the radio emission was much larger than the planet and roughly aligned with its equator.

As Jupiter rotates, it carries the magnetic fields and their trapped electrons with it. Since the radio signal of the electrons is beamed in a particular direction, it sweeps past the observer once every rotation, providing a precise measurement of the planet's rotation period: 9 hours 55 minutes 29.7 seconds, or 9.9249 hours.

radio emission indicate that the magnetic field rotates with a period of precisely 9 hours 55 minutes 29.7 seconds, or 9.9249 hours. Since the magnetic fields are generated deep within the planet, this is assumed to be Jupiter's rotation period; it differs from the rotation speed inferred from visible clouds that are blown in different directions and at various speeds by powerful winds.

In December 1973 and December 1974 *Pioneer 10* and *11* confirmed Jupiter's strong dipolar magnetic field and energetic trapped electrons. In 1979 *Voyager 1* and *2* obtained information about Jupiter's outer magnetosphere, as well as the magnetic interaction between Jupiter and Io.

The *Pioneer* data showed that the magnetic field at the cloud tops is 4.28×10^{-4} tesla, or about 14 times stronger than the Earth's equatorial magnetic field strength. As near the Earth's surface, the magnetic fields at Jupiter's cloud tops are strongest at the poles and weakest along the equator. The magnetic field can be described as a dipole with a magnetic axis that is tilted at 9.6 degrees with respect to the rotation axis, similar to the Earth's tilt of 11.7 degrees; but the magnetic poles are reversed in comparison to those of the Earth, so a north-seeking terrestrial compass would point south in the vicinity of Jupiter.

Like its terrestrial counterpart, Jupiter's magnetosphere contains electrons and protons that are supplied from outside by the variable solar wind. Unlike the Earth, the giant planet's magnetosphere is also fed from within, by ions erupted from the active volcanoes on its innermost large satellite, Io. The dominant ions are sulfur and oxygen, both products of Io's unique volcanic activity. Jupiter's belts of charged particles resemble the terrestrial Van Allen belts in shape, but the Jovian belts are up to a million times more densely filled with particles than those near the Earth are.

Ions and electrons within Jupiter's inner magnetosphere are accelerated by the spinning magnetic field of the planet, eventually reaching very high energies. The inner magnetosphere is a stiff, permanent structure that is tied to the planet and rotates with it. Once an electrically charged particle enters this region, the magnetic field picks the particle up and takes it for long rides around the planet. As the powerful magnetic field spins, it extracts rotational energy from the planet, lashing and accelerating the charged particles to nearly the speed of light.

Thus, the energy that populates and maintains the magnetosphere of Jupiter comes principally from the planet's rotation, as well as the tidal flexing of Io that results in its volcanoes. In contrast, the Earth's magnetosphere is principally energized by the solar wind. The numerous high-energy particles in Jupiter's magnetosphere are capable of destroying sensitive electronic circuits in spacecraft that pass near the planet.

The high-speed charged particles exert an outward pressure on Jupiter's magnetic field, inflating it like an air-filled balloon. Because the field is weakest in its equatorial plane, the forces and pressures associated with the rapid rotation stretch the equatorial regions outwards in the form of a thin, elongated disk.

The varying solar wind pressure buffets Jupiter's outer magnetosphere, so it expands and contracts. As varying solar activity pushed the magnetosphere in and out, approaching *Pioneer 10* and *11* and *Voyager 1* and *2* spacecraft crossed the bow shock several times. The changing shape and location of the magnetotail similarly caused the outward-bound spacecraft to cross it many times.

When gusts in the solar wind compress Jupiter's outer magnetosphere, some of its high-speed, electrically charged particles squirt out into interplanetary space with energies that exceed the typical energy of electrons and protons in the solar wind. The Jovian particles are continually replenished by acceleration within the planet's magnetosphere, spraying energetic electrons and protons throughout the solar system. Some of them reach the orbit of the Earth and even that of Mercury.

The discovery of Saturn's magnetosphere did not occur until September 1979 when *Pioneer 11* first crossed its bow shock, at 24 Saturn radii, closely followed by the *Voyager 1* and *2* encounters in November 1980 and August 1981, respectively. The magnetic field strength at Saturn's cloud tops is 70 percent of that at the Earth's equator, but spread over a much bigger volume. Saturn's dipolar magnetic field is almost precisely aligned with its poles of rotation. Like Jupiter, the magnetic poles of Saturn are reversed compared to those of the Earth. High-speed electrons in the spinning magnetic field give rise to periodic radio emission, and the inferred rotation period is 10 hours 39 minutes 22.3 seconds, or 10.6562 hours, and just 44 minutes longer than Jupiter's rotation period.

Saturn's sizeable satellites and its rings absorb energetic electrons and protons. Perhaps as a result, its magnetosphere does not contain a high density of high-energy electrons. The planet also does not have a volcanically active satellite to generate sulfur and oxygen ions. Its magnetic trap is instead permeated with low-energy ionized material, protons and oxygen ions, chipped or sputtered off the water ice in the planet's rings and on its satellite surfaces. A vast dense cloud of neutral, or un-ionized, hydroxyl (OH) molecules envelops the rings; it is also derived from the water ice H_2O.

The magnetic axes of Uranus and Neptune are tilted at a large angle with respect to their rotational axis, by 59 degrees for Uranus and 47 degrees for Neptune (Fig. 3.19),

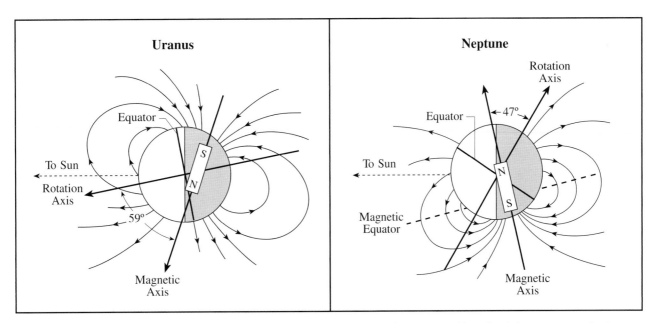

Fig. 3.19 Tilted magnetic fields   The magnetic fields of Uranus and Neptune can be represented by a simple bar magnet, or dipole, embedded in the planet, but with a magnetic axis that is tilted with respect to the rotation axis. For Uranus the tilt is about 59 degrees; Neptune has a tilt of 47 degrees. In contrast, the magnetic axes of Jupiter, Saturn and Earth are much more nearly aligned with their rotation axes. The arrow of the rotation axis points from the geographic south towards geographic north, and the magnetic axis similarly points from magnetic south to magnetic north. On Uranus and Neptune a terrestrial compass would point toward the southern hemisphere of the planet, while on Earth it points toward the geographic north pole. In addition to the dipole part of their magnetic field, Uranus and Neptune have a large additional component known as the quadrupole. A method of visualizing this is to imagine that the dipole has a magnetic center that is offset from the center of the planet. As shown here, the equivalent offset for Uranus is almost a third of the planet's radius, and there is a larger offset for Neptune of nearly half its radius. But such off-center dipoles are only useful as a picture of what the external field looks like and do not help in understanding how it is produced deep down.

and these planets have fully developed magnetospheres. The rotation periods of the magnetic fields of Uranus and Neptune, inferred from their periodic radio emission, are 17.24 and 16.11 hours, respectively.

All planetary magnetic fields are generated by the dynamo action of moving electrically conducting material in their interior. Internal rotation and convection produce the motions, somewhat like a spinning and boiling pot of water. Vigorous internal convection is powered by the decay of radioactive elements in the Earth and Mercury; the giant planets have retained much of their primordial heat to drive the internal convection and power their dynamo. The combination of convection and rotation concentrates the magnetism, amplifying its strength and regenerating the magnetic fields.

Mercury and the Earth have cores of molten iron alloys. At the high pressures inside Jupiter and Saturn, their most abundant ingredient, hydrogen, behaves like a liquid metal. Their strong magnetic fields are attributed to electrical currents driven by the fast rotation of their liquid metallic interiors. For Uranus and Neptune, water-rich material within their vast internal oceans most likely provides the electrical conductivity.

3.7 Aurora

Terrestrial aurora

Curtains of green or red light dance and shimmer across the night sky in the Earth's polar regions, far above the highest clouds (Figs. 3.20, 3.21). This light is called the *aurora* after the Roman goddess of the rosy-fingered dawn, a designation that has been traced back to Galileo Galilei (1564–1642). The aurora seen near the north and south poles have been given the Latin names *aurora borealis*, for Northern Lights, and *aurora australis*, for Southern Lights.

Most people never see the awesome lights, for aurora are normally confined to high latitudes in the north or south polar regions. But this does not mean that the aurora occur infrequently. Residents in far northern locations can see a faint aurora every clear and dark night.

The northern aurora borealis has been documented for centuries. Ancient Vikings (500–1500 AD) thought they were the spirits of fallen warriors being carried to Valhalla, the home of the gods. The southern aurora australis have never achieved a renown comparable to the Northern Lights, probably because the southern ones are not

Fig. 3.20 Northern Lights   Spectacular green curtains of light dance and shimmer across the northern sky. High-energy electrons are funneled down the Earth's polar magnetic field lines into the atmosphere, where they excite oxygen atoms that fluoresce green light, like a cosmic neon sign. The color is usually green, but the aurora can also have red bottoms, arising from excited nitrogen molecules. This photograph of the fluorescent Northern Lights, or *aurora borealis*, was taken over Fairbanks, Alaska. (Courtesy of Jan Curtis.)

Fig. 3.21 Twilight aurora   The form and brightness of the aurora will vary with time of night, appearing near twilight as bands or diffuse arcs, then rising and brightening as the night progresses and the display forms a curtain or drapery formation. The dark gaps located between the green bands, known as black aurora, have been attributed to negatively charged particles that are sucked out of the Earth's ionosphere along adjoining magnetic field lines, climbing to over 20 000 kilometers and lasting for several minutes. This photograph was taken over Fairbanks, Alaska. (Courtesy of Jan Curtis.)

usually located over inhabited land and are instead seen from oceans that are infrequently traveled.

Rare, brilliant aurora can extend down to the Earth's equator, becoming visible as far south as Athens, Rome or Mexico City. They were noted by the ancient Greeks. Plutarch (*c.* AD 46–120) reported one that occurred in 427 BC, but aurora do not extend down to Greece very often, perhaps every 50 or 100 years.

Since aurora become more frequent as one travels north from tropical latitudes, it was thought that they would become brighter and occur most frequently at the highest northern latitudes. Arctic explorers were therefore surprised to see that the intensity and frequency of aurora did not increase all the way to the poles and instead peaked in an oval-shaped band that encircles the North Pole. This northern aurora oval has a radius of about 2250 kilometers and is centered on the Earth's north magnetic pole, with an inner and outer radius separated by about 500 kilometers.

Nowadays we can use spacecraft to view both the northern and southern lights from space (Figs. 3.22, 3.23). The *Space Shuttle* has even flown right through the Northern Lights. While inside the display, astronauts could close their eyes and see flashes of light caused by the charged aurora particles speeding through the spacecraft walls and into their eyeballs.

When viewed from above, the aurora form a luminous oval centered at each magnetic pole, resembling a fiery halo (Fig. 3.23). The aurora oval is constantly in motion, expanding a little toward the equator or contacting a bit

Fig. 3.22 Aurora australis   The eerie, beautiful glow of auroras can be detected from space, as shown in this image of the *aurora australis*, or Southern Lights, taken from the *Space Shuttle Discovery*. The colored emission of atomic oxygen extends upward to between 200 and 300 kilometers above the Earth's surface. (Courtesy of NASA.)

toward the pole, and constantly changing in brightness. Such ever-changing aurora ovals are created simultaneously in both hemispheres and can be viewed at the same time from the Moon.

Visual auroras normally occur at 100 to 250 kilometers above the ground. This height is much smaller than either the average radius of the oval, at 2250 kilometers, or the radius of the Earth, 6380 kilometers. An observer on the ground therefore sees only a small, changing piece of the aurora oval, which can resemble a bright, thin, wind-blown curtain hanging vertically down from the Arctic sky.

Energetic electrons bombarding the upper atmosphere cause most aurora on the Earth. The reason that aurora are usually located near the polar regions is that the Earth's magnetic fields guide the energetic electrons there. Electrical currents as great as a million amperes can be produced along the aurora oval, and the electric power generated during the discharge is truly awesome — about ten times the annual consumption of electricity in the United States.

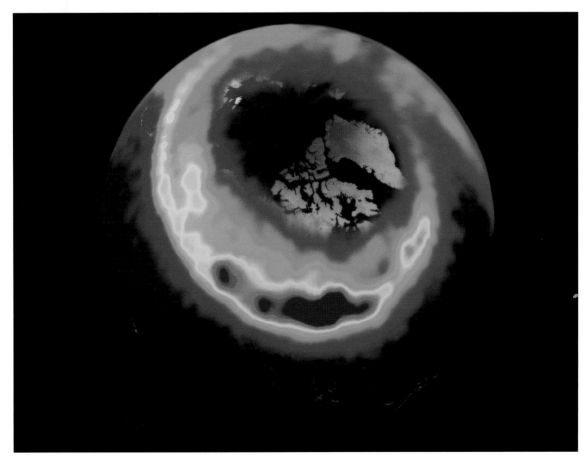

Fig. 3.23 The aurora oval   The *POLAR* spacecraft looks down on an aurora from high above the Earth's north polar region in February 2000, showing the Northern Lights in their entirety. The glowing oval is 4500 kilometers across. The most intense aurora activity appears in bright red or yellow. (Courtesy of NASA/U. Iowa.)

Table 3.6 Frequent spectral features in the aurora emission

Wavelength (nanometers)	Emitting atom, ion or molecule	Altitude (kilometers)	Visual color
391.4	N^+ (nitrogen ion)	1000	violet-purple
427.8	N^+ (nitrogen ion)	1000	violet-purple
557.7	O (oxygen atom)	90–150	green
630.0	O (oxygen atom)	>150	red
636.4	O (oxygen atom)	>150	red
661.1	N_2 (nitrogen molecule)	65–90	red
669.6	N_2 (nitrogen molecule)	65–90	red
676.8	N_2 (nitrogen molecule)	65–90	red
686.1	N_2 (nitrogen molecule)	65–90	red

When the fast-moving electrons slam into the upper atmosphere, at speeds of about 50 kilometers per second, they collide with the oxygen and nitrogen atoms or molecules there and excite them. The pumped-up particles quickly give up the energy they acquired from the electrons, emitting a burst of color in a process called fluorescence. It is something like electricity making the gas in a neon light shine or a fluorescent lamp glow.

The color of the aurora depends on which atoms or molecules are struck by the precipitating electrons, and the atmospheric height at which they are struck (Table 3.6). Low-altitude oxygen atoms produce green, a common aurora color, while the high-altitude oxygen atoms cause the rare all-red aurora. Nitrogen molecules create low-altitude red light, below the oxygen's green, while nitrogen ions can produce violet-purple light at high altitudes. The green oxygen emission appears at about 100 kilometers and the red oxygen light at 200 to 400 kilometers. At these heights, the aurora shines from the ionosphere, an electrically conducting layer in the Earth's upper atmosphere.

Even though changing conditions on the Sun may trigger exceptionally intense Northern and Southern Lights, we now know that the electrons that cause some of the everyday aurora arrive indirectly at the polar regions from the Earth's magnetic tail, and that these electrons can also be energized locally within the magnetosphere. As the solar wind flows past the Earth, terrestrial magnetic fields can capture and store energy from the winds, and the solar-wind magnetic fields can merge, or reconnect, with the terrestrial magnetic fields. The stretched-out magnetism eventually gets overloaded with too much energy, temporarily pinching off the Earth's magnetotail.

During this magnetic reconnection process, the magnetic fields heading in opposite directions – having opposite north and south polarities – break and reconnect at 140 000 to 160 000 kilometers downwind of Earth on its nightside. Electrons are pushed up and down the tail, and can be accelerated within the magnetosphere as they travel back toward the Earth and into its polar regions. The electrons that are thrown Earthward follow the path of magnetic field lines, which link the magnetotail to the polar regions and map into the aurora oval.

Magnetic reconnection between the solar-wind magnetic fields and the Earth's magnetic fields can also occur on the dayside of the Earth, facing the Sun, and this can also open a valve that lets the solar-wind energy cross into the magnetosphere. In this event, high-energy electrons also follow the path of magnetic field lines into the polar regions where they produce the aurora and create the aurora oval.

Although most aurora are caused by collisions between high-energy electrons and the atmosphere, protons can also be funneled down along the polar magnetic fields and sometimes cause aurora. The ultraviolet emission of such a proton aurora was, for example, recorded from the *IMAGE* spacecraft in 2000. Solar-wind protons, which enter through rare temporary openings in the Earth's magnetic field, cause them.

The rare bright aurora seen at low latitudes in more clement climates tend to occur when the Sun is near the peak of its 11-year magnetic activity cycle, when sunspots are most numerous; such exceptionally bright aurora occur less often at the minimum of the cycle when there are few sunspots. The sunspots do not themselves cause the intense aurora, but are instead a sign of strong magnetic fields on the Sun, which can explosively eject huge magnetic bubbles known as coronal mass ejections. At maximum activity, the Sun emits coronal mass ejections more frequently, and when they chance to hit the

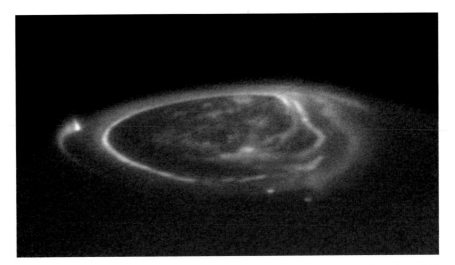

Fig. 3.24 Jupiter's aurora   The aurora oval or ring over the north polar region of Jupiter, as imaged from the *Hubble Space Telescope* in ultraviolet light in September 1997. High-energy electrons and ions cascade into Jupiter's upper atmosphere and create the bright ultraviolet aurora. Several of the bright spots are believed to mark the magnetic "footpoints" of three of Jupiter's largest moons. The footpoints are locations where powerful streams of electrons follow the magnetic fields of Jupiter's magnetosphere from the moons down into Jupiter's atmosphere. (Courtesy of NASA/STScI/U. Michigan.)

terrestrial magnetic field with the right magnetic orientation, they reconnect with it and compress the Earth's magnetic field. The aurora ovals then intensify and spread down as far as the tropics in both hemispheres. So it is really the Sun that controls the intensity of the brightest, most extensive aurora, like the dimming switch of a cosmic light.

Aurora on Jupiter and Saturn

Charged particles are funneled into the magnetic polar regions of Jupiter and Saturn, producing aurora ovals that shine in ultraviolet light, rather than Earth's green, red or violet. The atmospheres of these giant planets are primarily composed of hydrogen, unlike Earth's oxygen and nitrogen, and their ultraviolet auroras are emitted when charged particles stream into and excite the atomic and molecular hydrogen. Moreover, many, but not all, of the aurora at Jupiter and Saturn are caused by particles trapped in their immense magnetospheres, rather than being directly connected to solar-wind particles.

Like their terrestrial counterparts, the curtains of light on Jupiter are found in two oval-shaped regions circling the magnetic poles of the planet, just above the clouds. The aurora glows are produced in these high-latitude regions because that is where the magnetic fields direct electrically charged particles: electrons, protons and other ions. When these particles hit the planet's upper atmosphere, they collide with atoms and molecules there, leaving them

in an excited state. As on Earth, the atoms and molecules release the extra energy in the form of light, and return to their normal state. But unlike Earth's colored light show, Jupiter's aurora ovals were first observed from space at ultraviolet wavelengths (Fig. 3.24).

Jupiter's aurora is the most powerful in the solar system. At about 10^{14} watts, it is typically one thousand times more powerful than Earth's aurora. The Jovian lights are powered largely by energy extracted from planetary rotation, although there seems to be a contribution from the solar wind. Thus, internal processes seem to be the dominant source of power for Jupiter's aurora. This contrasts with Earth's aurora, which is mainly generated externally through the interaction of the solar wind and the terrestrial magnetosphere.

Jupiter's satellite Io affects the aurora on the planet. Electrons and ions spewed out by volcanoes on Io are captured by the intense, rapidly rotating magnetic field and spiral inward at high energies toward the planet's polar regions. As the rotating magnetic field sweeps past Io, an invisible current of charged particles flows along Jupiter's magnetic field lines into the polar regions, producing bright trails in the ultraviolet images (Fig. 3.24).

On Jupiter one can normally see a main ultraviolet oval, and in addition bright swirling streaks are sometimes detected both within and outside the oval (Fig. 3.24). They have been attributed to electric currents from three of the planet's large moons, Io, Ganymede and Europa. Very intense bursts of aurora activity have also

Fig. 3.25 Aurora on Saturn   High-energy electrons and ions are captured from the solar wind and funneled down into Saturn's upper atmosphere, creating aurora ovals at its northern (*upper left*) and southern (*lower right*) magnetic poles. This ultraviolet image was recorded from the *Hubble Space Telescope* in October 1997. The bright red aurora features are dominated by emission from atomic hydrogen, while the white regions within them are emitted by molecular hydrogen. (Courtesy of NASA/STScI/JPL.)

been detected; they are apparently regulated by the variable solar wind, perhaps because it affects the size of Jupiter's magnetosphere.

Saturn's ultraviolet aurora (Fig. 3.25) is most likely caused when the gusty solar wind sweeps over the planet, perhaps like the Earth's aurora. But unlike the Earth, Saturn's aurora oval has only been seen from spacecraft in ultraviolet light, at least so far. It could not be detected from beneath the Earth's atmosphere that absorbs the ultraviolet.

In Chapter 4 we turn our attention to our home planet, Earth, third rock from the Sun.

4   Restless Earth: third rock from the Sun

- Seismic waves generated by earthquakes have been used to look inside the Earth, determining its internal structure.

- There is a crystalline globe of solid iron at the center of the Earth that spins faster than the rest of the planet. This inner solid core is suspended in a much larger, fluid, outer core of molten iron, which is itself encased in a thick mantle of solid rock.

- The continents disperse and then reassemble, over and over again, roaming about the planet in an endless journey.

- Sound waves and gravitational data have been used to effectively empty the Earth's oceans and see their floors, revealing an underwater range of active volcanoes that snakes its way around the middle of the ocean floor.

- The bottom of the oceans remains in eternal youth as new floor spills out of mid-ocean volcanoes and old floor is pushed back inside the Earth, but the water above the floors has remained for billions of years, shifting about the globe as new oceans open up and old ones close.

- The outer part of the Earth is broken into a mosaic of large plates, like the cracked pieces of an eggshell; these plates move across the Earth at the rate of a few centimeters per year, or about as fast as your fingernails grow.

- Wheeling, churning motions deep inside the Earth's hot interior move continents sideways all across the planet.

- The Earth's moving plates squeeze oceans out of existence, grind against each other to create earthquakes, and dive into the Earth to produce volcanoes that make continents grow at their edges.

- Boston and Italy were once part of Africa, a glacier of ice once covered the Sahara Desert, and the Pacific Ocean once washed against the shores of Colorado.

- A colossal alp can erode away into a small, round knob of a hill in just a few hundred million years, while continents can also weld together to form new mountain ranges.

- The Earth's upper atmosphere is heated and ionized by the Sun's variable X-ray and extreme ultraviolet radiation.

- Ultraviolet radiation from the Sun creates the protective ozone layer in the stratosphere of the Earth's atmosphere.

- Synthetic chemicals called chloroflurocarbons (CFCs) have been destroying the thin layer of ozone that protects human beings from dangerous solar ultraviolet radiation. The production of these ozone-destroying chemicals was outlawed in 1987 by an international agreement named the *Montreal Protocol.*

- Invisible gases help to warm the Earth by trapping the Sun's heat and preventing some of it from being reflected back into space. This process is commonly known as the greenhouse effect.

- Warming of the Earth's surface and lower atmosphere by the greenhouse effect keeps the Earth from becoming a frozen ball of ice.

- Carbon dioxide and other heat-trapping gases, such as methane and nitrous oxide, have been increasing in the Earth's atmosphere for more than a century as the result of human activity.

- By burning coal and oil, humans have increased the amount of carbon dioxide in the Earth's atmosphere by 30 percent since the industrial revolution.

- Rising seas, retreating glaciers, melting ice caps, and increasing sea and air temperatures are all recent signs of global warming from increased emissions of heat-trapping gases.

- If current emissions of carbon dioxide and other greenhouse gases go unchecked over the next 100 years, global warming could produce agricultural disaster in the world's poorest countries, rising seas with coastal flooding throughout the world, and the spread of diseases carried by mosquitoes.

- An international agreement to limit the emission of heat-trapping gases was made in December 1997. Known as the *Kyoto Protocol,* it has had a limited effect on curbing global warming because it has not been ratified by China or the United States, two of the main climate-altering polluters.

- The world's most influential science academies have warned national leaders that global warming from emissions of carbon dioxide and other heat-trapping gases poses a clear and increasing threat.

- The 2007 Nobel Peace Prize was awarded to an Intergovernmental Panel on Climate Change and to Albert Gore Jr. for their contributions to knowledge about man-made climate change and for laying foundations to measures needed to counteract the change.

- The Copenhagen Summit in December 2009 sought international consensus on ways to combat global warming, but it did not result in any legally binding treaty on limiting carbon-dioxide emissions. Both China and the United States refused to accept such mandatory limits, but agreed with a hypothetical climate-change accord that has voluntary curbs and varying emission reductions for different countries.

- The major ice ages, which repeat every 100 000 years, are caused by astronomical rhythms that alter the angles and distances from which sunlight strikes the Earth.

- The Sun is slowly getting brighter as time goes on. It will become hot enough in 3 billion years to boil the Earth's oceans away, and 4 billion years thereafter our star will balloon into a giant star, engulfing the planet Mercury and becoming hot enough to melt the Earth's surface.

- Space weather refers to conditions on the Sun and in the Sun's winds, the Earth's magnetosphere, and the Earth's outer atmosphere that can influence the performance and reliability of space-borne and ground-based technological systems, and can affect human life and health.

- Explosive outbursts of solar flares and coronal mass ejections from the Sun can cripple spacecraft and seriously endanger unprotected astronauts that venture into outer space. Sun storms can also disrupt global radio communications and disable satellites used for navigation, military reconnaissance or surveillance, and communication, from cell phones to pagers, with considerable economic, safety and security consequences.

- Solar protons are the most energetic and therefore the most dangerous solar energetic particles. They can severely affect the health of unprotected astronauts traveling outside the Earth's magnetosphere, and they are capable of penetrating spacecraft to damage or disrupt sensitive technical systems. The strongest events produce radiation doses that might be lethal to astronauts fixing a spacecraft in outer space or taking a walk on the Moon or Mars.

- Interplanetary magnetic clouds travel behind interplanetary shocks, which are driven by coronal mass ejections. Such a magnetic cloud contains a well-organized, twisted magnetic flux tube, which can provide a "highway" for the transport of solar energetic particles.

- When encountering Earth with the right magnetic alignment, coronal mass ejections can trigger intense geomagnetic storms, accompanied by exceptionally bright aurora, and compress the magnetosphere, exposing geosynchronous satellites to the full force of the solar wind.

- Solar X-rays and extreme ultraviolet radiation both produce and significantly alter the Earth's ionosphere. The solar X-rays fluctuate in intensity by two orders of magnitude, or a factor of 100, during the Sun's 11-year magnetic activity cycle. Near activity maximum, greater amounts of X-rays produce increased ionization, greater heat, and expansion of the Earth's upper atmosphere, altering satellite orbits and disrupting communications.

4.1 Fundamentals

Our home planet Earth is larger and denser than the other terrestrial planets, and the only one with large oceans of liquid water. It revolves about the Sun once a year, at a mean distance of one astronomical unit, and rotates on its axis to view the same star every 23 hours 56 minutes and 4 seconds, the Earth's sidereal rotation period. Table 4.1 summarizes the physical properties of the Earth.

4.2 Journey to the center of the Earth

Looking inside the Earth's hidden interior

The internal structure of the Earth can be mapped with the help of earthquake waves. The Greek word for earthquake is *seismos*, meaning "to quake or tremor". Today, earthquake waves are often called *seismic waves*, and the study of earthquakes is known as *seismology*.

Earthquakes that originate in the planet's outer shell set the seismic waves in motion, and their velocities are determined by the density, temperature and chemical composition of the rocks they travel through. The waves become sluggish in hot, low-density rock, and they speed up in colder, denser regions. When moving between materials of differing physical properties, the seismic waves change their speed and direction of movement, enabling seismologists to determine boundaries between the Earth's internal layers.

The seismic investigations indicate that the Earth is layered inside like a peach. Its deeper layers are denser, and they are separated from one another in sharp transitions. There are three major parts: (1) the rocky *crust*, (2) a *mantle* of hot, plastic rock, and (3) the dense *core* (Fig. 4.1). They are the skin, pulp, and pit of the Earth, so to speak.

Table 4.1 Physical properties of the Earth

Mass	5.972×10^{24} kilograms
Mean radius	6371 kilometers
Bulk density	5513.4 kilograms per cubic meter
Sidereal rotation period	23 hours 56 minutes 4 seconds $= 8.6164 \times 10^4$ seconds $= 0.99727$ days
Sidereal orbital period	1 year $= 365.24$ days $= 3.1557 \times 10^7$ seconds
Mean distance from Sun	1.49598×10^8 kilometers $= 1.000$ AU
Orbital eccentricity	0.0167
Tilt of rotational axis, or obliquity	23.27 degrees
Age	4.6×10^9 years
Atmosphere	77 percent nitrogen, 21 percent oxygen
Surface pressure	1.013 bars at sea level
Surface temperature	288 to 293 kelvin
Magnetic field strength	0.305×10^{-4} tesla at the equator
Magnetic dipole moment	7.91×10^{15} tesla meters cubed

Table 4.2 The five most abundant elements in the Earth

Element	Symbol	Average abundance (percent by mass)
Iron	Fe	34.6
Oxygen	O	29.5
Silicon	Si	15.2
Magnesium	Mg	12.7
Nickel	Ni	2.4

The core has a liquid outer component and a solid inner one.

These different internal layers can be distinguished by the different chemical composition of their rocks. Most of the rocks of the mantle consist of minerals in which silicon (Si) and oxygen (O) are linked to other atoms. Such minerals are known as *silicates*. The core is composed mainly of iron (Fe) with some nickel (Ni). It is made up of an outer core of liquid molten iron and an inner core of solid iron. The abundance of these ingredients in planet Earth is given in Table 4.2.

Most earthquakes occur just beneath the Earth's surface at depths of no more than 700 kilometers, when massive blocks of rock grind, lurch and slide against one another. But they shake the Earth to its very center, at 6371 kilometers below the surface, causing the planet to vibrate and ring like a bell. The reverberations resemble ripples spreading out from a disturbance on the surface of a pond. These seismic waves move in all directions and their arrivals at various places on the Earth can be detected by seismometers. By combining the arrival times of different seismic waves that have traveled through the Earth's interior to various points on the surface, seismologists can determine the hidden interior structure of the Earth (Fig. 4.2).

Rock layers of different density and stiffness will propagate the waves at different speeds, much the way that a tightened violin string will sound at a higher pitch. As a

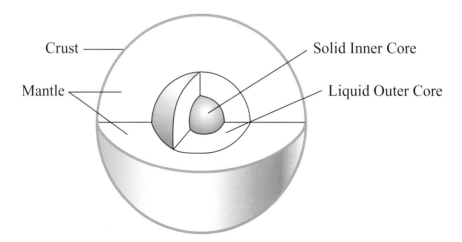

Fig. 4.1 Crust, mantle and core A relatively thin, rocky crust covers a thick silicate mantle. They overlie a liquid outer core, composed mainly of molten iron, and an inner core of solid iron. These nested layers have been inferred from seismic waves that travel through the Earth, changing velocity and direction at the layer boundaries.

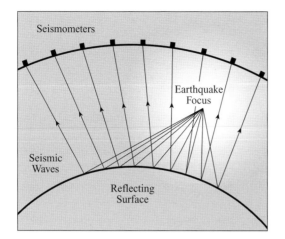

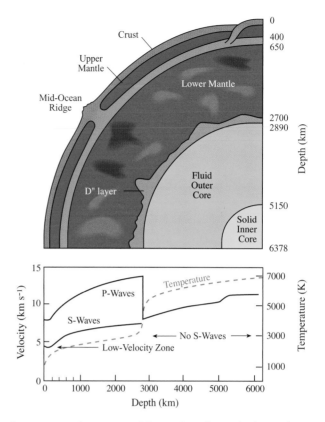

Fig. 4.2 Measuring earthquakes   When an earthquake occurs beneath the surface of the Earth, it becomes the focus of seismic waves that travel through the Earth. Seismometers on the surface of the Earth record the arrival of the waves, and locate the position of the boundaries between internal layers of different composition, density, pressure and temperature. Seismic waves known as S-waves (or "shear and shake" waves) cannot pass though a fluid, and are reflected by it. The reflected waves shown here mark the boundary of the Earth's liquid outer core.

Fig. 4.3 Layered structure of the Earth   The Earth's internal structure is determined by the varying velocity of earthquake waves. There are two types of waves that travel through the Earth. They are known as the compression P-waves (or "push and pull" waves) and the shear S-waves (or "shake" waves). The P-waves move almost twice as fast as the S-waves, and the P-waves pass through the fluid outer core, which the S-waves cannot do. The boundary between the mantle and core is marked by a precipitous drop in the velocity of the P-waves at a depth of about 2890 kilometers. The S-waves do not propagate beyond this boundary. The liquid outer core is separated from the solid inner core at a radius of 1220 kilometers where the P-waves increase in velocity.

result, the paths of seismic waves are bent and focused by their passage through the Earth's interior.

By careful mapping of the patterns of many earthquakes that travel to different depths, seismologists have peeled away the Earth's outer layers and looked at various levels within it (Fig. 4.3; Focus 4.1). It is similar to using an ultrasonic scanner to map out the shape of an unborn infant in a mother's womb. Seismology is also somewhat like using computed axial tomography (CAT) scans to derive clear views of the insides of living bodies from the numerous readings of X-rays that cross through the body from different directions.

The crust and mantle of the Earth

The outer skin of the Earth, its crust, is a thin veneer of rocky material that covers the planet like the lumpy and split crust of an apple pie. At the base of the Earth's crust lies the Mohorovicic discontinuity, a tongue-twisting name shortened by most geologists to "Moho".

The Moho separates the dense mantle from the light crust. The boundary lies at a depth of about 5 kilometers from the ocean bottom and 35 kilometers below most places on continents, but as much as 60 kilometers below mountains. So the crust is thinnest under the oceans and thickest under the continents. Because the continental and oceanic crusts are both less dense than the underlying material, they both tend to float on the mantle. High

mountains have deep crustal roots that provide buoyancy and keep them afloat, much the way icebergs float on the ocean.

The Earth's buoyant crust is made up of two different materials. There is the oceanic crust, which is made of the black, shiny, volcanic rock known as basalt, and the continental crust that contains granite. The ocean floor covers more than half the Earth's surface and has been produced by an outpouring of lava from volcanoes at the bottom of the sea. Volcanic islands like Hawaii and Iceland are also largely composed of basalt. The tough continental granites were formed in the fiery melts of magma, and include hard, colorless quartz.

The bulk of the Earth is in its mantle, the region that reaches down some 2890 kilometers, on average, from the

Focus 4.1   Taking the pulse of the Earth

Earthquakes produce three types of seismic waves. There are the *P-waves* and *S-waves* that travel into the Earth, and the *surface waves* that propagate around it. The P-waves consist of compression pulses through the Earth, expanding and compressing the rocks. They are analogous to sound waves in air, although the vibrations of the P seismic waves are much slower than audible sound. The P-waves arrive at monitoring stations before the S-waves, which set the Earth vibrating at right angles to the path of the waves, advancing like snakes. The P stands for "push and pull", while the S denotes "shear or shake".

The P-waves can propagate through every part of the Earth, even its center. A large portion of them penetrates the deep interior and then re-emerges toward the surface on the other side. The S-waves do not travel in a fluid; they propagate only in resilient, solid substances that have elastic resistance to twisting.

In 1906, the British geologist Richard D. Oldham (1858–1936) found that at a certain depth the P-waves slowed sharply and the S-waves couldn't propagate. These changes mark the bottom of the mantle and the top of the Earth's liquid outer core. The core–mantle boundary is sometimes called the *Gutenberg discontinuity*, after Beno Gutenberg (1889–1960), from the California Institute of Technology, who made the first accurate determination of its depth at an average of 2890 kilometers. In 1909, the Croatian geologist Andrija Mohorovicic (1857–1936) discovered that the speed of seismic waves increases at 35 to 60 kilometers below some continents. This *Mohorovicic discontinuity* marks the place where the crust ends and the mantle begins. The inner core, with a radius of about 1220 kilometers, was discovered in 1936 by the Danish seismologist Inge Lehmann (1888–1993).

thin crust to the top of the core. The difference between the crust and the mantle is one of chemical composition. Material brought up by volcanic eruptions, as well as eroded mountains, indicate that the upper mantle is composed of dense minerals known as olivine and pyroxene, which are silicates with a little magnesium or iron mixed in as minor constituents.

Lithosphere and asthenosphere of the Earth

The outermost parts of the Earth can be divided by their physical properties into the lithosphere and asthenosphere. The lithosphere is the solid region beneath the familiar oceans and mountains. It extends to depths of about 100 kilometers, which includes both the crust and upper mantle. The lithosphere consists of rocky crust and mantle down to a zone in the mantle that is lubricious enough to move. Beneath the lithosphere lies the warm and plastic asthenosphere that reaches to a depth of about 300 kilometers. Its material is revealed by the slowness with which it propagates seismic waves.

The distinction between the lithosphere and asthenosphere is one of stiffness. The lithosphere takes the root of its name from the Greek *lithos*, for "stone". The lithosphere is the solid "plate" of the plate tectonic theory mentioned later in this section. The word "asthenosphere" comes from the Greek *asthenos*, meaning "without strength" or "devoid of force".

The radioactive elements responsible for the warmth of the asthenosphere are too weakly concentrated to melt the rock, but they cause it to soften and behave like putty. Rock in the asthenosphere flows slowly when strained for a long time, like applying slow pressure to an open tube of toothpaste, but the asthenosphere responds like a solid when it is struck by an earthquake.

Two cores of the Earth

If you pick up a typical rock in the Earth's crust and determine its mass density, it will be roughly 3000 kilograms per cubic meter, or about three times that of water. By way of comparison, the mean mass density of the Earth is 5513.4 kilograms per cubic meter, which means that there must be high-density material located deep inside the Earth. The fact that the Earth is not homogeneous, with its densest parts concentrated inside, has been known since the time of Isaac Newton (1642–1727), from the varying gravitational pull measured by pendulums located at different places on the Earth's surface. The material with the greatest density is concentrated in the planet's core, with a mass density of about twelve times that of water.

The Earth's core reaches about halfway to the surface, implying a volume that is one-eighth that of the entire Earth. If the mass density of the Earth were uniform, the core would have an equal share, one-eighth, of the mass of the Earth, but its actual mass is nearly three times greater. This points to iron as the most likely core material, since it is also the most abundant heavy element in the Sun and in some meteorites.

Laboratory measurements also show that the densities and seismic-wave velocities of the core are more closely matched by iron than any other element. The seismic evidence indicates that the core is less dense than pure iron would be at the pressures there. Although the core is mostly iron, it must consist of an alloy of iron that includes light elements, one of which may be hydrogen.

By weight, the Earth is mostly iron, but relatively little of the metal is found in the Earth's crust. It is principally made of lighter elements like silicon and oxygen. Billions of years ago, during the planet's very early history, the Earth must have been molten, permitting the iron to sink to the interior because of its enormous weight. The planet would have then cooled from the outside, forming solid rocks in the crust and mantle that consist of lighter elements that are locked together and do not sink into the molten core.

Examination of earthquake waves has shown that there are two cores: an inner crystalline solid core and an outer fluid one. Compressed by the immense weight of the overlying material, the inner core solidified, while the outer core remained liquid. The two cores are very different in size. The solid inner core has a radius of about 1220 kilometers, which is slightly smaller than the Moon whose radius is 1740 kilometers. The outer fluid core is about 3490 kilometers in radius, or 55 percent of the Earth's radius.

The seismic evidence indicates the presence of a rugged, interactive zone, known as the D″ layer, where the liquid-metal outer core meets the lowermost part of the rocky mantle. This turbulent, irregular boundary contains troughs and swells, deeper than the Grand Canyon and higher than Mount Everest, spreading across continent-sized areas. It may represent material that was once dissolved in the underlying fluid core, dense material that sank through the mantle but could sink no further down, or material formed as a result of chemical reactions between the core and mantle. Seismologists speculate that irregularities in the D″ layer at the core–mantle boundary may channel heat flows to produce giant rising plumes of molten rock capable of penetrating the thick mantle and occasionally making their way to the Earth's surface.

The temperature of the deep core is difficult to determine, but we certainly know that the planet is hot inside. The central inner core appears to be about 6900 kelvin, which is a bit hotter than the visible disk of the Sun at 5780 kelvin. At first glance, this would seem to imply that the center of the Earth must be liquid, but this is contradicted by seismic evidence, which indicates a solid region in the deep interior. The clue to the apparent paradox is the high pressure at the center of the Earth, about 3.6 million times the pressure of the atmosphere at sea level. These pressures have been imitated in laboratory experiments, and they lead to a remarkable conclusion. At high pressures, iron can persist as a fairly rigid solid even at a temperature of thousands of kelvin.

Most liquids will solidify if the pressures are high enough and the temperatures are relatively low. Probably the entire core was once molten, but the drop in temperature associated with a loss of heat permitted the inner portion to solidify under the high pressures. The pressures are low enough, and the temperatures still high enough, to

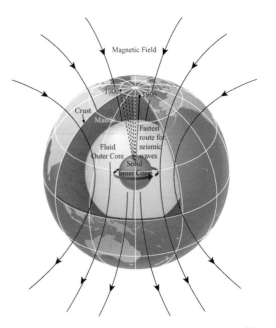

Fig. 4.4 The Earth's double core   The mantle and part of the crust have been cut away here to show the relative sizes of the Earth's fluid and solid cores. The outer fluid core is about 55 percent of the radius of the Earth, and the inner solid core is slightly smaller than the Earth's Moon. The Earth's magnetic field is thought to be generated and sustained by moving currents in the planet's electrically conducting, fluid outer core, which is composed of molten iron. Geophysicists have discovered that the route of the rapid polar (north–south) waves through the Earth's interior is gradually shifting eastward because the inner core is rotating slightly faster than the rest of the planet. The fast rotation of the inner solid core may help explain how Earth's magnetic field reverses polarity. (Courtesy of Paul Richards, Lamont-Doherty Earth Observatory.)

sustain a liquid outer core. It also remains liquid because iron alloys melt at lower temperatures than most rocks.

As our planet grows older and colder, the solid inner core is growing continuously at the expense of the liquid outer core. The iron liquid at the base of the fluid outer core is freezing, solidifying and snowing on the surface of the solid inner core, making it grow at the rate of about 0.01 meters every 100 years. At the same time, the rocky mantle may be slowly dissolving into the liquid metal of the outer core.

The Earth's inner core is a solid lump of iron suspended at the center of the much larger, fluid outer core, something like a golf ball levitated in the middle of a fish bowl (Fig. 4.4). The faint seismic vibrations that pierce the inner core move through it at different speeds that depend on their direction, faster on polar north–south paths than equatorial east–west ones. This directional dependence of seismic-wave velocities is explained by the crystalline structure of the inner core. The crystals give the solid inner core a texture with a preferred orientation, like the grain in wood. By lining up along the Earth's rotation axis, iron crystals make the inner core stiffer along this axis, thus making sound waves travel faster in this direction. Some scientists have even speculated that the inner core may be just one, single, gigantic crystal of iron atoms rather than a mass of tiny crystals, like a huge diamond suitable for an interplanetary engagement. In either case, each crystal probably takes its direction from either the stress generated by Earth's rotation or from the terrestrial magnetic field.

Recordings of weak earthquake rumbles, which have traveled through the central core of the Earth, indicate that it spins faster than the outer Earth, but that they both rotate in the same direction. The fast lane for seismic waves is tipped slightly with respect to the Earth's north–south axis, and it moves around it (Fig. 4.4). This shift in orientation means that the crystalline globe at the center of the Earth is turning slowly within its liquid metal enclosure. It is spinning with respect to the Earth's surface at between 0.2 and 0.3 degrees per year, completing one lap in between 1200 and 1800 years.

The Earth's magnetic field threads the solid inner core and could make it turn faster, much as a magnetic field turns the shaft of an electric motor. The magnetic coupling between the two cores could also account for reverses in the Earth's magnetic polarity; the planet switches its north and south magnetic poles a few times every million years. Currents in the electrically conducting, fluid outer core generate and maintain the magnetic fields, and turbulence in the fluid is always trying to toss the magnetism into a polarity reversal. The inner solid core exerts a stabilizing influence on this tendency, forcing the fields to stay in place, but the magnetic connection between the two cores is probably pulled apart as they rotate with respect to each other. The coupling eventually gives way and the magnetism flips.

Origin of the Earth's layered interior

The origin of the layered structure of the Earth's interior is still a geological mystery, but there are two extreme alternatives. According to one theory, the Earth accumulated rapidly (in 100 000 to 10 million years), and the kinetic energy of the impacting material that coalesced to form the Earth kept the planet hot and molten as it formed. If the rocks were molten as the Earth grew, its constituents would separate, with the dense, heavy material sinking toward the interior, creating the dense core, and the light substances rising to the surface to form the low-density mantle and crust.

In the alternative scenario, the Earth gathered itself together relatively slowly, in 100 million to 1 billion years, and the planet started out cold, homogeneous and solid.

The globe then became heated by emission from radioactive material that was uniformly distributed through the interior, and its temperature gradually rose to the melting point. When the planet melted, the heavy elements fell toward the center, forming a dense core, while the lighter elements rose toward the surface, producing chemically distinct layers.

In both the hot and cold theories, the layered internal structure of the Earth results from a process known as differentiation, in which gravity separates elements in a molten state, pulling the heavier ones down. A similar thing takes place in a blast furnace or smelter. Slag-forming rock is loaded into the furnace, and molten metal is tapped periodically from the bottom. Thus, in both theories the Earth was once molten and after a process of gravitational separation or differentiation, the Earth began cooling from the outside. The solid crust formed and then the mantle, and the basic layered structures remained a feature of the Earth since its early history.

4.3 Remodeling the Earth's surface

Earth's continents, oceans and ocean floors

There are two major types of terrain on Earth – the high, dry continents and the low, wet floor of the ocean (Fig. 4.5). Between them, and partially surrounding many continents, is a narrow strip of shallow ocean called the continental shelf. Today, the oceans cover 71 percent of the Earth's surface, and the world's continents amount only to scattered and isolated masses surrounded by water.

To those of us who are confined near the surface of the globe, the Earth seems rugged, with towering mountains rising several kilometers above the ocean (Fig. 4.6). But a scale model of the Earth would have to be quite smooth. The highest and lowest places reach only one-tenth of one percent, or 0.001, of the Earth's radius above and below the ocean surface. A basketball this smooth would have bumps no more than 0.0001, or 10^{-4}, meters high, roughly the size of the dot at the end of this sentence.

The smoothness of the Earth is due to the immense force of its gravity and the weight of its outer layers, which largely overcome the electrical force inside the solids making up the Earth and cause them to lie in concentric shells. In smaller bodies, such as asteroids less than a few hundred kilometers in diameter, the interior is strong enough to remain rigid and they retain their original irregular shapes and internal composition.

Moreover, ongoing erosion will wear down the world's highest mountains in just a few hundred million years, which is just a fraction of the Earth's age of 4.6 billion years. The Rocky Mountains were, for example, once

Fig. 4.5 Planet Earth from space   As illustrated in this image of Africa, the Arabian Peninsula, and the Indian Ocean, the Earth's surface consists of continents and oceans. Continents cover a little more than one-quarter of the Earth's surface, while ocean water covers almost three-quarters of the surface. Our home planet has a thin atmosphere with white clouds of water ice, and enough transparency that you can usually look right through it. The Antarctica ice cap gleams white at the bottom. *Apollo 17* astronauts took this image in December 1972 as they left the Earth en route to the Moon. (Courtesy of NASA.)

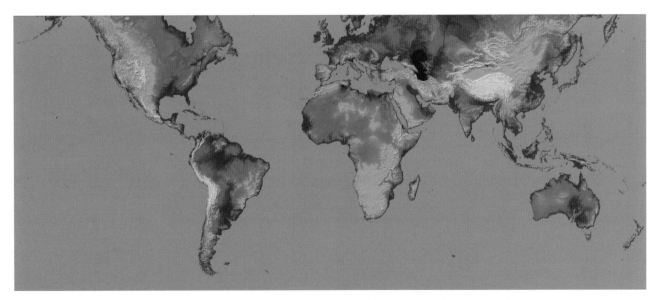

Fig. 4.6 Topography of the Earth's landmass   Global topographic image of land on the Earth is displayed by both colored height and shaded relief in this Mercator projection. Color-coding is directly related to topographic height, with green at the lower elevations, rising through yellow and tan, to white at the highest elevations. The shaded image displays topographic slopes in the northwest–southeast direction, with bright northwest slopes and dark southeast slopes. The elevation data were acquired from the Shuttle Radar Topography Mission (SRTM) aboard the *Space Shuttle Endeavor*, launched on 11 February 2000. (Courtesy of NASA/JPL/NIMA.)

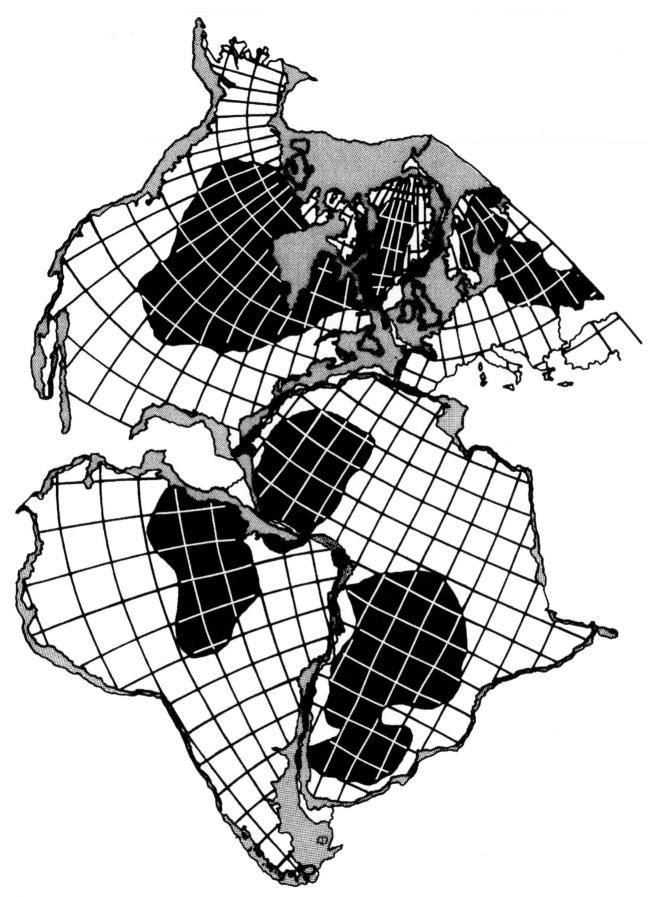

Fig. 4.7 Continental fit   The continents fit together like the pieces of a puzzle. Here the fit has been made along the continental slope at the depth of 910 meters, or 500 fathoms (*gray areas*). Within the present continents are ancient terrains between 1.7 and 3.8 billion years old (*black areas*). The close fit of the shorelines of the continents suggests that they once formed a single land mass known as Pangaea shown in Fig. 4.8.

twice as tall as they are today; they have been worn down through tens of millions of years of erosion by wind, rain and ice. If the planet were a perfectly smooth sphere, the oceans would cover the entire globe to a depth of 2.8 kilometers. So we can tell right away that high, dry land must be continuously recreated and pushed up out of the water.

There's water just about everywhere, but most of it is in the salty seas. Less than three percent of the Earth's water is fresh, and most of that is locked up in polar ice caps and glaciers. Lakes, rivers, and other sources of drinkable water make up less than one percent of the planet's total water. As time goes on, the demand for drinking water will increase, and the supply will diminish or even disappear in many dry places within a few decades or less.

Earth's drifting continents

The continents and oceans are not eternal, unchanging aspects of the Earth. Their appearance of permanence is an illusion caused by the brevity of the human lifespan. Just as an entire human lifetime is just a fleeting moment in the history of the Earth, today's map of the world is just a brief snapshot of its evolving, mobile, ever-changing surface. Over hundreds of millions of years, blocks of the Earth move about, producing drifting continents that completely alter our picture of the world.

The idea that continents have not always been fixed in their present positions was suggested more than three centuries ago, in 1596 by the Dutch mapmaker Abraham Ortelius (1527–1598) in his work *Thesaurus Geographicus*. However, the theory of moving continents was not developed into a thorough scientific hypothesis until the early 20th century, by the German meteorologist Alfred Wegener (1880–1930) in his influential and controversial book *Die Entstehung der Kontinente und Ozeane*, or *The Origin of Continents and Oceans*. Wegener noticed that the outlines of the continents exhibit a number of remarkable symmetries. For example, the eastern edge of South America would fit snugly into the western edge of Africa, a fit originally noticed by Ortelius. In fact, large parts of the east and west shores of the Atlantic are as well matched as the shores of a river (Fig. 4.7).

Wegener based his concept of continental drift not only on the similar shapes of the present continental edges, but also on the striking match of certain rocks and geologic formations, fossil creatures, and ancient climates along the borders of continents on opposite sides of the ocean. He concluded that all of the continents were once a part of a single landmass that fragmented and drifted apart (Fig. 4.8). If spacecraft had existed back then, their camera

eyes would have seen one large continent and a single ocean surrounding it.

This hypothetical super-continent is called *Pangaea*, a Greek word meaning "all lands" and pronounced *pan-gee-ah*. After all, if today's continents spread apart from their obvious puzzle fit, they had to have been together in the first place. This would also account for the Earth's curiously asymmetric face, in which the ocean waters dominate the southern hemisphere while the continents dominate the northern hemisphere.

Pangaea broke into pieces about 200 million years ago when large amphibians and reptiles ruled the land, leaving many fossils and forming the various smaller continents we see today. As the once-joined continents moved apart, the water rushed in to fill the gap caused by their separation. This led to the various smaller drifting continents that we see today.

Modern geologists have now pieced together the past, reconstructing the pieces of the Earth's moving jigsaw puzzle. They have shown that Boston and southern Florida are both former pieces of Africa, which were left behind when the Atlantic Ocean opened up. China used to be separated from Siberia by at least one ocean. Japan was once attached to Asia, and it may become part of Alaska in 800 million years.

Several super-continents formed and split before Pangaea. There was Rodinia, which covered much of the southern hemisphere 800 million years ago, and Gondwana that was located near the South Pole about 500 million years ago.

Today's continents will continue to move apart, and since the globe is round, all lands will eventually converge again. Thus, in about 250 million years, many of the continents will drift together and reposition themselves into another single, dominant landmass, forming a new Pangaea. And then, inevitably, another break-up will ensue as our restless planet continues to reform and reshape itself. In the process the continents will continue on their endless journey, forever roaming and wandering about the planet with no final destination.

Wegener's theory of continental drift was disparaged, ridiculed and even scorned by most geologists for at least half a century. In retrospect, their objections are hard to understand. The discovery of glacial deposits in Africa and of fossils of tropical plants, in the form of coal deposits, in Antarctica certainly meant that these two continents had once been located at different parts of the globe with climates much different from their present ones.

The main difficulty was understanding how the continents could move across the Earth and plow their way

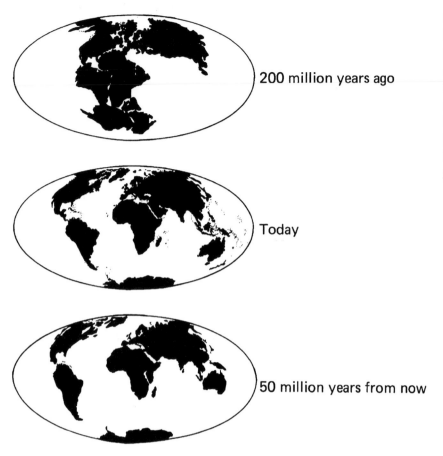

200 million years ago

Today

50 million years from now

Fig. 4.8 Earth's continental drift Two hundred million years ago all of the continents were grouped into a single super-continent called Pangaea and the world contained only one ocean (*top*). The continents then drifted away from Pangaea, riding on the back of plates to the positions they now occupy (*middle*). The bottom diagram depicts the world geography 50 million years from now.

through solid rock at the bottom of the ocean. A possible mechanism had been proposed by the Scottish geologist Arthur Holmes (1890–1965), who noticed that both the Earth's surface and interior could be in motion. Internal heat could drive churning motions that might propel the continents from below. But these prescient ideas, developed in the 1930s and revitalized in Holmes' 1944 classic *Principles of Physical Geology*, were not widely accepted. Exploration of the ocean floor by sound waves provided the first evidence that Wegener and Holmes were on the right track after all.

Sea-floor spreading

The bottom of the ocean is not flat. It contains underwater mountains and valleys that are as grand as those on any continent. Although we cannot see these features in the inky darkness of the deep sea, we can use sound waves to reach down and touch them. Their distance is determined by recording the time it takes for electrically generated sound signals, called pings, to travel from a ship to the floor and back.

The German Navy used such echo-sounding measurements to reveal the rugged sea-floor soon after World War I (1914–1918). They showed that a chain of submarine mountains runs right across the middle of the floor of the Atlantic Ocean. Known as the Mid-Atlantic Ridge, it extends about 2 kilometers above the adjacent sea-floor, which is at a depth of about 6 kilometers.

Nowadays, the United States Navy detects enemy submarines or ships with sonar, an acronym for sound navigation and ranging, transmitting a continuous train of pulsed sound waves and using the same echo technique to measure distance. Many modern ships, including warships and some commercial fishing boats, are also equipped with sonar to aid in navigation. Navy ships and research vessels can now use sonar to map a two-kilometer swath at the bottom of the ocean in a single ping of the sonar. Gravitational data, obtained from satellites that bounce radio beams off the sea surface (Focus 4.2), complement the sonar data, and they together result in highly detailed maps of the entire ocean floor. They have shown that the Mid-Atlantic Ridge is just a part of a global mid-ocean ridge that snakes its way across the bottom of the world's oceans (Fig. 4.9).

The global mid-ocean ridge is a gigantic network of underwater mountain ranges. The submerged mountains stand higher than the greatest peaks on land, and meander

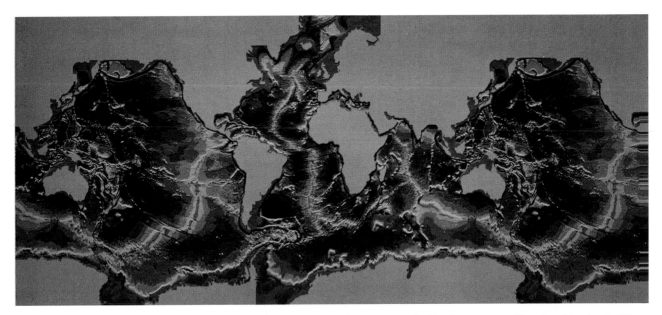

Fig. 4.9 Bottom of the Earth's oceans   This map of the world's ocean floors was acquired by the *Seasat* satellite. The Mid-Atlantic Ridge runs down the middle of the ocean floor separating Africa from North and South America. As shown here, a succession of great ridges runs through all of the world's ocean floors, although not always in the middle. (Courtesy of William F. Haxby, Lamont-Doherty Geophysical Observatory, Columbia University.)

Focus 4.2   Mapping the Earth's ocean floor from the top of the sea

The ocean depths have recently been charted with great accuracy by measuring the height of the sea surface. Through the action of gravity on water, seabed mountains produce swells at the surface and canyons or valleys produce dips. So the top of the ocean mimics the topography at its bottom.

A satellite is used to beam pulses of microwaves, or short radio waves, down at the ocean, and to determine the time for the reflected pulse to bounce back. Since the microwave pulse travels at the speed of light, the distance between the satellite and the top of the sea is half the product of that speed and the round-trip travel time. Because the separation between the satellite and the center of the

Earth is known from the satellite's orbit, one can establish the distance between the center and the top of the ocean by subtraction.

The topographical contours of the ocean top move up and down with features on the bottom, faithfully tracing out their highs and lows with an extraordinary precision of about one-tenth of a meter. This information helps submariners glide stealthily through the sea. Knowing the precise direction of gravity also improved the accuracy of guided missiles fired from the submarines.

With the cold war ending in the mid-1990s, the United States Navy declassified the data, obtained from the *Geosat* satellite, and it has been used with other satellite altimeter data, of the European Space Agency and NASA, to provide a full, detailed map of the sea-floor.

for more than 75 000 kilometers, creating the longest mountain chain on Earth. It is long enough to accommodate the total length of the Alps, Andes, Himalayas and Rockies. The mid-ocean ridge winds around the Earth, girdling the globe like the stitched seams of a baseball, not in simple lines but in offset segments. When the undersea mountains reach the surface they can form islands, like Iceland and its relatively new neighboring

island Surtsey, named after the Icelandic god of fire, Surtur (Fig. 4.10).

Even more remarkable are the deep canyons, collectively known as the Great Global Rift, that run along the mid-ocean ridges, splitting them as though they had been sliced with a giant's knife. Discovered in 1953 by the American scientists Maurice Ewing (1906–1974) and Bruce Heezen (1924–1977), the rift marks a line where

Fig. 4.10 Volcanic islands on Earth   Lava erupting from the volcanic island Surtsey on 19 August 1966, almost three years after it rose out of the sea near the coast of Iceland. The volcanic island of Jolnir is in the background. It disappeared back into the sea about one month after this picture was taken, but Surtsey is still visited for research purposes today. All of these volcanic islands, including Iceland, mark points where a mid-ocean ridge has risen out of the ocean. (Courtesy of Hjalmar R. Bardarson, Reykjavik, from his book *Ice and Fire*.)

much of the Earth's internal heat is released. It is filled with hot molten rock, or magma, coming up from inside the planet.

Amazing creatures live down there in the eternal darkness at the bottom of the sea, where life doesn't need the Sun. Giant clams, tubeworms and crabs are warmed and fed by the superheated water. They thrive without light by digesting sulfur minerals emitted from the hot vents, nutrients that other animals would find poisonous. Some of the heat-loving microbes breathe iron, and survive well above the boiling temperatures usually associated with sterilization.

The mid-ocean ridge is more accurately described as a tear in a sheet of paper rather than a cut, for it represents the place from which the ocean floor moves outward on both sides. It is as if the Earth was pulling itself apart and becoming unstitched.

Hot magma emerges from beneath the sea-floor, and oozes into the canyons of the Great Global Rift, filling them with lava. As the lava cools in the ocean water, it expands and pushes the ocean crust away from the ridge. More lava then fills the widening crack, creating new sea-floor that moves laterally away from the ridge on both sides, with bilateral symmetry.

If sea-floor is continuously created in the middle of the ocean, where does it go? If all of the new material kept on piling up, the Earth would grow bigger as time goes on, and that is not observed. The size of the Earth has not changed significantly during the past 600 million years, and very likely not since shortly after its formation 4.6 billion years ago. The Earth's unchanging size implies that the ocean floor must be destroyed at about the same rate as it is being created. The floor disappears back inside the Earth, where it is transformed by the heat and eventually recycles to rise again.

As it migrates away from the hot rift of its beginning, the new ocean floor grows colder and denser, subsiding to greater depths as it ages. After traveling across the Earth, in conveyer-belt fashion for many millions of years, the older, heavier floor bends and descends back into the Earth, often at the edges of continents, creating a deep-ocean trench in the underlying rock. Such trenches are found all around the edges of the Pacific Ocean, and they can sink as far below sea level as the tallest mountains rise above it.

The overall concept is known as sea-floor spreading, an idea introduced by the American geologist Harry H. Hess (1906–1969) in his 1962 paper entitled "History of Ocean

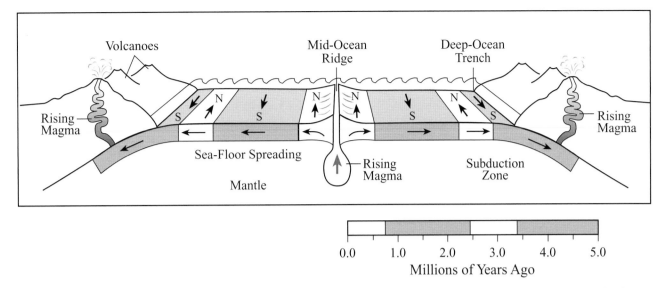

Fig. 4.11 Magnetic reversals and sea-floor spreading Radioactive dating of volcanic rocks on land have been used to determine the timescale of magnetic reversals on the Earth (*bottom*). They indicate that the Earth's magnetic field has flipped, or changed direction, several times during the past 5 million years. The data describe normal epochs (*white*) when compasses would have pointed toward the geographic north, as they do now, and reversed epochs (*gray*) when compasses would have pointed south. The pattern of magnetic reversals on both sides of the volcanic mid-ocean ridge (*top*) is the same, indicating that sea-floor spreading has carried the solidified lava away from the central ridge. The sea-floor is consumed at the other end, when it slides into a deep-ocean trench at a subduction zone.

Basins". In brief, new sea-floor is formed at a rift in the mid-ocean ridge, turning cold and heavy as it spreads away from its source in two directions; the sea-floor eventually sinks and disappears in a deep-ocean trench, where it is consumed. Its material is then recycled and born again as new floor emerges from the central ridge.

Sea-floor spreading accounts for the fact that the ocean floor and the sediments on it are both relatively young. Core samples recovered from the seabed during petroleum exploration show that the floor of the Atlantic Ocean is youngest in the middle, and grows progressively older with increasing distance from the mid-ocean ridge. Both the average age and the thickness of the sediment increase away from the ridge. Moreover, the thickness of the sediments indicates that none of them has been accumulating for more than 200 million years. The oldest fossils found down there are also no more than 200 million years old. From creation to extinction, from rift to trench, the ocean floor completely cleans house, erasing any record of its previous history in less than 200 million years.

In contrast, marine fossils in rock strata on land can be considerably older, including those found near the tops of the highest mountains, and the oldest known continental rocks have ages dating back to about 4 billion years. The sea most probably dates back to the formative stages of the young Earth, more than 4 billion years ago. Thus, young ocean floors have been replacing older ones, while the water above them has remained for billions of years.

Perhaps the most decisive evidence for sea-floor spreading was the discovery of regular magnetic-field patterns in the ocean floor. Magnetic detectors towed behind ships and carried in aircraft could measure very small differences in the Earth's magnetic field from place to place, known as magnetic anomalies. Positive magnetic anomalies are places where the magnetic field is stronger than expected, and negative ones are weaker than anticipated.

One of the motivations of these studies was to detect perturbations in the magnetism caused by enemy submarines, but the results had far greater consequences. The pattern of magnetic anomalies was symmetrically placed, or mirrored, on each side of the mid-ocean ridge (Fig. 4.11). Frederick Vines (1939–1988), then at Princeton University, and Drummond Matthews (1931–1997), working at Cambridge University, compared the magnetic irregularities with the known history of magnetic field reversals found on the flanks of volcanoes on land. These continental lavas show that every so often the Earth's magnetic field has changed its direction, or polarity, and the symmetric magnetic anomalies on the sea-floor exactly match these polarity reversals recorded on land (Fig. 4.11).

Each time the anomaly changes from positive to negative, the Earth's magnetic field turns upside down. Its magnetic poles flip, so the south magnetic pole switches from the north geographic pole to the south geographic pole or vice versa, and the north magnetic pole moves to the opposite geographic pole. Lava emerging at the present

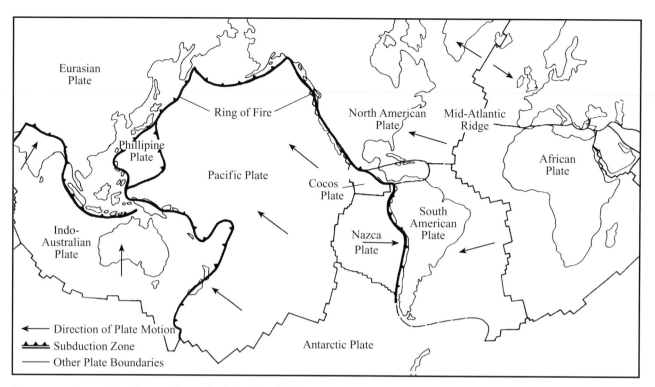

Fig. 4.12 Earth's moving plates   The Earth's lithosphere is broken into numerous plates. They move in the directions shown by the arrows at rates of about a tenth of a meter per year. The lithosphere dives into the underlying asthenosphere at zones of subduction. They are denoted by the thick line with triangles, forming the famous Ring of Fire around the edge of the Pacific and Nazca Plates. Most of the Earth's earthquake and continental volcanic activity is concentrated along the subduction zones.

time would have a positive magnetic anomaly, with the Earth's south magnetic pole located at the north geographic pole.

The orientation of the Earth's dipolar magnetic field imprints itself on the volcanic rocks at the time they form, whether on land or under the sea. When fresh molten lava pours out of them, magnetic minerals within the cooling lava become aligned with the Earth's magnetic field, and this orientation or polarity remains as a fossil magnetic record, locked into the rock when it solidifies. Vine and Matthews proposed that the lava on both sides of the mid-ocean rift solidified and moved away, freezing in the magnetic direction at the time, and when the Earth's poles flipped and reversed the lava flows preserved a set of parallel bands with opposite magnetic direction. Thus, the symmetric magnetic-anomaly stripes were recording the Earth's past magnetic field, and providing dramatic support for sea-floor spreading.

By radioactive dating of volcanic rocks on land, it is possible to tell when they solidified and to build up a chronology of the magnetic changes. This chronology can then put dates on the reversals found in the sea-floor, and from the distances traveled it is possible to compute the rate of sea-floor spreading, assuming that the floor has moved at a constant rate. These rates have been independently calibrated astronomically by comparison of the seabed sediments with the orbital parameters that govern climate changes recorded in fossil organisms in the sediments. They indicate that the sea-floor has indeed been spreading for the past 5 million years, moving away from the ridge at rates of 0.02 to 0.20 meters per year depending on the location, or just a little faster than your fingernails grow.

When sustained for 200 million years, the spreading sea-floor can push continents apart by between 400 and 4000 kilometers — entirely adequate to explain the widths of the great oceans. At the measured rate, it took just 150 million years for a slight fracture in an ancient former continent to widen into today's Atlantic Ocean.

Plate tectonics on Earth

The surface of the Earth is continually shaped and molded by plate tectonics, a process that does not occur on any other planet in the solar system. The rind of the Earth, its outer shell known as the lithosphere, is subdivided into a mosaic of large moving plates, each thousands of kilometers across (Fig. 4.12). They vaguely resemble the cracked pieces of an eggshell. The plates move horizontally atop a viscous layer of much hotter, softer, more malleable rock called the asthenosphere. Because of the high temperatures and immense pressures found there, the uppermost part of the asthenosphere flows along at the base of the lithosphere.

Plates are composed of the Earth's crust and the rigid upper mantle just beneath it. Most plates contain both continental and oceanic crust, and they all include oceanic crust. Six of the nine major plates are named for continents embedded in them: the North American, South American, Eurasian, African, Indo-Australian and Antarctic Plates. The other three are almost entirely oceanic: the Pacific, Nazca, and Cocos Plates. Accompanying them is a host of smaller plates.

Driven by heat from below, the plates move with respect to one another, accounting for most of our world's familiar surface features and phenomena, such as mountains, earthquakes and ocean basins. The continents are implanted within the moving plates, and continental drift is a consequence of the motion of plates carried along by the sea-floor spreading.

The rigid plates are in continual, relentless movement, and they deform at their boundaries. Like drops of olive oil gliding across a warm frying pan, the continents sometimes collide and coalesce, sometimes slide and rub against each other, and at other times break up and scatter. The transformations produced by these interactive motions are known as plate tectonics, from the Greek word *tectonic* for "carpenter or building". They are forever reconstructing the face of the Earth.

A mid-ocean ridge is a crack in the sea-floor that is filled in by magma from the underlying mantle as two diverging plates separate. So the ridge marks the boundary between two plates. As the plates move apart in opposite directions, a crack or rift opens up at the crest of the ridge, allowing more molten rock to move up and feed the spreading plates, like blood in an open wound that will not heal.

The spreading ocean floors are eventually pushed back down into the planet's hot interior, at subduction zones where two converging plates meet. When a moving oceanic plate encounters a light continental plate, or a younger, lighter oceanic one, the heavier oceanic plate plunges steeply into the Earth along a subduction zone, like a down-going escalator, producing a deep-ocean trench. Because the continental rock has the lowest density, it remains on the top, while the ocean floor slides underneath. The buried material is consumed within the Earth, only to re-emerge, recycled and transformed at some other location.

Magma is generated at the sinking subduction zones where dense oceanic plates are pushed under lighter continental ones, producing volcanoes that help us locate the edges of the plates. A dramatic example is the circular line of volcanoes that borders the Pacific Ocean. This active belt is known as the Ring of Fire because it is often the site of fiery volcanic eruptions (Fig. 4.12).

These volcanoes recycle atmospheric carbon dioxide, which was dissolved in rainwater and ingested by organisms in the sea. Their shells are deposited on the sea-floor, carried along by the diving oceanic plate, and released in the volcanic eruptions.

Scientists can track the plate motions using Very-Long-Baseline Interferometry (VLBI), and by Satellite Laser Ranging (SLR). In VLBI, radio receivers at widely separated telescopes record the strength and arrival times of cosmic radio signals from quasars, and the interference between the recorded data is used to determine the distance between the telescopes, with an accuracy of less than 0.01 meters. The SLR targets are satellites covered with tiny mirrors called corner cubes. Pulsed laser light, generated at stations on the ground, is bounced off the mirrors and the round-trip time for the light to return is used to establish the distances to the satellites. When combined with precise orbital information for the satellites, these distances can be used to determine the changing separation of the ground stations and the rate of tectonic plate motion.

The VLBI and SLR measurements indicate that the plates move laterally, or horizontally, across the Earth at rates of 0.02 to 0.20 meters per year depending on the plate, which is consistent with the rate of sea-floor spreading. When scientists extrapolate the current plate motions into the past, like running a movie backwards, they find that all of the continents converge, joining together into a single super-continent, Pangaea, about 200 million years ago. As suggested by the Canadian geophysicist J. Tuzo Wilson (1908–1993) in the 1970s, the heat-driven plate motions cause the continents to disperse and then reassemble over and over again, as oceans open and close.

The radio interferometer measurements indicate that the Pacific Plate is carrying Los Angeles northwest with respect to the plate that holds most of North America, at a velocity of 0.048 meters per year, while also producing earthquakes along the edge of the plate (Fig. 4.13). At this rate, Los Angeles will be a suburb of San Francisco in 10 million years. The interferometer observations also indicate that the Atlantic Ocean is widening by 0.017 meters per year, so it was 8.7 meters narrower when Columbus crossed it in 1492.

Earthquakes

An earthquake is a trembling or shaking of the ground caused by a sudden release of energy stored in the rocks below the Earth's surface. The devastating tremors and after-shocks can ravage large sections of the land, flattening

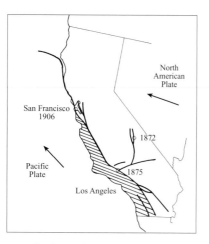

Fig. 4.13 Los Angeles is moving   The Pacific Plate is moving northwestward at the rate of 0.048 meters per year, or about 5 meters every century, slowly carrying Los Angeles towards San Francisco. The North American and Pacific Plates strike and rub against each other like immense grindstones, producing earthquakes along their boundary known as the San Andreas Fault. The dated circles denote places where very major earthquakes have occurred with magnitudes of 8 and over on the Richter scale.

entire cities, awakening dormant volcanoes and creating new ones, draining lakes and causing floods, avalanches and fires. Enormous tidal waves, called tsunami, can be generated by the quakes, racing across the ocean at high speed and wiping out everything on the shores they reach. Tens of thousands of people can be killed during a major earthquake, and hundreds of thousands more can be left homeless. Earthquakes are erratic, unpredictable and inevitable, with the power to merely shake you or completely destroy you.

Like volcanoes, the world's earthquakes do not occur just anywhere, but usually along the edges of plates. They occur most often where the ocean floor is being either created or destroyed along the mid-ocean ridges and the deep-ocean trenches. Since an earthquake in the middle of the ocean floor is not likely to disrupt human life, we are naturally most interested in earthquakes that occur near the edges of continents where cities are located.

In addition to diverging plates, which are moving apart at a mid-ocean ridge, and convergent plates, which are heading toward a collision, there is a third type of plate boundary proposed in 1965 by J. Tuzo Wilson. It is known as a transform fault, a place where plates move past one another, neither toward nor away, and this is where earthquakes can occur. When the two plates meet along a transform fault, they "transform" their encounter into a slipping, sliding horizontal motion, and a sudden lurch in this motion can produce an earthquake.

The two plates on each side of a transform fault bump, crush, grind, rub and slide against each other, without creating or destroying crust, like two high-speed cars sideswiping each other, but in slow motion. A famous and visible example is the San Andreas Fault in California that marks the meeting of the Pacific Plate with the North American Plate. In 1906 a great earthquake devastated San Francisco, which is located at the edge of this fault.

The plates on each side of a transform fault build up stress along the line where they meet, and the stress is greatest where they are most tightly locked. As the friction and strain accumulate and rise over the years, a moment comes when the rock can't take it anymore, as when a festering problem of a family can surface into a screaming fight. In effect, the strain surpasses the strength of the rock. The stress is pushed to the limit, the two plates cannot slide further, and the accumulated energy is released as an earthquake. That part of the fault line then lurches back to its original equilibrium position, waiting for the next big one.

Since the time between major earthquakes in a given location can be a little longer than a human lifetime, many imagine that the danger is over, but the ground beneath their feet can remain unstable. Thus people living near the San Andreas Fault should no longer be concerned about whether another earthquake will occur, but when it will happen.

An instrument called a seismograph can measure the relative amount of energy released by an earthquake, its magnitude. The earthquake magnitude is given on a numerical scale, named after the American seismologist Charles F. Richter (1913–1984) who established it. Each increase of one unit on the Richter scale represents a 32-fold increase in the intensity of an earthquake. Major earthquakes usually measure between 6.0 and 9.1 (the highest ever recorded) on the Richter scale. The 1906 San Francisco earthquake would have measured 8.3 on the Richter scale, and the one that occurred there in 1989 measured 6.9.

An even more powerful earthquake of magnitude 9.1 struck off the west coast of northern Sumatra on 26 December 2004, producing devastating tsunamis in the Indian Ocean, sending shock waves that rippled and ricocheted around the globe, and lifted the Earth's surface by a couple of centimeters halfway around the world. A strong earthquake of magnitude 7.0 hit Port-au-Prince, Haiti, on 12 January 2010 with devastating effects, killing more than 200 000 people and significantly damaging or destroying major buildings.

An earthquake of magnitude 8.8 occurred near the shores of Chile on 27 February 2010. Although loss of life and building destruction were not as great as the Haiti earthquake, the one in Chile shook the entire planet, shifting the Earth's axis by about 8 centimeters and shortening the length of an Earth day by about 1.26 microseconds, or 1.26 millionths of a second. The Sumatran earthquake should have shortened the length of day by 6.8 microseconds and shifted the Earth's axis by about 7 centimeters.

An international network established in the late 1990s monitors the globe for clandestine nuclear bomb blasts, and incidentally records the seismic din of earthquakes. Scientists can sometimes distinguish between the two by the form of their shock waves. A nuclear explosion begins with a sharp spike, while an earthquake starts with a gentle shaking that subsequently becomes more violent.

The Earth's internal heat engine

What pushes the tectonic plates across the globe? Like humans, most of the driving forces that transform the Earth's face are hidden below its surface. Or, as the saying goes, it's what inside that counts. Heat, bottled up deep inside the Earth, produces internal currents that move the plates and propel the drifting continents.

The Earth's internal heat is left over from the time of the planet's formation, within the liquid outer core, and augmented by the continued radioactive decay of elements such as uranium and thorium. As the internal heat tries to escape, it maintains a ceaseless, wheeling, churning and roiling motion, called convection, which turns and rolls over very slowly. Convection occurs when molten rock becomes swollen by heat and rises through the cooler overlying material of lower pressure, like the currents in a pot of thick soup or oatmeal about to boil.

The relatively low density of the hottest rock makes the material buoyant, so it ascends slowly; in contrast, the colder, denser rock sinks until heat escaping from the molten core warms it enough to make it rise again. Thus, the hot mantle rock flows in a circular pattern with hot rock rising in some places and cooler rock descending in others. The heated rock moves upward, spreads sideways, dragging portions of the lithosphere with it, and then cools and sinks, to be reheated and pushed upward again. And the crust rides passively atop these giant convection cells, like dirt on a conveyor belt.

Powerful motions deep inside the planet do not only push the plates horizontally; they also produce vertical changes at the surface. After all, the wheeling convection moves up and down, as well as sideways. Huge rising plumes of hot, buoyant, molten rock, originating and channeled at the rugged core–mantle boundary, can expand upward, piercing the mantle to lift and lower entire continents. The rising heat is now pushing South Africa up from below, and has been doing so for the past 100 million years.

The continents are poor conductors of heat, and therefore act like an insulating blanket that tries to block the heat's escape. The pent-up force of the trapped heat can become powerful enough to split a continent apart. A gigantic plume of hot magma may even have played a major role in the break-up of Pangaea.

Thus, the energy that drives the continents, spreads the sea-floor, sets off earthquakes and ignites volcanoes is ultimately derived from the hot interior of the Earth. From the inner core all the way to the surface of the Earth, the dynamical activity of the Earth is driven by heat, and like any heat engine the Earth must be gradually running down. When the Earth's internal heat becomes totally depleted, it will become a geologically dead planet, and erosion will gradually flatten the mountains. In the meantime, continents continue to be renewed as the result of the Earth's internal heat engine.

The Earth's impermanent face

Moving plates provide the tools for sculpting the Earth's surface and altering its landscape. They have profoundly changed the way we view the world. Its entire surface is continuously shifting about and changing in shape and form. It is something like a rich old lady who keeps going in for face-lifts, in a futile attempt to resurrect her youth. Indeed, nothing on the Earth's face resembles itself as it was even several millions of years ago.

As an ocean plate disappears into the trenches, great chains of volcanoes are created along the margins of continents. The descending slab of lithosphere causes underground rock to melt, and the magma generated rises buoyantly to widen the continents at their edges. The Andes are still growing higher in this way, as the floor of the Pacific Ocean plunges beneath the west coast of South America.

The Pacific Ocean once reached to Colorado, and the western United States has been grafted onto the continent. For most of the world's history, the land we call California did not exist. Where California has come to be there was only the deep blue sea reaching down to the spreading ocean floor. But the floor was moving into the Earth and being consumed under the ancient shoreline, creating volcanoes that rose to create new land.

On average, about 2 billion cubic meters of lava and ash are now being added to the continents by volcanoes each year. The rising material also brings valuable metals and precious stones to the surface. All the gold in California originated in this way, as did the famous copper deposits in *Cyprus*, the Greek word for "copper", rising with volcanic magma and spewing out on the surface with the lava flows.

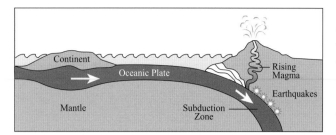

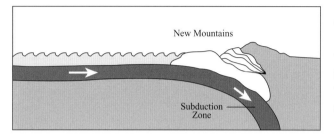

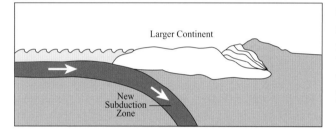

Fig. 4.14 Converging plates on Earth   Magma, volcanoes and earthquakes are generated at a subduction zone (*top*) where a dense oceanic plate is pushed under a lighter continental plate. When continents on two moving plates meet head on, new mountains are generated (*center*). In some situations, the advancing plate may become disrupted and the plate motion may stop. The two continents then become welded together forming a larger one, and a new subduction zone can be formed elsewhere (*bottom*).

Diamonds were also forged in the crucible of the Earth's hot interior. The crushing pressures and blistering heat far within the mantle worked in unison to squeeze carbon into diamonds, and some of them have risen from the deep, entrained in explosive volcanic eruptions, even in the middle of continents.

Eventually, a moving continent reaches an open trench and jams it shut (Fig. 4.14), like trying to shove an eggplant down a garbage disposal. Continents are too light and thick to be subducted, and when they arrive at a trench the suture is closed up.

When two continents meet, they buckle upward to form a range of mountains and help to hold the land above the sea. Both land and oceanic sediment, built up over many millions of years, are tossed into the sky. The magnificent Himalayan range was formed this way (Fig. 4.15), when the Indo-Australian plate, with India firmly embedded, ran into the Eurasian plate, like a head-on collision of two cars.

Slowly, the Himalayas shot up as India rammed into Asia, carrying the fossilized remains of ancient sea-creatures with them. Today the plate that carries India continues to slide beneath the Eurasian plate, widening the Indian Ocean and pushing the mighty Himalayas upward.

The European Alps have been fashioned in a similar way to the Himalayas, when the African Plate moved Italy up from Africa and collided with the Eurasian plate along Switzerland's former ocean shore. Today the African Plate continues pressing Italy northward and raising the Alps.

Like adolescents, the relatively young mountains in the Alps and Himalayas just can't stop growing, but there is a limit to how high they can stand. Gravity opposes the upward forces, and erosion wears away mountain summits as they are being pushed up from below.

Erosion provides the second major force, in addition to plate tectonics, for sculpting the Earth's face. As massive as it is, a range of mountains cannot resist eventual destruction by the erosion of wind, rain, flowing water, and ice. Old mountain ranges, such as the Appalachians in the United States, once stood as high as today's Himalayas, but they have eroded into gentle undulations and rounded knobs.

This erosion acting by itself would have worn away the continental mountains in a few hundred million years, and oceans would therefore have long ago covered the entire globe. But the mountains are constantly being rebuilt while enlarging the continents. Many of the world's present-day continents have indeed been assembled as former continents welded together.

The edges of plates are not the only place that land rises above the sea. Some oceanic islands are located thousands of kilometers from the nearest plate boundaries. Chains or strings of such isolated islands are attributed to hot spots, an idea introduced by the prolific J. Tuzo Wilson back in 1963. The hot spots are rising plumes of magma, or molten rock, anchored far beneath the ocean and deep within the mantle, even as far down as the core–mantle boundary. The relatively small, long-lasting and exceptionally hot regions provide a persistent source of magma capable of penetrating the mantle and piercing an overriding lithosphere plate, like a fixed blowtorch might melt holes in a steel plate moving by. As the plate glides slowly overhead at the rate of a few meters every century, it can leave a trail of islands that have risen out of the sea.

The Hawaiian Islands were formed in such a way, as the Pacific Plate moved over a deep, stationary hot spot at the slow rate of 0.13 meters per year (Fig. 4.16). Stretching to the north and west of the big island of Hawaii, they form a string of smaller islands, including Oahu and Midway, and submerged volcanoes, or seamounts, altogether extending about 6000 kilometers in length. Every

Fig. 4.15 Mount Kailash   This mountain is sacred to both Hindus and Buddhists. Tibetans know it as Kang Rinpoche (The Precious Snow Mountain). It was formed by the collision of two former continents, now welded together in a seam known as the Himalayan mountain range. Buddhists consider it to be the palace Chakasamvara (Wheel of Supreme Bliss) and Hindus consider it to be the dwelling place of Shiva. There are also many nearby caves where famous hermits like Milarepa meditated for years. Pilgrims have been visiting and circumambulating Mount Kailash, in western Tibet, for thousands of years. (Courtesy of Matthieu Ricard, Shechen Monastery, Katmandu, Nepal.)

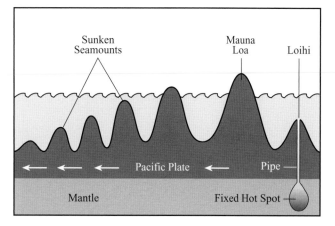

Fig. 4.16 Hot spot forms the Hawaiian Islands   A hot spot that is anchored deep within the Earth has recently fed molten lava through a long pipe to Mauna Loa on the big island of Hawaii. The moving Pacific Plate has carried three other volcanic islands away from the hot spot. As the plate moves on, wind and water erode the peaks, reducing these now-extinct volcanoes to sunken islands known as seamounts. An underwater volcano, named Loihi, is now forming over the hot spot; it should rise above the ocean to become another Hawaiian island in about 50 000 years.

one of these islands and seamounts was formed in the exact place where Hawaii now stands. The plume pushed the first Hawaiian island up above the ocean surface in this location about 70 million years ago.

Kiluea, the world's largest active volcano, is still rumbling because Hawaii has yet to move completely off the hot spot. At the same time, the underwater volcano, Loihi, is being formed as the Pacific Plate moves steadily on, continuing its relentless journey over the hot spot (Fig. 4.16). In about 50 000 years Loihi should grow high enough to form the next Hawaiian island.

But why aren't the Hawaiian Islands just one long extended island, eroded away on the oldest end and standing tallest at the youngest end? Although the lower parts of the hot-spot plumes are shaped like the thin stem of a wine glass, their tops flare out into mushroom-shaped reservoirs of molten rock that pools under the lithosphere and overriding crust, slowly melting them and gathering enough strength for penetration. It takes thousands of years to break on through to the other side, just as a welder's torch takes a while to burst through a steel plate. The hot rock breaks though the lithosphere and overlying ocean floor sporadically rather than in a continuous stream, forming a succession of oceanic islands.

Most of the hot spots lie under oceans and give birth to island chains, but some of them penetrate the mantle under the continents. Such a hot spot has created the hot springs, boiling mud and geysers of the Yellowstone National Park in Wyoming. As the North American Plate

moved above this hot spot, it created a long line of volcanoes that are now extinct, with ages ranging from 0.6 to 6 million years old. A semi-dormant volcano now rests under Yellowstone.

If a plate carrying a continent comes to rest over a hot spot, the heat and pressure from the upwelling magma will weaken and stretch the overlying material. And when the continental crust is stretched beyond its limits, cracks or rifts will form in it. The magma rises and squeezes through the widening cracks, forming volcanoes. If the upwelling is short-lived, the result is merely a rift scar, such as the Rhine Valley. If it persists, the rift widens and a continent can literally be split in two. In time, the gap reaches a coastline, permitting seawater to flow in and a new ocean is created (Fig. 4.17).

Hot spots are now tearing Africa apart at its seams. A Great Rift Valley stretches from Ethiopia to Tanzania; as it widens the continent will break apart and the sea will eventually enter. The African and Arabian Plates have already pulled apart in another location, forming the Red Sea and the Gulf of Aden. They are developing into an ocean that may eventually rival the Atlantic Ocean in size (Fig. 4.18). At the same time, the Mediterranean Sea is narrowing as Africa moves toward Europe.

Thus a new dynamic picture of the Earth has emerged. Continents are growing in size by accumulating volcanic material at their edges, when oceanic plates plunge under them, or by colliding continental plates that can weld together and create some of the world's largest mountains, but wind, rain, flowing water and ice inevitably cut them down to size. Earthquakes are bringing vast destruction as two plates grind together; strings of volcanic islands are rising from the ocean's depths; converging continents are squeezing oceans out of existence while new ones open up where continents are splitting apart; and the ocean floor remains in eternal youth as new floor spills out of the mid-ocean ridges and old floor is pushed back into the Earth. As we shall next see, our atmosphere is also in a perpetual state of flux, forever changing in sometimes dangerous ways.

4.4   The Earth's changing atmosphere

Our Sun-layered atmosphere

Our thin atmosphere is pulled close to the Earth by its gravity, and suspended above the ground by molecular motion. Its height is about 1 percent of the planet's diameter. Compared to the size of the Earth, the thickness of the air is something like the width of a window on a big building.

Because air molecules are mainly far apart, our atmosphere is mostly empty space, and it can always be

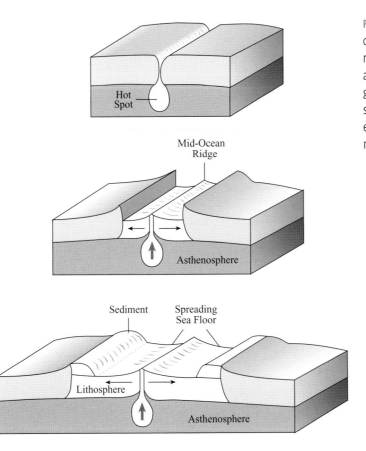

Fig. 4.17 The rifting of a continent   A continental rift begins when molten lava rises up from deep in the Earth's interior and splits a continent open. As the fissure grows and widens, a future ocean floor spreads away from the ridge. Water should eventually flow into the cavity, making a new ocean.

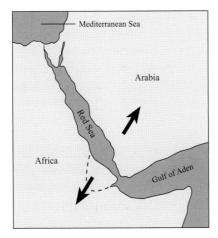

Fig. 4.18 An infant ocean   An ocean is being born where the Arabian peninsula and the African continent are moving apart, a process that began about 20 million years ago. In a few hundred million years, the Red Sea could be as wide as the Atlantic Ocean is now.

Not only does the atmospheric pressure decrease as we go upward; the temperature of the air also changes. It decreases steadily with increasing height in the lowest region of our atmosphere, called the troposphere from the Greek *tropo* for "turning". The troposphere extends from the Earth's surface to about 12 kilometers above sea level.

All of the Earth's weather takes place in the troposphere, a thin planetary skin just one-thousandth the diameter of the Earth. Yet about 80 percent of the atmosphere's total mass and almost all its water are concentrated in the troposphere.

The temperature falls at increasing heights in the troposphere because this layer of our atmosphere is heated from the warm ground below, and because the air expands in the lower pressure at higher altitudes and becomes cooler. The average air temperature drops below the freezing point of water, at 273 kelvin, about 1 kilometer above the Earth's surface, and bottoms out at roughly 10 times this height.

But the temperature is not a simple fall-off with height. It falls and rises in two full cycles as we move off into space (Fig. 4.19). The temperature increases are produced by the Sun's invisible radiation.

Different types of radiation differ in their wavelength, though they propagate at the same constant speed — the velocity of light. They range from long radio waves, about

squeezed into a smaller volume. The atmosphere near the ground is compacted to its greatest density and pressure by the weight of the overlying air. At greater heights there is less air pushing down from above, so the compression is less and the density and pressure of the air falls off into the near vacuum of space.

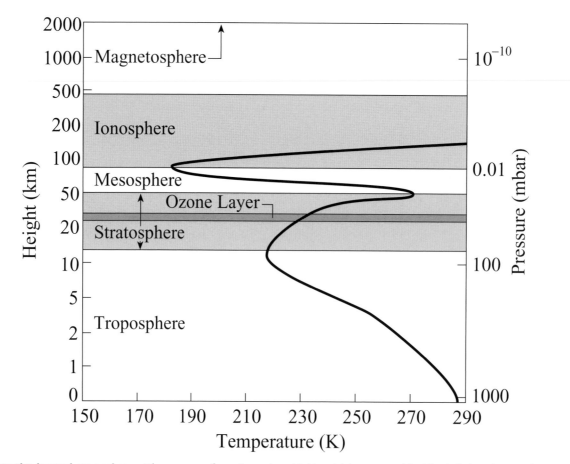

Fig. 4.19 Earth's layered atmosphere   The pressure of our atmosphere (*right scale*) decreases with altitude (*left scale*; note that both the pressure and altitude scales are logarithmic). This is because fewer particles are able to overcome the Earth's gravitational pull and reach higher altitudes. The temperature (*bottom scale*) also decreases steadily with height in the ground-hugging troposphere, but the temperature increases in two higher regions that are heated by the Sun. They are the stratosphere, with its critical ozone layer, and the ionosphere. The stratosphere is mainly heated by ultraviolet radiation from the Sun, and the ionosphere is created and modulated by the Sun's X-ray and extreme ultraviolet radiation.

a meter in length, to short X-rays whose wavelengths are roughly a billionth, or 10^{-9}, of a meter. The radiation at most of these wavelengths goes unseen by humans. Our eyes are only sensitive to a narrow range of visible colors, with wavelengths between 4 ten-millionths and 7 ten-millionths, or 4×10^{-7} and 7×10^{-7}, meters. Ultraviolet radiation is on the short-wavelength side of blue light, with wavelengths of about 2 ten-millionths, or 2×10^{-7}, meters. The wavelengths of X-rays are about a hundred times shorter than the ultraviolet rays.

The most intense radiation from the Sun is emitted at visible wavelengths, and our atmosphere permits it to reach the ground. That is the colored sunlight that our eyes respond to. The Sun emits lesser amounts of invisible, short-wavelength radiation, which is partially or totally absorbed in the atmosphere.

Even though the total amount of invisible solar radiation is substantially less than the visible emission, the individual short-wavelength rays are more energetic. That is why we get sunburns from the ultraviolet radiation that manages to get through the atmosphere, and why we need to be protected from the Sun when climbing at high altitudes where the air is thinner and more ultraviolet penetrates the atmosphere. The greater energy of radiation at shorter wavelengths also explains why X-rays, generated by machines, can see through your skin and muscles to detect your bones.

When absorbed in our air, the invisible short-wavelength radiation from the Sun transfers its energy to the atoms and molecules there, causing the temperature to rise. There is, for example, a gradual increase in temperature just above the troposphere, within the next atmospheric layer named the stratosphere (Fig. 4.19). This layer is located between 10 and 50 kilometers above the Earth's surface. Its name is coined from the words "stratum" and "sphere".

The mesosphere, from the Greek *meso* for "intermediate", lies just above the stratosphere. The temperature declines rapidly with increasing height in the mesosphere, reaching the lowest levels in the entire atmosphere.

The temperature then begins to rise again with altitude in the ionosphere, a permanent spherical shell of electrons and ions, reaching temperatures that are hotter than the ground. The ionosphere is created and heated by absorbing the extreme ultraviolet and X-ray portions of the Sun's energy. This radiation tears electrons off the atoms and molecules in the upper atmosphere, thereby creating ions and free electrons that are not attached to atoms.

The ionosphere was postulated in 1902 to explain Guglielmo Marconi's (1874–1937) transatlantic radio communications. Since radio waves travel in straight lines and cannot pass through the solid Earth, they get around the planet's curvature by reflection from electrons in the ionosphere.

The Sun's invisible rays make the Earth's ozone layer

Solar ultraviolet radiation is largely absorbed in the cold and barren stratosphere, where it helps make ozone. When ultraviolet rays strike a molecule of the ordinary diatomic oxygen that we breathe, denoted by O_2, they split it into its two component oxygen atoms, or two O. Some of the freed oxygen atoms then bump into, and become attached with, an oxygen molecule, creating an ozone molecule, abbreviated O_3, that has three oxygen atoms instead of two. The Sun's ultraviolet rays thereby produce a globe-circling layer of ozone in the stratosphere.

Solar ultraviolet radiation also heats the statosphere, making the molecules of ozone move faster. The main reason for the decreasing temperatures in the overlying mesosphere is the falling ozone concentration and decreased absorption of solar ultraviolet.

Although the ozone is present to the extent of only about 10 parts per million, the ozone layer is critical to life below. It protects us by absorbing most of the Sun's ultraviolet emission and keeping its destructive rays from reaching the ground. If there were no ozone shield, plants, animals and humans could not even exist on land.

The amount of ozone in the stratosphere resembles the level of water in a leaky bucket. When water is poured into the bucket, it rises until the amount of water poured in each minute equals the amount leaking out. A steady state has then been reached, and the amount of water in the bucket stops rising. It will stay at the same level as long as you keep pouring water in at the same rate. However, if you pour the water in at a different rate, or punch a few more holes in the bucket, the steady-state level of water in the bucket changes.

Solar ultraviolet radiation supplies ozone to the stratosphere from above, like pouring water into a bucket, at a rate that varies slightly with the changing ultraviolet output of the Sun. We have recently been punching holes in the ozone layer from below, with chemicals used in everyday lives.

Synthetic chemicals are destroying the Earth's ozone layer

Man-made chemicals, called chloroflurocarbons, are consuming the protective ozone layer, eating holes in it and making it thinner. They are synthetic chemicals, entirely of human origin with no counterparts in nature. The name of the chemicals is a giveaway to their composition. Each molecule has been constructed in company laboratories by linking atoms of chlorine, fluorine and carbon.

The shorthand CFC notation abbreviates some of them. A number sometimes follows, providing a complex description of the number of atoms in each molecule, the most widely used being CFC-11 and CFC-12.

Beginning in 1930, the biggest producer of CFCs, the Du Pont Company, manufactured and marketed them under the name Freons. They have been widely used in refrigerators, plastic foams, spray-can propellants, automobile air-conditioning systems, and the cleaning of circuit boards used in televisions and computers.

The hardy substances don't interact chemically to form other ones. They are so inert and stable that once entering the atmosphere the CFC molecules can survive for more than a century, permitting them to drift and waft up into the ozone layer in the stratosphere. Although more than 20 million tons of CFCs have been released into the air, their combined concentration isn't very significant, only about one CFC molecule for every two billion molecules in the air. Yet even these seemingly insignificant amounts can have enormous impact.

In 1974, Mario J. Molina (1943–) and F. Sherwood Rowland (1927–), then at the University of California at Irvine, showed that the chlorine in the CFCs could destroy enormous amounts of ozone. Once arriving in the stratosphere, the Sun's ultraviolet rays will split chlorine atoms out of the CFCs, and the liberated chlorine sets off a self-sustaining chain reaction that destroys the ozone. A single chlorine atom will react with an ozone molecule, taking one oxygen atom to form chlorine monoxide; the ozone is thereby returned to a normal oxygen molecule. Moreover, when the chlorine monoxide encounters a free oxygen atom, the chlorine is set free to strike again. Each

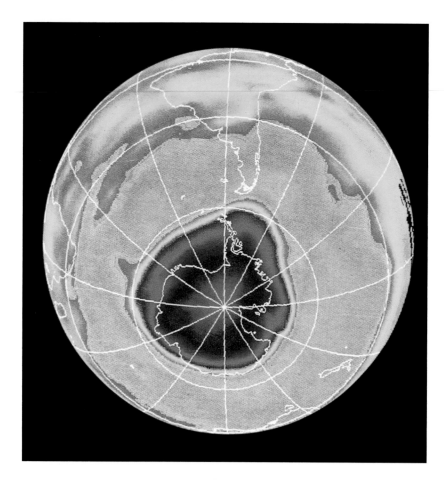

Fig. 4.20 Hole in the sky A satellite map showing an exceptionally low concentration of ozone, called the ozone hole, which forms above the South Pole in the polar spring (September–November). In October 1990 it had an area larger than the Antarctic continent, shown in outline below the hole. Eventually spring warming breaks up the polar vortex and disperses the ozone-poor air over the rest of the planet. (Courtesy of NASA.)

chlorine atom thus acts as a catalyst, destroying about 10 000 ozone molecules before it finally combines permanently with hydrogen in the air.

Molina and Rowland were awarded the Nobel Prize in Chemistry in 1995 for their "contribution to our salvation from a global environmental problem that could have catastrophic consequences". They shared the prize with the German chemist Paul Crutzen (1933–), who showed how the rate of ozone depletion could be accelerated by other chemical reactions in the atmosphere.

The ozone layer is itself invisible. But you can determine its ozone content by measuring the amount of solar ultraviolet radiation getting through the layer and reaching the ground. When there is more ozone, greater amounts of ultraviolet are absorbed in the stratosphere and less reaches the ground, and when the ozone layer is depleted, more of the Sun's ultraviolet rays strike the Earth's surface.

The British scientist G. M. B. (Gordon Miller Bourne) Dobson (1889–1976) pioneered measurements of the air's ozone content more than half a century ago. When his instrument was installed at Halley Bay, Antarctica, in 1957–58, Dobson found that the ozone abundance in polar spring (September–November) was noticeably less than that above other parts of the world.

Other British scientists continuously monitored the southern polar skies for 27 years; always detecting a springtime loss that became steadily larger as the years went on. By 1985 the ozone loss above Antarctica had nearly doubled when compared to the earlier measurements in the 1960s, and it extended all the way to the tip of South America, where another British monitoring station detected the ozone depletion. A continent-sized hole had opened up in the sky – the ozone hole (Fig. 4.20).

This unexpected discovery astounded space-age scientists who had not detected any ozone hole using satellites that had been monitoring the ozone layer from above. Their computers had been programmed to automatically reject large ozone depletions, apparently because their models did not predict such huge losses. So the now-famous ozone hole had been discarded as an anomaly, perhaps caused by an instrumental error. After reanalyzing the satellite data, the scientists confirmed the existence of an ozone hole in the local springtime above the South Pole.

We now know that strong winds concentrate ozone-destroying chemicals, the CFCs, within a vast towering vortex above Antarctica, resembling the eye of an immense hurricane. Each year the gaping hole opens up during Antarctic spring when the sunlight triggers ozone-destroying chemical reactions; the hole starts to close up in the early polar fall when the long sunless winter begins. Ozone-depleted air is dispersed globally, and the ozone is slowly restored, filling the hole until the cycle repeats in the following year.

Doing something about ozone depletion

The sudden and frightening discovery of an enormous ozone hole in 1985 sparked public awareness of the fragile ozone layer. In the meantime, the scientific community had been actively investigating Molina and Rowland's theory that the CFCs could be destroying the ozone layer. Although global models of the expected ozone depletion initially led to widely varying estimates of the potential threat, affecting the scientists' credibility and dampening public concern, a coordinated international investigation eventually led to a unified assessment of the problem.

A group of approximately 150 scientific experts reported in 1986 that atmospheric accumulations of CFC-11 and CFC-12 had nearly doubled from 1975 to 1985. The continued release of the synthetic chemicals at the 1980 rate could, they said, deplete the ozone layer by about 9 percent on a global average by the last half of the 21st century, with even greater seasonal and latitudinal differences. As a result, higher levels of dangerous ultraviolet radiation could reach heavily populated regions of the northern hemisphere.

Who cares if chemicals are punching a few holes in the sky and letting a little more sunlight reach the ground? The United States Environmental Protection Agency (EPA) cared. In 1986 it published a report of the many serious consequences of ozone depletion. A thinner ozone layer lets more solar ultraviolet radiation through to the ground, where it can produce severe biological harm. The most energetic ultraviolet rays will reduce the effectiveness of the human immune system, increasing human vulnerability to infections and cancer.

The EPA estimated that there could be over 150 million new cases of skin cancer in the United States alone among people currently alive or born by the year 2075, resulting in over 3 million deaths. The dangerous ultraviolet would also produce eye cataracts, distorting the vision of about 18 million people in the same population and blinding many of them. Added to this was the potential of widespread genetic damage to crops and forests, if nothing was done to stem the production of ozone-destroying chemicals.

Faced with the evidence of vanishing ozone, the global increase of atmospheric CFCs, and the prospect of widespread skin cancer and eye cataracts, international diplomats forged an accord in 1987 to limit and eventually ban the production of the substances that deplete the ozone layer. The treaty, known as the *Montreal Protocol*, has led to substantial reductions in ozone destroyers.

The *Montreal Protocol* was the first international agreement to protect the global environment. The treaty also marked the first time that the governments of the industrial nations agreed to help developing countries with environmentally safe substances and technology. It was further hoped that the precedent would pave the way for international agreement on global warming.

The *Montreal Protocal* was gradually strengthened over the years. A variety of ozone-depleting substances were added to those already banned, as amendments to the initial agreement made at meetings held in London, Copenhagen, Montreal and Beijing in 1992, 1994, 1999 and 2002, respectively, and the number of participating countries grew to nearly 200, including the vast majority of both the producers and consumers of the dangerous substances. This rapid and comprehensive accord was undoubtedly eased by the development of substitutes for CFCs in refrigerators, air conditioners, foaming, and cleaning solvents. In fact, the biggest producer, Du Pont, unilaterally stopped making the chemicals even before the *Protocol* required it.

Although production of ozone-destroying substances has been substantially curtailed under international agreement, more than 20 million tons of them have already been dumped into the atmosphere, and this damage cannot be undone. Because of their long lifetime and slow diffusion into the stratosphere, the synthetic chemicals that are already in the air will keep on destroying the ozone layer for about a century and full recovery will probably not occur until about 2070.

Scientists continue to use satellites to monitor the circulation, composition and temperature of the stratosphere, while also keeping a close eye on the Sun's varying ultraviolet output, which modulates ozone production. Computer models that forecast the future of ozone depletion take into account the Sun's varying ultraviolet radiation, and current measurements of the amounts of different ozone-destroying substances in the stratosphere, including the identification of new gases that might dominate ozone depletion, such a nitrous oxide. In 2006, for example, an analysis of 25 years of ozone observations at different altitudes in the stratosphere indicated that the Earth's protective ozone layer outside of polar regions stopped thinning about 10 years after the *Montreal Protocol* was first enacted.

The computer projections indicate that the Earth's ozone hole ought to eventually close, but it is going to take a while. The models suggest that the ozone hole will remain open longer than initially expected, forecasting that it will not substantially shrink until about 2020. An uneven ozone recovery is now anticipated due to the way global warming is changing the circulation of stratospheric air masses from the tropics to the poles, with over-recovery at mid-latitudes and depletion remaining over the tropics beyond 2100. This therefore brings us to the related scientific, and political, topic of global warming by heat-trapping gases dumped into the atmosphere by humans.

Global warming by human emission of heat-trapping, "greenhouse" gases

Our planet's surface is now comfortably warm because the atmosphere traps some of the Sun's heat and keeps it near the surface. The thin blanket of gas acts like a one-way filter, allowing visible sunlight through to warm the Earth's lands and oceans, but preventing the escape of some of the heat into the sink of space. Much of the ground's heat is reradiated out toward space in the form of longer infrared waves that are less energetic than visible ones and thus do not pass through the atmosphere's gas as easily as sunlight.

Some of the infrared heat radiation is absorbed in the air by "greenhouse gases", such as carbon dioxide, and some of the trapped heat is reradiated downward to warm the planet's surface and the air immediately above it.

The greenhouse effect is literally a matter of life and death. If the Earth had no atmosphere, it would be directly heated by the Sun's light to only 255 kelvin, which is well below the freezing point of water at 273 kelvin. Fortunately for life on Earth, the greenhouse gases in the air warm the planet to as much as 288 kelvin, and this extra heat can keep the oceans, lakes and streams from turning into ice. This "natural" greenhouse warming comes from minor ingredients of our atmosphere, such as water vapor and carbon dioxide, and it is perfectly normal and beneficial.

Of course, you can have too much of a good thing, like lying in the Sun all day in the summer or eating too much ice-cream too fast. For hundreds of years, humans have been filling the sky with carbon dioxide. The invisible waste gas is dumped into the air by burning fossil fuels – coal, oil and natural gas. When these materials are burned, their carbon atoms, denoted C, enter the air and combine with oxygen atoms, O, or oxygen molecules, O_2, to make carbon dioxide, CO_2.

About a century ago, in the middle of the industrial revolution, it was mainly coal that fueled factory boilers and warmed city houses, releasing carbon into the atmosphere to make carbon dioxide. Since the gas is colorless and odorless, it could not be seen or smelled, but it was detected indirectly by the noxious fumes emitted as a byproduct of burning high-sulfur coal. It has blackened entire cities,

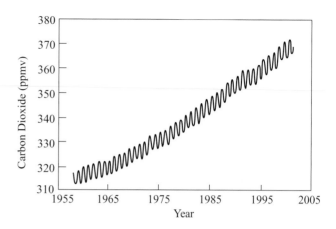

Fig. 4.21 Rise in Earth's atmospheric carbon dioxide   The average monthly concentration of atmospheric carbon dioxide (CO_2) in parts per million by volume (ppmv) of dry air plotted against time in years observed since 1958 at the Mauna Loa Observatory, Hawaii. It shows that atmospheric amounts of the principal waste gas of industrial societies, carbon dioxide, have risen steadily for more than forty years. The up and down fluctuations superimposed on the systematic increase reflect a seasonal rise and fall in the absorption of carbon dioxide by trees and other vegetation. Summertime lows are caused by the uptake of carbon dioxide by plants, and the winter highs occur when the plants' leaves fall and some of the gas is returned to the air. (Courtesy of Dave Keeling and Tim Whorf, Scripps Institution of Oceanography.)

such as London, described in 1854 by Charles Dickens (1812–1870) in *Hard Times*. His account is an eerie foreboding of the dark, polluted skies that have recently been found in Beijing and Mexico City.

In the 20th century, the perfection of the internal combustion engine and the mass production of automobiles made oil one of the most important sources of atmospheric carbon, and therefore of carbon dioxide. Around the globe, cars, sports utility vehicles and trucks have been releasing huge amounts of the potentially dangerous heat-trapping gas into the air.

Every time we drive a car, use electricity from coal-fired power plants, or heat our homes with oil or natural gas, we release carbon into the lower atmosphere. The burning of forests, whose trees hold much carbon dioxide, has also contributed.

Just a few decades ago, no one knew if any of the carbon dioxide stayed in the atmosphere or if it was all being absorbed in the oceans. Then in 1958 Charles D. Keeling (1928–2005) began measurements of its abundance in the clean air at the Mauna Loa Observatory in Hawaii, showing that the amount of carbon dioxide in the atmosphere has been increasing non-stop at an accelerating rate over half a century (Fig. 4.21). Superimposed on the relatively small annual fluctuations, due to the growth and loss of plant leaves and associated carbon dioxide intake, there was a large systematic increase over the entire period of observation. Year by year the total measured concentration of carbon dioxide grew, as inexorably as the expansion of the world's population and human industry.

Moreover, studies of ice deposits in Antarctica indicate that the amount of CO_2 has been increasing at an exponential rate ever since the beginning of the industrial revolution in the mid-18th century. Air bubbles that are trapped in the ice act like time capsules, conserving the atmosphere of the past. The air was sealed in the bubbles when the ice was laid down, and can be extracted from cores drilled deep within the layered ice deposits. They indicate that the atmospheric concentration of carbon dioxide has increased 31 percent during the past two and a half centuries, a mere blink in the eye of cosmic time.

The atmosphere now contains almost 800 billion tons of carbon dioxide. Humans continue to release about 7 billion tons of it each year. In other words, each person on Earth is, on average, dumping about a ton of carbon dioxide into the air every year. Once added to the air, carbon dioxide spreads throughout the entire atmosphere. The sea will absorb about half of that carbon dioxide, but it will take decades and even centuries to disappear. So future generations will have to contend with our present activities.

Roger Revelle (1909–1991) and Hans E. Suess (1909–1993) realized the threat decades ago. They argued that the oceans might not readily absorb all of the carbon dioxide being released into the air, and that the amount of atmospheric carbon dioxide would steadily increase as the fuel and power requirements of our worldwide civilization continued to rise. With prophetic insight, they wrote in 1957 that the increase might alter our weather and climate.

The increase in carbon dioxide isn't the entire story. During the past several decades other heat-trapping gases have been accumulating noticeably in the atmosphere, such as methane (CH_4) and nitrous oxide (N_2O). Even though the total emissions of these molecules are quite small when compared with those of carbon dioxide, they are much more efficient at trapping infrared heat radiation. As a result, they can together contribute about as much global warming as carbon dioxide alone.

Methane is the same natural gas that we use at home for cooking and heating. When found in swamps, methane is known as marsh gas; it sometimes ignites spontaneously, producing flickering blue flares called will-o'-the-wisps. Some of it also escapes from coalmines, natural gas wells and leaky pipelines.

Most of the atmospheric methane does not, however, come from gas wells. It is produced by agricultural activities such as growing rice and raising cattle. The gas is emitted by bacteria that thrive in oxygen-free places like rice paddies and the stomachs of cows. Since pre-industrial times, the atmospheric concentration of methane has increased more than 110 percent. Although methane is about 200 times less abundant than carbon dioxide, each incremental molecule of methane has about 20 times the heat-trapping power as each additional molecule of carbon dioxide.

Nitrous oxide, or laughing gas, is also building up in the air, although not as rapidly as methane. The current rate of increase is about 0.2 percent a year, primarily as the result of nitrogen-based fertilizers but also from the burning of fossil fuels in cars and power plants.

The chlorofluorocarbons (CFCs) are very effective heat-trapping molecules. The addition of one CFC molecule to the air can have the same greenhouse effect as the addition of 10 000 molecules of carbon dioxide to the present atmosphere. Fortunately, the warming effect of these industrial chemicals may soon be leveling off since they have been banned on the basis of their ozone-destroying capability.

Contrary to popular misconception, however, ozone depletion and global warming are not the same thing. The CFC molecules that destroy ozone also trap heat, but the thinning of the ozone layer does not by itself make the Earth's surface hotter.

Possible consequences of Earth's rising temperature

The inexorable, irreversible build-up of carbon dioxide and other heat-trapping gases is raising temperatures across the Earth, giving it a rising fever. The Swedish chemist Svante Arrhenius (1859–1927) saw it coming, setting out in 1896 to find out what would happen if the amount of carbon dioxide were changed from the amounts then in the air. He concluded that a doubling of the atmospheric carbon dioxide would boost the Earth's temperature by a few degrees, and that the temperature would drop by almost the same amount if the gas decreased by half. Although obtained without the use of modern satellite observations or extensive computer models, Arrhenius' estimate of the global warming by doubling the amount of carbon dioxide in our air is comparable to modern estimates. He also pointed out that industrial activity was then noticeably increasing the amount of atmospheric carbon dioxide, and that humans were therefore altering the temperature of the globe.

There can be no doubt that the temperatures are already rising. The evidence comes from direct measurements of rising surface air temperatures and subsurface ocean temperatures, as well as retreating glaciers, increases in average global sea levels, and changes to many physical and biological systems. Taken alone, these events are no proof of global warming, but in combination they provide strong evidence for a warmer climate.

As the water in the sea gets warmer, it will expand as most substances do when heated. The sea will then ascend to higher levels, in much the same way that heating the fluid in a thermometer causes the fluid to expand and rise up. This is because warm water or other fluids occupy a greater volume than cold ones. Measurements indicate that the global sea level increased somewhere between 10 and 25 centimeters during the 20th century. However, you could not have noticed the change, for the sea level was only rising between 1.0 and 2.5 millimeters per year.

The melting of ice that now covers land, such as mountain glaciers, also contributes to the sea-level rise and most likely results from global warming. By the end of the 20th century, glaciers were retreating throughout the world, and those in Alaska were typically becoming about a meter thinner every four or five years. The loss of ice in Greenland and Antarctica is now increasing at an accelerated pace, and the melting of Arctic ice may eventually shut down the Gulf Stream, which carries tropical heat away from the mid-Atlantic.

Melting glacial ice releases water into streams and rivers, which add to the sea. Such melt-waters from mountain glaciers boosted the sea level between 2 and 5 centimeters in the 20th century.

Contrary to popular belief, the melting of floating icebergs will not raise the level of the surrounding sea. When ice cubes in your drink at home melt, they similarly do not cause any change in its level, for the melted ice produces the same volume of water as it displaces.

No one can see much advantage in the rising seas, which are one of the most certain effects of the warming projected during the coming decades. There is just one indirect advantage – the meltdown of polar ice, that contributes to the sea rising, could result in a permanent ice-free passage in the Arctic Ocean, providing a new shipping route between Europe and Asia.

The climate experts predict a rise in sea level of between 0.09 and 0.9 meters (3 inches to 3 feet) over the next 100 years if nothing is done to curtail the emission of greenhouse gases. The resultant flooding will seriously disrupt coastal areas where more than a quarter of the world's population now lives. In the worst-case increase, Venice and Alexandria will be inundated, as will many cities on the Atlantic and Gulf coasts of the United States, including Boston and New York City. Residents in South Florida will not have to worry about the sweltering heat;

their homes will be flooded with seawater. Thirty million people in Bangladesh could be displaced by a 0.9-meter increase in sea level, and the rising waters would most likely force the evacuation of 70 million Chinese.

Salt water could move several kilometers inland at the mouths of rivers, invading coastal drinking-water systems. The Nile, Yangtse, Mekong and Mississippi deltas are all at risk. Island nations will suffer severe flooding or completely disappear under the rising waters; they include the Bahamas, many of the Caribbean islands, Cyprus and Malta in the Mediterranean, and several archipelagos around the Pacific Ocean (Fig. 4.22).

Even a modest rise in sea level will wipe many of the world's beaches out of existence. Flooding isn't the problem; it's the removal of sand by waves. Even a 0.3-meter (1-foot) rise in sea level creates wave action that might erode away up to 50 meters (150 feet) of some beaches. So people who live by the seashore had better sell their homes, and you can forget winter vacations in parts of Florida and the Caribbean islands.

The modest increase in temperatures at mid-northern latitudes, where most people live, will be welcome. There will be longer summers, shorter winters and warmer nights. Residents of cities like Boston will suffer fewer colds, experience fewer heart attacks from shoveling snow, and spend less on heating, snowplowing and road salting. On the other hand, summer air-conditioning will cost more, the winter ski slopes may turn to slush, and the colorful fall foliage could disappear as future generations of trees move away from the heat to the north.

Even if global warming is at the upper end of the predictions in 100 years, many humans should be able to adapt without much difficulty. After all, the predicted rise in temperature is less than the average daily temperature difference between New York City and Atlanta, Georgia, or between Paris and Naples. And those who live in the colder, northern locales are already used to a seasonal temperature increase between winter and summer that can be three times greater than the largest predicted heating over the next 100 years. Moreover, life has thrived during past periods when the planet was substantially warmer than it is now, and the Arctic was free of ice.

So the good thing is that humans are adaptable. But the bad thing is also that humans are adaptable. As long as the climate changes occur slowly, we can adapt without realizing what is happening, but some very uncomfortable things can happen at the top part of the expected warming in 100 years.

There is an applicable proverb about frogs. If you put a frog in boiling water, it will jump out and save itself. But if you gradually increase the heat, with the frog in the water, it will die.

Fig. 4.22 Independent State of Samoa The topography of Savai'i (*background*) and Upolu (*foreground*), the two large islands of the Independent State of Samoa. The highest volcano rises 1.9 kilometers. The low-lying coastal lands were inundated on 29 September 2009 by a tsunami generated by a major earthquake located about 200 kilometers to the south. Villages located on the coasts are also in danger of flooding by rising seas expected from future global warming. Color-coding is directly related to topographic height, with green at the lower elevations, rising through yellow and tan, to white at the highest elevations. Bright and dark shades display topographic slope in the northeast-southwest direction. The Shuttle Radar Topography Mission aboard the *Space Shuttle Endeavor* acquired the elevation data used in this image on 11 February 2000. (Courtesy of NASA/JPL/NGA.)

Very hot will be decidedly unwelcome in many places. Within deserts, entire cities will be immobilized under the heat. The wealthy will move out of Palm Springs, and Las Vegas could become a ghost town. Residents in many other large cities should experience severe heat waves, making them feel like the world is melting down in a pool of sweat. As the climate becomes hotter and drier, drought will probably become more severe in areas prone to it, and supplies of freshwater will dwindle. More frequent bouts of extreme weather will also be expected, with widespread flooding and intense hurricanes.

Agriculture in some regions will be better than other regions. The longer growing season and increasing carbon dioxide will foster plant growth, making much of the developed world greener. Agriculture will likely become more productive in Canada, northern Europe, Russia and the northern United States. As droughts turn some mid-American farms to dust, both agricultural production and population centers in the United States will shift north, and the same thing will probably happen in Europe.

The world's poorest countries are, on the other hand, highly vulnerable to agricultural disaster, for they are already located in arid and semi-arid regions. A further rise in temperature will almost certainly reduce crop yield in south Asia and sub-Sahara Africa, where expanding deserts will additionally claim more land.

As environmental conditions change over time, plants and animals will migrate, as they have throughout geological history – moving up and down in latitude as the globe warms and cools. The recent increase in temperatures has already caused some species to move north, and the accelerated heating could wipe out many of them in the future. Some plants and animals might not be able to move fast enough to keep pace with the rapid rate of temperature change, and climate-sensitive habitats could

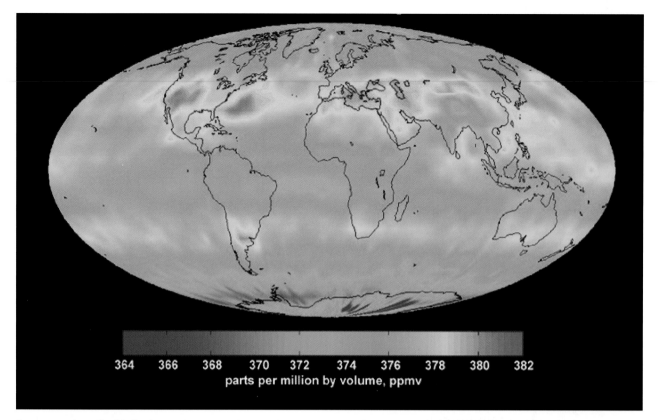

Fig. 4.23 Carbon dioxide in the Earth's troposphere   The concentration of carbon dioxide about 8 kilometers above the Earth's surface in July 2003. The regional distribution can still be seen by the time that the gases reach this height in the mid-troposphere, with large concentrations above parts of the United States and China. This image was acquired from the Atmospheric Infrared Sounder Experiment aboard the *Aqua* spacecraft. (Courtesy of NASA/JPL.)

be destroyed altogether, hastening the extinction of some species.

And there are other catastrophes that might result from significant global warming in the future. Hurricanes will become stronger and wetter; water supplies will be reduced and forest fires will become more common; more species will become extinct; drought will be intensified within the interiors of many continents; the American Midwest might become a colossal dust bowl; and power companies will be unable to air-condition sweltering cities.

The last factor in our catalog of likely consequences of severe global warming is increased health risk – a topic close to the hearts of most people and nearly every politician. Many diseases might spread dramatically as the temperatures head upward, especially those carried by mosquitoes. As the world warms, mosquitoes will move north, into regions where the winter cold used to kill them, bringing malaria, dengue fever, yellow fever and encephalitis with them. Extremely hot weather may also directly terminate the lives of a lot of people, particularly among the very old and very young, the poor and weak, and those with cardiovascular and respiratory disease.

Doing something about global warming

Although the most severe consequences of global warming are not likely to be noticed by you or your children, we've already initiated changes that will affect future generations. Once in the atmosphere, about half of the carbon dioxide stays there for centuries, so our grandchildren and their children will have to contend with the consequences of our present actions. The invisible waste gases that we have already dumped in the air will slowly change the climate of the Earth regardless of future actions.

The scientists are doing all they can to improve our understanding of global warming. They continue to monitor signs of the rising temperatures, such as the meltdown of Antarctica, retreating glaciers and rising sea levels. Instruments aboard orbiting satellites are now monitoring the global distribution and temporal changes of heat-trapping gases in the atmosphere (Figs. 4.23, 4.24). And climate scientists continue to refine their models, using measurements of increasing accuracy to predict the future in intricate computer calculations.

When you strip away the rhetoric, the experts are uncertain about the exact future severity of climate change and

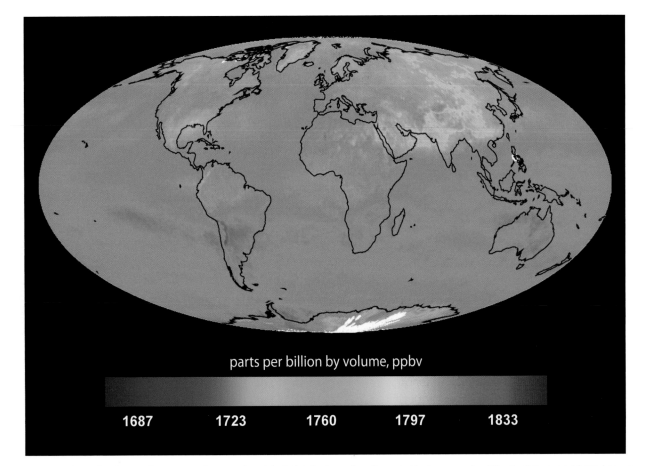

Fig. 4.24 Methane in the Earth's troposphere   The global distribution of methane in the lower parts of the Earth's atmosphere, known as the troposphere. This image was taken from the Atmospheric Infrared Sounder Experiment aboard the *Aqua* spacecraft in August 2005. Such images will help identify sources of this heat-trapping greenhouse gas, as well as its seasonal and multi-year variation and its transport around the globe. (Courtesy of NASA/JPL.)

even less about the future physical impact in particular countries or regions. After decades of research, the model builders cannot say precisely what will happen to the climate as the result of the atmospheric build-up of heat-trapping gases. They just don't know enough about the atmosphere, clouds or the oceans to predict accurately the future global climate. Despite the lack of precision, the general trends have nevertheless become obvious, with uncontestable observations of global warming.

The majority of scientists are therefore now in agreement that we ought to do something to curtail human emissions of carbon dioxide and other heat-trapping gases. The chances of serious consequences are high enough to take action to stop their probable effects. In 2005, for example, the world's most influential scientific academies warned national leaders that they can no longer ignore the clear and increasing threat posed by global warming, and that the scientific understanding of climate change was then sufficiently clear to justify nations taking prompt action. The unprecedented joint statement included the heads of

the scientific academies of Brazil, Canada, China, France, Germany, India, Italy, Japan, Russia, the United Kingdom, and the United States.

Then in 2007 the Nobel Peace Prize was awarded, in two equal parts, to the Intergovernmental Panel on Climate Change (IPCC), and Albert Arnold "Al" Gore Jr. (1948–) for their efforts to build up and disseminate greater knowledge about man-made climate change and to lay the foundations for the measures that are needed to counteract such change. The Nobel presentation speech recognized the importance of understanding and containing man-made global warming that will decisively impact our existence on Earth.

Most scientists support prudent steps to curb the continued build-up of heat-trapping gases, even asserting that the evidence warrants a sense of urgency. Yet, whether we like it or not, global warming has become politicized, the subject of a contentious debate. It has entered the arena of world politics, a shadowy realm of diplomacy, economic interests, political alliances, and national security.

So what's being done about the problem? In December 1997, representatives of the world's nations met in Kyoto, Japan, to establish, for the first time, specific legally binding targets and timetables for the emission of heat-trapping gases. The treaty, called the *Kyoto Protocol*, calls for mandatory reductions in the emissions of greenhouse gases such as carbon dioxide and methane. But it has not been signed by the United States, which contributes about 18 percent of the total emissions with just 4 percent of the world's population. And developing nations are not bound by the treaty restrictions even though their emissions are expected to surpass even the unrestrained emissions of the richer nations in a few decades. China, for example, has already overtaken the United States as the world's largest carbon-dioxide emitter on an annual basis, and the combined increase in greenhouse gases contributed by these two countries will outstrip any reductions agreed to by other countries.

More recently, heads of state and government of more than a hundred nations gathered in Copenhagen, Denmark, between 7 December and 18 December 2009 to seek a consensus on an international strategy for fighting global warming and associated climate change. Despite the serious intent of global leaders and meaningful proposals for reductions of carbon-dioxide emissions, if a binding agreement was made, the United Nations' sponsored Copenhagen Summit fell short of even modest expectations.

No legally binding treaty was ratified by any nation. A draft *Copenhagen Accord* asks countries to submit emission targets, but it was not adopted or passed unanimously, and it did not contain any legally binding commitments for reducing carbon-dioxide emissions.

The main difficulty in arriving at a more meaningful agreement lies in a long-standing division between the rich and poor nations. The United States, whose emissions of carbon dioxide over the past 200 years has helped cause global warming, has long refused to accept any binding limits on its greenhouse-gas emissions. Fast-developing China, a "poor" nation, is also strongly opposed to mandatory ceilings on the emissions. For China, and also India, the main concern is economic growth, which delivers stability and prosperity, and keeps the government in power. Since that growth involves burning more coal, which releases large amounts of carbon dioxide into the atmosphere, these countries are going to find it difficult to substantially reduce their emissions. Coal provides roughly half of the electricity in the United States and most of that in China, and the inexpensive fuel will help move the populations of China and India from poverty to middle-class prosperity.

Both the United States and China probably find it unacceptable to approve a treaty that might cause serious economic damage. Wealthy countries fear that mandatory limits on the emissions of heat-trapping gases could cause a recession in their prosperous economies, while the poor ones are afraid that such limits will destroy the economic growth needed for the very survival of their people. And since neither side in the debate will compromise, a consensus is impossible.

To put it another way, no one likes to be told what to do, and voluntary steps to improve the climate will be opposed by the vested interests of governments, corporations and other powerful institutions. As a result, no comprehensive, ratified, international global-warming treaty exists, at least so far. As always, the greatest hope is in the young, who are contributing to a mass movement on behalf of climate action.

Something important may nevertheless result from the future interactions of the calm, compromising diplomats and politicians. Most major nations – including the United States, the 27 nations of the European Union, China, India, Japan and Brazil, for example, restated, in January 2010, their earlier pledges to curb emission of heat-trapping gases by 2020, some by promising absolute cuts, others by reducing the rate of increase.

As the traditional Negro jubilee spiritual goes:

He's got the whole world in his hands,
The whole wide world in his hands,

but the "he" is not a supernatural deity – it's us. We humans have modified the atmosphere, warming the globe, and we are starting to do something about it. But it is bound to be only a temporary fix.

In the long run nature will take over the weather and climate once again. A hundred million years ago, when the dinosaurs roamed the Earth, there were no ice caps and tropical plants flourished near the South Pole. Deep cold nearly turned the Earth into a ball of ice about 10 000 years ago, when the planet was in the depths of an ice age. In just a few million years from now, entire continents and oceans can be destroyed or created new, changing the flow of air and ocean currents and altering global weather patterns. And even if it is pretty warm right now, the die is cast for the next ice age and the glaciers will come again.

Ice and fire

During the past million years the Earth has undergone a series of warm and cold periods. During the cold periods, called ice ages, huge ice sheets build up on the continents and in the polar seas. The growing layer of continental ice flows towards the equator, scouring and covering large areas of land. Then the climate warms and the ice retreats.

Each glacial ice age lasts about 100 000 years. There is a relatively short interval of unusual warmth between the ice ages that lasts 10 000 or 20 000 years. During such an interglacial interval, the world's climate becomes more pleasant and serene. We now live in such a warm time, called the Holocene period, which has enabled human civilization to flourish.

The recurrent ice ages and warm intervals are caused by variations in the amount and distribution of sunlight reaching the Earth, but not by any intrinsic fluctuations in the amount of light radiated by the Sun itself. Three astronomical cycles combine to alter the angles and distance at which sunlight strikes the far northern latitudes of Earth, triggering the ice ages. This explanation was fully developed by the Serbian astrophysicist Milutin Milankovitch (1879–1958) from 1920 to 1941, so the astronomical cycles are now sometimes called the Milankovitch cycles.

When there is less sunlight being received in far northern latitudes, the winter temperatures are colder there, and the summer temperatures are milder. So less polar ice melts in the summer, and over time the winter snows are compressed into ice to make the glaciers grow.

The varying gravitational forces produced by the other planets, whose distances from Earth change, produce a rhythmic stretching of the Earth's orbit. These planetary perturbations periodically change the shape of the Earth's orbit from circular to slightly elliptical and back again, over a period of 100 000 years. As its path becomes more elongated, the Earth's distance from the Sun varies more during each year, intensifying the seasons in one hemisphere and moderating them in the other.

Shorter cycles are due to repetitive changes in the wobble and tilt of the Earth's rotational axis, which vary over 23 000 and 41 000 years respectively. The greater the tilt, the more intense the seasons in both hemispheres, with hotter summers and colder winters.

Successive layers of frozen atmosphere have been laid down in Greenland and Antarctica, providing a natural archive of the Earth's past climate over the past 420 000 years. Bubbles of air trapped in falling snowflakes and entombed in ice are deposited every year, building up on top of each other like layers of sediment. When extracted in deep ice cores, they reveal secrets about the ancient climate. Such cores strongly support the idea that changes in the Earth's orbit and spin axis cause variations in the intensity and distribution of sunlight arriving at Earth, which in turn initiate natural climate changes and trigger the ebb and flow of glacial ice.

The current Holocene interglacial, which has already lasted 11 000 years, may not continue for more than a few thousand years, and we could then enter an ice age. The next time it happens, the advancing glaciers might bury Copenhagen, Detroit and Montreal under mountains of ice, and because of the drop in sea level people might then walk from England to France, from Siberia to Alaska, and from New Guinea to Australia.

Perhaps global warming by human activity will help counteract a coming ice age – no one knows for sure. And there is no way out in the long run, for the Sun will inevitably fry the Earth. Well-accepted models of stellar evolution indicate that the Sun began its life about 4.5 billion years ago shining with about 70 percent of the brightness it has today, and that it has been slowly increasing in brightness ever since. As the Sun continues to brighten, the planet will eventually become a burned-out cinder, a dead and sterile place.

Astronomers calculate that the Sun will become hot enough in 3 billion years to evaporate the oceans away, and 4 billion years thereafter the Sun will balloon into a giant star, engulfing the planet Mercury and melting the Earth's surface. Thus, our very remote descendants are destined to an end in fire, consumed by the Sun that once nurtured us.

So our long-term prospects aren't all that great, and we might as well concentrate on protecting, improving and experiencing the magnificent world that we are so privileged to inhabit. And to get back to more immediate concerns, the Sun provides dangers whenever humans or their spacecraft venture into space.

4.5 Space weather

The Sun is the ultimate power source. It warms the ground we walk on, lights our days, sustains life, and provides directly or indirectly most of the energy on our planet. And it is solar heat that powers the winds and cycles water from sea to rain, the source of our weather and arbiter of our climate. Nowadays, and in all former times, it is the Sun-driven seasons that dominate weather on Earth.

Once it was realized that the space between the Sun and Earth is not empty, and just more rarefied than our transparent atmosphere, it was natural to suppose that the Sun also powered space weather. The term refers to conditions on the Sun, in the Sun's winds, and near the Earth that can affect space-borne and ground-based technological systems and human life and health.

As our civilization deploys ever more sophisticated technology, it becomes increasingly at the mercy of storms in space. Its gusts and squalls, the cosmic equivalent of terrestrial blizzards or hurricanes, are related to explosive outbursts on the Sun, and to dynamic processes in interplanetary space, in near-Earth space and in the magnetosphere.

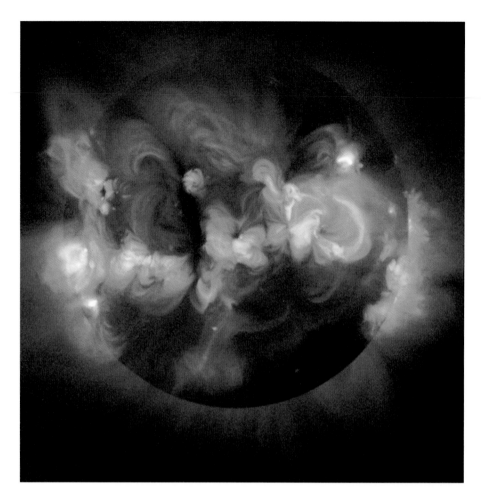

Fig. 4.25 The Sun in X-rays The bright glow seen in this X-ray image of the Sun is produced by ionized gas at a temperature of a few million kelvin. It shows magnetic coronal loops that thread the corona and hold the hot gas in place. The brightest features are called active regions and correspond to the sites of the most intense magnetic field strength. The Soft X-ray Telescope (SXT) aboard the Japanese *Yohkoh* satellite recorded this image of the Sun's corona on 1 February 1992, near a maximum of the 11-year cycle of solar magnetic activity. Subsequent SXT images, taken about five years later near activity minimum, show a remarkable dimming of the corona when the active regions associated with sunspots have almost disappeared, and the Sun's magnetic field has changed from a complex structure to a simpler configuration. (Courtesy of Gregory L. Slater, Gary A. Linford, and Lawrence Shing, NASA, ISAS, the Lockheed-Martin Solar and Astrophysics Laboratory, the National Astronomical Observatory of Japan, and the University of Tokyo.)

Down here on the ground, we are shielded from much of this space weather by the Earth's atmosphere and magnetic fields, keeping us from bodily harm. But out in deep space there is no place to hide, and both humans and satellites are vulnerable. Energetic protons accelerated by explosions on the Sun can cripple spacecraft and seriously endanger unprotected astronauts that venture into outer space. Sun storms can also disrupt global radio communications and disable satellites used for navigation, military reconnaissance or surveillance, and communication, from cell phones to pagers, with considerable economic, safety and security consequences. This technology has become part of our everyday lives, enhancing our vulnerability to space weather and increasing the importance of understanding and predicting it.

High-flying humans at risk from solar explosions

Powerful bursts of radiation, as well as energetic charged particles and magnetic fields, are being hurled into interplanetary space by solar explosions. These outbursts come in two main varieties: solar flares, and coronal mass ejections (CMEs). Both kinds of solar activity are powered by the Sun's magnetic energy, and they both vary in step with the Sun's 11-year cycle of magnetic activity. Solar flares and CMEs are more frequent and tend to be more powerful during the maximum in the activity cycle.

All of the solar flares, and most of the fastest coronal mass ejections, with the largest amount of energy,

Fig. 4.26 Mass ejection from the Sun   A huge coronal mass ejection is seen in this coronagraph image, taken on 5 December 2003 with the Large Angle Spectrometric COronagraph (LASCO) on the *SOlar and Heliospheric Observatory* (*SOHO*). The solid red circle corresponds to the occulting disk of the coronagraph that blocks intense sunlight and permits the corona to be seen. An image of the singly ionized helium, denoted He II, emission of the Sun, taken at about the same time, has been appropriately scaled and superimposed at the center of the LASCO image. The full disk helium image was taken at a wavelength of 30.4 nanometers, corresponding to a temperature of about 60 000 kelvin, using the Extreme-ultraviolet Imaging Telescope, or EIT for short, aboard *SOHO*. (Courtesy of the *SOHO* LASCO and EIT consortia. *SOHO* is a project of international cooperation between ESA and NASA.)

originate from places of intense magnetism on the Sun. Known as solar active regions, they contain closed magnetic fields that are rooted in sunspots and whose loops constrain the intense X-ray-emitting gas of the quiescent, non-exploding Sun (Fig. 4.25).

Solar flares are brief catastrophic outbursts that flood the solar system with intense radiation and high-speed electrons and protons. In just a few minutes they can release an explosive energy of up to 10^{25} joule, equivalent to 20 million 100-megaton terrestrial nuclear bombs, raising the temperature of Earth-sized regions on the Sun to tens of millions of kelvin. The other type of solar explosive activity, the CMEs, expand away from the Sun at speeds of hundreds of kilometers per second, becoming larger than the Sun and removing up to 50 billion tons, or 5×10^{13} kilograms, of the Sun's atmosphere (Fig. 4.26).

At any given phase of the solar cycle, intense solar flares are as much as 100 times more frequent than mass ejections, but the CMEs energize particles on a grand scale that covers large regions in interplanetary space (Fig. 4.27). They move straight out of the Sun and flatten everything in their path, like a gigantic falling tree or a car out of control. Fast CMEs plow into the slower-moving solar wind and

act like a piston that drives shock waves ahead of them, accelerating electrons and protons as they go, like ocean waves propelling surfers.

High-energy protons from a solar flare or coronal mass ejection can easily pierce a spacesuit, causing damage to human cells and tissues. The explosive solar emissions can endanger the health and even the lives of astronauts when they venture into outer space to construct a space station, repair a spacecraft, or walk on the Moon or Mars.

Solar astronomers, and employees of national space-weather forecast centers, therefore keep careful watch over the Sun during space missions, to warn of possible solar activity occurring at just the wrong place and time. Space flight controllers can then postpone space walks during solar storms, keeping astronauts within the heavily shielded recesses of a satellite or space station. The astronauts would also be told to curtail any strolls on the Moon or Mars, and to move inside underground storm shelters.

Failing to communicate on Earth

Eight minutes after the outburst of an energetic flare on the Sun, a strong blast of X-rays and extreme ultraviolet radiation reaches the Earth and radically alters the structure of the planet's upper atmosphere, the ionosphere, by producing an increase in the amount of free electrons that are no longer attached to atoms. Even during moderately intense flares, long-distance radio communications can be temporarily silenced over the Earth's entire sunlit hemisphere. The radio blackouts are particularly troublesome for the commercial airline industry, which uses radio transmissions for weather, air traffic and location information; the United States Air Force and Navy are also concerned about this solar threat to radio communications.

The Air Force operates a global system of ground-based radio and optical telescopes and taps into the output of national, space-borne X-ray telescopes and particle detectors in order to continuously monitor the Sun for intense flares that might severely disrupt military communications and satellite surveillance.

Space-weather interference with radio communication can be avoided by using short-wavelength, ultra-high-frequency signals that pass right through the ionosphere to satellites that can relay the transmissions to other locations. But the telecommunications industry is also threatened by the loss of their satellites due to disabling solar outbursts.

Earth-orbiting satellites in danger

Solar energetic particles arising from solar flares or coronal mass ejections can degrade, disrupt or destroy a satellite.

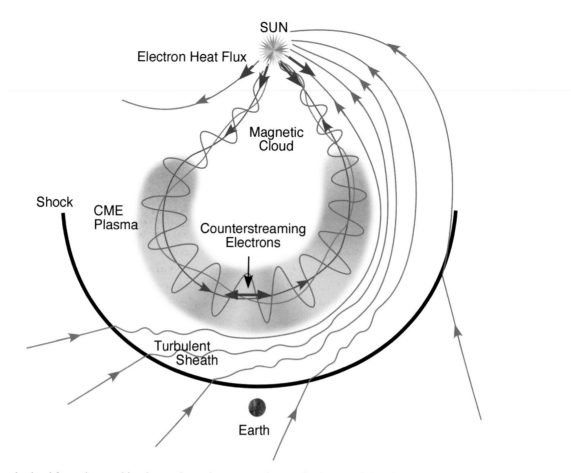

Fig 4.27 Magnetic cloud from the Sun hits the Earth   When a coronal mass ejection travels into interplanetary space, it can create a huge magnetic cloud containing bidirectional, or counterstreaming, beams of electrons that flow in opposite directions within the magnetic loops that are rooted at both ends in the Sun. The magnetic cloud also drives an upstream shock ahead of it. Magnetic clouds are only present in a subset of observed interplanetary coronal mass ejections. (Courtesy of Deborah Eddy and Thomas Zurbuchen.)

And there are now roughly 1000 of them in daily use by governments, corporations and ordinary citizens. Geosynchronous satellites, which orbit the Earth at the same rate that the planet spins, stay above the same place on Earth to relay and beam down signals used for aviation and marine navigation, cellular phones, global positioning systems, national defense, and internet commerce and data transmission. Other satellites whip around the planet, scanning air, land and sea for environmental change, weather forecasting and military reconnaissance. All of these spacecraft can be temporarily or permanently disabled by solar energetic particle events, causing engineers to design spacecraft with greater shielding and increased redundancy in their components.

Geosynchronous satellites, for example, are endangered by the coronal mass ejections that cause intense geomagnetic storms. These satellites orbit our planet once every 24 hours at an altitude of 35 786 kilometers, or a distance of about 6.6 Earth radii from the planet's center, and thus remain at constant longitude above the Earth. A powerful coronal mass ejection can compress the magnetosphere from its usual location at about 10 Earth radii to below the satellites' geostationary orbits, exposing them to the full brunt of the gusty solar wind and its charged, energized ingredients.

Infrequent, anomalously large eruptions on the Sun can hurl very energetic protons toward the Earth and elsewhere in space, interfering with satellite instruments (Fig. 4.28). The solar protons can easily enter a spacecraft to produce single-event upsets in electronic components by ionizing a track along parts of their circuits. The ionized tracks can occur in transistors and memory devices, producing erroneous commands and crippling their microelectronics. Such single-event upsets have already destroyed at least one weather satellite and disabled several communications satellites.

Space weapons can also wipe out a satellite; so if you didn't know the Sun was at fault, you might think someone was trying to shoot down the satellites. But error-correcting software has been developed to decrease damage by

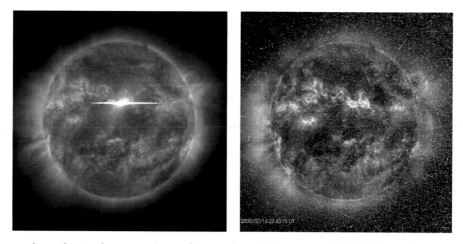

Fig. 4.28 Solar flare produces threatening space storm of energetic particles A powerful solar flare (*left*), occurring at 10 hours 24 minutes Universal Time on Bastille day, 14 July 2000, unleashed high-energy protons that began striking the *SOHO* spacecraft near Earth about 8 minutes later, continuing for many hours (*right*). Both images were taken from the Extreme-ultraviolet Imaging Telescope (EIT) on the *SOlar and Heliospheric Observatory* (*SOHO*). (Courtesy of the *SOHO* EIT consortium. *SOHO* is a project of international cooperation between ESA and NASA.)

single-event upsets to military satellite operations. Commercial satellites, which are less expensive than military ones to build, have less protection and are more vulnerable.

Solar threats to electrical power systems on Earth

While altering the Earth's magnetic field, a colliding coronal mass ejection can produce strong electric currents in nearby space. If these currents connect to long-distance power lines on the ground, they can blow circuit breakers, overheat and melt the windings of transformers, and cause massive failures of electrical distribution systems. They can plunge major urban centers, like New York City or Montreal, into complete darkness, causing social chaos and threatening safety. The threat is greatest in high-latitude regions where the currents are strongest, such as Canada, the northern United States and Scandinavia.

Forecasting space weather

Our technological society has become increasingly vulnerable to explosions on the Sun. They emit energetic particles, intense radiation, powerful magnetic fields and strong shocks that can have enormous practical implications when directed toward Earth. The solar emissions can disrupt navigation and communication systems, pose significant hazards to humans in space, destroy Earth-orbiting satellites, and create power surges that can black out entire cities. Recognizing our vulnerability, national centers and defense agencies continuously monitor the Sun from ground and space to forecast threatening activity. An example is the Space Environment Center (SEC) of the United States National Oceanic and Atmospheric Administration. It collects and distributes space weather data, using satellites and ground-based telescopes to monitor the Sun and interplanetary space.

With adequate warning, operators can power down sensitive electronics on navigation and positional satellites, putting them to sleep until the danger passes. Airplane pilots and cellular telephone customers can be warned of potential communication failures. The launch of manned space flight missions can be postponed, and walks outside spacecraft or on the Moon or Mars might be delayed. Utility companies can reduce load in anticipation of induced currents on power lines, in that way trading a temporary "brown-out" for a potentially disastrous "black-out".

What everyone wants to know is how strong the storm is and when it is going to hit us. Most of the more energetic coronal mass ejections come from magnetic explosions in active regions with sunspots, producing a flare in tandem with the ejections. So a good place to begin our space weather forecasts is to know when a threatening active region, with its sunspots and strong magnetic fields, is on the Sun.

Active regions appear more frequently near the maximum of the 11-year sunspot cycle, as do solar flares and coronal mass ejections. So long-term solar activity can be forecast in a general way using this cycle.

On a shorter timescale of weeks, we can use helioseismology to detect large solar active regions on the hidden backside of the Sun. The technique of helioseismology uses observations of solar pulsations to infer the

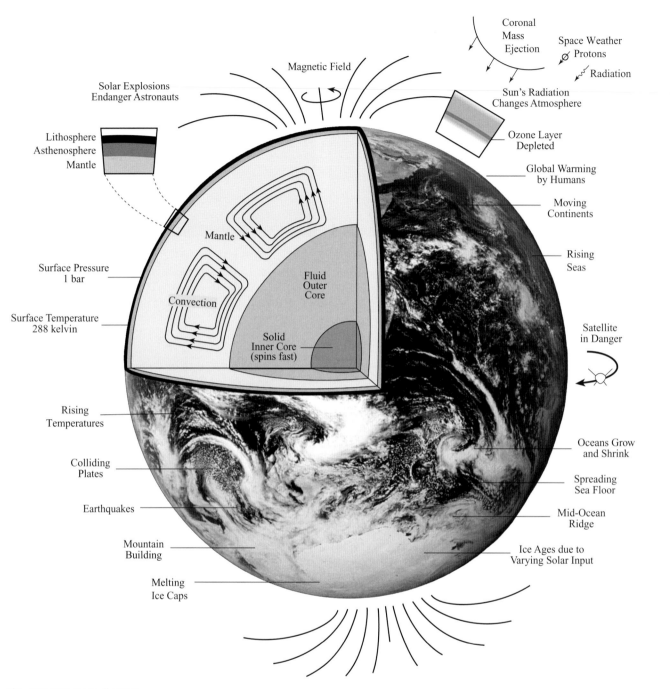

Fig. 4.29 Summary diagram

trajectories of sound waves within the Sun, including those that see through it. Since the solar equator rotates with a period of 27 days, when viewed from Earth, the detection of a magnetically complex and strong active region on the far side of the Sun can give more than a week's warning before it swings into view to threaten the Earth. Daily images of the unseen, far side of the Sun are available on the web at http://soi.stanford.edu/data/full_farside and http://gong.nso.edu. Like winter storms on Earth, some of the effects of space weather can be predicted days in advance. A coronal mass ejection arrives at the Earth one to four days after leaving the Sun, and solar astronomers can watch them happen days in advance. Solar flares are another matter. As soon as you can see a solar flare on the Sun, its radiation and fastest particles have already reached us, taking just 8 minutes to travel from the Sun to Earth. Dangerous but less energetic particles might take an hour to get here.

The ultimate goal of space weather forecasters is to predict when the Sun is about to unleash its pent-up energy, before a solar flare or coronal mass ejection occurs. One promising technique is to watch to see when the Sun's magnetism has become twisted into a stressed situation, for it may then be about to explode. Observing the X-ray emission of an active region to determine when the magnetic fields are sheared and twisted can do this. The signature of an immanent explosion might be found deeper down, under the visible disk of the Sun. The techniques of local helioseismology have demonstrated that the strength of flares from active regions is correlated with the amount of circulating flows beneath them.

However, some regions that exhibit magnetic shear and twist never erupt, so contorted magnetism may be a necessary but not sufficient condition for solar flares or coronal mass ejections. And the Sun's sudden and unexpected outbursts often remain as unpredictable as most human passions. They just keep on happening, and even seem to be necessary to purge the Sun of pent-up frustration and to relieve it of twisted, contorted magnetism.

And to be honest, scientists have not solved the question of what exactly initiates a solar flare or coronal mass ejection, igniting the explosion from stressed coronal magnetic fields. They think the storms might be triggered when magnetized coronal loops are pressed together, driven by motions beneath them, meeting to touch each other, merging to break open the magnetic fields and release free magnetic energy. But no one has identified a signature that allows prediction of exactly when such an outburst might occur. So far, we only have signs of a possible solar storm; it's something like seeing that dark storm cloud but not knowing if it's going to rain.

5 The Earth's Moon: stepping stone to the planets

- When the Moon moves into the Earth's shadow, the full Moon turns blood red; when the Earth travels into the Moon's shadow it can become dark during the day.

- The full Moon looks bigger near the horizon than directly overhead, but its changing size is an illusion.

- The Moon spins on its axis with the same period in which it revolves around the Earth, at 27.3 days, keeping its far side forever hidden to Earth-bound observers.

- The near side of the Moon contains light, rugged, cratered regions called highlands and dark smooth lava flows dubbed maria; the far side of the Moon is mostly highlands and has very few maria.

- For more than two centuries, lunar craters were attributed to volcanoes on the Moon, but they are now widely known to be due to the explosive impact of interplanetary projectiles, known as meteors when in space and meteorites upon hitting the surface of a moon or planet.

- More than thirty years ago, twelve humans roamed the surface of the Moon and brought back nearly half a ton of rocks.

- Because the Moon has almost no atmosphere, its sky remains pitch black in broad daylight and there is no sound or weather on the Moon.

- Two modest spacecraft, named *Clementine* and *Lunar Prospector*, chalked up an impressive list of accomplishments in the 1990s, including evidence for a lunar core and for water ice at the poles of the Moon.

- Rocks returned from the Moon contain no significant amounts of water, but there is evidence for small quantities of water in some places such as the permanently shaded regions at the lunar poles. Comets may have deposited the water.

- Space agencies from China, Europe, India, Japan, and the United States have all sent spacecraft to the Moon in the early 21st century, obtaining detailed information about the altitude, geological, chemical and gravity characteristics of the lunar surface and sending their spacecraft into controlled impact with the Moon.

- High-resolution maps acquired from lunar orbit are being used to specify potential landing sites and resources for future human exploration of the Moon.

- Humans might return to the Moon to create unique astronomical observatories, and establish a permanent base and way station for trips to Mars; but that is not likely to happen in the near future.

- Moonquakes, which are much weaker than earthquakes, indicate that the Moon has a small dense core, probably surrounded by a partially molten zone. The core has been confirmed by gravity measurements from the orbiting *Lunar Prospector* spacecraft, and laser-ranging measurements have confirmed the molten zone.

- There is no life on the Moon, and there apparently never was any.

- Earth rocks and Moon rocks are similar in their mix of light and heavy oxygen isotopes, but the Moon rocks contain relatively little iron and few volatile elements common on Earth.

- Impact basins excavated by cosmic collision produce as much topographical relief on the Moon as there is on the Earth due to ongoing tectonic processes.

- Vast blocks of the lunar surface are magnetized, but they do not combine into an overall global dipole like the Earth's magnetism. Some of the ancient lunar magnetism has been concentrated on the other side of the Moon from large impact basins.

- Radioactive dating indicates that the oldest rocks returned from the Moon are about 4.6 billion years old, which is about the same age as the Earth.

- During its early youth, between 4.4 and 4.6 billion years ago, a global sea of molten rock covered the Moon, but now a layer of fine, powdery Moon dust covers it.

- A heavy bombardment cratered the highlands until about 3.9 billion years ago, when the large impact basins were formed; lunar volcanism subsequently filled these basins to create the maria between 3.2 and 3.9 billion years ago.

- Most of the features we now see on the Moon have been there for more than 3 billion years.

- The Moon's gravity draws the Earth's oceans into the shape of an egg, causing two high tides as the planet's rotation carries the continents past the two tidal bulges each day.

- The Moon acts as a brake on the Earth's rotation, causing the length of the day to steadily increase and the Moon to move away from the Earth.

- The Moon provides a steadying influence to the Earth's seasonal climatic variation, anchoring and limiting the tilt of the planet's rotation axis.

- The Moon was most likely born during the ancient collision of a Mars-sized body with the young Earth; the giant impact dislodged material that would become the Moon that we know.

5.1 Fundamentals

The Earth has one Moon, which is a natural satellite. To distinguish it from the moons of other planets, we denote our Moon with a capital M and also sometimes call it the Earth's Moon. The mass, size, density and other physical properties of our Moon are given in Table 5.1.

The Earth's Moon is a unique satellite within the solar system, the largest relative to its planet. Mars is the only other terrestrial planet to have a moon, and its two satellites are very small. The giant planets have extensive satellite systems, but these moons are usually composed of low-density rock-ice mixtures unlike our high-density rocky Moon.

Table 5.1 Physical properties of the Moon[a]

Mass	7.348×10^{22} kilograms $= 0.0123\ M_E$
Mean radius	1737.5 kilometers $= 0.2725\ R_E$
Bulk density	3344 kilograms per cubic meter
Sidereal rotation period	27.322 days $=$ fixed star to fixed star
Sidereal orbital period	27.322 days $=$ fixed star to fixed star
Synodic month	29.53 days $=$ new Moon to new Moon
Mean distance from Earth	3.844×10^{8} meters
Increase in mean distance	0.0382 ± 0.0007 meters per year
Mean orbital speed	1023 meters per second
Angular radius at mean distance (geocentric)	15 minutes 32.6 seconds of arc
Angular radius at mean distance (topocentric)	15 minutes 48.3 seconds of arc
Age	4.55×10^{9} years

[a] Here M_E and R_E respectively denote the mass and radius of the Earth. The Earth to Moon mass ratio is 81.300 587.

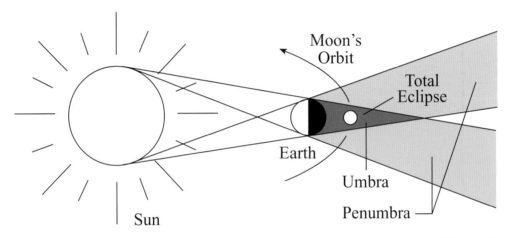

Fig. 5.1 Lunar eclipse   During a lunar eclipse the initially full Moon passes through the Earth's shadow. A total lunar eclipse occurs when the entire Moon moves into the umbra. Because no portion of the Sun's visible disk can be seen from the umbra, it is the darkest part of the Earth's shadow. Only part of the Sun's disk is blocked out in the larger penumbra. A partial lunar eclipse occurs when the Moon's orbit takes it only partially through the umbra or only through the penumbra.

5.2 Eclipses of the Moon and the Sun

Once or twice in a typical year, the Moon's orbital motion carries it through the Earth's shadow. This is an eclipse of the Moon, when the Sun's illumination of the Moon has been removed. The word *eclipse* is derived from the Greek term for "abandonment".

A lunar eclipse can be seen from half of the Earth. There are two regions in the Earth's shadow at the time of a lunar eclipse: the *umbral* region where there is no direct sunlight, and the *penumbral* region where the Sun's light is partially shadowed (Fig. 5.1). The umbral shadow is darker, and it is in the shape of a narrow cone pointing away from the Earth.

The full Moon turns a deep red when in the umbral shadow of the Earth (Fig. 5.2). Ancient Hebrew writers often used this appearance as a metaphor to describe the end of the world. For instance, the prophet Joel declared that the Lord:

> will shew wonders in the heavens and in the Earth, blood, and fire, and pillars of smoke. The Sun shall be turned to darkness, and the Moon into blood...
>
> (Joel 2:30, 31; KJV)

And then there are the lyrics to some of Bob Dylan's (1941–) songs that include "the Moon rising like wildfire", and "when there was blood on the Moon".

A total eclipse of the Sun occurs when the Moon passes between the Earth and the Sun, and the Moon's shadow falls on the Earth. In an incredible cosmic coincidence, the Moon is just the right size and distance to blot out the visible solar disk when properly aligned and viewed from

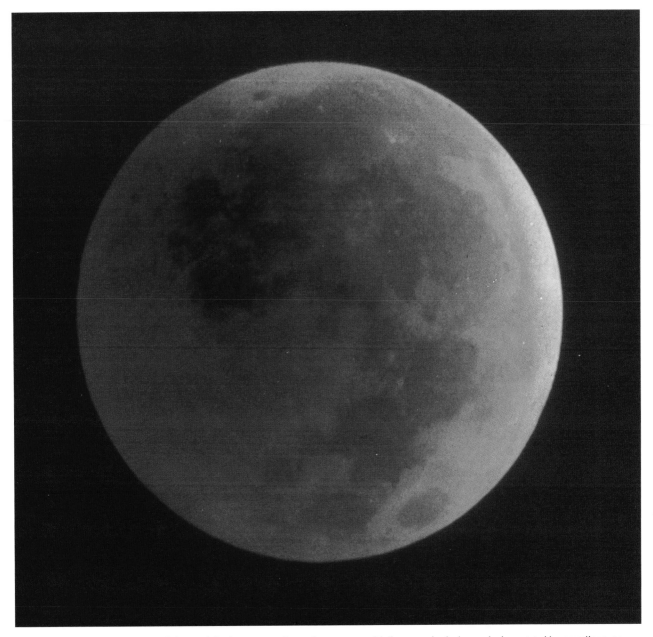

Fig. 5.2 The blood-red Moon   If the Earth had no atmosphere, the Moon would disappear in darkness during a total lunar eclipse. As shown here, the Moon actually becomes dark red for an hour or so. This is because the Moon is illuminated by sunlight that is bent part way around the Earth and is reddened in passing through the Earth's atmosphere, just as the Sun is reddened at sunset. If the Earth is heavily clouded, the sunlight is obstructed and the Moon is particularly dark during a lunar eclipse. (Courtesy of Eric Mandon, Observatoire Populaire de Rouen.)

the Earth. In other words, the apparent angular diameter of the Moon and the visible solar disk are almost exactly the same, about 30 minutes of arc, so that under favorable circumstances the Moon's shadow can reach the Earth and cut off the light of the Sun.

The outer atmosphere of the Sun, known as its corona, becomes momentarily visible to the unaided eye when the Moon blocks out the Sun's bright disk and it becomes dark during the day. The corona is then seen at the limb, or apparent edge, of the Sun, against the blackened sky as a faint, shimmering halo of pearl-white light (Fig. 5.3). But be careful if you watch an eclipse, for the light of the corona is still very hazardous to human eyes and should not be viewed directly.

Since the Moon and the Earth move along different orbits whose planes are inclined to each other (Fig. 5.4),

Fig. 5.3 Gossamer corona   The Sun's corona as photographed during the total solar eclipse of 26 February 1998, observed from Oranjestad, Aruba. To extract this much coronal detail, several individual images, made with different exposure times, were combined and processed electronically in a computer. The resultant composite image shows the solar corona approximately as it appears to the human eye during totality. Note the fine rays and helmet streamers that extend far from the Sun and correspond to a wide range of brightness. (Courtesy of Fred Espenak.)

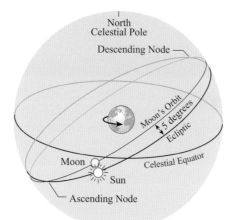

Fig 5.4 Celestial paths of the Moon and Sun   The orbit of the Earth's Moon is tilted 5 degrees to the Sun's route across the sky, the ecliptic, allowing these paths to cross at two nodes. These are the only points at which eclipses can occur. During a lunar eclipse the Moon and Sun are located at opposing nodes, so that the Moon can move into the Earth's shadow cast by the Sun. A solar eclipse occurs when the Moon and Sun cross paths at the same node.

a total eclipse of the Sun does not happen very often. The Moon only passes between the Earth and the Sun about three times every decade on average. Even then, a total eclipse occurs along a relatively narrow region of the Earth's surface, where the tip of the Moon's shadow touches the Earth (Fig. 5.5). At other nearby places on the Earth, the Sun will be partially eclipsed, and at more remote locations you cannot see any eclipse of the Sun.

The Moon's orbital motion carries its shadow rapidly eastward across the ground at about 1600 kilometers per hour. As a result, the longest total eclipse of the Sun observed at a fixed point on the ground lasts just under eight minutes.

If the Moon is at a distant part of its orbit at the time of solar eclipse, the Moon appears smaller than the Sun, and the tip of the Moon's shadow does not quite reach the Earth. The bright ring of the Sun's disk is then seen around the edge of the Moon. This is an annular eclipse, and it has none of the darkness and excitement of a total eclipse.

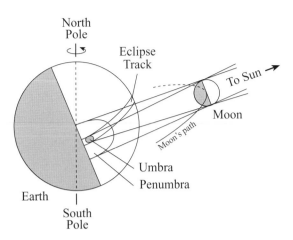

Fig. 5.5 Solar eclipse During a solar eclipse, the Moon casts its shadow upon the Earth. No portion of the Sun's photosphere can be seen from the umbra region of the Moon's shadow (*small gray spot*); but the Sun's light is only partially blocked in the penumbra region (*larger half circle*). A total solar eclipse, observable only from the umbra region, traces a narrow path across the Earth's surface.

5.3 The Moon's face

When a full Moon rises or sets, it is a captivating sight. It looks huge, dwarfing everything in the foreground (Fig. 5.6), and as the song goes, "when the Moon hits your eye like a big pizza pie, that's amore". But appearances can be deceiving. The Moon is no bigger when it is close to the horizon than when it is high in the sky. Its changing apparent size may be an illusion caused by comparing the Moon to other objects when it is viewed along the ground.

This so-called Moon illusion seems to arise from the way that the brain deals with apparent distance, not size. When people view the Moon near the horizon, there are large foreground objects, such as trees, buildings and hills, for comparison, so the Moon looks very far away and huge. When the Moon is overhead, alone in an otherwise empty sky, there are no other objects to gauge its distance; the Moon then appears to be closer and we think it is smaller than at the horizon (Fig. 5.7). Our perception of the dome of the background sky may also play a role in the Moon illusion.

Artists often portray the Moon's face in all of its round fullness, and the full Moon is the subject of all kinds of myths and superstitions (Focus 5.1).

Although it appears bright in contrast to the night sky, the Moon's face is as dark as the asphalt on highways and darker than most rocks on Earth. The fraction of incoming sunlight that is reflected from the lunar surface is known as its *albedo*, and values range from 5 to 10 percent for the darkest regions and 12 to 18 percent for the brightest ones. On average, the lunar surface reflects just 12 percent

Fig. 5.6 An enormous Moon In this awesome picture, a man and child seem enveloped by the Earth's Moon, which looks huge in comparison to the tree in the foreground. When the Moon is overhead, alone in an otherwise empty sky, there are no other objects to gauge its distance; we then think the Moon is smaller than at the horizon. (Courtesy of der Foto-Treff.)

Fig. 5.7 Moon illusion We make decisions about size because of our perceptions of distance. The two black disks in this figure are the same size, but we see the bottom one as smaller because we think it is closer. The top disk seems larger because it appears to be farther away. The Moon on the horizon is similarly thought to be huge because comparisons with objects on the ground make us think it is far away – also see Fig. 5.6. When people look straight up at the Moon, in an otherwise empty sky, they no longer have land clues to compute the Moon's distance and it is perceived as being closer and smaller.

of the sunlight that strikes it, which makes it one of the least shiny objects in the solar system.

Earth-bound observers always see the same side of the Moon. We call this the near side of the Moon, in contrast to the far side, not visible from Earth. The far side of the Moon is not its dark side, for it is illuminated by sunlight in the same way as the near side.

Gravitational interaction between the Earth and its Moon tie the two together, and lock the Moon's rotation into synchronism with its orbital motion. Our planet's greater gravitational pull on the near side of the Moon brakes the Moon's rotation and holds it in place like an invisible string, so the Moon's sidereal rotational period is precisely equal to its sidereal orbital period of 27.322 days, the time it takes for the Moon to return to a given position among the stars. In other words, the Earth's gravity has synchronized the Moon's rotation with its orbital motion, so the Moon rotates on its axis once each orbit. This condition, in which the spin of one body is precisely equal to, or synchronized with, its revolution around another body, is known as a synchronous orbit.

You can demonstrate synchronous rotation by holding a ball at arm's length and slowly turning around. As your body completes one rotation you always see the same side of the ball, but the ball has completed one rotation while revolving once about your body. But you can't

Focus 5.1   Full Moons

Ancient Greeks thought marriages consummated during a full Moon would be prosperous and happy. In England, a distinction was made between lunacy and insanity; the former happened only during a full Moon, while the latter was permanent. Yet there is no scientific evidence that people become abnormally crazy at the time of full Moon. The Navajo Indians believed that a woman is more likely to give birth during a full Moon because of its pull on the amniotic fluid. It has even been rumored that a male child is more likely to be conceived when aided by the extra gravitational pull of a full Moon. Of course, the pull of lunar gravity depends only on the Moon's mass and distance, and has no direct connection with the amount of sunlight that we see illuminating it.

A full Moon is considered unlucky on Sunday; the Sun's day, but lucky on Monday, whose name is derived from Moon Day. The phrase "once in a blue Moon" refers to the second full Moon in a single month, which typically occurs every few years. The reason for the rarity is that the 29.5306 days between full Moons is just slightly shorter than the average month of 30.4369 days in the average year of 365.425 days.

A blue Moon is also the fourth full Moon in a season, which normally has three, and by the way, the last time a month elapsed without at least one full Moon was in February 1866, an event that will not repeat itself for 2.5 million years.

At the time of harvest Moon, the full Moon rises at sunset, providing extra time for farmers to harvest their crops. According to folklore, the harvest Moon also appears bigger than other full Moons, but this is because it stays close to the horizon. A full Moon that illuminates the landscape all through the night is known as a hunter's Moon. All full Moons look bigger near the ground than directly overhead, a visual effect known as the Moon illusion.

The yellow color of a rising full Moon is due to scattering of light in the great thickness of air near the direction of the horizon; the haze and humidity of summer air can provide an orange color. It has even been suggested that the term "honeymoon" derives from the amber-colored full Moons of June, but the origin of the word "honeymoon" dates back to the 16th century, when the first month of marriage was said to be the sweetest.

Fig. 5.8 The gibbous Moon   The gleaming light of a gibbous Moon is shown in this digital combination of several high-resolution Earth-based images and a representative background star field. Though not visible to the eye, even with a telescope, the color differences at various places on the lunar surface are real. They correspond to regions with different chemical compositions that have been carefully mapped using spectrometers in satellites orbiting the Moon. (Courtesy of Noel Carboni.)

watch someone else do this; you have to demonstrate it to yourself.

Although the same side of the Moon always faces the Earth, this doesn't mean that one side of the Moon is always dark. Like the Earth, the Moon gets its light from the Sun, and sunlight always illuminates one half of the Moon. As the Moon orbits the Earth we see varying amounts of its illuminated near side. When we see a full Moon, the near side is in sunlight and the far side is dark. And when a new Moon is seen from Earth, the near side is dark and the far side is in full sunlight. In between new and full Moon we see either a crescent Moon (concave) or a gibbous Moon (bulging out from the sunlit side; Fig. 5.8). Thus, although there is a "dark side of the Moon", it is not equivalent to the "far side".

The Moon is the only planetary body that can be distinguished with the unaided eye as a globe, and even without a telescope you can tell that its surface is not uniform. Its face contains large irregular features of light and dark material (Fig. 5.9), familiarly known as the "Man in the Moon".

It wasn't until Galileo Galilei (1564–1642) turned his primitive telescope to the Moon that it became clear that our satellite is rugged and mountainous like the Earth. When he looked closely at the division between light and shadow − day and night − on the globe of the Moon, Galileo discovered that the dividing line was ragged and that he could see high mountain peaks casting long pointed shadows. When sunlight strikes the lunar surface obliquely, every mountain, hill or valley is sharply delineated.

Since he had clear evidence that the Moon was not the perfectly smooth crystalline sphere that had been proclaimed in the writings of Aristotle (384–322 BC), Galileo could write in 1610 that:

> the surface of the Moon is not perfectly smooth, free from inequalities and exactly spherical, as a large body of philosophers considers with regard to the Moon and other [heavenly] bodies, but on the contrary, it is full of inequalities, uneven, full of hollows and protuberances. It is like the surface of the Earth itself, which is varied everywhere by lofty mountains and deep valleys.

Fig. 5.9 The full Moon   Our Moon glows by light it reflects from the Sun, orbiting the Earth about once a month, where the "mon" in month is short for moon. The gleaming light of the full Moon, shown here, has beckoned humanity since ancient times, pulling us closer to the heavens and out into the cosmos. It lies suspended in space, always apart yet inextricably linked to the Earth. Terrestrial observers always see the same near side of the Moon. This view of the near-side full Moon enhances the contrast between the dark maria and the bright craters. The dark circular Mare Imbrium (Sea of Rains) is prominent in the northwest (*upper left*), immediately above the bright rays of craters Copernicus and Kepler (*middle left*). The dark circular Mare Serenitatis (Sea of Serenity) lies to the east (*right*) of Imbrium. (Photo courtesy of UCO/Lick Observatory.)

But do not be deceived by these Earthly comparisons. The conspicuous mountain ranges on the Moon were thrown up about 4 billion years ago as rims of impact basins gouged out by immense cosmic collisions, and the lunar mountains have nothing to do with the plate tectonics that created the much younger terrestrial mountains.

The Moon's rough terrain is mostly confined to the brighter regions that Galileo called *terrae*, Latin for "lands"; they are now known as the highlands because they are higher than the dark regions (Fig. 5.10, left).

Galileo also discovered that the dark patches are smooth and level, resembling seas seen from a distance.

He called them *maria*, the Latin word for "seas"; *mare*, pronounced "MAHrey", is the singular for "sea" (Fig. 5.10, right). However, we now know there are no substantial amounts of water in the maria. The dark maria cover about 17 percent of the lunar surface. When spacecraft were sent past the Moon to look at its averted face, they found that the far side contains very few maria (Fig. 5.11). Altogether, The heavily cratered highlands cover more than 80 percent of the Moon's total surface.

Chemical examination of rock samples returned from the Moon has shown that the maria are ancient volcanic outflows composed of dark lava. This material flowed out from inside the Moon to fill large impact basins that were formed at about the same time as the lunar highlands (Fig. 5.12; Table 5.2). One of them, the Imbrium Basin that contains Mare Imbrium, now forms the "eyesocket" in the face of the "Man on the Moon"; it has a diameter of 1500 kilometers.

Craters form one of the most striking features of the Moon's landscape (Fig. 5.13). There are at least 30 000 of them with a diameter greater than one kilometer. The word *crater* is derived from the Greek word for "cup or bowl", and it is a good description of the bowl-shaped depressions. They are just beyond the limit of visibility with the unaided eye, but a pair of binoculars will reveal a few of the larger ones. When seen through a telescope, the bright highlands are resolved into an enormous number of overlapping craters that have been visible to generations of telescope-using observers.

At around the time of full Moon, a pair of binoculars will also show bright streaks that radiate from several craters like the spokes of a wheel. These are the lunar rays, and the debris of crater formation produced them. Some of the rays go more than one-quarter of the way around the Moon (Fig. 5.14).

The ubiquitous craters were thought to be of volcanic origin throughout the 19th century and well into the 20th century, but they are now widely known to be due to the explosive impact of interplanetary projectiles (Focus 5.2). Although the maria are filled with ancient volcanic outpourings of molten rock, it spread out rapidly and did not build up in one place. So there are no large volcanoes or calderas on the Moon.

Unlike the Earth, where erosion and tectonic processes tend to obscure the effects of impact and to destroy its ancient surface rocks, the surface of the Moon preserves a pristine record of an ancient bombardment extending back several billion years. Even the youngest rocks on the Moon are as old as some of the oldest rocks found on Earth, about 3.2 billion years. Many of the planets and other satellites in the solar system bear the scars of a similar ancient rain of debris, providing a common element in their history.

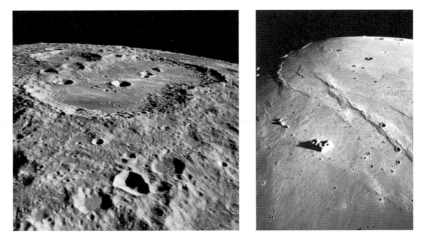

Fig. 5.10 Rough highlands and smooth maria on the Moon   (*Left*) The heavily cratered lunar highlands are illustrated in this *Apollo 17* image of the Van de Graaff crater. It is located on the far side of the Moon, at the northeast edge of Mare Ingenii. This unusual crater formation seems to be composed of two merged craters with no intervening rim, and the surrounding region is saturated with craters upon craters formed during an intense bombardment of the Moon about 3.9 billion years ago.

(*Right*) Lunar volcanism is seen frozen into place in this *Apollo 15* image of Mare Imbrium, the Sea of Rains. The oblique perspective enhances individual lava flows and the relief of lunar maria ridges. The lunar maria contain relatively few craters when compared with the lunar highlands. The maria formed a secondary crust on the Moon, when lava filled the large impact basins over a period of several hundred million years ending around 3.2 billion years ago. The fluid spread rapidly, creating thin extensive sheets rather than piling up to form volcanoes. (Courtesy of NASA.)

Fig. 5.11 Far side of the Moon
Locked into synchronous rotation by tidal interaction with the Earth, our Moon always presents its familiar near side to us. The far side, which remains invisible from Earth, is seen from Moon-orbiting spacecraft. This image, taken from *Apollo 16*, shows the eastern edge of the near side (*left*) and the rough, heavily cratered far side of the Moon, which contains fewer smooth, dark lunar maria than the near side. This is most likely because the far-side crust is thicker than the near-side crust, so molten material, or magma, have greater difficulty in flowing to the surface to form smooth maria on the far side. (Courtesy of NASA.)

Table 5.2 Large impact basins and maria on the Moon[a]

Maria (Latin)	Seas (English)	Basin diameter (kilometers)
Oceanus Procellarum	Ocean of Storms	3200
Mare Imbrium	Sea of Rains	1500
Mare Crisium	Sea of Crises	1060
Mare Orientale	Eastern Sea	930
Mare Serenitatis	Sea of Serenity	880
Mare Nectaris	Sea of Nectar	860
Mare Smythii	Smyth's Sea	840
Mare Humorum	Sea of Moisture	820
Mare Tranquillitatis	Sea of Tranquility	775
Mare Nubium	Sea of Clouds	690
Mare Fecunditatis	Sea of Fertility	690

[a] Dating of rocks returned from the Moon indicate that the Imbrium, Serenitatis and Nectaris impacts occurred 3.85, 3.87 and 3.92 billion years ago.

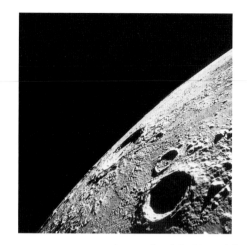

Fig. 5.13 Lunar craters Copernicus and Reinhold Bright ejected material radiates outward from the crater Copernicus near the lunar horizon. It is one of the youngest lunar craters on the near side of the Moon, with an estimated age of 800 million years and a diameter of 93 kilometers. The craters in the foreground are Reinhold A and B. (Courtesy of NASA.)

Fig. 5.12 Orientale impact basin on the Moon The Orientale impact basin, shown here in a *Lunar Orbiter IV* image, is nearly 1000 kilometers across. It was probably created during an intense bombardment of the Moon, about 3.9 billion years ago. The collision caused ripples in the lunar crust, resulting in three concentric circular rings, standing a few kilometers high. Molten basaltic lava from the Moon's interior subsequently flowed into the impact site, most likely about 3.2 billion years ago, creating the dark, smooth floor in the center of the basin. Much smaller, sharper, younger craters have impacted this ancient basin in more recent times. Located on the extreme western edge of the Moon, the Orientale basin is difficult to see from the Earth. (Courtesy of NASA.)

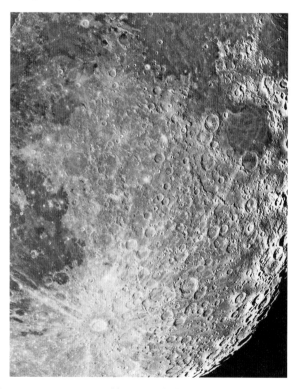

Fig. 5.14 Lunar rays White rays splash out across the Moon from crater Tycho (*bottom center of this image*). Tycho is a large, young crater with a diameter of 85 kilometers and an age of 108 million years. Only relatively recent craters retain their white rays, for those of older craters are darkened and worn away by continued meteorite impact. The dark, flat circular feature in the upper right is Mare Nectaris (Sea of Nectar). This clear image was produced using the unsharp masking technique that permits high contrast and fine resolution. (Anglo-Australian Telescope © 1976. Photo prepared by David F. Malin.)

Focus 5.2   Lunar craters – volcanoes or bombs?

Early interpretations of the lunar craters suggested they were formed by volcanic activity. At the time, volcanic craters were the only Earth craters known, and no impact craters had been recognized on Earth. In addition, for more than two centuries reputable astronomers reported seeing smoke and even fire coming from volcanic eruptions on the Moon.

Gradually, the evidence began to favor the idea that the craters are formed by meteoritic impact. It was found that the floors of most craters are slightly depressed below the surrounding level, in contrast to volcanic caldera that appear at the summits of volcanoes. In addition, large lunar craters contain flat floors and central peaks that are not often found in volcanic craters on the Earth. The central peaks of large craters on the Moon are created by the rebound of the underlying lunar surface following the collision of a big meteorite.

The round shape of nearly all the craters on the Moon can also be explained by the impact hypothesis. Because they strike at great speed, projectiles from space disintegrate in the explosive impact, producing a circular crater regardless of the direction at which the projectile struck. The lunar craters were also seen to resemble impact craters because the amount of material piled in the rims is nearly equal to the material excavated from the interior.

Ralph B. Baldwin (1912–2010) marshaled additional evidence for the explosive origin of lunar craters in his influential book *The Face of the Moon*, first published in 1949. He connected the relationship between the depth and diameter of craters on the Moon to the one describing the shell and bomb craters created during World War II, additionally noting that these man-made explosions have a circular form regardless of the angle of impact. Baldwin also argued that the dark, smooth maria occupy huge basins that were gouged out by rare, powerful impacts that punctured holes in the thin lunar crust, permitting lava to well out into them from the molten lunar interior.

The impact origin for lunar craters was confirmed when rocks were returned from the Moon. Samples from the highland craters and larger basins are conglomerates of pre-existing rocks that have been welded together by impact. Dating of these rocks indicate that the battered highland crust is a museum of impact scars created during an ancient bombardment about 4 billion years ago.

Although the lunar maria were filled during ancient episodes of volcanism, the Moon has apparently been volcanically inactive for 3 billion years. The supposed volcanic outbursts that were reported by several astronomers may have been explosive flashes of light generated during the impact of small meter-sized meteorites on the dark side of the Moon.

5.4   *Apollo* expeditions to the Moon

Race to the Moon

The Soviet Union triggered the Space Age about half a century ago, by launching the first artificial Earth satellite, *Prosteyshiy Sputnik*, the simplest satellite, in 1957, hurling the *Luna 3* spacecraft past the invisible far side of the Moon two years later, and sending the first human, cosmonaut Yuri A. Gagarin (1934–1968), into Earth orbit aboard the *Vostok 1* capsule two years after that.

Soviet officials cited there early accomplishments as evidence that communism is a superior form of social and economic organization. And the United States feared that a missile gap existed between it and its adversary, which seemed to verify the threat that the Soviet Union posed to world peace.

Stimulated by the worldwide excitement generated by the first human flight in space, the visionary President, John Fitzgerald Kennedy (1917–1963), decided that the United States had to defeat the Soviets at their own game, and deliver an American to the Moon.

Thus, on 25 May 1961, just six weeks after the Gagarin flight, Kennedy delivered his now-famous address to a joint session of Congress, including the declaration: "I believe that this nation should commit itself to achieving the goal, before the decade is out, of landing a man on the Moon and returning him safely to Earth". The president's call to action struck a responsive chord in the American public and was galvanized under the newly created National Aeronautics and Space Administration (NASA). In less than nine months, on 20 February 1962, John H. Glenn Jr. (1921–) became the first American to orbit the Earth, and the race to the Moon was in full tilt.

For several years the two superpowers traded accolades. The Russians sent the first woman – Valentina Tereshkova (1937–) – into space, and they were the first to orbit three men in the same spacecraft. On 18 March 1965 the Russian cosmonaut Aleksei A. Leonov (1934–) was the first to walk in space, from the *Vostkod 2* capsule, closely followed by the American astronaut Edward H. White (1930–1967) who took the first United States spacewalk on 3 June 1965 from the *Gemini 4* spacecraft.

During 1965 and 1966 the United States launched 10 successful flights of the two-man *Gemini* spacecraft, including the first rendezvous of two spacecraft, *Gemini 6* and *7*, and it was well prepared to embark on the *Apollo* program to land men on the Moon. It began with an ill-fated flight simulation on 27 January 1967, when faulty wiring ignited a flash electrical fire that asphyxiated and incinerated three astronauts on the ground. Yet, in just 22 months after this tragic setback, the manned *Apollo 8* spacecraft entered lunar orbit, all but ending the race to the Moon. It opened the way for the historic *Apollo 11* mission seven months later, when Neil Armstrong (1930–) planted the American flag on lunar soil. The achievement was a spectacular triumph, the ultimate space first in the global geopolitical competition with the Soviet Union.

From 1960 to 1970 the Soviet Union sent several unmanned spacecraft to the Moon, including roving vehicles and sample returns to Earth, but they had lost the race to put a man on the Moon. Personal rivalries, shifting political alliances and bureaucratic inefficiencies had apparently bred failures and delays. After the triumphs of *Apollo 8* and *11*, the Russian lunar program faded into oblivion, and they turned their attention to long-duration missions in Earth-orbiting space stations.

But to be honest, the driving factor in the race to the Moon was not scientific. It was the Cold War rivalry between the United States and the Soviet Union. And the American achievement was a spectacular political triumph. The dissipation of the Soviet Union's lead in space tarnished the image of Soviet competence and diminished their status in world affairs. In contrast, landing men on the Moon and conquering the frontier of space taught the American people that nothing is impossible if they set their sights high enough; with resolve and willpower you can accomplish anything, especially in a democratic nation that stresses individual freedom. And it was surely a contributing factor to the idea that success can result from technological superiority.

The *Apollo* program to land men on the Moon

Before the United States accomplished manned landings, three types of robot spacecraft were sent to reconnoiter and answer two main questions for the proposed lunar landing. The first concerned the danger of encountering rocky terrain, where it would be impossible to land without capsizing. The second was the prediction, by some astronomers, that a thick layer of dust covers the lunar surface, perhaps as deep as a kilometer, which would make travel impossible. In fact, the astronauts might sink into the dust, suffocate and vanish into the Moon, like sinking into quicksand on Earth. After all, the lunar surface has been battered, churned and worn down by a hail of meteorites over the eons, creating loose debris of rocks, pebbles, grains, soil and dust.

To start resolving these uncertainties, three *Ranger* spacecraft crashed into the Moon, transmitting television pictures back to Earth as they rapidly approached the lunar surface. Watching these pictures was a dizzying experience, and the transmission of the final frames was interrupted by the crash itself. These were followed by five *Lunar Orbiters* that mapped most of the Moon's surface to locate potential landing sites, missing only the polar regions. The final stage of preparation involved soft landings by the *Surveyors 1*, *3*, *5* and *7* that tested the detailed physical and chemical properties of the lunar surface and certified the safety of the initial *Apollo* landing sites. While the ground-control crews watched anxiously, the feet of the three-legged *Surveyor* robots sank only a few centimeters into the lunar soil, showing that there was no thick dust layer and people could walk on the Moon without sinking in over their heads.

The *Apollo* spacecraft was designed to carry three men into orbit around the Moon. A small, Spartan landing craft, the *Lunar Excursion Module* (*LEM*), would ferry two of the crewmen from lunar obit to the Moon's surface and then back to the mother ship, while the third astronaut remained orbiting the Moon in the larger *Command and Service Module* (*CSM*).

On 21 December 1968 three *Apollo 8* astronauts became the first humans to break free of the Earth's gravity. Although the crew would only orbit the Moon and not land on it, the unprecedented voyage provided the first sight of the Earth seen from afar – a radiant blue-and-white sphere rising beyond the battered face of the Moon in the dark void of space (Fig. 5.15). We then saw our home world in a new perspective, beautiful and vulnerable, a tiny, fragile oasis shimmering all alone in the vast deep chill of outer space. The sheer isolation of the Earth became plain to every person on the planet. It stimulated a worldwide awareness of the Earth as a unique and vulnerable place, fostering the ecology movement and helping us to get a better feeling for the planet's place in our lives and the Universe.

On 20 July 1969, the spindly-legged *Lunar Module Eagle* carried two *Apollo 11* astronauts to the lunar surface. While an estimated half-billion people watched, Neil Armstrong took the controls to avoid a hazardous crater, and radioed the first words from another world: "Houston, Tranquility Base here. The *Eagle* has landed."

With Buzz Aldrin (1930–) at his heels, Armstrong groped cautiously down the ladder to the surface. He stood firmly on the fine-grained surface, and an ancient dream

Fig. 5.15 Earthrise   In 1968, the *Apollo 8* spacecraft carried the first humans on a journey around the Earth's Moon. When they reached the far side of the Moon, the crew looked back toward the Earth along the lunar horizon, watching our planet rise as the spacecraft continued its orbit around the Moon. They helped create a new image of the Earth as a blue and turquoise ball suspended alone in dark space, light and round and shimmering like a bubble, flecked with delicate white clouds. (Courtesy of NASA.)

had come true – man had set foot on another world and humans were no longer confined to their native planet.

As Armstrong put it: "That's one small step for man, one giant leap for mankind." Although he forgot the "a" in front of "man", everyone knew what Armstrong meant. Moments after his initial footstep, Aldrin gazed out at the Sea of Tranquility and said simply "magnificent desolation". The next day, the Italian newspapers put it more succinctly: "Fantastico!"

After the historic landing, it was time to return to Earth. The two astronauts flew the ascent stage of *Eagle* back to the Moon-orbiting *CSM*, where Michael Collins (1930–) was waiting to take them home (Fig. 5.16).

By the time of the *Apollo 17* mission in 1972, the lunar landings had become so commonplace that astronaut Eugene A. Cernan (1934–) muttered "Let's get this mother out of here" as he blasted off the Moon in his *Lunar Module Challenger*. That's more like Sir Edmund Hillary's (1919–2008) comment when descending from the summit

of Mount Everest on 29 May 1953 – "We knocked the bastard off."

Still, the first human visit to the Moon was a historic occasion. Even now, there is a sense of participation, a feeling that our lives were enriched and made memorable by the landing.

In all, twelve humans have walked on the lunar surface, to gather samples, take photographs and make other scientific measurements (Table 5.3). All of the landing sites were on the near side and close to the lunar equator because these were the only places the astronauts could go safely (Fig. 5.17). Direct radio contact with Earth would be lost if they landed on the far side of the Moon. Sites near the equator were chosen to always be able to get astronauts back from the lunar surface quickly in case something bad happened on the Moon. A landing near the edge or limb of the Moon, as viewed from Earth, was ruled out if the spacecraft was to return to Earth in daylight. Within these constraints, the landing sites were chosen to provide samples

Table 5.3 *Apollo* missions to the Moon

Mission	Launch date[a]	Landing site	Accomplishments	Sample (kilograms)
Apollo 8	21 Dec. 1968	Lunar Orbiter	First humans to orbit Moon	
Apollo 10	18 May 1969	Lunar Orbiter	Test Lunar Excursion Module	
Apollo 11	16 July 1969	Mare Tranquillitatis	First human landing	21.6
Apollo 12	14 Nov. 1969	Oceanus Procellarum	First ALSEP[b]	34.3
Apollo 13	11 Apr. 1970	Flyby	Landing aborted	–
Apollo 14	31 Jan. 1971	Fra Mauro, highland	First highland landing	42.6
Apollo 15	26 July 1971	Hadley-Apennine	First lunar rover	77.3
Apollo 16	16 Apr. 1972	Descartes	Highland landing	95.7
Apollo 17	07 Dec. 1972	Taurus-Littrow	Last flight	100.5

[a] The spacecraft landed on the Moon four or five days after launch.
[b] ALSEP is an acronym for Apollo Lunar Surface Experiments Package.

Fig. 5.16 Going home from the Moon
The ascending *Lunar Module* of the *Apollo 11* mission, carrying Neil Armstrong (1930–) and Buzz Aldrin (1930–) back from the first human landing on the Moon. They were returning to the Moon-orbiting *Command and Service Module*, when its pilot Michael Collins (1930–) took this photograph, on 21 July 1969. (Courtesy of NASA.)

of a wide variety of terrain, from the smooth maria to the heavily cratered highlands.

Apollo 11 landed on the smooth plains of Mare Tranquillitatis, and *Apollo 12* settled down on a mare site near the edge of the vast Oceanus Procellarum. Rocks returned from these first missions confirmed the volcanic basalt nature of the maria and established their antiquity, with ages greater than 3 billion years. The *Apollo 14* landing site was located in highland terrain near the crater Fra Mauro, an area thought to be covered with debris thrown out by the impact that formed the Imbrium Basin. Dating of material obtained from this site indicated that the basin-forming impact occurred 3.85 billion years ago. *Apollo 15* was the first mission to employ a roving vehicle; it was sent to the Hadley-Apennine region containing both mare and highland units. The so-called Genesis

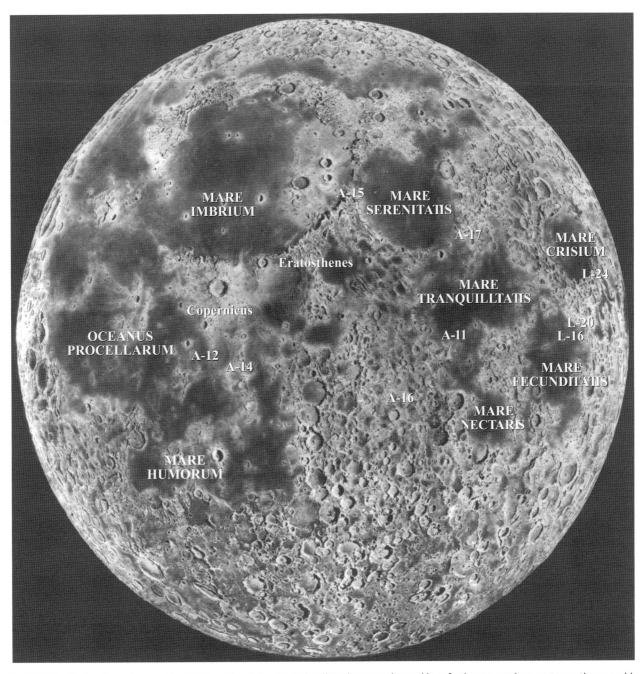

Fig. 5.17 Apollo landing sites on the Moon   The six *Apollo* (A) landing sites were located in safe places near the equator on the near side of the Moon. Within this constraint, the sites were designed to obtain samples from a wide variety of terrain. *Apollo 11* and *12* respectively landed on Mare Tranquillitatis and Oceanus Procellarum. The spot chosen for *Apollo 14* was the Fra Mauro Formation, which is covered with material ejected during the ancient impact that created the Imbrium Basin. By landing at a point just inside the Apennine Mountains, the *Apollo 15* astronauts could sample highlands, maria and the Hadley Rille. The *Apollo 16* mission sampled the highlands near crater Descartes, while *Apollo 17* landed near Mare Serenitatis. The location of the three Soviet *Luna* (L) unmanned sample-return sites are also shown.

rock found during this mission is a primitive chunk of highland anorthosite dating back some 4.5 billion years. *Apollo 16* landed on the Descartes highlands near the rim of the Nectaris basin, blasted out 3.92 billion years ago, and *Apollo 17* was sent to the Taurus-Littrow site, at the

edge of the Serenitatis basin, excavated 3.87 billion years ago.

What happened to *Apollo 13*? Its three astronauts almost lost their lives when an oxygen tank exploded aboard the spacecraft on the way to the Moon. Sealed

Focus 5.3 Black skies on the Moon

The lunar astronauts stepped into a stark but beautiful world. With no clouds, dust, moisture or haze to obscure the view, distant details stood out clearly against the deep black background. Because the Moon has no atmosphere to speak of, the sky was pitch black in broad daylight, there were no sounds to disturb the eternal stillness, and the Sun's true light could be seen, unfiltered by any air. And since there is no air to breathe on the Moon, visiting astronauts were bundled in oxygen cocoons known as spacesuits.

By way of comparison, incident sunlight contains all the colors, but the Earth's air molecules scatter blue light more strongly than red light, making the overhead sky appear blue. The Sun's rays being bent by the atmosphere cause the twilight zone between the night's darkness and sunrise or sunset on Earth. And when the Sun rises or sets, most of the blue light is scattered out before reaching us, so the light of the setting Sun is reddened; airborne dust helps this effect. In contrast, astronauts orbiting the Moon saw no twilight, and no colorful sunrise or sunset.

How do we know that the Moon has no appreciable atmosphere? The earliest evidence came from the abrupt vanishing of stars behind the edge of the Moon during lunar occultations. The word *occult* means "to hide". If the Moon had a thick atmosphere, the starlight would gradually dim during a lunar occultation, but this vanishing actually takes less than one second. It is in sharp contrast to the gradual fading of the Sun, stars, and planets when they set behind the horizon on Earth. The lack of a significant atmosphere also follows from the Moon's relatively small mass and gravity, together with its proximity to the warming Sun.

Instruments carried to the Moon by the *Apollo* astronauts identified the barest wisp of helium and argon atoms, and subsequent imaging observations from Earth revealed just a whisper of sodium and potassium atoms, emitting a detectable fluorescent glow when exposed to sunlight. But this is not a permanent atmosphere in the sense that ours is, for the lunar "exosphere" is continuously being created, lost and replaced every few hours or weeks, depending on the atom.

The Moon's atmosphere is 100 trillion, or 10^{14}, times more tenuous than the Earth's air, and so thin and rarefied that its constituent particles hardly ever hit each other. So the Moon has no atmosphere in any practical sense.

---

inside, the crew was in danger of dying by re-breathing their carbon dioxide. They survived by converting the tiny *Lunar Module*, with its intact, breathable air and fuel, into a lifeboat, canceling the planned lunar landing and heading home. *Apollo 13* was very nearly a catastrophe, which has been shown in the captivating 1995 Oscar-winning movie *Apollo 13*. If the tank had exploded earlier, there would not have been enough electric power and water to go around the Moon and get home again. And if it had occurred later, when the astronauts were on their way down to land on the Moon, there would not have been enough fuel left in the *Lunar Module* to go home.

The other lunar astronauts recorded an eerie wasteland below a blackened sky (Focus 5.3), battered and scarred with craters of all sizes and covered with dust. It clung to the astronauts' clothing and equipment and showed the sharp outline of their footprints (Fig. 5.18); but there were no clouds of dust above the airless surface. Walking on the lunar surface was like walking on plowed soil or wet sand, and most of the finer dust had evidently been plowed down into the Moon by the churning of the meteorites.

Armstrong and Aldrin never strayed more than a hundred meters from their lander, like timid children testing the water when entering a lake or sea for the first time. The astronauts of the next two missions (*Apollo 12* and

Fig. 5.18 Boot prints on the Moon On 20 July 1969, Neil Armstrong (1930–) became the first human to walk on the Moon. His boot print, shown here, reveals a thin layer of Moon dust, about 0.01 meters thick. Because there is no atmosphere or weather on the Moon, the footprint will probably remain for 1 or 2 million years. By that time, the constant rain of micrometeorites will have erased it. Altogether, twelve astronauts have left boot prints on the Moon. (Courtesy of NASA.)

Fig. 5.19 Driving on the Moon *Lunar Rovers* were used to travel across the Moon's rugged terrain, gathering rocks from a wide variety of locations. In this image, *Apollo* 15 astronaut James Irwin (1930–1991) prepares to take a *Lunar Rover* for a drive. The *Lunar Module* "Falcon" is at the left side of this image. The St. George Crater is located about 5 kilometers behind Irwin, and the lunar mountains Hadley Delta and Appennine Front are in the background at the left. The *Rovers* were left on the Moon. Free from wind, rain and rust, they will remain intact for millions of years; one might even imagine a returning astronaut using one that was discarded hundreds of years before. (Courtesy of NASA.)

14) had greater confidence and took longer moonwalks. During the last three missions (*Apollo 15, 16* and *17*) astronauts roamed as far as 7 kilometers from the landing site, visiting some of the most spectacular places on the Moon in a battery-powered car called the *Lunar Rover* (Figs. 5.19, 5.20). Still, the visits were always short, with the entire missions lasting between one and two weeks.

Unlike the early landings on the smooth lunar maria, the last three *Apollo* flights visited mountainous areas: the Appenines, the Descartes highlands and the Taurus Mountains. The tops of all the mountains were rounded off into gentle hills without sharp peaks or steep cliffs. Although it looked as if the Moon had been sandblasted smooth by eons of meteorite bombardment, the main reason for the

Fig. 5.20 Moonwalk   Charles Duke (1935–) strolls across the lunar surface during the *Apollo 16* mission in April 1972. Small impacting particles have sandblasted the lunar surface, producing smoothed, undulating layers of fine dust and rounding the surfaces of lunar rocks. Larger meteorites have pounded and churned the surface, producing a layer of ground-up rocky debris. (Courtesy of NASA.)

gradual lunar slopes is that there is no water or ice erosion, as on Earth, to cut deep valleys and shape mountain crags, or tectonic activity to toss up crumpled mountain ranges.

The astronauts left behind the Apollo Lunar Surface Experiments Package (ALSEP). This nuclear-powered array of instruments included seismometers to monitor vibrations of moonquakes and meteorite impact, magnetometers to measure possible magnetic fields, and other instruments to analyze gases and charged particles streaming from the Sun to the Moon. The astronauts also brought lunar soil and rocks back home with them, altogether 382 kilograms and not an ounce of cheese (Fig. 5.21).

Mirrors were also left on the Moon, to reflect laser light fired from Earth. Every reflector contained 100 small mirrors, each in the shape of the three-sided corner of a box. These corner cubes reflect light directly back toward its point of origin. Observations of pulsed laser light, sent to the lunar mirrors and back, has permitted astronomers to measure the Moon's distance with an accuracy of two centimeters, or to better than one part in 10 billion, showing that the Moon is moving very slowly away from the Earth.

Sophisticated experiments were also performed from the *Command Module*, mapping the magnetic fields, chemical composition, surface radioactivity and terrain from a distance as the mother ship circled the Moon.

Returning to the orbiting craft, the astronauts jettisoned the landing *Lunar Module* and headed for Earth, arriving home about three days later. Biologists felt there was a chance that the astronauts, or the returned rock samples, might infect the human race with some deadly lunar virus. The astronauts from the first three lunar landing missions, *Apollo 11*, *12* and *14*, were therefore placed in quarantine for three weeks after their return. They remained in fine health, and the crews of the last three missions, *Apollo 15*, *16* and *17*, did not have to suffer through the quarantine.

The achievement of landing humans on the Moon was a spectacular American triumph in the cold-war confrontation with the Soviet Union, in an incredible, warlike mobilization of scientists, engineers, and technology with an optimistic, can-do spirit. Imaginative thinkers viewed it as the stepping stone to permanent lunar bases, giant space stations and the colonization of Mars. But the dreams quickly dissipated and the sense of mission disappeared. The "age of *Apollo*" was short-lived, and public interest quickly waned.

The goal had been reached, the crisis was over, and the enemy had been conquered. The *Apollo* program lost its political appeal in the face of growing public indifference, its budget shriveled, and the program was abruptly cut

Fig. 5.21 Moon rock Harrison Schmitt (1935–) about to walk behind Split Rock during the *Apollo 17* mission in December 1972. Eugene Cernan (1934–) had already scooped up samples from the debris on the front side of the boulder. The huge rock rolled down about a billion years ago, splitting into five pieces during the fall. The total length of the boulder, when reassembled, is about 20 meters. *Apollo 17* was the last of the six missions that landed humans on the Earth's Moon and returned them safely. It investigated the dark terrain at the Taurus-Littrow landing site, deployed explosives to be used with seismographs to examine the lunar interior, and returned rocks to the Earth. (Courtesy of NASA.)

short with the cancellation of the *Apollo 18, 19* and *20* lunar missions, largely for political reasons.

The excitement must have also quickly dissipated for the men who visited the Moon. They had trained their whole lives to stay for just a few days on the Moon, waiting for years for that opportunity. A few moments of fame followed, but there was no encore. There wasn't anything comparable to do, and many of the astronauts had some hard adjusting to do after their return from the Moon.

5.5 Inside the Moon

Moonquakes

As the Earth has earthquakes, so the Moon has moonquakes which were first detected by the sensitive seismometers placed by the *Apollo* astronauts at four widely spaced locations on the lunar surface — at the *Apollo 12, 14, 15* and *16* landing sites. Because interfering winds, sea waves, and road traffic do not shake the Moon, the lunar seismometers can detect moonquakes that are relatively weak by terrestrial standards.

More than 12 500 seismic events were recorded over the eight years the seismometers were used. As expected, they were able to record occasional tremors of the Moon caused by the impact of small meteorites, as well as a shuddering from the deliberate crash landing of *Lunar Modules* near the end of some *Apollo* missions. But a great many more events were generated inside the Moon.

A few of these moonquakes, 28 in all, were shallow, emanating from the upper mantle, the largest about 5.0 on the Richter scale. But more unexpected were the numerous tiny moonquakes, several a day on average, which occurred further down, about halfway to the center of the Moon and deeper than any earthquake.

These deep moonquakes are much smaller than even mild earthquakes. If you stood directly over most moonquakes you would not even feel your feet shake. They almost never exceed a magnitude of 2 on the Richter scale. Although earthquakes of this magnitude are recorded on Earth, they are not felt by humans and produce no damage to buildings.

The moonquakes are not only gentler than earthquakes; they also have distinctly different behavior. While tremors on the Earth start suddenly and persist for only a few minutes, the moonquake waves build up gradually and continue for more than an hour, suggesting that the body of the Moon is an almost perfect medium for the propagation of seismic waves.

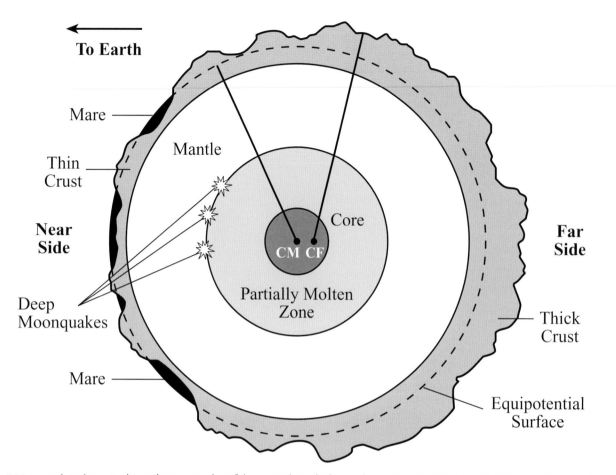

Fig. 5.22 Lunar interior A schematic cross-section of the Moon shows its internal structure. The lunar crust is thinner on the near side that faces the Earth, and thicker on the far side. Fractures in the thin crust have allowed magma to reach the surface on the near side, where the lava-filled maria are concentrated. The Moon has an iron-rich core with a radius of about 20 percent of the Moon's average radius of 1738 kilometers. A partially molten layer is believed to encircle the Moon's core, out to depths of about 1000 kilometers. The Moon's center of mass (CM) is offset by 2 kilometers from its center of figure (CF), so an equipotential surface, which experiences an equal gravitation force at all points, lies closer to the lunar surface on the hemisphere facing Earth. Therefore magmas originating at equipotential depths will have greater difficulty reaching the surface on the far side.

Moreover, nearly identical seismograph records were obtained again and again, indicating that certain regions, known as *nests*, are repeatedly generating moonquakes in the same way. The largest nest emitted 323 moonquakes during the period of observation.

The rate of the moonquakes, in general, seemed to rise and fall every 27 days, the time it takes the Moon to circle the Earth, suggesting that they were caused by the tidal pull of the Earth.

The seismograph records have permitted the construction of a model for the lunar interior, in much the way that geologists have modeled the inside of the Earth. The Moon is slightly asymmetrical in bulk form, with a thicker crust on the far side. Most of the volcanic maria occur on the near side where the crust is thinner (Fig. 5.22). On average, the crust on the near side of the Moon is about 50 kilometers thick, or less. It is only a few tens of kilometers thick beneath the mare basins. In contrast, the far side crust is believed to be about 15 kilometers thicker than the near side crust. As a result, the Moon's center of mass is offset from its geometric center by about 2 kilometers in the direction of the Earth.

A small lunar core

Scientists from the *Apollo* era were unable to agree whether the Moon has an iron-rich core; but they were certain that it had to be much smaller than the core of the Earth, which is 55 percent of the radius of the planet and 32 percent of the planet's mass. Data from the *Lunar Prospector* spacecraft, obtained in 1998–99, have been used to gauge the size of the lunar core. They indicate that the Moon's core has a radius of about 350 kilometers, or 20 percent of the satellite's radius, and only about 2 percent of the body's mass.

Radio telescopes on Earth were used to measure small Doppler-effect changes in the *Lunar Prospector*'s radio signal as the spacecraft moved toward or away from the Earth, thereby identifying slight variations in the craft's velocity as it orbited the Moon. Since these velocity changes are caused by the varying gravitational pull of the Moon, they could be used to construct a full gravity map of the near and far sides of the Moon, from pole to pole. The resulting map revealed the distribution of mass within the Moon, and showed that it has a small, dense, metallic core, about 350 kilometers in radius if mostly iron.

A second method studied the weak magnetic field induced within the Moon when it passes through the tail of the Earth's magnetosphere each month. This technique confirmed the presence of a lunar core of about the same size as that inferred from the gravity data.

The relatively small core of the Moon has profound implications for its origin. If the Moon and Earth coalesced independently, their cores might be expected to occupy a similar fraction of their volume; instead the Moon seems to have coalesced from material that has been blasted out of the young Earth. The giant impact might have taken place when the Earth was still forming. In that case, most of the Earth's iron might have already sunk to its core, but there could be enough iron-rich rock, expelled into space from the Earth and impacting object, to build a lunar core.

Internal, partially molten zone in the Moon

The *Apollo* seismic data suggested that the outer half of the Moon is cold and solid, but that it might be warm and partially molten in a lower zone. The moonquake waves lost energy if they went deeper than 1000 kilometers, about halfway to the center of the Moon at 1738 kilometers down. The *Apollo* scientists argued that the deep moonquakes might be generated at the boundary between the outer solid shell and the inner molten zone.

Evidence for a partially molten zone has been obtained from accurate measurements of the Moon's distance using laser ranging. The laser beams are sent to the Moon from telescopes on Earth and reflected from corner mirrors left on the Moon by the *Apollo* astronauts. Measurements of the round-trip travel time of a pulse of laser light yields twice the distance between the Earth and the Moon, by multiplying the time by the velocity of light, with an accuracy of 0.02 meters. After more than 30 years of such determinations, scientists have concluded that the Moon's surface moves in and out by as much as a tenth of a meter, or 10 centimeters, every 27 days, in response to the shifting gravitational tugs of the Earth. This elastic yielding suggests that the interior is pliable, with a larger, partially molten layer surrounding the core.

Mascons on the Moon

Precise radio tracking of *Apollo* spacecraft on the near side of the Moon showed that their orbits are gravitationally deflected toward the circular maria. The spacecraft acted as though the maria contained mass concentrations, abbreviated as *mascons*, which pulled at the spacecraft and changed their velocity when passing overhead. Virtually all the maria on the near side showed this unexpected feature, and the excess mass in each is about 10^{18} kilograms, or 1/70 000 the total mass of the Moon. Because radio tracking of the orbiting spacecraft was not possible when they passed to the far side, it was not known from the *Apollo* missions whether there are mascons on the far side of the Moon.

The *Lunar Prospector* spacecraft has been used to detect mascons from gravity data for the entire Moon, discovering several mass concentrations beneath the floors of large impact basins, including at least four on the far side. The nearside basins have been filled with mare lava, but the farside basins remain unfilled, so it isn't the lava that is providing the extra gravitational pull.

What are the mascons? The most likely explanation is that they represent an upward bulging of high-density mantle rocks that rose in the aftermath of basin-forming impacts. The impact that formed the largest basins has weakened the crust so much that the dense mantle has moved up beneath them, raising and fracturing the basin floors.

5.6 The lunar surface

Rocks from the *Apollo* missions

During the *Apollo* landings from 1969 to 1972 a dozen people roamed the Moon taking hundreds of rock samples, placing them in labeled bags, and returning 382 kilograms of Moon rocks to Earth in sealed containers. These specimens from another world have permitted scientists to decipher the composition of the lunar crust, and to reconstruct our satellite's history.

Since contact with the Earth's atmosphere would alter the composition of the lunar samples, they are kept in cabinets filled with a dry, oxygen-free atmosphere of nitrogen and are manipulated with long gloves sealed to the walls of the cabinets. When not under investigation, the rocks are kept in a massive steel-lined vault at the Lunar Receiving Laboratory of NASA's Johnson Space Center at Houston, Texas.

Scientists spent years examining the rocks brought back from the Moon in search of water, but none was found. For more than 40 years the Moon was therefore thought to be bone dry. As discussed in Section 2.4, orbiting lunar spacecraft nevertheless detected signs of surface water on the Moon in the late 1990s and early 2000s, and this probably led to a re-examination of rocks collected during the *Apollo* missions. An analysis of the lunar samples in 2008 showed evidence of very small amounts of water, at about 50 parts per million, in volcanic glasses formed on the Moon about 4 billion years ago. The chemical signatures of water on the Moon were confirmed in 2010, in the form of hydroxyl ions, each containing one atom of oxygen and one atom of hydrogen. The hydroxyl was locked up in mineral crystals as water-bearing magma cooled on the Moon's surface soon after it formed.

Although scientists have now found 100 times more water in the Moon's minerals than previous limits, the concentrations are very low and until recently impossible to detect. Moreover, the Moon was never soaking wet, and both its interior and surface are still considered drier than the driest desert on Earth.

The Moon is also lifeless. Extensive testing revealed no evidence for life, past or present, among the lunar samples. They contain no living organisms, fossils or native organic compounds. Thus the Moon is a desolate place, barren of life.

From its low mean mass density, we would expect the bulk of the Moon to be composed of silicates, or minerals in which atoms of silicon and oxygen are linked to other elements. Laboratory investigations of the lunar samples indicate that the lunar crust is indeed composed of such minerals, just as the Earth's crust is.

The surface of the Moon has been bombarded by meteorites for billions of years, breaking the lunar crust into rock fragments and fine-grained material known as the lunar *regolith*. It is the loose debris that has fallen back to the Moon after eons of meteorite bombardments. The regolith is the Moon's version of soil, which has organic connotations here on Earth, but the Moon's soil contains no organic material.

The regolith covers the entire lunar surface to depths as great as 20 meters. It is thickest in the highland regions that have been exposed to meteoritic bombardment longest. The regolith in the maria is 2 to 8 meters deep.

The Moon rocks are roughly divisible into three types: anorthosites, basalts and breccias, and they all exhibit important differences from terrestrial rocks. The lunar samples are all much older than most rocks found on Earth, and they are composed of material that has been previously melted (anorthosites), erupted through magma–lava outflow (basalts), and crushed by meteorite impacts (breccias).

The *anorthosites* are the oldest rocks ever found, dating back to 4.5 billion years ago. They are found in the light-colored lunar highlands, and contain a type of mineral known as plagioclase feldspar, commonly found on Earth, but with a difference. The lunar anorthosites were melted more than 4 billion years ago.

Basalts are dark lava that fills mare basins, forming a secondary crust. These thin volcanic veneers were created after heat from radioactive decay accumulated in the Moon, leading to the rise of magma and the eruption of basaltic lava about 3 billion years ago. The surface of Venus and the Earth's ocean floor are also secondary crusts formed in this way. But most of Venus was resurfaced about 750 million years ago, and the Earth's ocean floor is still being created.

The Moon contains none of the type of rocks that were generated deep inside the Earth during plate tectonic processes, such as the continental granites.

After the lunar rocks solidified, they were broken up, flung about and pulverized by meteorite impacts. Energetic impacts, powerful enough to excavate meter-sized craters, have compacted and welded the regolith into aggregates called *breccias*. They retain compositional information from the era in which they formed.

Scanning the surface of the Moon from spacecraft in the 1990s

The *Apollo* rock and soil samples came from only six sites on the near side, chosen mainly to be safe and easy to get to. A global perspective of the Moon's surface composition therefore had to wait until the *Clementine* spacecraft surveyed the unexplored regions on both the near and far sides in the 1990s.

After the *Apollo* missions, no one had even a single glimpse of the Moon's far side for nearly two decades, and then it was obtained by the *Galileo* spacecraft on its way to explore Jupiter's realm. In order to reach the giant planet, *Galileo* gained speed by twice swinging past the Earth, passing by the Moon in the process. In 1990 and 1992 the instruments on *Galileo* obtained images of the lunar limb and far side from vantage points not previously obtained.

Composites of *Galileo* images taken in three colors – violet, red and near infrared – have been used to depict compositional variations of the lunar surface (Fig. 5.23). They have been calibrated by *Apollo* sample returns that specify the chemistry at specific sites on the near side of the Moon. Some mare basalts are rich in titanium, while many others are relatively low in titanium but rich in iron and

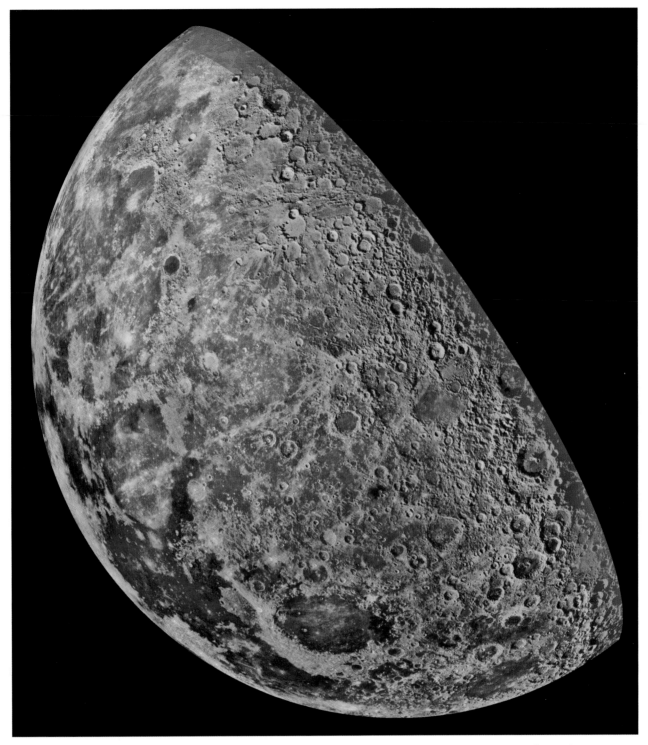

Fig. 5.23 Compositional variations on the Moon's surface   This mosaic of images, taken through three spectral filters, shows exaggerated false-color differences in reflected sunlight in order to specify composition differences on the lunar surface. The image shows volcanic flows with relatively high titanium content (*blue*), volcanic flows that are low in titanium but rich in iron and magnesium (*green, yellow and light orange*), and heavily cratered highlands that are typically poor in titanium, iron and magnesium (*pink and red*). In this view, taken by *Galileo* on 7 December 1992, bright pink highlands surround the lava-filled Crisium impact basin (*bottom*), and the dark blue Mare Tranquillitatis (*left*) is richer in titanium than the green and orange maria above it. The youngest craters have prominent blue rays extending from them. (Courtesy of NASA/JPL.)

magnesium. The heavily cratered highlands are typically poor in titanium, iron and magnesium.

In early 1994, the United States Department of Defense placed a small spacecraft in orbit about the Moon. In sharp contrast to the eight-year, $25 billion *Apollo* program, the tiny unmanned satellite required only two years and $75 million to build and launch.

Because one of the mission's byproducts was to prospect the surface mineral content of the Moon, the spacecraft was given the name *Clementine*, after the miner's darling daughter in the old Gold Rush ballad. However, this was not the main purpose of the spacecraft. It was built primarily as a military test of lightweight electronic imaging sensors that could detect the launch and track the flight of enemy ballistic missiles, possibly for use in a future star-wars missile shield.

Like its namesake, the spacecraft was "lost and gone forever" after orbiting the Moon for two months, but not without first chalking up an impressive list of accomplishments. Unlike the *Apollo Command Modules* that circled the Moon in low near-equatorial orbits, *Clementine* orbited across the lunar poles, permitting a global perspective as different regions rotated into view (Fig. 5.24).

Its ultraviolet, visible and infrared cameras took pictures at eleven different wavelengths used to identify different types of minerals. Since various rock-forming minerals reflect and absorb incident sunlight at different wavelengths, the *Clementine* data could be used to infer the chemical composition of most of the lunar surface, and rock samples could be used to calibrate the global data when it overlapped the *Apollo* landing sites.

The *Clementine* global data was used to map the abundance and distribution of iron on the Moon, showing that the dark, nearside maria consist of iron-rich lava, containing up to 14 percent iron by weight. In contrast, iron is practically absent in the nearside highland crust and across vast tracts of the far side, at about 3 percent iron by weight. These regions of very low iron content are dominated by aluminum-rich anorthosite.

The highland crust on both the near and far sides is just what one would expect if the entire Moon were once covered in liquid rock at least several thousand kilometers deep. The heavy iron sank into this magma "ocean", while low-density feldspar (anorthosite) grains accumulated into floating "rockbergs" that coalesced and cooled to form the eventual constituents of the Moon's highlands. Some of the iron subsequently resurfaced when the Moon heated up inside and magma flowed up to the nearside maria.

Large meteorite impacts dig holes into the lunar crust, exposing deeper material and revealing its composition. For instance, the floor of the South Pole – Aitken basin on

Fig. 5.24 *Clementine* **observes the Moon, Sun, and Venus** In this image, taken from the *Clementine* spacecraft in 1994, the Sun is just behind the Moon's limb, or edge, so most of the lunar surface is illuminated by light reflected from the Earth, known as Earthshine. Light from the Sun's outer atmosphere, the solar corona, produced the bright glow on the lunar horizon. The planet Venus is seen at the top of the frame. (Courtesy of NASA/JPL/USGS.)

the far side has an iron abundance of nearly 10 percent by weight.

Until *Clementine*, we also had no global map of the topography of the Moon. The laser altimeter on the spacecraft fired pulses of light at the Moon once every second and timed how long it took for the light beam to travel down to the lunar surface and back again. This enabled scientists to determine the distance to the surface, over and over again, with an accuracy of 50 meters. When these distances were combined with knowledge of the spacecraft orbit, maps of the elevation, or topography, of the entire lunar surface were obtained (Fig. 5.25).

The new global maps showed an unexpected range of heights, over 16 kilometers and comparable to that seen in the geologically different Earth. The wide range of relief on the Moon is caused by the presence of large impact basins. The huge South Pole – Aitken basin, which is over 12 kilometers deep and about 2600 kilometers across, dominates the farside topography. The near side is relatively

Table 5.4 The *Clementine* and *Lunar Prospector* missions to the Moon

Mission	Launch date	Lunar orbit	Accomplishments
Clementine	25 Jan. 1994	19 Feb. 1994 to 3 May 1994	Global surface composition, global topography, map of South Pole – Aitken basin, possible water ice at poles
Lunar Prospector	6 Jan. 1998	15 Jan. 1998 to 31 July 1999	Global elemental abundance, global magnetic field maps, global gravity maps, detection of lunar core, water ice at poles.

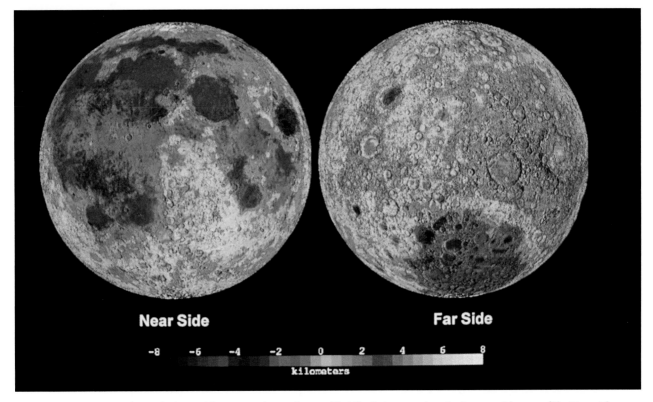

Fig. 5.25 Lunar topography The laser altimeter on *Clementine* provided the first comprehensive topographic map of the Moon. The contour interval is 500 meters, and the altitude in kilometers is coded by color from blue to red (*bottom*). The near side (*left*) is relatively smooth and low (*blue* and *purple*), primarily because of the prominent impact basins, including Imbrium, Crisium and Nectaris, which are at least partly filled with mare basalt. In contrast, the far side (*right*) shows high relief (*red*) and extreme topographic variation comparable to that of the Earth. The Moon's wide altitude range is attributed to ancient impact basins that have been preserved for about 3.9 billion years, while the Earth's wide range stems from ongoing mountain building by colliding tectonic plates. The large circular feature on the southern far side (*right bottom*) is the South Pole – Aitken basin, 2600 kilometers in diameter and 12 kilometers deep. (Courtesy of Paul D. Spudis, Lunar and Planetary Institute.)

smooth, with typical relief of about 5 kilometers, primarily because its impact basins have been filled with mare basalt. As substantiated by *Clementine* gravity data, a thicker crust blocks the outward flow of magma on the far side.

In the face of dwindling budgets and growing public interest in problems here on Earth, scientists and engineers found less-expensive ways of exploring space. NASA adopted a new "smaller, faster, cheaper" mode of operation, in which several cost-effective, high-risk spacecraft rather than a few major, expensive and low-risk ones do science. The *Lunar Prospector* spacecraft is an example. Launched on 6 January 1998, it was designed to obtain global data on elemental abundance, magnetic fields and gravity fields, and it has also achieved an impressive array of accomplishments (Table 5.4).

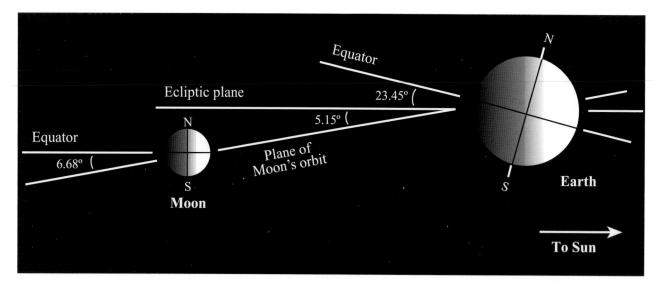

Fig. 5.26 Dark, cold lunar poles   The near-vertical orientation of the Moon's north–south rotation axis to the ecliptic plane creates permanent night and deep freeze at the floor of craters located at the lunar poles. These regions might be reservoirs of water-ice, delivered there by comets. The angle between the Earth's equator and the ecliptic, or the plane of the Earth's orbit around the Sun, is 23.5 degrees, and this tilt produces the seasons. The Moon provides a steadying influence for the Earth's tilt, keeping it from varying widely and producing dramatic climate variations. Also note that the plane of the lunar orbit falls neither in the Earth's orbital plane nor in the ecliptic.

The Moon is not completely dry

Bright radar echoes, returned to *Clementine* from the south pole of the Moon, suggested that this region might contain radar-reflective water ice. Instruments aboard the *Lunar Prospector* spacecraft then strengthened the possibility of water ice at the south pole, and also discovered what appears to be additional ice near the north pole. During its passes over the poles, an instrument on *Lunar Prospector* detected substantial quantities of hydrogen, which mission scientists attributed to water ice found in permanently shaded areas near both poles. They estimate that there could be as much as 6 billion tons, or 6×10^{12} kilograms, of water ice located in the polar regions.

The Moon's rotation axis is orientated nearly perpendicular to the ecliptic plane (Fig. 5.26), so the lunar poles are never tilted toward the Sun by more than a very small amount. This means that the bottoms of craters at the poles are in constant shadow and in a perpetual deep freeze with temperatures of 50 to 70 kelvin. Any ice deposited in these frozen reservoirs would be preserved indefinitely in the eternal dark and cold.

Hopes for water on the Earth's Moon were strengthened in 2009 when the Moon Mineralogy Mapper aboard India's *Chandrayaan-1* spacecraft revealed the infrared signatures of water surrounding a small, fresh impact crater on the far side of the Moon (Fig. 5.27). An infrared mapping instrument aboard the *Cassini* spacecraft, on its way to Saturn, confirmed the existence of lunar water, and the *LCROSS* spacecraft bombed a shadowed polar crater, discovering water in the ejected plume. Unfortunately, none of the observations reveal large quantities of water. Even the dampest possibility leaves the lunar surface drier than almost any place on Earth's surface.

But how could there be water ice on the Moon when the rocks returned from the Moon show no signs of ever being exposed to significant amounts of water? The rocks were taken from near-equatorial regions where intense sunlight would boil away any liquid water or water ice, and the Moon's fiery origin seems to have removed volatile elements during the satellite's formation.

Water ice could have been delivered to the Moon by comets, which are essentially big balls of dirty ice. Comet-borne water could have been deposited in the cold traps at the top and bottom of the Moon for more than 4 billion years, ever since the Moon's rocky crust formed, slowly accumulating in amount.

If there were a source of water on the Moon, it would make it more attractive for human outposts. Water could be purified to drink, or it could be chemically split into hydrogen, to burn as a rocket propellant, and oxygen to breathe. This would make it easier to establish a colony on the Moon, or to build a fueling station on it for interplanetary spacecraft. Nevertheless, most of the water on the Moon is not immediately accessible. It is incorporated in the rocky interior of the Moon. Moreover, the concentrations of water on the lunar surface are so low that its extraction is impracticable.

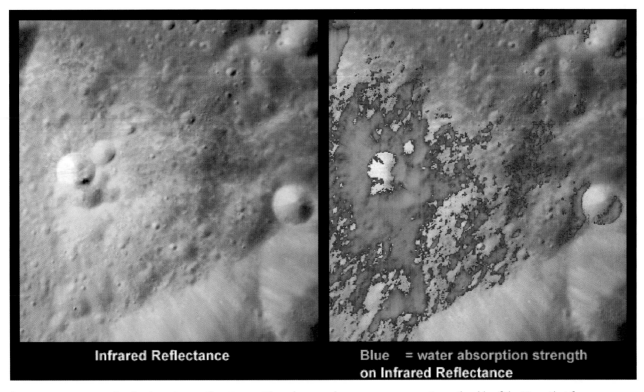

Fig. 5.27 Water around a fresh crater on the Moon   These images show a very young crater or the side of the Moon that faces away from the Earth, as viewed from NASA's Moon Mineralogy Mapper instrument aboard the Indian Space Research Organization's *Chandrayaan-1* lunar-orbiter spacecraft. Sunlight reflected at infrared wavelengths (*left*) shows a very young, fresh crater on the far side of the Moon. On the right, the distribution of water-rich minerals, seen in absorption at a specific infrared wavelength (*false-color blue*), is shown around the crater. Small amounts of both water and hydroxyl-rich material were found to be associated with material ejected from the crater, but the amount of water is very small. (Courtesy of ISRO, NASA, JPL-Caltech/USGS/Brown University.)

The magnetized Moon

The Moon has no overall dipolar magnetic field, at least none that is strong enough to be detected. Its magnetic moment is at least 10 million times weaker than the Earth's. Yet some of the lunar rocks returned to Earth are magnetized. They have survived since the time that molten rocks covered the Moon and solidified 3 to 4 billion years ago, preserving fossilized remnants of ancient magnetic fields.

The *Apollo 15* and *16* missions each carried small subsatellites designed to measure the Moon's magnetism from lunar orbit. They found localized regions on the Moon with surface magnetic fields of between one-hundredth and one-thousandth that of Earth; the Earth's equatorial magnetic field strength is 3×10^{-5} tesla. Large blocks of the lunar crust, as broad as 100 kilometers, are magnetized, but they do not combine into an overall global pattern.

Lunar Prospector measurements have shown that the largest concentrations of strong magnetic fields are on the lunar far side, located diametrically opposite to the Imbrium, Serenitatis, Crisium and Orientale impact basins on the near side. The basin rock is itself weakly magnetized, suggesting that the large basin-forming impacts demagnetized the crust at the impact side, while simultaneously magnetizing the crust on the opposite side of the Moon.

According to one explanation, the young Moon may have had its own global magnetic field generated early in its history when molten metal circulated in a small core. Large impacts, like the one creating the Imbrium basin, would create an ionized fireball racing around the Moon, piling the magnetic field up and concentrating it at the point opposite the impact. As the rocks cooled, the strongest, localized magnetic fields survived as fossils after the Moon lost its global magnetic field.

This is not the only explanation for lunar magnetism, for there are hundreds of smaller magnetized regions scattered over the entire Moon, in regions that are not located opposite to a large impact site. Perhaps they swept up, amplified and concentrated the magnetized winds that flow from the Sun. On the other hand, the ancient magnetism may have been incorporated from surrounding

Table 5.5 Voyages to the Moon in the early 21st century

Country/Region	Mission	Launch date	Orbit entry	Accomplishments
Europe	*SMART-1*	27 Sept. 2003	15 Nov. 2004	Tested propulsion technology, mapped lunar surface composition, impacted Moon on 3 Sept. 2006
Japan	*Kaguya*	14 Sept. 2007	3 Oct. 2007	Obtained altitude, geological and gravity data, impacted Moon on 10 June 2009
China	*Chang'e-1*	24 Oct. 2007	5 Nov. 2007	Obtained images of landforms, geological structures, and distribution of chemical elements, impacted Moon on 1 March 2009
India	*Chandrayaan-1*	22 Oct. 2008	8 Nov. 2008	Sent probe into lunar south polar region on 14 Nov. 2008, discovered small amounts of water on lunar surface, mission terminated on 29 August 2009
United States	*Lunar Reconnaissance Orbiter* and *LCROSS*	18 June 2009	23 June 2009	Obtained maps of the lunar surface intended to specify potential landing sites and resources for future human exploration of the Moon. Simultaneous launch of *LCROSS* with impact probe that confirmed trace amounts of water ice and vapor in south polar region of the Moon

material during the early stages of formation, or the Moon could have been magnetized by the Earth, at a time when the two bodies were closer together.

This is one of the remaining mysteries yet to be solved by the future space-age exploration of the Moon, already begun in the early 21st century.

5.7 Return to the Moon

A new space race to the Moon

Space agencies from China, Europe, India, Japan and the United States have all sent spacecraft to the Moon in the early 21st century (Table 5.5), with extended plans to return. Altruistic scientific goals are not the main impetus this time around, any more than they were during the first flights. Each agency is eager to demonstrate its technological skills, and to elevate their international status. In some instances, there might also be close ties between the developing space program and defense or military objectives. But there is a genuine spirit of international collaboration in these new lunar missions, including instrument sharing between countries.

The European Space Agency's first-ever trip to the Moon began in 2003 with the launch of the low-cost *SMART-1* spacecraft, the first of its *Small Missions for*

Advanced Research in Technology. As the name suggests, its primary goal was flight-testing of new technologies, such as solar-powered electric propulsion. After a long testing cruise, lasting more than a year, *SMART-1* entered lunar orbit and its instruments mapped and studied the Moon in great detail for more than a year and a half. They surveyed lunar resources, mapping the surface composition with then unrivalled resolution, and investigated potential landing sites and outposts for future missions. The spacecraft was intentionally sent into a controlled impact with the lunar surface on 3 September 2006 in order to study the ejected material.

The next modern lunar mission, *Kaguya*, was launched on 14 September 2007 by Japan. Kaguya is the name of a princess from the Moon found in a bamboo thicket by an old couple, and described in the ancient Japanese folktale *The Tale of the Bamboo Cutter*. This name replaced the more austere one of *SELENE*, an acronym for *SELenological and ENgineering Explorer*. The spacecraft's objectives were to study the origin and evolution of the Moon and to serve as an engineering test for future deep-space missions. Its instruments obtained improved lunar topographical maps and geological data using a laser altimeter and X-ray and gamma-ray spectrometers. After 1 year and 8 months of lunar orbit, *Kaguya* was intentionally impacted on the lunar surface, on 10 June 2009, for an examination of its impact plume by Earth-based telescopes. *Kaguya* also carried and released two smaller satellites to study the Moon's gravity and ionosphere, named *Okina* and *Ouna*, also from Japanese folklore.

The People's Republic of China entered the modern race to the Moon on 24 October 2007 with the launch of *Chang'e-1*, a lunar orbiter designed: to make three-dimensional imagery of the Moon; to study the chemical elements and surface composition of the lunar surface, including the terrestrially-rare isotope helium-3; and to investigate the prospect of mining the Moon. The spacecraft is named for an angel in a Chinese fairy tale who takes a magic potion and flies to the Moon.

Chang'e-1 entered lunar orbit on 5 November 2007 and was operated until 1 March 2009, when it was deliberately taken out of the orbit and made to impact the Moon. During orbit, the spacecraft obtained images of the landforms and geological structures of the entire lunar surface, and analyzed and mapped the distribution of chemical elements. As the number 1 suggests, it is the first of an ambitious series of Chinese spacecraft, including the launch of a sister orbiter, *Chang'e-2*, in 2011, and the landing of a rover on the Moon and a sample-return mission during the subsequent decade. China has also expressed interest in a manned Moon landing.

India's national space agency, the Indian Space Research Organization (ISRO) launched the lunar orbiter *Chandrayaan-1*, or first moon craft, on 22 October 2008, entering lunar orbit on 8 November 2008. A few days after that, it released an impact probe, making India the fourth country to touch down on the lunar surface. The main orbiting spacecraft was originally intended for two years of mapping the chemical and mineralogical characteristics of the lunar surface, as well as its geological and topographical details. Unfortunately, the failure of the spacecraft's star sensors and problems with overheating led to the termination of the mission after 10 months in space. Many of its primary objectives were nevertheless obtained, including the discovery of trace amounts of water on the lunar surface and confirmation that the Moon was once completely molten. ISRO is planning a *Chandrayaan-2*, and hopes to land a motorized rover on the Moon in 2012 as part of this second mission. Manned spaceflights and a manned mission to the Moon are also part of their future agenda.

In 2004 President George W. Bush (1946–) announced plans to send American astronauts back to the Moon and later Mars. Summoning the spirit of discovery, exemplified by the Lewis and Clark exploration of the vast American West two centuries before, Bush called on the human need to venture into space, "for the same reason we were once drawn into unknown lands and across the open sea",

urging NASA to return people to the Moon no later than 2020.

NASA's first step back to the Moon was achieved with the launch of its *Lunar Reconnaissance Orbiter* (*LRO*) on 18 June 2009. Instruments aboard the spacecraft are providing scientists with a detailed global topographical map of the Moon, specifying its impactor populations as well as potential landing sites, and with global high-resolution infrared maps of the Moon, enabling detection of previously unseen compositional differences. The *Lunar Crater Observation and Sensing Satellite* (*LCROSS*) was launched with the *LRO* mission. *LCROSS* included a small probe that was sent into a controlled impact within a crater in the south polar region of the Moon. Scrutiny of the ejected material confirmed that small amounts of water were present.

This bold plan to return Americans to the Moon was grounded by President Barack Obama (1961–) in February 2010, after NASA had already spent 9.1 billion dollars on the program. Obama instead proposed 6 billion dollars over five years to encourage private companies to build spacecraft that NASA could rent to ferry astronauts to the *Space Station* – after the *Space Shuttle* program is terminated.

Nevertheless, the idea of a human return to the Moon still has its attractive aspects. Technology can now be used that wasn't even imagined when humans first walked on the Moon, more than 40 years ago. The *Apollo* spacecraft, for example, carried handheld cameras to photograph the lunar surface. Now high-resolution imaging and remote sensing from lunar orbit are common.

NASA's cancelled plans included the future use of robotic lunar rovers, similar to those used on Mars, to pave the way for riskier human missions. Eventually a lunar outpost was envisaged, as a permanent base and way station for trips to Mars. With only one-eightieth of the Earth's mass, the lower gravitational pull of the Moon would make it easier and less costly to launch missions to other parts of the solar system, using spacecraft assembled and provisioned on the Moon with its abundant resources.

Although the prospects look dim right now, the United States might eventually establish a permanent human settlement on the Moon. It could serve as a support station for exploration of the rest of the solar system, including manned trips to Mars. Oxygen, hydrogen and metals might be extracted from the lunar soil to produce water, air, fuel and construction material. The Moon might even be mined for scarce resources on the Earth, such as helium-3: the gas that makes balloons float and might be used in fusion reactors in the future. The global reserves of helium-3 on the entire Earth amount to about 100 kilograms, but there is perhaps 10 million times as much helium-3 on the Moon's surface, implanted there by the Sun's winds. And since the Moon has no appreciable atmosphere, some promising astronomy could be accomplished from it.

Astronomy from the Moon

A manned scientific station on the Moon would provide excellent opportunities for astronomy in the future. With no significant atmosphere, every wavelength of radiation, from the longest radio waves to the shortest gamma rays, streams down without absorption to the lunar surface. Telescopes in artificial Earth satellites are now used to observe cosmic ultraviolet and X-ray emission that is partially or totally absorbed in the Earth's atmosphere, but such instruments are constrained by their weight, size, complexity and cost. Telescopes constructed on the rock-solid surface of the Moon would rest on a more stable platform than an orbiting satellite or space station, permitting high-resolution observations of the Universe at wavelengths that cannot be seen from the ground, including X-rays and gamma rays.

The near absolute-zero temperatures and airless environment on the Moon would also avoid the problem of image blurring caused by atmospheric turbulence that usually limits angular resolution at visual wavelengths to 1.0 seconds of arc. A lunar optical telescope as small as 1 meter in diameter would have a resolution of about 0.1 seconds of arc, while also avoiding interruption caused by cloudy weather and daytime. Relatively small telescopes could be linked together electronically, using the techniques of interferometry to achieve angular resolutions at visible wavelengths that are thousands of times better than those currently available from either space or the ground. The great increase in resolution might, for example, be used to investigate Earth-like planets around nearby stars, or to obtain focused infrared observations for weeks and months at a time, detecting the most distant galaxies as they are forming.

The Moon's far side, permanently shielded from the noisy, interfering Earth, is an ideal site for future radio astronomy. It could also be used to open a new window on the Universe at wavelengths longer than about 20 meters that are reflected by the Earth's ionosphere and do not reach the ground.

5.8   The Moon's history

The age of the oldest rocks — Moon, Earth and meteorites

The time at which different features on the Moon originated can be determined from rocks returned from them. These relics have remained unaffected by the erosion that removed the primordial record from most terrestrial rocks. The ages of the lunar rocks can be determined by examining unstable radioactive elements and their stable decay products.

What matters are not the actual quantities of each element present, but the proportions — the ratio of stable elements, like lead, to unstable ones like uranium and thorium. When this ratio is combined with the known rates of radioactive decay, the time since the rock solidified and "locked in" the radioactive atoms is found.

The method is known as *radioactive dating*, and it works this way. Certain types of nuclei, known as unstable parent isotopes, decay at a constant rate into stable lighter isotopes known as daughters. By measuring the amount of daughter material and knowing the rate of decay, the age of the rock can be estimated. The detailed mathematical treatment is given in Focus 5.4. The method is something like determining how long a log has been burning by measuring the amount of ash and watching a while to determine how rapidly the ash is being produced.

The daughter isotopes must be trapped in the rock and not escape or the estimated age will be too short. In fact, the daughters can escape quite easily when the rock is molten; only when it cools and solidifies do the daughters start to accumulate. For this reason, the ages determined for the rocks are really the times since the rock became solid. And if the rock is re-melted, say by the impact of a meteorite, its radioactive clock is reset, and the age will measure the time since the last solidification.

The radioactive dating method has been used to study rocks returned from the Moon. The oldest lunar samples, returned from the light, rugged highlands, indicate an age of nearly 4.6 billion years, and the lava flows that created the dark, nearside maria are dated at 3.2 to 3.9 billion years ago.

The oldest rocks found on Earth are about 3.9 billion years old, and the oldest known terrestrial minerals are found in crystals of zircon that have been dated to between 4.1 and 4.3 billion years old. Erosion by wind, water and geological processes has wiped out the oldest terrestrial rocks. The deep-ocean sediments, that are least affected by continuing geological activity on Earth, have ages of about 4.55 billion years.

Primitive meteorites known as carbonaceous chondrites have an age of 4.566 billion years, with an uncertainty of about 0.002 billion years. These meteorites are thought to date back to the earliest days of the solar system.

Rounding off the numbers and allowing for possible systematic errors, we can say that the Earth, Moon and primitive meteorites solidified at about the same time some 4.6 billion years ago, with an uncertainty of no more than 0.1 billion years. If the solar system originated as one entity, then this should also be the approximate age of the Sun and the rest of the solar system.

Focus 5.4 Radioactive dating

Radioactive elements can be used to clock the age of rocks on the Earth's surface, meteorites, and lunar rock samples. The number, N, of radioactive atoms in a rock changes with the time, t, since its solidification according to the differential equation:

$$\frac{dN}{dt} = -\lambda N$$

where λ is the decay rate. This equation integrates to give the number of radioactive atoms, N_t, at time t:

$$N_t = N_0 \exp(-\lambda t) = N_0 \exp\left(\frac{-0.693\,t}{\tau_{1/2}}\right)$$

where N_0 is the number of atoms at time $t = 0$, the time of solidification. The radioactive decay constant λ is $0.693/\tau_{1/2} = \ln(2)/\tau_{1/2}$, and $\tau_{1/2}$ is the half-life of the radioactive species. Half-lives for the decay of radioactive isotopes are given in Table 5.6.

The number of radioactive atoms in the rock will be halved in a time equal to the half-life. Radioactive uranium ^{238}U decays, for example, into lead ^{206}Pb, which is stable with a half-life of about 4.47 billion years; so every 4.47 billion years the amount of uranium-238 in a rock will be halved. We can apply the equations to ^{238}U, and express the abundance in terms of another kind of lead, ^{204}Pb, that is not a radioactive decay product. If a terrestrial rock, lunar sample or a non-terrestrial meteorite became a closed system at time $t = 0$, then the present abundance of lead and uranium are related by the equation:

$$\left(\frac{^{206}\text{Pb}}{^{204}\text{Pb}}\right)_t = \left(\frac{^{238}\text{U}}{^{204}\text{Pb}}\right)_t \left[\exp(\lambda_{238}\,t) - 1\right] + \left(\frac{^{206}\text{Pb}}{^{204}\text{Pb}}\right)_0$$

where the subscripts t and 0 denote the present and initial abundance, respectively.

If all of the rock samples have the same initial ^{206}Pb / ^{204}Pb abundance, and if all of them have the same age, t, then a plot of $(^{206}$Pb/ ^{204}Pb$)_t$ against $(^{238}$U/ ^{204}Pb$)_t$ should lie in a straight line of slope $[\exp(\lambda_{238}\,t) - 1]$. Such a plot is called an *isochron*. If a system formed t years ago and initially contained no lead, then a curve of the ratios ^{207}Pb / ^{206}Pb and ^{238}U / ^{206}Pb also provides the age t.

These and similar methods have been used to show that the Earth, Moon and meteorites are 4.6 billion years old, with an uncertainty of no more than 0.1 billion years, where 1 billion years = 10^9 years = one Gyr.

Table 5.6 Radioactive isotopes used for dating

Radioactive parent [name (symbol) mass no.]	Stable daughter [name (symbol) mass no.]	Half-life (millions of years)
Rubidium (Rb) 187	Strontium (Sr) 87	48 800
Rhenium (Re) 187	Osmium (Os) 187	44 000
Lutetium (Lu) 176	Hafnium (Hf) 176	35 700
Thorium (Th) 232	Lead (Pb) 208	14 050
Uranium (U) 238	Lead (Pb) 206	4 470
Potassium (K) 40	Argon (Ar) 40	1 270
Uranium (U) 235	Lead (Pb) 207	704
Samarium (Sm) 146	Neodymium (Nd) 142	100
Plutonium (Pu) 244	Thorium (Th) 232	83
Iodine (I) 129	Xenon (Xe) 129	16
Palladium (Pd) 107	Silver (Ag) 107	6.5
Manganese (Mn) 53	Chromium (Cr) 53	3.7
Aluminum (Al) 26	Magnesium (Mg) 26	0.72

Formation of the highlands and maria

The retrieved lunar rocks have taken us back into time, to the formative stages of the Moon. They record events from the earliest history of the solar system that have been erased on Earth by water, wind and geologic activity. Radioactive dating of Moon rocks and primitive meteorites indicates, for example, that the Moon was assembled a mere 50 million years after the solar system itself was born 4.6 billion years ago.

The cataclysmic bombardment associated with the final stages of the Moon's formation was so energetic that the globe was melted to depths of several hundred kilometers. In this global magma ocean, lighter mineral

Table 5.7 Lunar timescales

Period	Age[a] (billions of years)	Characteristics
Pre-Nectarian	4.6 to 3.92	Moon accumulated 4.5 to 4.6 billion years ago in Earth orbit, newly formed Moon wrapped in molten rock, a magma ocean, until 4.4 billion years ago, solidification of lunar crust and formation of oldest impact basins
Nectarian	3.92 to 3.85	Nectaris basin probably formed 3.9 billion years ago, Serenitatitis, Crisium and other impact basins are most likely this old, as are most lunar rocks
Imbrian	3.85 to 3.15	Period of lunar volcanism when most maria were formed, Imbrium basin excavated 3.85 billion years ago and the last big basin, Orientale, created 3.8 billion years ago; these basins were filled with lava to create the lunar maria up until 3.15 billion years ago
Eratosthenian	3.15 to about 1.0	Craters that are slightly degraded but without rays; most of lunar surface remains unchanged, but some mare volcanism and large impacts
Copernican	About 1.0 to present	Youngest craters formed, most of which have preserved rays; crater Copernicus excavated 0.85 billion years ago and crater Tycho created just 0.10 billion years ago; most of lunar surface remains unchanged

[a] These ages are estimates based on inferences of the geological setting of the lunar samples.

species floated to the top and formed the Moon's crust, and the denser material sank to the interior. The lunar highlands still contain the remains of these early low-density rocks, the anorthosites.

The magma ocean gradually cooled and crystallized between 4.6 and 4.4 billion years ago, forming a thin, low-density lunar crust. Portions of this crust are today's highlands, on both the near and far sides of the Moon, rich in light elements such as aluminum and poor in heavy ones like iron.

When the radioactive dating method is applied to highland rocks retrieved from the Moon, there are two significant results. Highland rocks that crystallized from internally generated magmas, such as the anorthosite rocks, date from the earliest times, 4.6 to 4.1 billion years ago – a span of 500 million years after the origin of the Moon. In contrast, highland rocks that assembled from pre-existing rocks by impact, the breccias, all date from 3.9 to 3.8 billion years ago. The two results attest to prolonged igneous evolution on the early Moon, followed by an apparently short, "cataclysmic" impact bombardment about 3.9 billion years ago. This hail of meteoritic debris cratered the lunar highlands and obliterated most of the direct evidence of the first half billion years of lunar history.

The large impact basins were formed during the final stages of this heavy bombardment. The Nectaris basin was created 3.92 billion years ago, and the Imbrium impact took place an estimated 3.85 billion years ago. These impacts have been used to mark key events in the lunar history (Table 5.7). In pre-Nectarian time, the Moon's crust solidified. The major impact basins were gouged out during the heavy bombardment of Nectarian time. When the Imbrium basin was excavated, it marked the start of the Imbrian period of lunar volcanism that created the maria.

As the external cratering rate was declining rapidly, internal processes set to work. The radioactive decay of long-lived unstable elements, such as uranium and thorium, produced heat that gradually warmed up the lunar interior. There followed an era of volcanism, lasting for 700 million years, from 3.9 to 3.2 billion years ago. The outer zone of solid rock gradually cooled from the outside in, becoming thicker, and lava worked its way from deeper and deeper in the Moon. The magma flow may have stopped 3.2 billion years ago, since the youngest lunar lava samples are this old, but some mare basalts could be as young as 1 billion years old.

As molten basaltic rock welled up from the interior, it penetrated the thin crust beneath the great impact basins on the near side of the Moon, flooding them with lava and producing the dark circular maria that can be seen today (Fig. 5.28). Successive lava flows set their marks in some maria, showing that they were not formed in a single quick pulse of volcanism, but by repeated outpourings that gradually filled the nearside basins.

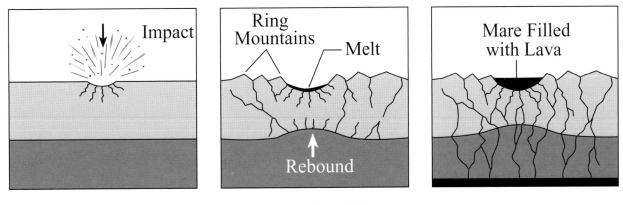

Fig. 5.28 Mare formation on the Moon   Disintegrating and vaporizing as it strikes, a meteorite blasts a huge impact basin out of the lunar surface (*left*), while the associated shock waves create fractures in the rock beneath the basin. The blast hurls up mountain ranges around the basin (*middle*), and the underlying rock adjusts to the loss of mass above it by rebounding upward. The uplifted mantle causes additional fractures in the rock, while a pool of shock-melted rock solidifies in the basin. All the major impact basins on the Moon were created in this way between 4.3 and 3.9 billion years ago. Later, interior heat from radioactivity caused partial melting inside the Moon, and magma rose along the fractures, filling the basin with lava to form a dark mare (*right*). The lunar maria were filled by this volcanic outpouring between 3.9 and 3.1 billion years ago.

The liquid lava moved quickly away from its vents, covering the sources rather than piling up, so no volcanic mountains were created. Flowing with about the consistency of motor oil, the lava spread for hundreds of kilometers before hardening into a thin veneer, only a few hundred meters or less in thickness. And since the crust on the far side of the Moon was relatively thick to begin with, the molten rock had greater difficulty in penetrating it, explaining why there are so few maria on that side of the Moon.

The lava inundated all craters in its path, wiping the slate clean of previous impacts and preparing a fresh surface to record new impacts, which, by this time, had greatly diminished in intensity. Thus the maria are relatively unscarred and most of their craters are small and relatively young.

Despite its violent beginnings, the Moon then settled down into a long quiet life. The airless Moon's face has remained largely unchanged for 3.2 billion years. By that time, the Moon had cooled so much that magma could no longer break through and erupt, and the pummeling by meteorite impacts continued at a much lower rate. Impacts occasionally blasted out craters like Copernicus, some 800 million years old, and Tycho, dated at 108 million years, but the ongoing rain of lesser impacts mainly churned up the lunar surface, covering it with rock fragments. This rubble forms a dusty covering to a magnificent, ancient museum of craters and basins formed billions of years ago.

5.9 Tides and the once and future Moon

The pattern of the tides

Walking along the ocean beach some morning, we might notice that the waves seem to be reaching farther and farther up the sand. The tide is flooding the beach. A few hours later, it hesitates and then begins to ebb, retreating onto the flats where the clams may often be found. The high tides occur simultaneously and symmetrically on opposite sides of the Earth; they return every 12 hours 25 minutes in each location, although not precisely to the same height. The time between consecutive high tides is slightly more than half a day because the Moon's revolution around Earth is in the same direction as the Earth's rotation on its axis, so Earth needs an extra 25 minutes of rotation to out-race the Moon and get into position.

The Moon creates two high tides because the gravitational force of the Moon draws the ocean out into an ellipsoid, or the shape of an egg. We can understand this by remembering that the gravitational force decreases with distance, so the Moon pulls hardest on the ocean facing it, and least on the opposite ocean; the Earth between is pulled with an intermediate force. As a result, the water directly beneath the Moon is pulled up away from the Earth's center, and the Earth's center is pulled away from the water on the opposite side, causing another high tide. Thus the differences of the gravitational attraction of the Moon on opposite sides of the Earth produce two tidal bulges — one facing the Moon and one facing away (Fig. 5.29).

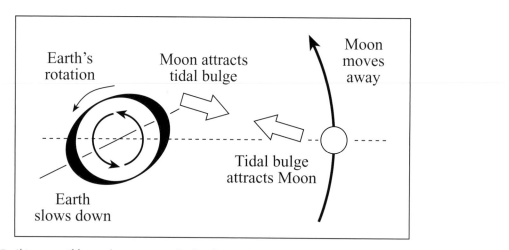

Fig 5.29 Cause of the Earth's ocean tides   The Moon's gravitational attraction causes two tidal bulges in the Earth's ocean water, one on the closest side to the Moon and one on the farthest side. The Earth's rotation twists the closest bulge ahead of the Earth–Moon line (*dashed line*), and this produces a lag in time between the time the Moon is directly overhead and the highest tide. The Moon pulls on the nearest tidal bulge, slowing the Earth's rotation down. At the same time, the tidal bulge nearest the Moon produces a force that tends to pull the Moon ahead in its orbit, causing the Moon to spiral slowly outward.

As the Earth's rotation carries the continents past the tidal humps, we experience the rise and fall of water. In mid-ocean the tide is only 0.01 to 0.30 meters in height and usually goes unnoticed. But, when a shore blocks the tide, it often runs 2 or 3 meters high. Tides can also resonate in estuaries of the right shape, amplifying and building up the tides to 10 or 20 meters in height, as in the Bay of Fundy in Nova Scotia. You can create a similar effect by sloshing the water in your bathtub back and forth at just the right rate.

On a slowly rotating planet without continents, the tide would be highest along the line joining the centers of the planet and its moon – that is, when the moon is overhead. This is not the case for the Earth. The friction of the continents and the rapid rotation of the Earth carry the ocean's tidal bulge forward so it precedes the Earth–Moon line by about 3 degrees (Fig. 5.29). This means that in the open ocean the high tide actually occurs about 12 minutes after the Moon is overhead.

There are further delays for tides at the continental shores. When the flood tide moves in from the ocean, it may have to work its way among islands and peninsulas and along channels; this twisted path will delay the arrival by amounts that vary with location and with the time of the month. This time delay is called the *establishment of the port*, and the result is that high tide usually occurs an hour or two after the Moon is overhead, and occasionally more. Similar delays can be noted in tidal pools. They are lowest long after low tide.

Most people think that the Moon alone causes the tides, but that is not the case. The Sun and the Moon both contribute to the formation of the tides, but the major portion

of this rhythmic ebb and flood is driven by the Moon, whose tide is 2.2 times as high as the Sun's. In the course of a month, the changing alignment of the Sun and Moon causes the tides produced by these two bodies to alternately reinforce and interfere, leading to the cycle of spring tides and neap tides. Once a month the Sun and Moon are aligned, both on the same side of the Earth, producing the spring tide that swells exceptionally high.

The spring tide occurs near new and full moon, when the Sun and Moon reinforce each other's tides, and the neap tide occurs near first and third quarter, when they interfere with each other (Fig. 5.30). The spring tides can be two or three times as high as the neap tides. On new moon nights, the spring tide can pull a boat much farther out than the average quarter-moon neap tide. The Sun's tides also vary by a small amount over the year as the Earth travels around its eccentric orbit and the Sun–Earth distance changes, with the greatest solar tides when the Earth is nearest the Sun.

The days are getting longer

As the Earth rotates, the bulge raised on its surface by the Moon's gravity is always a little ahead of the Moon rather than directly under it. The Moon pulls back on the bulge, and in the process it slows the whole planet down. In other words, our planet meets resistance in its daily rotation caused by the tidal interaction of the Moon with the Earth.

As the ocean tides flood and ebb, they create eddies in the water, producing friction and dissipating energy at the expense of the Earth's rotation. The motion of the

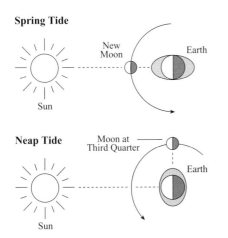

Fig. 5.30 Earth's spring and neap ocean tides   The height of the tides and the phase of the Moon depend on the relative positions of the Earth, Moon and Sun. When the tide-raising forces of the Sun and Moon are in the same direction, they reinforce each other, making the highest high tides and the lowest low tides. These spring tides (*top*) occur at new or full moon. The range of tides is least when the Moon is at first or third quarter, and the tide-raising forces of the Sun and Moon are at right angles to each other. The tidal forces are then in opposition, producing the lowest high tides and the highest low tides, or the neap tides (*bottom*). The height of the tides has been greatly exaggerated in comparison to the size of the Earth.

tides heats the ocean water ever so slightly and the Earth's rotation is slowed. The tides therefore act as brakes on the spinning Earth, slowing it by friction in much the way that the brakes of a car slow its wheels and become warm. As the result of this tidal friction, the rotation of the Earth is slowing down and the day is becoming longer at a rate of 2 milliseconds, or 0.002 seconds, per century (Focus 5.5). In other words, the days are getting longer at the rate of one second every 50 000 years, and tomorrow will be 60 billionths of a second longer than today.

This tidal effect on the Earth's rotation can help us understand the ancient astronomical records, and conversely those records also help us understand the effect of the tides. If the current, slow rotation rate is used to wind the Earth backward 2500 years to the time of an eclipse, the Earth would rotate about a quarter-turn too little, putting those predicted eclipse paths several thousand kilometers west of their actual locations. These paths would then conflict with the reported occurrences. As we shall see, the length of the month is also increasing; this reduces the discrepancy considerably, so the full story is a bit complicated.

Indirect historical measures of the Earth's rotation have been made by paleontologists through studies of fossil corals. The growth patterns of these corals consist of annual bands and fine daily ridges, produced by the effect of seasonal and daily changes of water temperature on the growth rate. The days were shorter in the past, but the year was the same, so the number of days per year increases as we go back in time. Ancient corals confirm this, and they show a greater number of daily ridges per annual band than modern corals. Careful counting reveals that the day was only 22 hours long when we look back 400 million years. Studies of daily grown increments have been extended to fossilized algae called stromatolites, which indicate that the day may have been only 10 hours long 2 billion years ago.

Aside from such historical and paleontological determinations, this change of the Earth's rotation is imperceptible to humans, and it has not yet been measured directly. It is also mixed in with an erratic rate produced by the vagaries of the weather and the seasons, so all in all the Earth's rotation is no longer the best choice of a clock. Astronomers now prefer to rely on atomic clocks due to their stability and continued accuracy.

The planets Mercury and Venus have exceptionally slow rotation periods, of 58.646 and 243 Earth-days, respectively. The youthful energy of their fast initial rotation has probably been tempered by tidal interaction with the massive nearby Sun. These would be tides in the solid body of the planets, for there are no oceans on Mercury or Venus. Such a tidal interaction is suggested by the fact that Mercury spins on its axis exactly three times during two full revolutions about the Sun, so its rotation period is exactly two-thirds of Mercury's orbital period of 87.969 Earth days.

Earth's tidal influence on the Moon

The Moon pulls the Earth's oceans, and the oceans pull back, in accord with Newton's third law that every action has an equal and opposite reaction. The net effect is to swing the Moon outward into a more distant orbit. This is because the tidal bulge on the side facing the Moon is displaced ahead of the Moon and this bulge pulls the Moon forward.

As the Earth slows down, the angular momentum it loses is transferred to the Moon, which speeds up in its orbit around us. It is not hard to see that this will swing the Moon away from the Earth if we look at the key equations (Focus 5.6). When we do the arithmetic, we find that the change of 0.002 seconds per century in the length of the day implies an outward motion of the Moon amounting to about 0.04 meters per year. Small as it is, this value is just measurable with the laser reflectors planted on the Moon by the *Apollo* astronauts. The lunar laser ranging data indicate that the Moon is moving away from the Earth at a rate of 0.0382 ± 0.0007 meters per year.

Focus 5.5   Tidal friction slows the rotation of the Earth

In most of the ocean, the tidal currents are confined to the top of the deep sea, never reaching its bottom. Most of the tidal energy is therefore dissipated in shallow seas near land, where the turbulent tidal water reaches the ocean bottom, at depths of 100 meters or less.

When the tide moves toward a beach at velocity V, the frictional energy ΔE dissipated by tidal currents on the sea bottom per unit time Δt and unit area ΔA is

$$\frac{\Delta E}{\Delta t \, \Delta A} = \gamma \rho V^3$$

$$\approx 2 \text{ joule per second per square meter}$$

where the density of seawater is $\rho \approx 1000$ kilograms per cubic meter, a typical velocity is $V \approx 1$ meter per second, and the stress on the sea bottom is $\gamma \rho V^2$ with an empirical drag coefficient $\gamma \approx 0.002$ for wind stress on the ground and a river's stress on its bed as well as tidal currents in the bottom of the sea.

In 1919, Sir G. I. Taylor (1886–1975), a British expert on turbulence in air and water, used this equation to obtain $\Delta E/\Delta t \approx 5 \times 10^{10}$ joule per second for the Irish Sea alone, and in the following year Sir Harold Jeffreys (1891–1989) estimated that the total energy dissipated by tidal friction in the shallow seas surrounding Europe, Asia, and North and South America is

$$\text{Energy lost by tidal friction} = \frac{\Delta E}{\Delta t}$$

$$\approx 10^{12} \text{ joules per second}$$

This is comparable to the estimate obtained by considering the flux of energy convected into the shallow seas by tidal currents.

The lost energy comes from the Earth's rotational energy, which is equal to

$$\text{Rotational energy} = \frac{1}{2} M_E \, V_{rot}^2 = \frac{2\pi^2 \, M_E \, R_E^2}{P_E^2}$$

where the mass of the Earth M_E is 5.972×10^{24} kilograms, the mean radius of the Earth R_E is 6.371×10^6 meters, and the rotation period of the Earth P_E is 24 hours = 8.616×10^4 seconds. For a period change ΔP in time interval Δt, the loss in rotational energy is

$$\text{Energy lost by rotation} = \frac{4\pi^2 \, M_E \, R_E^2 \, \Delta P}{P^3}$$

Setting this equal to $\Delta E/\Delta t$ and collecting terms, we obtain:

$$\frac{\Delta P}{\Delta t} = \frac{P^3}{4\pi^2 \, M_E \, R_E^2} \frac{\Delta E}{\Delta t}$$

$$\approx 10^{-12} \text{ seconds per second}$$

$$\approx 0.003 \text{ seconds per century}$$

where one century equals 3.156×10^9 seconds.

Will the Moon's outward motion carry it away from the Earth altogether? Probably not, because there is not enough energy in the Earth–Moon system for these bodies to overcome their binding energy and go their separate ways. Only the intrusion of a massive third body could achieve that, or some fantastic project to attach enormous rockets to the Moon and launch it into space.

What will ultimately happen is the following. The combination of the slowing Earth and the receding Moon means that the Earth's day will eventually catch up with the length of the month. When the day and the month are equal, the Moon-induced tides will cease moving; from then on the oceans will rise and fall much more gently under the influence of the Sun. The Moon will hang motionless in the sky, and will be visible from only one hemisphere. At that stage the recession of the Moon will stop.

Then, billions of years from now, the Sun's tidal action will take over; slowing the Earth's rotation even further, until the day becomes longer than the month. At this point, angular momentum will be drawn from the Moon, and it will begin approaching the Earth, heading on a course of self-destruction until it is finally torn apart by the tidal action of the Earth. Perhaps it will form a ring around our planet. In any case, it will probably end its years where it apparently began – close to the Earth. By this time, however, the brighter Sun will have boiled the oceans away, and the Earth will have become a dry and barren place.

Stabilizing the Earth

The orientation of the Earth's rotation axis causes the annual seasonal variations of our climate, and small variations in its orientation contribute to the advance and retreat of the ice ages. The obliquity of the Earth, the angle that its rotation axis makes with the perpendicular to its orbital plane, is now a modest 23.5 degrees. This is sufficient to bring summer and winter as the northern or southern

Focus 5.6   Conservation of angular momentum in the Earth–Moon system

One of the fundamental, unbreakable laws of physics is the law of conservation of angular momentum, and this means that the angular momentum that the Earth loses in slowing down will be transferred to the Moon. That is, the product of mass M, velocity V and radius R is unchanged in a closed system, which is not subject to an outside force. Thus:

Conservation of angular momentum $= M \times V \times R$

$$= \text{constant}$$

For the Earth, the angular momentum is rotational, with $V = 2\pi R_E / P_E$, where P is the Earth's rotation period of one day and the subscript E denotes the Earth. So, we have

Earth's rotational angular momentum $= \dfrac{2\pi M_E R_E^2}{P_E}$

Since the length of the Earth's day is increasing as time goes on, the Earth's rotational angular momentum is decreasing by the amount

Decrease in rotational angular momentum

$$= \frac{2\pi M_E R_E^2 \, \Delta P_E}{P_E^2}$$

The loss has to be made up by an equivalent gain somewhere else in order to conserve angular momentum. This is done by an increase in the Moon's orbital angular momentum, which is given by:

Moon's orbital angular momentum

$$= M_M \times V_M \times D_M$$

where the subscript M denotes the Moon, M_M is the mass of the Moon, D_M is the distance between the Earth and

the Moon, and the orbital velocity of the Moon can be estimated by the escape velocity of the Earth at the Moon's distance, or by

$$V_M = \left(\frac{2 G M_E}{D_M} \right)^{1/2}$$

where G is the gravitational constant. Substituting this velocity expression into the angular momentum relation, we obtain

Moon's orbital angular momentum

$$= M_M D_M (2 G M_E / D_M)^{1/2}$$

Since the mass of the Moon and the Earth do not change, the Moon's distance has to increase by an amount ΔD_M to provide an increase in the angular momentum:

Increase in orbital angular momentum

$$= M_M \, \Delta D_M \left(\frac{2 G M_E}{D_M} \right)^{1/2}.$$

Setting the loss in rotational angular momentum equal to the gain in orbital angular momentum and collecting terms, we obtain

$$\Delta D_M = \frac{2\pi M_E R_E^2 \, \Delta P_E}{M_M P_E^2 \left(\dfrac{2 G M_E}{D_M} \right)^{1/2}}$$

$$\approx 1.3 \times 10^{-9} \text{meters per second}$$

$$\approx 0.04 \text{meters per year}$$

where $M_E = 5.972 \times 10^{24}$ kilograms, $R_E = 6.371 \times 10^6$ meters, $\Delta P_E = 0.002$ seconds per century $\approx 0.66 \times 10^{-12}$ seconds per second, $M_M = 7.348 \times 10^{22}$ kilograms, $P_E = 24$ hours $= 8.616 \times 10^4$ seconds, $G = 6.673 \times 10^{-11}$ m^3 kg^{-1} s^{-2}, $D_M = 3.844 \times 10^8$ meters and 1 year $= 3.156 \times 10^7$ seconds. This is the amount measured by sending laser pulses from the Earth to the corner reflectors left on the Moon by astronauts.

---

hemisphere is tilted toward or away from the Sun. Variation in the Earth's obliquity as small as ±1.3 degrees, around a mean value of 23.3 degrees, may contribute to, or trigger, the ice ages.

The climate forecast for a Moon-less Earth would be a lot bleaker. The gravitational pull of our large Moon acts as an anchor, limiting excursions in the Earth's rotation axis and keeping the climate relatively stable (Fig. 5.31). Without the Moon, the tilt of Earth's rotation axis would vary chaotically between 0 and 85 degrees. Such large variations in the planet's obliquity would result in dramatic changes in climate. With an obliquity of 0 degrees, there

would be no seasonal variation in the distribution of sunlight on Earth. At 85 degrees, the Earth's axis would be tipped completely over. The equatorial tropics could then be permanently in cold winter snows, and the poles would be alternately pointed almost directly at or away from the Sun over the course of a single year. Such wide climate changes might be hostile to many forms of life on Earth.

The nearby massive Sun holds the tilt of Mercury and Venus in place, but there is no help for Mars. Located far from the Sun and with two puny satellites, the obliquity of the red planet exhibits wild variations from 0 to

Fig. 5.31 Steadying influence of the Moon on the Earth The brightly colored, sunlit half of the Earth contrasts strongly with the darker, subdued colors of its Moon, which reflects only about one-third as much sunlight as our world. The Moon holds the Earth upright in space, stabilizing its orientation and keeping the planet from tilting over. Without the Moon's influence, chaotic forces could tip the Earth's rotation axis down so far that its poles are pointing at or away from the Sun, producing wild swings in the Earth's climate. This image of the Moon and Earth was taken from a distance of 6.2 billion meters, by the *Galileo* spacecraft on 16 December 1992, soon after swinging around the Earth on its way to Jupiter. (Courtesy of NASA/JPL.)

60 degrees on timescales of about 5 million years, with profound changes in the Martian climate.

5.10 Origin of the Moon

What models of the Moon's origin must explain

There are several facts that must be explained by a successful account of how the Moon originated. Some of them have been known for more than a century, while others result from laboratory investigations of the rocks returned from the Moon.

Any origin theory must, for example, explain why the Earth has a relatively massive Moon when Mercury and Venus have no known moons, and Mars only has two miniscule ones that may be captured asteroids. In other words, our Moon is an unusual event in the formation process of the rocky terrestrial planets.

A satisfactory theory for the origin of the Moon must also explain the Moon's peculiar orbit, which lies neither in the plane of the Earth's orbit around the Sun, the ecliptic plane, nor in the Earth's equatorial plane. Our satellite revolves around the Earth inclined about 5 degrees to the ecliptic plane, which is itself tilted 23.5 degrees with respect to the Earth's equatorial plane.

Perhaps even more important is the mean mass density of the Moon, just 3344 kilograms per cubic meter, much lower than the Earth's mean mass density of 5513 in the same units. The Moon's overall mass density is much closer to the terrestrial mantle than that of the Earth as a whole, which includes its dense iron core.

If we extrapolate the outward motion of the Moon back into the past, assuming a constant rate of 0.04 meters per year, we see that the Moon was 1.8×10^8 meters closer to the Earth 4.6 billion years ago, when the Earth and Moon were formed. That's more than half the current distance to the Moon, of 3.844×10^8 meters, suggesting that the Moon might well have formed near or even out of the Earth in the distant past, for stronger tidal interaction probably propelled the Moon outward at a quicker rate in the past.

Comparison of the lunar rocks to terrestrial rocks provides further constraints on the Moon's parentage (Table 5.8). The oldest rocks on the Moon solidified 4.5 billion years ago, which means that the Moon is about as old as the Earth. One important distinction comes from the similar quantities of oxygen isotopes, or light and heavy oxygen atoms, in Moon rocks and Earth rocks, indicating a close kinship and suggesting a common ancestry, instead of the Moon forming elsewhere and then being

Table 5.8 Constraints on models for the origin of the Moon

Constraint	Implication
No massive moons on other rocky planets	Moon formation is an unusual process
Moon orbit tilted to Earth's equator	Origin process must explain peculiar orbit
Low mean mass density of Moon	Moon has no large iron core
Moon moving away from Earth	Moon once closer to Earth
Some Moon rocks 4.5 billion years old	Moon about as old as Earth
Oxygen isotope ratios	Earth and Moon formed nearby
Depletion of volatiles	Moon formed at high temperature
Enrichment of refractories	Condensation at high temperature
Depletion of metals	Removal of iron prior to formation

captured by the Earth's gravity. Objects formed in other parts of the solar system exhibit different oxygen isotope ratios. This indicates that the Earth and Moon formed in roughly the same part of the primeval solar nebula, unlike all of the meteorites and planetary samples found to date.

A second key constraint is in the compositional differences between the Earth and the Moon. The Moon rocks lack any detectable water-bearing minerals. They are also missing other kinds of volatile elements, with low melting points, that could have been boiled out into space at high temperatures. Relative to Earth rocks, the Moon rocks are also highly depleted in siderophile, or "metal-loving", elements such as cobalt or nickel, which tend to occur in rocks containing iron.

Yet when compared to the Earth our satellite is enriched in non-volatile substances. Called refractories, these elements are the opposite of volatiles; they have high melting points and remain solid at high temperatures and require extraordinary heat to vaporize.

Early origin hypotheses

There are three classical hypotheses for the origin of the Moon that have been advocated for more than a century. They are the fission, capture and accretion models, nicknamed the daughter, pickup and sister theories (Fig. 5.32). But as Sherlock Holmes said in *The Adventure of Silver Blaze*, "I am afraid that whatever theory we state has very grave objections to it." So we will briefly discuss the advantages and flaws of each of these hypotheses, and then move on to the more successful giant impact theory.

The fission hypothesis supposes that the Earth had no satellite in its earliest youth, but that it was once spinning so fast that a large fraction of its mass tore away to create the Moon. If this occurred after the Earth's iron had settled to the center, the Moon would naturally be depleted in metals and would have a low mean mass density characteristic of the outer layers of the Earth. Once the Moon had separated, tidal friction caused it to move slowly away toward its present orbit.

The fission hypothesis does not easily account for the compositional differences between the Moon and the Earth, such as the depletion of volatiles on the Moon. There are also two dynamical difficulties. First, the primordial

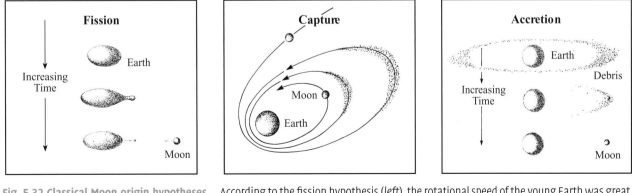

Fig. 5.32 Classical Moon origin hypotheses   According to the fission hypothesis (*left*), the rotational speed of the young Earth was great enough for its equatorial bulge to separate from the Earth and become the Moon. In the capture hypothesis (*middle*), a vagabond Moon–sized object once passed close enough to be captured by the Earth's gravitational embrace. We have pictured disruptive capture, with subsequent accretion, but the Moon might have been captured intact. The accretion hypothesis (*right*) asserts that the Moon formed from a disk near the young Earth.

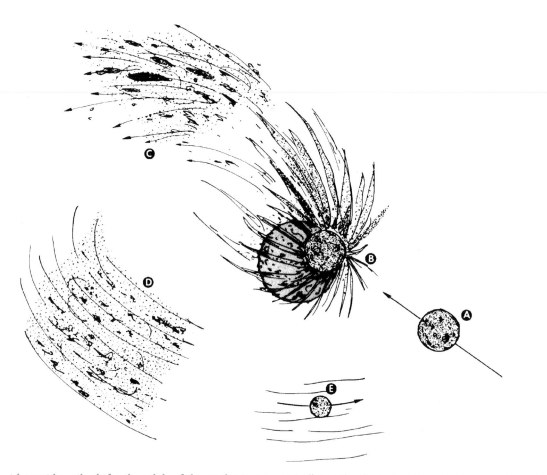

Fig. 5.33 Giant impact hypothesis for the origin of the Earth's Moon   According to the giant impact hypothesis, a massive projectile (A), about the size of Mars, struck the young, still-forming Earth (B) in a catastrophic, glancing blow nearly 4.6 billion years ago, resulting in a tremendous explosion and the jetting outward of both projectile and Earth mass. Some fraction of this mass remained in Earth orbit (C), while the rest escaped Earth or impacted again on Earth's surface. A proto-Moon began to form from the orbiting material (D), accreting neighboring matter, and finally became the Moon (E). It may be mostly derived from the crust and mantle of the Earth and/or the impacting object, accounting for the Moon's relatively low mean mass density and lack of iron when compared with the Earth. The Moon accumulated so rapidly that the outer crust was molten, helping to account for the relative lack of water and other volatile elements. Then, as the crust cooled, the newborn Moon swept up the remaining objects nearby, blasting out impact basins and pockmarking the surface with numerous craters. (Courtesy of Alan P. Boss, Carnegie Institution of Washington.)

Earth would have had to rotate exceptionally fast, once every 2.5 hours, if it were to throw off the material that became the Moon. Second, the Moon's orbital plane is tilted from the equatorial plane of the Earth; if the Moon spun off the planet's equator, the two planes ought to be coincident. There are ways around both of these problems, but the fission theory loses its attractive simplicity when it is doctored in these contrived ways.

If the Moon was not plucked out of the Earth, perhaps our satellite is a maverick that formed elsewhere, strayed too near the Earth, and was captured in orbit − either intact or as fragments torn apart by our planet's strong gravity. The principal advantage of this capture hypothesis is that it easily permits compositional differences between the Earth and Moon since they were formed in different locations within the solar system. The main obstacle is understanding how the capture could have taken place. A passing body would either collide with the Earth or receive a gravitational boost that would hurl it away from the Earth in slingshot fashion. In order to go into orbit about the Earth, the approaching Moon would have to slow down, and the chances of this happening are exceptionally low.

The third classical hypothesis suggests that the Moon and Earth formed concurrently from a cloud of gas and dust through a process not unlike the probable formation of the planets around the Sun. The raw materials for the Moon came from a disk of material in orbit around the Earth, and the planets originated in a similar disk orbiting the Sun.

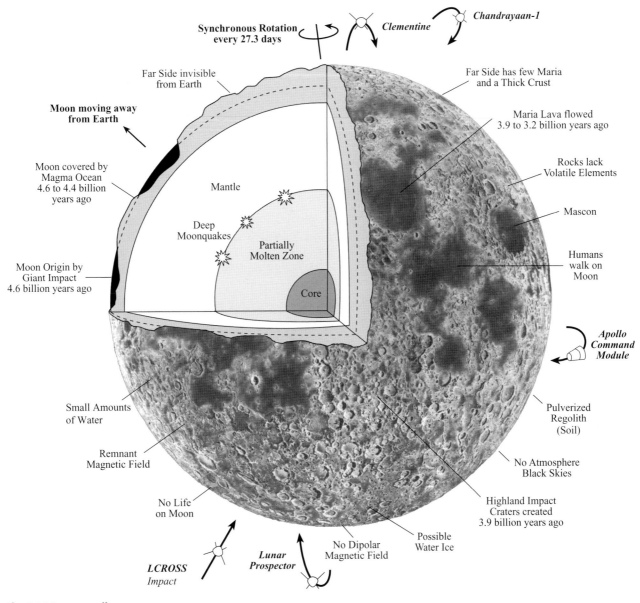

Fig. 5.34 Summary diagram

Such a model seems to apply nicely to giant gaseous planets, such as Jupiter, that have families of satellites resembling the solar system. But where does that leave the rocky terrestrial planets that have no known moons, Mercury and Venus? And what about Mars, with only two tiny satellites? If the process that formed our massive Moon is the natural way of things, we have difficulty understanding these other planets. And why should the chemistry of the Earth and Moon be so different, and how were the volatile elements driven out of the Moon if it always orbited the Earth? Special assumptions can help extricate the accretion theory of its difficulties, but it also loses its appeal when these special assumptions are introduced.

The giant impact hypothesis

Today many astronomers favor a hybrid of the fission and capture theories, with a violent and catastrophic lunar birth. According to this newer, giant impact hypothesis, a massive rogue projectile, perhaps 2.5 to 3.0 times the mass of Mars, sideswiped the Earth and dislodged the material that would become the Moon. The collision may have knocked the Earth away from its original upright position, giving the planet its current axial tilt of 23.5 degrees.

This glancing, planet-shattering blow occurred almost 4.6 billion years ago, during a heavy bombardment that marked the last stages of the solar system's formative period. At this time, iron had already sunk to the core of the Earth, and a rocky crust was beginning to congeal around the partially molten planet.

The giant impact mechanism permits the Moon to form initially in the same part of the solar system as the Earth and to undergo a process that explains both the dearth of metals and volatile elements, as well as an enrichment of refractory elements, before solidification. It might have shattered the colliding object to smithereens and vaporized parts of the iron-poor upper layers of the Earth, blasting off a mix of terrestrial and impactor material into orbit where it coalesced within about a century to form the Moon (Fig. 5.33). The intense heat of the collision vaporized water and most volatile materials from the cast-off material, which clumped together and reassembled into the orbiting Moon.

The glancing blow would have knocked the collision debris into the Moon's current tilted orbit, and if the satellite included material in the mantle of the impactor, this could also help explain the compositional differences between the Moon and Earth. The searing heat of such a collision would explain why the Moon holds no appreciable amounts of water and few volatile elements. All were boiled away. The new hypothesis therefore seems to explain the facts with a minimum of assumptions. As one astronomer stated, "it requires no magic, no special pleading, no extra twiddling and no *deus ex machina*".

Another reason that this model became at least imaginable was the discovery of extremely large impact basins on the Moon, such as the South Pole – Aitken basin on the far side. It was a short mental leap from very large impact basins to a planetary collision.

Thus, exploration of the Moon has resulted in the probable solution of the ancient mystery of the Moon's origin. It was most likely born in a fiery cataclysm out of the infant Earth, the result of an enormous, off-center collision during the early days of the solar system when such events were common.

The giant impact that gave rise to the Moon is a natural consequence of planet formation, making astronomers more aware of impact catastrophes in solar system history. Similar collisions with larger or smaller projectiles could explain major planetary anomalies, such as the removal of Mercury's low-density mantle, the off-kilter, backward spin of Venus, the global crustal dichotomy of Mars, the planet Uranus' bizarre, sideways orientation, and even the demise of the dinosaurs that redirected the course of life on Earth 65 million years ago.

The voyage to the Moon became the stepping-stone to outer space, resembling the first, tentative steps of a child testing the water before learning to swim. It opened a path to the rest of the solar system, to an ongoing, close-up exploration of the planets and their satellites, which have been mapped and surveyed with a detail surpassing that of most countries on our home planet Earth.

Our satellite was the first port of call in our captivating voyage to the planets, to which we now turn, beginning with Mercury whose composition probably resulted from a giant impact of its own.

6 Mercury: a dense battered world

- Because of its close proximity to the Sun, the innermost planet Mercury cannot be studied from Earth against the dark night sky; many astronomers and most people have never seen the elusive planet.

- During the daytime, Mercury's ground temperature reaches 740 kelvin, hot enough to vaporize water and melt lead; at night it plunges to a freezing 90 kelvin.

- Although Mercury is one of the Earth's nearest planetary neighbors, only two spacecraft have ventured near Mercury. They are the *Mariner 10* spacecraft in 1974–75 and the *MErcury Surface, Space ENvironment, GEochemistry, and Ranging* (*MESSENGER*) spacecraft in 2008–11.

- There is a simple three-to-two resonance between Mercury's rotation period of 58.646 Earth days and its orbital year of 87.969 Earth days. This spin–orbit coupling is produced by solar tides in the solid planet.

- The interval from sunrise to sunset at a given location on Mercury is 87.969 Earth days, and the night lasts 87.969 Earth days more, so the day on Mercury lasts 175.938 Earth days and is twice Mercury's year.

- Mercury's rotation axis is aligned perpendicular to its orbital plane, so there are no seasons on the planet, and its polar regions never receive the direct rays of sunlight. Radar echoes suggest that water ice may reside in permanently shaded regions within deep craters near Mercury's poles.

- Mercury has highland craters and impact basins that resemble those found on the Moon. The craters and basins on both objects were most likely formed during a late heavy bombardment by meteorites 3.9 billion years ago.

- An ancient period of volcanic flow, during the late heavy bombardment of Mercury 3.9 billion years ago, obliterated small craters, partially filled larger craters, and created intercrater plains that are not found on the Moon.

- Smooth volcanic plains have filled old craters and impact basins after they formed, and covered approximately 40 percent of the surface of Mercury.

- Irregularly shaped depressions surrounded by bright material have been attributed to volcanic vents on Mercury.

- Long, winding cliffs, or rupes, are found on Mercury, and not on the Moon. They are attributed to the contraction of the young planet as it cooled.

- Relative to its size, Mercury has the biggest iron core of all the terrestrial planets. Mercury's core is much larger than the core of the Moon.

- Mercury may have been blown apart by an ancient collision with a planet-sized object, removing its low-density rocky mantle.

- Mercury has a dipolar magnetic field with a magnetic axis closely aligned with the planet's rotation axis.

- The magnetosphere of Mercury can be opened on its dayside by magnetic reconnection during interaction with the magnetic fields emanating from the Sun.

- Rotational twists discovered by radar observations of Mercury suggest that it has a liquid core in which the planet's magnetic field might be generated.

- More than a century ago, astronomers found that Mercury did not appear in its expected place, leading Einstein to develop a new theory of gravity in which the Sun curves nearby space.

6.1 Fundamentals

Mercury is the smallest of the four rocky, terrestrial planets. It is the planet closest to the Sun, moving around it with the fastest speed and shortest year of any planet. Mercury spins with a slow rotation period of 58.646 Earth days, just two-thirds of its year. It has practically no atmosphere at all, but retains an unexpectedly strong magnetic field.

6.2 A tiny world in the glare of sunlight

Mercury revolves closer to the Sun than any other known planet, with a mean distance from the Sun of just 0.3871 AU. It also has the most eccentric orbit of any major planet, with a distance varying from 0.3072 AU at perihelion, its closest approach to the Sun, to 0.4667 AU at aphelion, when it is furthest from the Sun.

Table 6.1 Physical properties of Mercury[a]

Mass	3.301×10^{23} kilograms $= 0.0553\ M_E$
Mean radius	2439.7 kilometers $= 0.382\ R_E$
Bulk density	5427 kilograms per cubic meter
Sidereal rotation period	58.6462 Earth days
Sidereal orbital period	87.97 Earth days $= 0.240\,846\,7$ Earth years
Mean distance from Sun	5.79×10^{10} meters $= 0.387$ AU
Age	4.6×10^9 years
Exosphere	Hydrogen, helium, sodium, potassium atoms
Surface pressure (exosphere)	10^{-12} bars
Surface temperature	90 to 740 kelvin
Magnetic field strength	0.0033×10^{-4} tesla at the equator $= 0.01\ B_E$
Magnetic dipole moment	5.54×10^{13} tesla meters cubed

[a] The symbols M_E, R_E and B_E respectively denote the mass, radius and magnetic field strength of the Earth.

Mercury also has the shortest year — about 88 Earth days — and the highest orbital speed of any planet. Like a moth about a flame, Mercury races around the Sun at an average speed of 48 kilometers per second. Its rapid motion explains why Mercury is named after the wing-footed messenger of the gods in Roman mythology.

Any planet so close to the Sun is subject to intense sunlight, and therefore has to become very hot. Mercury's maximum daytime surface temperature, when closest to the Sun at perihelion and on the equator, is 740 kelvin. This is hot enough to melt tin, lead and even zinc. Because there is no atmosphere to hold in the heat, the surface temperature plummets to about 90 kelvin when the sunlit side rotates into the Sun's shadow during the planet's long night.

As planets go, Mercury is a tiny world, with the smallest size of any terrestrial planet and slightly smaller than Jupiter's moon Ganymede and Saturn's moon Titan. Mercury's linear radius is easy to measure from its angular radius and distance. Its radius is 2439.7 kilometers or about 1.4 times the radius of the Earth's Moon. Mercury's mass has been a more elusive quantity to determine, because the planet has no satellites. The mass was first estimated from Mercury's gravitational influence on the orbital motion of Venus and passing comets, and improved by the planet's gravitational deflection of the space probes *Mariner 10* and *MESSENGER*. The planet's mass is 3.301×10^{23} kilograms.

Mercury is surprisingly massive for its size. Its volume is only slightly larger than the Moon's and yet it has four times the Moon's mass. This implies a mean mass density of 5427 kilograms per cubic meter, which is nearly as high as that of the Earth, 5513 in the same units, and a little more than Venus at 5243.

The planet's small apparent size and its proximity to the Sun make it difficult to see from Earth. The innermost planet never wanders more than 27.7 degrees in angular separation from the Sun (Fig. 6.1). This angle is less than that made by the hands of a watch at one o'clock. From Earth's perspective, Mercury's tight orbit never reaches into the dark night sky, and it can thus be observed only during the day.

Mercury is visible to the unaided eye only in twilight when it is low in the sky and must be seen through a thick layer of air. Astronomers have therefore taken to observing Mercury near mid-day, when it is far from the horizon and can be seen through a relatively thin layer of air. At such times, Mercury can only be seen with telescopes.

But the terrestrial atmosphere limits the resolution of even the best Earth-based telescopes to features on Mercury that are a few hundred kilometers across or wider — a resolution far worse than that for the Moon with the

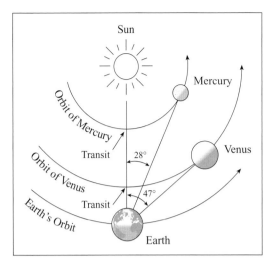

Fig. 6.1 Maximum elongation   Unlike the planets that orbit the Sun beyond the Earth, Mercury and Venus can never be seen high in the sky in the dead of night. Because they are close to the Sun and inside Earth's orbit, these planets are always seen soon after sunset or shortly before sunrise, and they show phases much as the Moon does. The elongation of Mercury and Venus, or their angular distances from the Sun as viewed from the Earth, never exceeds 28 degrees and 47 degrees, respectively.

unaided eye. Moreover, space-borne telescopes with better resolution, such as the *Hubble Space Telescope*, cannot point at Mercury, because even stray light from the nearby Sun could damage their sensitive instruments.

Most people have never seen Mercury. Even Copernicus complained that it eluded him. So it is not surprising that little was known about this enigmatic planet until it was probed by terrestrial radar (short for "radio detection and ranging"), and scrutinized during close encounters with two spacecraft.

6.3 Space-age investigations of Mercury

Close-up scrutiny of Mercury was first carried out from the *Mariner 10* spacecraft, which was launched on 3 November 1973, and obtained data during its three flybys of the planet on 29 March and 21 September 1974 and 16 March 1975. Instruments aboard *Mariner 10* imaged about 45 percent of the surface at an average resolution of 1 kilometer, and indicated that the planet retains an unexpectedly strong magnetic field. As far as we know, *Mariner 10* is still moving by Mercury every six months, but without obtaining any observational data.

Ground-based radar observations had already demonstrated in 1965 that Mercury spins with a slow rotation period of 58.646 Earth days, just two-thirds of Mercury's year of 87.97 Earth days. Detailed radar investigations of the rotation in 2007 provided new information about the planet's liquid core, and radar echoes from Mercury's polar regions indicated that water ice might be trapped in its permanently shadowed craters.

On 3 August 2004, the *MESSENGER* spacecraft was launched to explore Mercury close-up for the first time in more than 30 years. The name *MESSENGER* is an acronym of *MErcury, Surface, Space ENvironment, GEochemistry, and Ranging*. The name also reminds us that to the ancients Mercury was the messenger of the gods.

After an Earth flyby and two Venus flybys, *MESSENGER* passed nearby Mercury on three occasions, on 14 January and 6 October 2008 and 29 September 2009. During the first flyby, its instruments imaged 20 percent of Mercury's surface not previously seen by spacecraft, and it imaged an additional unseen 30 percent during the second flyby. Measurements of the planet's magnetic field, exosphere, surface composition, and gravitational field were also made.

After seven years and six planetary encounters, *MESSENGER* will have slowed enough to insert it into orbit around Mercury, on 17–18 March 2011, using a conventional retro-rocket. The spacecraft would have been traveling at such a high speed during a direct flight, such as the four-month *Mariner 10* trip, that it could not have been directed into orbit around the planet.

6.4   Radar probes of Mercury

The halting spin of old age

Astronomers once supposed that solar tides in the body of Mercury would cause the planet to rotate on its axis once every 87.97 Earth days, in step with its sidereal orbital period. (The sidereal rotation or orbital period is the time taken to complete one rotation or orbit relative to the "fixed" stars.) Just as the Moon always presents the same face to the Earth, it was thought that one side of Mercury was always turned toward the Sun. To test this idea, the Italian astronomer Giovanni Schiaparelli (1835–1910) monitored Mercury's surface markings seen though his 0.46-meter telescope, and he concluded that the same side of the planet did, indeed, always face the Sun. For three-quarters of a century, telescopic observers agreed with his conclusion. All of these astronomers were wrong!

In 1965, Mercury's true rotational period was determined with radio signals that rebounded from the planet. The world's largest radio telescope, located in Arecibo, Puerto Rico, was used to transmit megawatts of pulsed radio power at Mercury, and to receive the faint echo (Fig. 6.2). This technique is known as radar, and it is also used to locate and guide airplanes near airports.

Each pulse was finely tuned, with a narrow range of wavelengths. Upon hitting the planet, its rotation de-tuned the pulse, slightly spreading the range of wavelengths (Fig. 6.3). One side of the globe was rotating away from the Earth, while the other side was rotating toward our planet. These motions produced slight changes in the wavelength of the echo; from these changes, the speed of the surface and the rotational period were calculated, using the well-known expression for the Doppler effect (Focus 6.1; Fig. 6.4).

The result came as an unexpected surprise. The sidereal rotation period was 58.646 Earth days, or exactly two-thirds of the 87.969-day period that had been accepted so long. Thus, with respect to the stellar background, Mercury spins on its axis three times during two full revolutions about the Sun, so we say that there is a 3-to-2 spin–orbit resonance in Mercury's rotation. This relationship follows from $3 \times 58.646 = 2 \times 87.969$, and it is technically known as spin–orbit coupling. In comparison, the Moon has a one-to-one spin–orbit resonance in which its rotation period is equal to its orbital period.

Cause of Mercury's spin–orbit resonance

But why are Mercury's day and year related to each other in such a simple 3-to-2 ratio? The answer lies in the Sun's varying tidal forces as Mercury revolves about its elongated orbit. The solar gravity pulls hardest on Mercury when the planet is closest to the Sun, at perihelion, and least at the opposite side of its eccentric orbit, at aphelion. This extra gravitational pull of the Sun at perihelion gives an abrupt twist to Mercury's non-spherical body, speeding up the rotation rate and forcing it into synchronism at perihelion with the 3-to-2 resonance. If Mercury's orbit around the Sun were much closer to a circular shape, like the nearly round orbit of the Moon around the Earth, then the Sun's tidal forces would have slowed Mercury's rotation into synchronism with its orbital motion, in a one-to-one resonance with a rotation period equal to its 87.969-day orbital period.

Long, hot afternoons on Mercury

The days are certainly long on Mercury, longer than the planet's year. Mercury's solar day, the time from sunrise to sunset and back to sunrise, is two Mercury years long (Fig. 6.5). At any given point in its orbit, the same hemisphere does not face the Sun during each orbit, but during every other orbit.

Any markings at a given location on Mercury's surface will return to the sunlit side after two orbital revolutions, and this may have misled early astronomers. After two of

Fig. 6.2 Arecibo Observatory   The world's largest radio telescope is nestled into the hills near Arecibo, Puerto Rico. Its metal reflecting surface has a spherical shape with a diameter of 305 meters. The reflected radio signals are focused to detectors suspended on the triangular structure hanging from the three towers. The facility can also transmit powerful radio pulses, sending them off the metal surface into space. Such pulses, sent and received from this giant telescope, first measured Mercury's rotation period in 1965. Until then it had been wrongly thought that Mercury kept one side permanently facing the Sun, with a rotation period equal to its orbital period.

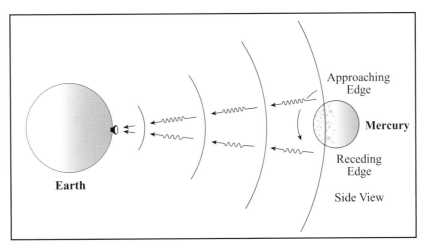

Fig. 6.3 Radar probes of Mercury   A radio signal spreads out as a spherical wave, and Mercury intercepts only a small fraction of them. As the wave sweeps by the planet, it is reflected in spherical wavelets whose wavelengths are Doppler-shifted by the rotational motion of Mercury's surface. The waves from the receding side are red-shifted towards longer wavelengths and those from the approaching side are blue-shifted to shorter wavelengths. The total amount of wavelength change, from red to blue, reveals the speed of rotation, and the rotation period can be obtained by dividing the planet's circumference by this speed.

Focus 6.1   The Doppler effect

Just as a source of sound can vary in pitch or wavelength, depending on its motion, the wavelength of electromagnetic radiation shifts when the emitting source moves with respect to the observer. This Doppler shift is named after the Austrian scientist, mathematician and schoolteacher Christian Doppler (1803–1853), who discovered it in 1842. If the motion is toward the observer, the shift is to shorter wavelengths, and when the motion is away the wavelength becomes longer (Fig. 6.4). You notice the effect when listening to the changing pitch of a passing ambulance siren. The tone of the siren is higher while the ambulance approaches you and lower when it moves away from you.

If the radiation is emitted at a specific wavelength, $\lambda_{emitted}$, by a source at rest, the wavelength, $\lambda_{observed}$, observed from a moving source is given by the relation

$$\text{Redshift} = z = \frac{\lambda_{observed} - \lambda_{emitted}}{\lambda_{emitted}} = \frac{V_r}{c}$$

where V_r is the radial velocity of the source along the line of sight away from the observer, and c is the velocity of light, $c = 2.9979 \times 10^8$ meters per second. The parameter z is called the redshift since the Doppler shift is toward the longer, redder wavelengths in the visible part of the electromagnetic spectrum. When the motion is toward the observer, V_r is negative and there is a blue shift to shorter, bluer wavelengths.

A rotating object will produce a blueshift on the side spinning toward an observer, and a redshift on the opposite side. Their combined effect will broaden a finely tuned radio pulse at wavelength λ by an amount $\Delta\lambda$ given by the expression

$$\frac{\Delta\lambda}{\lambda} = \frac{V_{rot}}{c}$$

where V_{rot} denotes the rotation velocity.

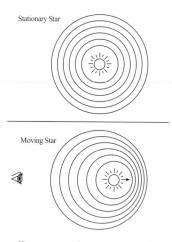

Fig. 6.4 Doppler effect   A stationary source of radiation (*top*) emits regularly spaced light waves that get stretched out or scrunched up if the sources moves (*bottom*). Here we show a star moving away (*bottom right*) from the observer (*bottom left*). The stretching of light waves that occurs when the source moves away from an observer along the line of sight is called a redshift, because red light waves are relatively long visible light waves; the compression of light waves that occurs when the source moves along the line of sight toward an observer is called a blueshift, because blue light waves are relatively short. The wavelength change, from the stationary to moving condition, is called the Doppler shift, and its size provides a measurement of radial velocity, or the velocity of the component of the source's motion along the line of sight. The Doppler effect is named after the Austrian physicist Christian Doppler (1803–1853), who first considered it in 1842.

its orbital periods they would see the same markings on the sunlit side and would find no disagreement with the 87.969-day period that they expected, probably ignoring or missing conflicting observations during the intervening orbit.

Possible water ice at the poles of Mercury

Radar astronomers have found evidence for water ice in both the north and south polar regions of Mercury. Both the intensity and orientation, or polarization, of the bright radar echoes suggest the presence of highly reflective water

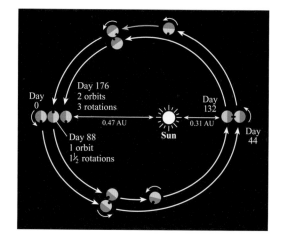

Fig. 6.5 The days are long on Mercury   Because of its spin–orbit coupling, Mercury rotates once every 58.6 Earth days, and orbits the Sun in 88.0 Earth days. After two orbits the planet has rotated three times, but from Mercury's surface the Sun appears to have moved only once around the planet. So sunrise is repeated at a given point on the planet's surface (*black dot*) every two orbits, and a solar day on Mercury lasts two of the planet's years and 176 Earth days. The labels on this figure refer to Earth days.

Fig. 6.6 Mercury's south pole   During its first flyby, the *MESSENGER* spacecraft obtained high-resolution images of a side of Mercury's south polar region not previously seen by spacecraft (*top*), while the opposite side, which was previously seen from *Mariner 10*, was imaged during *MESSENGER*'s second flyby (*bottom*). Radar-reflection data suggest that water ice might be present in the cold, dark, permanently shadowed floors of craters in the polar regions of Mercury. (Courtesy of NASA/JHUAPL/CIW.)

ice. The polar radar characteristics are similar to those returned from the large ice-covered moons of Jupiter and the residual water-ice cap of Mars.

The prospect of a planet so close to the Sun having any water ice seems preposterous. Yet the ice could reside at the bottoms of polar craters (Fig. 6.6). Mercury's rotation axis is very nearly perpendicular to the plane of its orbit, with an obliquity of zero degrees, so the planet does not experience seasons and its polar regions never receive the direct rays of sunlight. In other words, the planet's rotation axis does not tilt, so the Sun never gets very high in the sky at Mercury's poles. Moreover, the atmosphere of Mercury is so thin that heat from the Sun-baked areas does not spread to colder ones. That is why the temperatures during Mercury's night are so cold, down to 90 kelvin, in spite of long dayside highs of about 700 kelvin.

The floors of deep craters near the poles are never exposed to sunlight and ought to always be colder than about 112 kelvin, well below the freezing point of water at 273 kelvin. Moreover, the radar-bright features are concentrated in specific, fresh-looking polar craters with high rims and permanently shadowed floors; the bottoms of degraded, shallow craters are not in perpetual shadow and do not exhibit bright radar signals.

The water could have been delivered to the crater floors by the impacts of comets, subsequently remaining cold-trapped within the craters for as long as billions of years, depending on when the collision occurred.

The shaded polar craters on Mercury might nevertheless contain other volatile substances, such as sulfur, which could produce strong radar echoes but have a higher melting temperature than water ice. So, we may not definitely know if there is water ice at the top and bottom of Mercury until inquisitive robot spacecraft land there and make the appropriate tests. In the meantime, we turn to the startling results of the *Mariner 10* and *MESSENGER* spacecraft.

6.5 A modified Moon-like surface

At first glance, Mercury closely resembles the Moon, for both worlds are small, heavily cratered, and without a significant atmosphere to cause erosion. They both contain large impact basins, ubiquitous craters upon craters, and vast smoother places attributed to volcanic flows. Yet there are differences, for volcanism on Mercury differs from the lunar variety, and the planet exhibits long cliffs that traverse the surface for hundreds of kilometers and are not found on the Moon.

Craters on Mercury

In 1974–75 the *Mariner 10* spacecraft penetrated the glare surrounding Mercury, providing images of about half of Mercury's surface, and showing features thousands of times smaller than those seen from the best telescopes on Earth. These close-ups revealed a landscape that had never been seen before (Fig. 6.7), and they gave us a glimpse of the planet's past. Then in 2008–09 the *MESSENGER* spacecraft returned fascinating images taken at close range (Fig. 6.8) for most of the other half of Mercury that had not been seen before.

Like the Moon, the planet Mercury has highlands that are pockmarked with impact craters, ranging in diameter from impact basins 1000 kilometers across to craters only 100 meters in diameter. As on the Moon, there are small bowl-shaped craters, larger craters with terraces and central peaks, relatively young craters with bright rays, and huge impact basins on Mercury. Once identified, the

Fig. 6.7 Moon-like surface of Mercury   After passing the dark side of Mercury, the *Mariner 10* spacecraft looked back at the sunlit hemisphere of the planet and photographed these images that have been assembled into a mosaic. It shows large tracts of smooth plains, which may be due to extensive volcanism. Partially visible along the day–night terminator (*left*) is half of the Caloris basin (*above center toward the top*), a gigantic multi-ringed impact scar. (Courtesy of NASA/JPL.)

Focus 6.2   Names of surface features on Mercury

The impact craters on Mercury have been mostly named after famous deceased authors, artists, and musicians, such as Dickens, Matisse or Michelangelo, and Mozart. The biggest ones, in order of decreasing size, are named Rembrandt, Beethoven, Dostoevskij, Tolstoj, Goethe, Shakespeare, Raphael and Homer; they range from 720 to 314 kilometers in diameter.

New craters found in *MESSENGER* images and approved by the International Astronomical Union (IAU) include, for example, Calvino, Gibran, Hemingway and Poe, after the authors Italo Calvino, Kahil Gibran, Ernest Hemingway, and Edgar Allan Poe.

Cliffs (*rupes* – Latin for "cliff") take the names of famous ships, such as the British naval vessel *Beagle*, which the naturalist Charles Darwin (1809–1882) traveled on, or *Discovery*, Captain James Cook's (1728–1779) ship on his last voyage to the Pacific, and *Santa Maria*, Christopher Columbus' (1451–1506) flagship on his first voyage to America in 1492. Plains, or plaitiae, take the names of Mercury, as a planet or god, in various cultures.

The most prominent impact basin viewed by *Mariner 10* has a unique designation, the Caloris Basin or Basin of Heat, so-named because it nearly coincides with one of the hottest locations on the planet.

A complete description of all the naming details can be found in the "Gazetteer of Planetary Nomenclature" of the IAU Working Group for Planetary System Nomenclature, available on the Internet at http://planetarynames.wr.usgs.gov/.

craters are mostly named after famous writers, painters, and composers (Focus 6.2).

The ubiquitous craters on Mercury strongly resemble their lunar counterparts, indicating that they were formed by meteoritic impact, and most of the Mercurian craters probably record the period of late heavy bombardment by meteorites of all size, which formed the lunar highlands about 3.9 billion years ago.

Because the force of gravity on the surface of Mercury is about twice that on the Moon, material ejected from a crater on Mercury is thrown about half as far. Thus, secondary craters are closer to the primary crater rim on Mercury than they are on the Moon. The comparatively high gravity on Mercury also helps flatten crater walls, perhaps explaining why Mercury's craters appear to be shallower than similarly sized craters on the Moon (Fig. 6.9).

The freshest craters on Mercury have extensive ray systems, some of which extend over 1000 kilometers (Fig. 6.10). The bright rays are thrown out during a crater-forming explosion when the meteorite impact occurs on the surface of an airless body like Mercury or the Moon. But the rays fade with time as tiny meteorites and particles from the solar wind strike the surface and darken the rays. So the rayed craters were formed relatively recently, on the order of 1 billion years ago or less.

Some craters on Mercury have dark rims or nearby dark "halos" surrounding them (Fig. 6.11). There are two possible explanations for these features. A subsurface layer of dark material might have been excavated when the crater-producing impact penetrated to the right depth. Alternatively, the heat of impact might have melted some of the surface, and the molten rock splashed to the edge of the crater where it re-solidified as a dark, glassy substance.

Fig. 6.8 Previously unseen side of Mercury   A mosaic of images collected as the *MESSENGER* spacecraft viewed Mercury obliquely on its second approach on 6 October 2008. The Rembrandt impact basin, 715 kilometers in diameter and also shown in Fig. 6.14, is seen at the center as night was falling across the planet's eastern (right) edge. (Courtesy of NASA/JHUAPL/CIW.)

Fig. 6.9 Shallow craters on Mercury
This mosaic image was acquired from the *MESSENGER* spacecraft during its first flyby of Mercury. Many of the impact craters appear to be shallower than similarly sized craters on the Moon. The comparatively high gravity of Mercury helps flatten tall structures like crater walls. Smooth volcanic plains are seen near the center of the image. The large crater to the lower right is 200 kilometers wide. The shadowed area to the right of this crater marks the boundary between the sunlit day side and dark night side of the planet. (Courtesy of NASA/JHUAPL/CIW.)

Fig. 6.10 Bright rayed crater on Mercury   A relatively young crater, about 80 kilometers in diameter, has splashed bright rays far around Mercury. The bright rays are produced by impacts that excavate and eject relatively unweathered subsurface material. Space-weathering effects, such as the continued bombardment by small meteorites or solar-wind particles, will eventually darken the ejected material and erase the rays from view. Bright-rayed craters are also found on the Moon, and *MESSENGER* images, such as this one, indicate that Mercury has more of them, perhaps due to more violent impacts caused by Mercury's greater mass and proximity to the Sun. (Courtesy of NASA/JHUAPL/CIW.)

Fig 6.11 Dark halo craters on Mercury Two of the larger craters in this *MESSENGER* image appear to have darkened crater rims and partial "halos" of dark material immediately surrounding them. The explosive impact that produced the dark-haloed craters could have excavated darker subsurface material or melted part of the surface, splashing it across the surface where it re-solidified as dark material. (Courtesy of NASA/JHUAPL/CIW.)

The Moon also has dark-haloed craters; Tycho is a well-known example. In the lower gravity of the Moon, dark material will be ejected a greater distance from the crater than on Mercury.

The heavily cratered terrain on Mercury also exhibits a paucity of smaller craters when compared to the lunar highlands, most likely due to volcanic flows during the late heavy bombardment that obliterated the small craters on Mercury, but not in the lunar highlands. Mercury's craters are also less densely packed than their counterparts on the Moon, possibly due to volcanism that obscured the older craters on Mercury and created extensive intercrater plains not found in the lunar highlands.

Old intercrater plains on Mercury

In many important respects, Mercury's resemblance to the Moon is superficial. The planet's most densely cratered surfaces are not as heavily cratered as the lunar highlands, and Mercury does not contain regions of overlapping large craters and basins. Also unlike the Moon, the heavily cratered terrain on Mercury includes large regions of gently rolling, intercrater plains that were discovered in *Mariner 10* images (Fig. 6.12).

These plains obliterate or partially fill older craters, but the plains have other craters superposed upon them, suggesting that lava flows created the intercrater highland plains during the late heavy bombardment roughly 3.9 billion years ago. This widespread volcanic resurfacing covered the older craters, erasing them from view, and the

planet's heavily cratered terrain was then excavated out of the cooled and solidified lava near the end of the heavy bombardment recorded on the Moon.

Active volcanic past on Mercury

Widespread areas of Mercury are covered by relatively flat, sparsely cratered terrain called the smooth lowland plains (Fig. 6.13). They are younger than the intercrater highland plains, and they are about 2 kilometers lower. Unlike the dark maria on the Moon, the smooth lowland plains on Mercury have about the same brightness or color as the heavily cratered terrain and intercrater plains in the highlands of Mercury.

Detailed investigations from the *MESSENGER* spacecraft indicate an active volcanic past for Mercury. The Rembrandt impact basin (Fig. 6.14) has, for example, apparently been covered by effusive volcanism after the basin's formation, in multiple stages of smooth volcanic flow and surface contraction and deformation.

Smooth plains are found within and near craters and impact basins all over Mercury. They are attributed to gentle, effusive volcanic flows that occurred after the late heavy bombardment that excavated the older craters, but before the craters superimposed on them. The smooth plains are distributed across the planet, covering about 40 percent of its surface.

High-resolution imaging from *MESSENGER* has also revealed several non-circular, irregularly shaped depressions on Mercury, which are surrounded by bright, highly reflective halos with a distinctive color (Fig. 6.15). The depressions are interpreted as explosive volcanic vents, and the surrounding bright material as deposits ejected during volcanic eruptions at the vents.

Cliffs, or rupes, on Mercury

Mercury has another type of surface terrain not found on the Moon. They are the remarkable winding cliffs, or *rupes* – Latin for "cliffs" (Fig. 6.16), which are widely distributed over the planet and occur on a global scale. Individual cliffs range from 20 to more than 500 kilometers in length, and have heights from about 300 meters to about 3 kilometers. The cliffs cut across craters, and few craters are superimposed on them. This indicates that the cliffs were formed relatively late in early Mercury history, after the formation of the heavily cratered regions and the intercrater plains.

The long cliffs on Mercury look like the wrinkled skin of a shriveled apple, and the shrinking of the planet as the interior cooled probably caused them. The total shrinkage in the radius of the planet implied by the cliff heights is

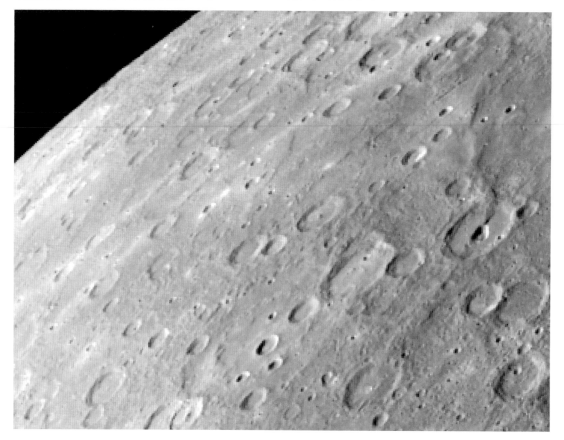

Fig. 6.12 Intercrater plains on Mercury The heavily cratered regions of Mercury exhibit abundant craters, just as the lunar highlands do, but unlike the Moon intercrater plains are found surrounding the craters in Mercury's highlands. This image, taken from the *Mariner 10* spacecraft, also exhibits long, dark cliffs, known as scarps or rupes, and a long ridge that runs along the right side of the image, cutting across a large crater about 80 kilometers in diameter. (Courtesy of NASA/JPL/Northwestern University.)

Fig. 6.13 Smooth plains on Mercury A large expanse of smooth plains on Mercury imaged from the *MESSENGER* spacecraft. Craters in these volcanic plains appear to have been significantly flooded with lava, leaving only their circular rims preserved. (Courtesy of NASA/JHUAPL/CIW.)

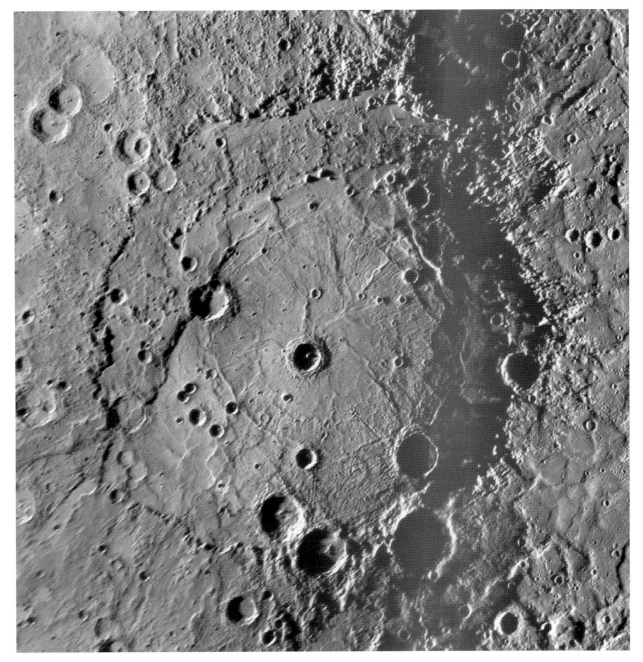

Fig. 6.14 Rembrandt impact basin   This well-preserved impact basin was discovered in images taken from the *MESSENGER* spacecraft. Named Rembrandt, after the Dutch painter, it is about 715 kilometers in diameter. Scientists estimate that it was formed about 3.9 billion years ago, near the end of the period of heavy bombardment of the inner solar system. Although ancient, the distributions of smaller craters on its surface indicate that Rembrandt is younger than many other impact basins on Mercury. The basin and some of the small craters have also been flooded by effusive volcanic flows with a smooth appearance. Multi-colored images of the crater floor indicate areas with unusually high amounts of iron and titanium. (Courtesy of NASA/JHUAPL/SSI/CIW.)

estimated to be about 1 kilometer. Since the cliffs formed after most of the craters, but before some of the smaller craters found on them, Mercury probably shrank many hundreds of millions of years after the solidification of the crust, during the cooling of the planet's underlying mantle and partial solidification of its internal core. As the young planet solidified from the outside in, great blocks of its crust shifted up on one side and down on the other, thrusting the cliffs up into the sky.

There is no evidence of global shrinking on the Moon. This is probably because the Moon has a small metallic core, while Mercury has a large one. Mercury's entire globe probably shrank when at least some of its large iron core cooled and solidified.

Fig. 6.15 Volcanic activity on Mercury   This enhanced-color view, taken during the third flyby of Mercury by the *MESSENGER* spacecraft, enhances compositional differences in the features observed. The bright yellow area near the top right is centered on an irregular depression that is possibly an explosive volcanic vent. The double-ring basin in the center of this image may be filled by the flow of effusive volcanic lava; the basin is 290 kilometers in diameter. Smooth plains, thought to be the result of volcanic activity at earlier episodes than the others shown here, are located in much of the surrounding area. (Courtesy of NASA/JHUAPL/CIW.)

6.6 An iron world

The terrestrial bodies — the Moon, Mars, Venus and Earth — exhibit a fairly linear relationship between size and mean mass density, in which bigger objects have greater density, but Mercury does not conform to this relationship (Fig. 6.17). Although it is less than half the size of the Earth and not much bigger than our Moon, the bulk mass density of Mercury is 5427 kilograms per cubic meter, only slightly less than the Earth's bulk mass density of 5513 in the same units.

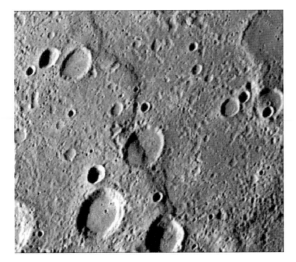

Fig. 6.16 Santa Maria rupes   Mercury's surface is distinguished from those of the other terrestrial planets and the Earth's Moon by having enormous cliffs, or rupes, that cut across its surface. The dark lobate cliff that diagonally crosses this *Mariner 10* image has been named Santa Maria rupes, after Christopher Columbus' (1451–1506) flagship on his first voyage to America in 1492; "rupes" is the Latin word for "cliff". This long cliff was probably created when the planet cooled and contracted. The 90-km diameter crater in the upper right hand corner has been filled, or embayed, by intercrater plains. (Courtesy of NASA/JPL/Northeastern U.)

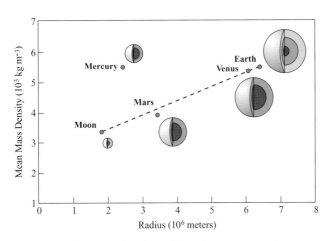

Fig. 6.17 Radius, mass density and interior structure of terrestrial bodies   Mercury is an anomaly in this comparison of size and mean mass density. Mercury is just slightly bigger than the Moon and a little smaller than Mars, but its mean mass density is comparable to that of the larger terrestrial planets, Earth and Venus. As shown in the internal cross-sections, Mercury's unexpectedly large mass density is due to a large dense core extending up to 75 percent of its radius. The Moon's core occupies just 20 percent of its radius. The Earth's inner and outer core take up 55 percent of its radius, while that of Mars is smaller.

The most natural explanation of Mercury's high mass density is that it contains an unusual amount of iron, which is cosmically the most abundant heavy element. The dense iron core may take up to 75 percent of the planet's radius, or some 42 percent of its volume, so Mercury is mostly iron core surrounded by a relatively thin silicate mantle. In comparison, the Earth's iron core has a radius of 54 percent of the planet's radius and occupies just 16 percent of the planet's volume.

Soon after its formation, Mercury probably remained molten long enough for the heavy substance to settle at its center, just as iron drops below slag in a smelter. The high weight of the iron atoms would have slowly carried them down into the interior, leaving the lighter silicates to form a rocky mantle on top.

Why does Mercury have so much iron and so little rock? Some astronomers think the silicate rock was blasted off long ago, when Mercury collided with another planet-sized object. The collision might have removed much of Mercury's silicate mantle, leaving the large iron core largely intact. Or both colliding objects could have had thick rocky mantles that were completely vaporized in the collision, while their iron cores clumped together under their mutual gravitational pull. The light vaporized rock particles might have spiraled into the massive Sun, while the iron planet remained in orbit.

Such cataclysmic collisions were apparently common about 4.5 billion years ago, when a giant impact of a Mars-sized object with the nascent Earth created its Moon, with a lot of rock and little metal. A similar collision might have knocked Venus into its slow retrograde spin, in the opposite direction to both the rotation and orbital motion of all the other major planets. Another large impact in the final growth stages of Mars could additionally account for the large-scale dichotomy in its crustal topography.

In a second possibility, the intense radiation and particle emission from the active young Sun could have removed the silicate mantle from once-larger Mercury. In a third possibility, called selective accretion, the accumulation of heavy iron, as opposed to light silicate substances, was favored during the planet's formation; such an enrichment would not occur at greater distances from the Sun. There is not enough evidence to decide between the three possibilities.

6.7 A mysterious magnetic field

An instrument aboard *Mariner 10* showed that Mercury is a magnetized planet, with a magnetic field emanating from it. The increasing strength of this field as the spacecraft approached the planet indicated that magnetism at Mercury's surface would be 0.0033×10^{-4} tesla, about

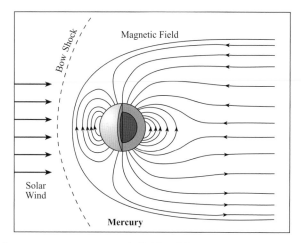

Fig. 6.18 Mercury's magnetic field and large core   The magnetic field of Mercury is a miniature version of Earth's magnetic field, complete with bow shock, magnetosphere and magnetotail. Mercury's magnetic axis is closely aligned with its rotation axis, and its polarity is the same as the Earth's, with magnetic south corresponding to geographic north. This magnetic field is probably generated within the planet's large iron core, but the exact mechanism for creating the field remains a mystery. The electrified solar wind compresses the magnetic field into a bow shock on the sunlit side and draws it out into a tail on the opposite side.

1 percent or 0.01 times the strength of Earth's equatorial magnetic field.

Mariner 10's encounter provided magnetic data only for Mercury's eastern hemisphere, but during its second flyby *MESSENGER*'s magnetometer measured the western hemisphere, showing that the planet's magnetism is highly symmetric, with a dipolar magnetic field whose axis is closely aligned, within 2 degrees, with the planet's rotation axis.

The relatively weak magnetic field on Mercury is still strong enough to hold off the solar wind, forming a bow shock on the sunlit side of the planet and carving out an elongated magnetosphere with a tail pointing away from the Sun (Fig. 6.18). It is a scaled-down, miniaturized version of the Earth's magnetosphere, with the planet Mercury occupying a larger fraction of its magnetosphere. That is, Mercury's magnetosphere is relatively small, with a bow shock located at about 1.5 Mercury radii above the planet's surface; the bow-shock distance for Earth's magnetosphere is about 10 times larger, in planetary radii.

Because of Mercury's weak magnetic field strength and the small size of its magnetosphere, the outer boundary of its magnetosphere, known as the magnetopause, is frequently penetrated during interaction with the interplanetary magnetic fields emanating from the nearby Sun. When the interplanetary and planetary magnetic fields are pointing in opposite directions on contact, they merge together and reconnect into a new magnetic configuration, converting magnetic energy into kinetic energy. This opens Mercury's magnetosphere, exposing the planet to energetic charged particles in the Sun's winds.

The magnetometer aboard *MESSENGER* has demonstrated that such perforations of the dayside magnetosphere are large and frequent, with openings of up to 900 kilometers across and with a reconnection rate about 10 times that typical of Earth. The magnetic flux transfer has been likened to magnetic twisters dancing on the magnetopause, rooted deep in the planetary interior and carried into interplanetary space by the solar wind.

The discovery of Mercury's magnetic field was completely unexpected. Its presence implied the existence of an electrically conducting, molten core in which the magnetism is sustained by dynamo action. But back in 1974–75, at the time of the *Mariner 10* encounters with Mercury, most scientists thought that Mercury's core would have solidified long ago.

The argument for a solid core goes something like this. Small objects like Mercury have a high proportion of surface area to volume; that proportion decreases with increasing size of the object. Since internal heat is generated throughout the volume of a satellite or planet, and radiated to space only from the surface, smaller bodies radiate their energy into space faster and cool more rapidly. Since the Earth and Mercury are both the same age, and the much larger Earth has a solid inner core, retaining only a liquid outer core, the interior of Mercury should have completely solidified over its 4.6-billion-year lifetime.

Nonetheless, Mercury does have a global magnetic field; about 1 percent as strong as that found on Earth, suggesting that at least some molten material might be circulating inside the planet, and detailed radar measurements provided strong evidence that the planet does have a molten core. By 2007, radar astronomers had measured the planet's rotation with an accuracy of one-thousandth of a percent, discovering rotational twists back and forth that were double what would be expected for a completely solid body. This meant that the core, or at the very least a thin outer core, is at least partially molten, and decoupled from the overlying solid mantle. The liquid core is not forced to rotate with the mantle, and as the liquid sloshes around it alters the planet's overall rotation with a twisting first in one direction and then in another. Maintaining a molten iron core over up to 4.6 billion years may require that the core also contains an element lighter than iron, such as sulfur, to lower the melting temperature of the core material. When *MESSENGER* begins to orbit Mercury in 2011, it will provide additional evidence about the planet's core and its effect on the spacecraft's orbit.

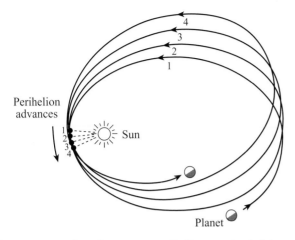

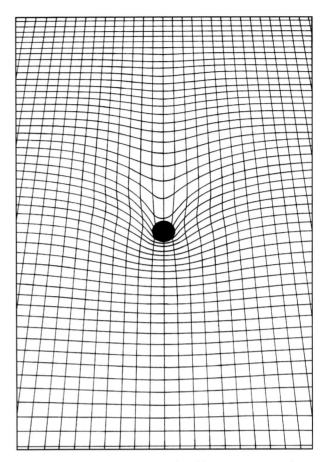

Fig. 6.19 Precession of Mercury's perihelion   Instead of always tracing out the same ellipse, the orbit of Mercury pivots around the focus occupied by the Sun. The point of closest approach to the Sun, the perihelion, is slowly rotating ahead of the point predicted by Newton's theory of gravitation. This was at first explained by the gravitational tug of an unknown planet called Vulcan thought to be revolving about the Sun inside Mercury's orbit, but we now know that Vulcan does not exist. Mercury's anomalous motion was eventually explained by Einstein's new theory of gravity in which the Sun's curvature of space makes the planet move in a slowly revolving ellipse.

Fig. 6.20 Space curvature   A massive object creates a curved indentation upon the flat Euclidean space that describes a world which is without matter. Notice that the amount of space curvature is greatest in the regions near the object, while further away the effect is lessened.

6.8 Einstein and Mercury's anomalous orbital motion

For nearly two and a half centuries, the solar system appeared to behave according to Newton's law of gravitation, which was used to predict the paths of the planets with great precision, but there was something wrong with Mercury's motion. Newton's theory failed to provide the expected connection between old and new measures of the planet's position, and its trajectory could not be precisely specified.

Instead of returning to its starting point to form a closed ellipse in one orbital period, Mercury moves slightly ahead in a winding path that can be described as a rotating ellipse. As a result, the point of Mercury's closest approach to the Sun, the perihelion, advances by a small amount, 43 seconds of arc per century, beyond that which can be accounted for by planetary perturbations using Newton's law (Fig. 6.19). This unexpected effect is known as the anomalous precession of Mercury's perihelion.

The unexplained twist in Mercury's motion was discovered more than one and a half centuries ago, as the result of watching the planet pass in front of the Sun every decade or so. This resulted in accurate determinations of the planet's orbital position. An analysis of these transits led the French mathematician Urbain Jean Joseph Le

Verrier (1811–1877) to report the anomalous precession to the French Academy of Sciences in 1849. He subsequently attributed the anomalous motion to the gravitational pull of an unknown planet orbiting the Sun inside Mercury's orbit and moving ahead of it. The hypothetical planet was named Vulcan, and extensive searches were conducted for it. No such planet was ever reliably detected.

The cause of Mercury's anomalous motion remained a mystery until 1915 when Albert Einstein (1879–1955) explained it using a new theory of gravity in a paper entitled "Explanation of the Perihelion Motion of Mercury by Means of the General Theory of Relativity". According to Einstein's theory, space is distorted and curved in the neighborhood of matter (Fig. 6.20). In effect, space has both content and shape. The curved shape is molded into space by its content, the massive objects. That bending, twisting and distortion of space is gravity. But such curvature effects are only noticeable in extreme conditions very close to an exceptionally massive object, and

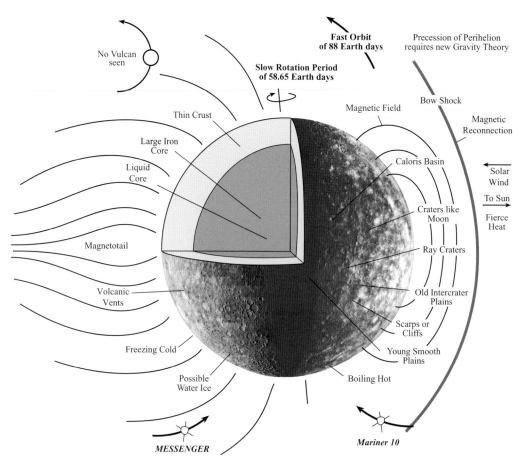

Fig. 6.21 Summary diagram.

the differences between the Newton and Einstein theories are indistinguishable in everyday life.

The curvature of space in the neighborhood of the massive Sun guides the motion of Mercury. The planet is traveling as straight as it can along an invisible curved track in space, like a ball spinning in a roulette wheel, making Mercury overshoot its expected perihelion position by precisely the observed amount of just 43 seconds of arc per century. Because the amount of space curvature produced by the Sun falls off with increasing distance, the perihelion advances for the other planets are much smaller than Mercury's.

The accord between Einstein's calculations and the observed motion of Mercury depends on the assumption that the Sun is a nearly perfect sphere. If the interior of the Sun is rotating very fast, it will push the equator out further than the poles, so its shape ought to be somewhat oblate rather than perfectly spherical. The gravitational influence of the outward bulge will provide an added twist to Mercury's orbital motion, shifting its orbit around the Sun by an additional amount and lessening the agreement with Einstein's theory of gravity. Fortunately, scientists have

used sound waves, detected as five-minute pulsations of the Sun's visible disk, to see inside the Sun and show that the differences in internal rotation rates is not enough to produce a substantial asymmetry in the Sun's shape. So we may safely conclude that measurements of Mercury's orbit confirm Einstein's explanation of the planet's motion.

Unlike some of today's theoretical physicists, Einstein realized that his new theory needed to be verified by definitive predictions of other consequences. He noticed that light has energy and therefore mass, and that the Sun's gravity will pull the light toward it. The path of light passing near the Sun will therefore be bent by the curvature of space, and Einstein predicted that the apparent positions of stars would therefore be displaced when passing near the Sun. When Einstein's predicted value, of 1.75 seconds of arc for a Sun-grazing light ray, was confirmed by observations of stellar positions during a solar eclipse on 29 May 1919, it brought him international recognition.

The solar curvature of nearby space has been measured with increasingly greater precision for nearly a century. In one test, radio telescopes on opposite sides of the Earth combine their observations of the changing positions of remote galaxies or quasars when they pass behind the Sun. Another test measures the time for a radio signal to travel from a spacecraft home. When the line of sight passes near the Sun, the radio waves travel along a curved path and take slightly longer to return to Earth. The measurements require extremely precise clocks, for the extra time delay caused by the Sun's curvature of nearby space amounts to only one-ten-thousandth of a second.

In 2003, measurements using radio links with the *Cassini* spacecraft confirmed the predicted curvature to one part in ten thousand or to the fourth decimal place. Einstein's theory has now been verified by so many experiments and to such precision that it has become widely accepted as a brilliant contribution to our understanding of nature — begun by his attempts to account for Mercury's unexplained motion.

7   Venus: the veiled planet

- When visible, Venus is the brightest planet in the sky. It orbits the Sun inside Earth's orbit, appearing in the evening or morning hours and never in the middle of the night.

- No human eye has ever gazed on the surface of Venus, which is forever hidden by a thick overcast of impenetrable clouds.

- *Venera* spacecraft have parachuted through the clouds of Venus, surviving long enough to measure the properties of its torrid surface and even photographing it.

- The deadly efficient greenhouse effect of a thick, carbon-dioxide atmosphere has scorched Venus's surface, raising its temperature to 735 kelvin, even hotter than Mercury's average dayside temperature.

- In size, density and composition, Venus is almost identical to the Earth, but radar signals and space probes have penetrated its clouds to reveal an unearthly surface without a trace of liquid water or life.

- The pale yellow clouds of Venus are composed of concentrated sulfuric-acid droplets.

- The surface of Venus lies under a crushing atmosphere whose surface pressure is 92 times that on Earth.

- It takes only 4 Earth days for the high-flying clouds to move once about Venus, from east to west, blown by fierce, rapid winds, but the slow winds near the surface rotate with the planet, once every 243 Earth days in the same backwards, retrograde direction.

- The high, rapid winds on Venus spiral toward its poles, producing a huge, whirling polar vortex at both poles of the planet.

- There is no detectable magnetic field on Venus, but its dense atmosphere deflects the solar wind.

- Venus has a day longer than its year. The planet rotates once every 243 Earth days, in the opposite, retrograde direction from other planets except Uranus, and it takes 224.7 Earth days for Venus to orbit once about the Sun.

- The radar instrument aboard the *Magellan* spacecraft spent more than four years mapping out the surface of Venus in unprecedented detail, revealing rugged highlands, smooth plains, volcanoes, and sparse, pristine impact craters.

- About 85 percent of the surface of Venus is covered by smooth, low-lying volcanic flows of lava, and much of the remaining 15 percent is high-standing with towering volcanoes.

- The entire surface of Venus was probably covered by rivers of outpouring lava roughly 750 million years ago, wiping out all previous craters and about 90 percent of the planet's history; volcanic activity has continued at a reduced level up to the present.

- Tens of thousands of volcanoes now pepper the surface of Venus; some of the volcanoes could now be active.

- High volcanic rises on Venus are kept up by active motions below.

- Vertical motions associated with upwelling hot spots have buckled, crumpled, deformed, fractured and stretched the surface of Venus.

- Venus exhibits every type of volcanic edifice known on Earth, and some, called arachnoids and coronae, which have never been seen before.

- The surface of Venus moves mostly up and down, rather than sideways.

- Liquid water is non-existent on Venus, and the lack of water could be why Venus does not have moving plates similar to those found on Earth.

7.1 Fundamentals

Venus is the planet most like the Earth in size and mass. Its radius of 6051.8 kilometers is 94.9 percent of the Earth's radius, and its mass is 81.5 percent of the Earth's mass. Like the Earth, the planet Venus is a dense, rocky world, one of the four terrestrial planets. Venus spins in the backward direction, and so slowly that its day is longer than its year. The planet's surface lies under a hot and heavy atmosphere, with a high temperature and pressure, but no magnetic field has been observed on the planet.

7.2 Bright, beautiful Venus

Brilliant torch of the heavens

When visible, Venus is the most brilliant of the planets; it is the brightest object in the night sky, after the Moon. The stunning beauty of Venus must have been known since the dawn of human history. Our name Friday is, for example, derived from the Anglo-Saxon Frigadaeg, combining Friga, or Venus, and daeg, or day.

A female association has been common since the beginning of civilization. As Euripides (484–407 BC) put it:

Table 7.1 Physical properties of Venus[a]

Mass	4.86732×10^{24} kilograms = $0.815\,M_E$
Mean radius	6051.8 kilometers = $0.949\,R_E$
Bulk density	5243 kilograms per cubic meter
Sidereal rotation period	-243.018 Earth days, retrograde
Sidereal orbital period	224.7 Earth days = 0.615 197 3 Earth years
Mean distance from Sun	1.08157×10^{11} meters = 0.723 AU
Age	4.6×10^9 years
Atmosphere	96.5 percent carbon dioxide, 3.5 percent nitrogen
Surface pressure	92 bars
Surface temperature	735 kelvin
Magnetic field strength	Less than 3×10^{-9} tesla or $10^{-5}\,B_E$

[a] The symbols M_E, R_E, and B_E denote respectively the mass, radius and magnetic field strength of the Earth.

"Venus, the eternal sway, all race of men obey". In another example, the Chinese named Venus *T'ai-pe* — "the Beautiful White One".

The name Venus is that of the ancient Roman goddess of love and beauty; the Greek equivalent was Aphrodite. The Greeks worshipped Aphrodite on the island of Cythera, and therefore the adjective "Cytherean" has often been applied to the planet.

The oldest recorded observations of Venus are those of the Babylonians, who called the planet Ishtar, "the bright torch of heaven" — the embodiment of all things womanly, the Mother of the Gods, and the goddess who evoked the power of dawn. The Maya built their calendar around the appearances and disappearances of Venus, though for them it was more fearsome than alluring.

There are other non-female names for the planet. The Judeo-Christian Devil is also known as Lucifer, which was originally a Latin name for Venus as a morning star. For the Mayan civilization, which flourished between 300 and 900 AD, Venus was the Sun's brother, the male god named Kukulkan, who preceded the Sun in rising from the underworld of night. The Mayan astronomer-priests could accurately predict Venus' appearance for over a hundred years, but they also got a bit carried away and made human sacrifices to the planet.

The view of Venus from Earth

Venus is visible at the edges of night, lingering near either dawn or dusk. It hangs low and bright in the morning or evening sky, sometimes near the crescent Moon. Because Venus's greatest angular distance from the Sun, known as its maximum elongation, is 47 degrees, it appears as the "evening star" just after sunset or as the "morning star" just before sunrise, but never as both an evening and morning star on the same day.

The brightest planet is never seen in the middle of the night or at midday. In the dark black of midnight, we are on the opposite side of the Earth from the Sun and Venus. At noon the planet is hidden in the full, bright glare of the Sun.

Venus is the second planet from the Sun, the world next door and the nearest planet to us in space. Every 19 months the planet swings to within 100 times the distance of the Moon. At closest approach, Venus is only 0.28 AU, or 38 million kilometers, from us, and its angular diameter is 64 seconds of arc.

Venus moves around the Sun in a nearly circular path once every 224.7 Earth days, like a runner on the inside track, at a mean distance of 0.723 AU. Since the Earth orbits the Sun at 1.00 AU in the same direction as Venus and with a slightly slower rate, it takes 584 Earth

days or about 19 months for Venus to catch up with us. That is, every 19 months Venus passes between the Sun and us.

During each 19-month circuit, Venus is visible from the Earth for approximately 260 days as an evening star on one side of the Sun, and for about 260 days as a morning star on the other side of the Sun. Between its evening and morning appearances, Venus disappears from view; it then passes between us and the Sun or moves behind the star.

The approximate 260-day length of a Venus appearance in the morning or evening coincides closely with the average length of a human pregnancy. So the cosmos certainly seems to be in tune with female cycles. Most of us are familiar with a woman's monthly lunar cycle, and there is another Sun-related one that is not so well known. The Sun pulsates, moving in and out, every 5 minutes. This is the average length of a woman's contractions during childbirth; at least it was when my kids were born.

When viewed through a telescope, Venus brightens and fades, and also changes in apparent size, during its dance around the Sun (Fig. 7.1). As noticed by Galileo Galilei (1564–1642) in 1610, the planet exhibits a complete sequence of Moon-like phases, which means that Venus should orbit the Sun rather than the Earth. Its apparent illumination goes from a full round disk to a narrow crescent and back to rotundity again every 19 months. Venus also appears to grow when it approaches us in its orbit and shrinks as it recedes. When Venus is farthest from the Earth, on the opposite side of the Sun, it is fully illuminated and smallest. As the planet comes closer to Earth, it looks partly illuminated and larger.

Venus's exceptional brightness is partly due to its highly reflective clouds, which reflect about 65 percent of the incident sunlight back into space. By way of comparison, Mercury and the Earth's Moon reflect roughly 10 percent of the sunlight reaching them.

The clouds that help make Venus the brightest of planetary worlds also perpetually hide its surface from view (Fig. 7.2). No features can be seen beneath the unbroken layer of clouds by the human eye, or even with a telescope on the ground or in space. The high-flying clouds whip around the planet from east to west every four Earth days, in the same direction as the planet rotates but in the opposite, retrograde direction to its orbital motion.

Spectroscopic observations have been more rewarding than casual visual inspection, showing in the 1930s that the planet's upper atmosphere is mainly composed of carbon dioxide (CO_2). We now know that its thick, massive atmosphere is 96 percent CO_2, and that it contains about 300 000 times as much CO_2 as is present in our air.

FIVE PHASES OF VENUS

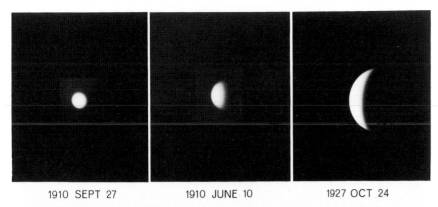

1910 SEPT 27   1910 JUNE 10   1927 OCT 24

1919 SEPT 25   1964 JUNE 19

Fig. 7.1 The phases of Venus When fully illuminated, Venus looks small and far away; its apparent size is about seven times larger when the crescents are narrowest and Venus is nearest the Earth. After observing similar phases with his small telescope in 1610, Galileo wrote "Cynthiae figuras aemulatur mater amorum", or "The mother of love (Venus) emulates the figure of Cynthia (Moon)". The phases and variation in the apparent size of Venus provided important early evidence that the planets revolve around the Sun rather than the Earth. (Lowell Observatory photographs.)

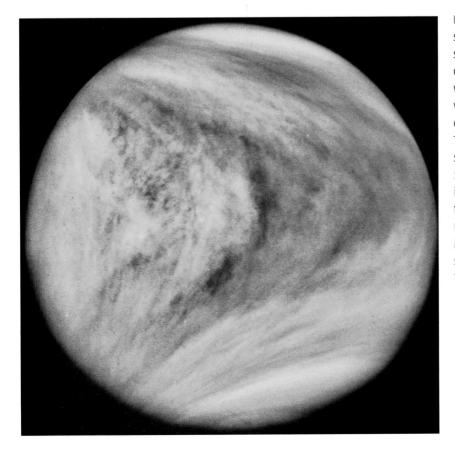

Fig. 7.2 Cloud tops of Venus Clouds of sulfuric acid permanently shroud Venus' surface from view. The horizontal Y-shaped clouds near the equator move from east to west along with high-speed winds at velocities of up to 100 meters per second, circling the planet every four Earth days. This image was taken from the *Mariner 10* spacecraft when it encountered Venus on 5 February 1974, using its gravitational influence to be flung on to Mercury. Similar features have been seen in violet or ultraviolet light from the *Mariner 10*, *Pioneer Venus*, *Galileo*, and *Venus Express* spacecraft, as well as from the *Hubble Space Telescope*. (Courtesy of NASA.)

7.3 Penetrating the clouds of Venus

Venus has a hot and heavy atmosphere

In many ways, Venus is the Earth's twin sister, with almost the same weight and waistline. Its mass is 19 percent less than the Earth's mass, and its mean radius of 6051.8 kilometers is just 5 percent less than the Earth's radius. So the feel of gravity on the planet's surface is like that on Earth. The two planets are also alike in mass density and composition, so at formation they must have been very similar to each other. And because Venus is just a little closer to the Sun than Earth, the climate beneath Venus's clouds was once thought to be warm and temperate.

Since no one could see the surface under the clouds, scientists were free to speculate about what might be found there, and fascinating creatures were imagined to flourish in the warm, wet environment. Throughout the 19th and most of the 20th century, our sister planet was pictured with a verdant, swampy surface, perhaps with oceans of water. It might even be teeming with life.

But space-age scientists have drastically altered this romantic vision. Their radio telescopes and space probes have penetrated the clouds and uncovered the elusive planet's hidden secrets. They revealed a truly hellish and sterile surface, without a trace of flowing water or a sign of life. Beneath her pure, gleaming clouds, the planet of love is a torrid inferno!

The first hint of the inhospitable environment came in 1958 from ground-based radio astronomers when they measured unexpectedly large amounts of microwave radiation emitted by Venus. If the radiation was coming from the planet's surface, it had a temperature of hundreds of kelvin, hotter than a microwave oven turned up high and hot enough to boil away any oceans on Venus.

This explanation was not universally accepted. Some scientists thought the microwave radiation might be coming from the atmosphere, where lightning could generate the emission. But then the United States launched the first interplanetary spacecraft, *Mariner 2*, which flew by Venus in 1962. Instruments aboard *Mariner 2* pinpointed the source of the microwave radiation, showing that it comes from the scorching surface of Venus rather than from its atmosphere. As one scientist put it, Venus is no place to raise the kids.

Diving into the inferno

Of course, the only way to be certain about what lies beneath the clouds was to send a space probe through them. After some initial failures, the former Soviet Union mastered the technique. In 1967 their *Venera 4* spacecraft was the first to enter the atmosphere of another planet, make measurements there, and radio home the results. It confirmed that the atmosphere of Venus is about 96 percent carbon dioxide, and recorded increasing temperatures and pressures on its way down, at least until the spacecraft was either burnt up or crushed to pieces. Venus contains about 300 000 times as much carbon dioxide in its atmosphere as the Earth does.

The day after *Venera 4* made its historic entry into the atmosphere of Venus, *Mariner 5* flew by the planet. Measurements of the way the atmosphere changed *Mariner 5*'s radio signal, when it passed behind the planet and reappeared on the other side, provided a clear profile of the temperature and pressure and confirmed their high values near the ground.

The next two probes, *Veneras 5* and *6*, penetrated further, and then in 1970 *Venera 7* survived the heat and pressure of descent to become the first spacecraft to land on the surface of another planet. *Venera 8* repeated this feat in 1972.

Veneras 7 and *8* measured the atmosphere's temperature and pressure all the way down to the ground, showing that the surface temperature is a sizzling 735 kelvin. That is about as hot as a self-cleaning oven and hot enough to boil the ground dry, and to incinerate any humans that might visit the planet. Venus is the hottest planet in the solar system, with a surface temperature above the average 700 kelvin of Mercury's daytime surface.

The high surface temperature on Venus is maintained by the greenhouse effect, driven by the small percent of solar energy that reaches the ground. The thick atmosphere of Venus traps the Sun's heat near the surface, raising the ground temperature to about three times what it would be without an atmosphere. Although carbon dioxide is the most significant greenhouse gas on Venus, the presence of sulfur dioxide and clouds of sulfuric acid enhance its action.

The massive atmosphere imposes a pressure that is 92 bars, or 92 times the air pressure at sea level on Earth. It would crush you out of existence. The surface pressure is comparable to that experienced by a submarine 500 fathoms, or 1000 meters, below the surface of our terrestrial oceans. So it's a marvel that *Veneras 7* and *8* could withstand the pressure and send back information. None of the landers lasted more than a few hours before succumbing to the destructive heat and pressure.

Five years later, *Veneras 9* and *10* obtained surface photographs, and the Soviets subsequently parachuted seven more entry probes to the surface, determining among other things the composition of the rocks. Two more landers, *Veneras 11* and *12*, descended to the surface in December 1978.

Table 7.2 Some important American, European, and Soviet missions to Venus[a]

Mission	Arrival date[b]	Accomplishments
Mariner 2	14 Dec. 1962	Flyby of Venus, first successful planetary flight, confirmed intense microwaves from surface of Venus, measured solar wind in interplanetary space
Venera 4	18 Oct. 1967	First entry probe of another planet's atmosphere, confirmed that carbon dioxide is its main ingredient
Mariner 5	19 Oct. 1967	Flyby, confirmed high temperature and pressure at surface of Venus
Veneras 5, 6	16, 17 May 1969	Penetrated farther than *Venera 4*
Venera 7	15 Dec. 1970	First probe to land on the surface of another planet, measured surface temperature, pressure and radioactive elements
Venera 8	22 July 1972	Landed on surface
Veneras 9, 10	23, 26 Oct. 1975	First photographs of surface
Pioneer Venus	4 Dec. 1978	First global radar maps of topography, first map of gravity field, five atmospheric probes
Veneras 11, 12	21, 25 Dec. 1978	Landers
Veneras 13, 14	1, 5 Mar. 1982	Chemical analysis of rocks
Veneras 15, 16	10, 14 Oct. 1983	Orbiters, radar images of surface
Vegas 1, 2[c]	11, 15 June 1985	Landers, balloon atmosphere probes
Magellan	10 Aug. 1990	Radar-imaging orbiter, global high-resolution maps of features
Venus Express	11 Apr. 2006	Measurement of atmosphere and cloud patterns, including polar vortex

[a] America's NASA missions to Venus are *Mariner*, *Pioneer* and *Magellan*, Europe's ESA mission is *Venus Express*, and the Soviet missions are named *Venera* and *Vega*.
[b] The *Venera* spacecraft were sent to Venus when the planet was near the Earth, taking about four months to get there.
[c] The main *Vega 1* and *2* spacecraft continued on to comet Halley after dropping probes at Venus.

Two American spacecraft, together known as the *Pioneer Venus* mission, arrived at Venus in the same month as *Venera 12* landed. The *Pioneer Venus Orbiter*, also known as *Pioneer 12*, circled the planet and sent back useful data for 14 years. By sending down radio signals and measuring their echoes, its radar instrument made the first topographic map of the surface, revealing an exceptionally smooth world with just a few high places. Other instruments aboard the *Orbiter* scrutinized the atmosphere, clouds and ionosphere, the electrically charged layer between the atmosphere and outer space.

The *Pioneer Venus Multiprobe*, also called *Pioneer 13*, carried four craft, one large probe and three small ones, which plunged into the atmosphere at both high and low latitudes and on both the daylight and night-time sides, providing a comprehensive picture of the atmospheric structure.

In the late 1960s, the Soviet Union began to hurl spacecraft toward Venus at 19-month intervals, every time the planet's orbital motion carried it nearest to the Earth (Table 7.2). A spacecraft launched near that time requires the least energy and fuel to reach Venus, taking about four months. The Soviets sent them in steadily increasing numbers for two decades; the American launches were fewer, but they included more technologically sophisticated instruments.

Clouds of concentrated sulfuric acid on Venus

What accounts for the unbroken layer of pale yellow clouds that form a thick impenetrable layer at altitudes of between 50 and 80 kilometers above Venus's surface? The billowing white clouds on Earth are formed when water vapor rises about 12 kilometers into the cold atmosphere. But because of the high surface temperature on Venus, there can be no liquid water on its surface and its atmosphere is extremely dry compared to the Earth. It possesses only a hundred-thousandth as much water as the Earth has in its oceans. If all of Venus's water could somehow be condensed onto the surface, it would make a global puddle only a couple of centimeters deep. So there are no water clouds on Venus, and no water rain.

Detailed study of the sunlight reflected from the uppermost clouds indicates that the reflecting cloud particles have a spherical shape, implying that the particles are liquid droplets rather than ice crystals. Water and other plausible liquids were ruled out because they have the wrong reflecting and refracting properties.

Baffled astronomers found the answer in the 1970s. A combination of spectroscopy and polarimetry, or how the cloud droplets polarize light, showed that the clouds of Venus are composed of concentrated sulfuric acid! That is the same sulfuric acid that is commonly used in car batteries and contributes to the eye-stinging quality of smog in some industrial cities, especially near smelters.

But where does the acid come from? Instruments aboard *Pioneer Venus* showed that it is derived from gaseous sulfur dioxide and water vapor in the atmosphere. Chemical reactions involving sulfur dioxide (SO_2) and water vapor (H_2O) form the sulfuric acid (H_2SO_4). A similar series of chemical reactions forms terrestrial smog.

When the concentrated sulfuric acid droplets fall into the warmer atmosphere on Venus, they evaporate, and the resulting gas rises again to the cloud layers. Thus, the acid rain on Venus never reaches the surface, and it is not removed from the atmosphere.

The *Pioneer Venus* investigations also showed that related substances contribute to the greenhouse effect on Venus. Although carbon dioxide is the most significant greenhouse gas on the planet, its action is enhanced by the presence of sulfur dioxide, water vapor, carbon monoxide, and cloud particles. This mixture prevents most of the heat radiation from escaping into space, yielding the torrid surface temperatures.

Venus is certainly too hot for liquid water to exist on its surface, but large amounts of water vapor are bound up with the sulfur dioxide in its clouds. Volcanoes probably supply these gases and maintain the clouds. Solar ultraviolet radiation will decompose any water vapor or sulfur dioxide molecules that rise to high levels in the atmosphere. The hydrogen is continuously lost into space, after water is torn apart. Sulfur and oxygen are too heavy to escape and they react with other atmospheric constituents. The thick clouds that are observed today could only be present if volcanoes supplied their constituents within the past 30 million years. Such volcanoes probably stay active for tens of millions of years, so they may still be active today.

On Earth, water rain efficiently removes sulfur dioxide and other sulfur gases from the atmosphere, so they are only present in very small amounts. Any sulfur gases that are discharged into the air by factories or volcanoes dissolve in our white clouds of water ice, to form droplets of sulfuric acid that are quickly washed out by water rain, sparing us

the corrosive acid clouds on Venus. But the sulfuric acid that does rain down to the Earth's surface can damage forests and lakes, so it is still of concern.

Glimpse of a volcanic surface on Venus

Veneras 9 and *10* touched down in October 1975, surviving long enough to send back one picture each, the first from the surface of any other planet. Six years later, *Veneras 13* and *14* parachuted down to the surface and transmitted more images, and analyzed the rocks. Since the sulfuric-acid clouds reflect most of the incident sunlight, and absorb almost all of the rest of it, Soviet astronomers once thought that there might not be any light at the surface. They therefore equipped their earlier spacecraft with floodlights to illuminate the scene when they reached the ground, but it turned out that there was enough natural light to take the historic pictures.

The surface of Venus is always bathed in a diffuse light under a heavy overcast. So the landers found that daylight on Venus resembles a dark, smoggy day in Los Angeles or Mexico City. The pictures are sometimes colored orange, for this is the main color to make it through the thick clouds.

Some scientists thought that the high temperatures and pressures would melt, flatten and chemically weather the surface into a featureless plain. However, the surface photographs showed fresh-appearing rock without eroded edges (Fig. 7.3). The sharp-edged, slab-like rocks are probably formed from flowing lava. As the molten lava spread across the surface and cooled, it would produce the thin, fractured layers of rock that we see in the surface photographs.

Analysis of the surface rocks showed that they have a composition and mass density that resemble basalt, the type of dark, fine-grained lava that lines the Earth's ocean floors and fills the mare basins on the Earth's Moon. The basalt rocks were identified at several landing sites on Venus, suggesting that much of the planet is encrusted by lava, covering its original surface. The basaltic crust was most likely extracted from a differentiated interior, when molten rock rose up through the crust, forming volcanoes that once poured lava over much of the planet's surface.

Circulation of the atmosphere of Venus

At ground level, the heavy atmosphere is sluggish and turgid, hardly moving at all. The speeds measured from the *Venera* landers and the *Pioneer Venus* probes are between 0.3 and 1.0 meters per second, or about the walking speed of a tired old man. A fast wind would have a tremendous

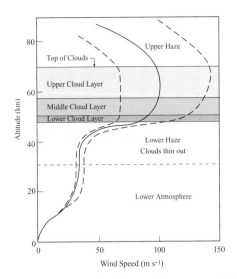

Fig. 7.4 **Wind speeds and cloud layers of Venus**   At all altitudes in the thick atmosphere of Venus, the dominant winds blow westward with a speed that increases with height. From a gentle breeze near the ground, the wind speed increases to 100 meters per second at great heights. Space probes that penetrated the planet's clouds have been able to detect three distinct layers in the opaque sulfurous clouds. The top layer contains small droplets of sulfuric acid; the middle layer contains larger but fewer particles. The bottom layer is the densest and contains the largest particles; it is comparable to bad city smog in visibility. Beneath the lowest layer, the atmosphere is hot enough to vaporize all particles, so it is relatively clear down there.

Fig. 7.3 **Surface rocks on Venus**   On 5 March 1982 the *Venera 14* lander touched down on Venus at 13 degrees south latitude and 310 degrees east longitude, where it survived for just one hour before succumbing to the planet's heat. That was long enough to radio back these photographs of the surface of Venus, which include part of the lander and a mechanical arm at the bottom. The thin, plate-like slabs of rock could be due to molten lava that cooled and cracked. The composition and texture of these rocks is similar to terrestrial basaltic lava. (Courtesy of Iosif Shklovskii.)

force in the dense atmosphere, disrupting the surface. Because of the slow winds, the bottom of the atmosphere moves along with the surface.

The wind speed increases with altitude, rising to about 100 meters per second in the clouds at about 60 kilometers in height (Fig. 7.4). The high-flying clouds race around the planet once every four Earth days, from east to west in the same backwards direction that the planet rotates. So the top of the atmosphere is blown around Venus more than 100 times faster than the planet rotates; such a rapid motion is sometimes called super-rotation. These high-speed zonal (east–west) winds are partly driven by the rotation of the solid planet beneath them, but the exact mechanism for maintaining the flow is not well understood.

The atmosphere and winds have transformed the impact craters on Venus, which are unlike those seen on any other world. The dense atmosphere affects the impact debris, changing it into fluid-like flows, and the material ejected during impact is moved by the winds. Some fresh craters are surrounded by radar-bright halos, streamlined hoods and tail-like wind streaks that act like wind vanes, pointing downwind at the time of impact (Fig. 7.5). The wind streaks indicate that the winds just above the surface were blowing toward the equator from the northern and southern hemisphere.

The atmosphere redistributes heat from one part of Venus to another, thereby moderating temperature differences. Most of the sunlight falling on Venus is either reflected by the clouds or absorbed in them. And because the Sun's rays fall directly on the equator and obliquely at the poles, the equatorial clouds are initially warmer than the polar ones. But this temperature difference generates winds that transfer heat in a single large Hadley cell (Fig. 7.6), named after George Hadley (1685–1768) who first proposed such a circulation for the Earth's atmosphere.

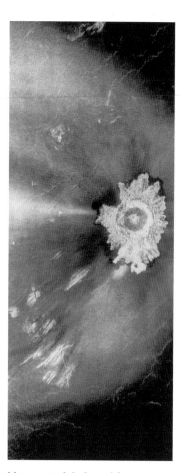

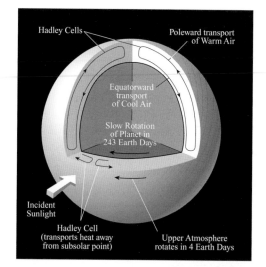

Fig. 7.6 **Hadley cells in the atmosphere of Venus**   Incident solar energy drives a Hadley-cell circulation of the atmosphere, which keeps heat from building up at any one location. Air rises in warm regions near the equator on the sunlit side of Venus, where the planet is heated most by the Sun. Some of the warm air flows towards cooler zones near the poles, and sinks and returns to warm equatorial regions again. Strong winds also blow around the planet in the direction that Venus rotates, from east to west.

Fig. 7.5 **Winds blow material ejected from a crater on Venus** A radar image of Avidar Crater is surrounded by streamlined, horseshoe or parabolic shapes (*top and bottom right*) and a bright, jet-like streak (*center left*) that extends over the surrounding plains. These unusual features, seen only on Venus, are believed to result from the interaction of ejected debris and high-speed winds in the upper atmosphere, blowing from the east (*right*). The crater is 30 kilometers in diameter, and is named for the Turkish educator and author Halide Adivar (1883–1964); it is located at 9 degrees north latitude and 76 degrees east longitude, just north of the western Aphrodite highland. (Courtesy of NASA/JPL.)

As on the Earth, the warm air rises at the equator to the cloud tops, where winds propel it toward both poles. After warming the poles, the circulating atmosphere sinks and flows back towards the equator at lower levels near the base of the clouds, completing the cell. The stronger zonal (east to west) circulation on Venus combines with this weaker Hadley (north–south) circulation, giving rise to a wind vortex that carries the clouds in a slow spiral toward the poles.

The turbulent atmosphere of Venus has been studied in great detail using ultraviolet and infrared detectors aboard the *Venus Express* spacecraft, which began orbiting the planet on 11 April 2006. These instruments have provided new insights to the planet's global cloud patterns that were first detected at ultraviolet wavelengths. The equatorial areas on Venus that appear dark in ultraviolet light are regions of relatively high temperature, where intense convection brings up dark material from below. In contrast, the bright regions at mid-latitudes are areas where the temperature in the atmosphere decreases with depth. The *Venus Express* observations have also confirmed that lightning is produced in the sulfuric acid clouds.

At the polar regions, gases that were moving toward the poles are deflected sideways by the planet's rotation, creating a huge whirling vortex that resembles the eye of a gigantic hurricane (Fig. 7.7). An immense, spinning vortex spirals around both poles. Each vortex draws the atmosphere downward, like water running down a bathtub drain, and they change shape in a matter of days, even forming a double eye in one polar region on occasion.

The dense lower atmosphere transports heat so efficiently from one part of the globe to another that the ground temperature varies by no more than a few degrees between the equator and poles or from day to night, so there is no place to escape the heat.

The changing distribution of sunlight on the Earth causes stormy weather and produces the changing winds that propel clouds across the globe. In contrast, Venus has a steady atmospheric circulation pattern almost devoid of

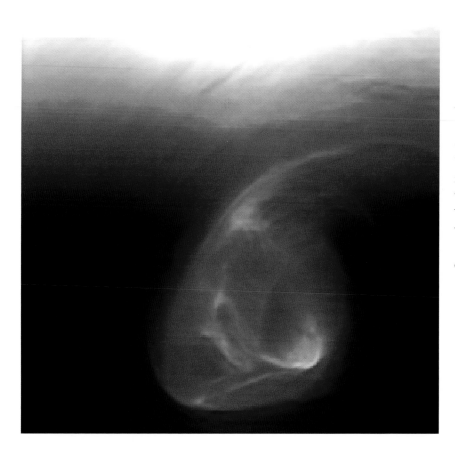

Fig. 7.7 South polar vortex at Venus
The core of a whirling polar vortex shines brightly in this infrared image taken on 9 August 2007 from the *Venus Express* spacecraft. The vortex, which resembles the eye of a hurricane on Earth, is located near the south pole of Venus. The 2000-kilometer-wide vortex was first discovered from the *Mariner 10* flyby of Venus on 5 February 1974. There is a similar structure near the planet's north pole, which was observed from *Venus Express*. The central part of the southern vortex is very dynamic, changing shape within a matter of days, including the formation of a double-eye. (Courtesy of ESA.)

weather. Since the planet's orbit is nearly a perfect circle and its rotation axis is hardly tipped at all, there are no noticeable seasons on Venus.

Missing magnetism on Venus

Unlike the Earth, Mercury and the four giant planets, Venus has no significant global magnetic field, for reasons not fully understood. Measurements from *Pioneer Venus Orbiter* showed that if Venus has any magnetic field its strength is less than one hundred-thousandth that of Earth, or less than 3×10^{-9} tesla. Lacking a magnetic field, Venus has no belts of trapped particles such as the Van Allen belts near Earth.

The weakness of this magnetic field is rather surprising because Venus and the Earth have a similar size and mass, and they might be expected to have similar interiors. The Earth has a molten outer core and a solid inner one, and by analogy Venus ought to possess similar cores. However, Venus does not show the magnetic field that would be produced by currents within a molten outer core.

Contrary to popular belief, the slow rotation of Venus is not responsible for the lack of a significant magnetic field, but something in its core is out of whack. Some investigators argue that the core is now completely solid, while others suggest that core solidification has not yet commenced. In either case, Venus would lack both a molten outer core for currents to flow in and the heat given off by the creation of a solid inner core, which would help maintain the flow.

The planet's lack of an appreciable magnetic field exposes the upper atmosphere to the continuous hail of charged particles from the Sun. The intrinsic magnetic fields of the Earth and other planets fend off and deflect this electrically charged solar wind. Nonetheless, the solar wind is prevented from reaching the surface by Venus's dense atmosphere and ionosphere.

Energetic ultraviolet sunlight ionizes some of the atoms and molecules in the outer atmosphere above the clouds, forming an electrified layer similar to the Earth's ionosphere, and this layer helps shield the ground from the solar wind. The ions provide conduction paths for electrical currents that produce forces that counter the wind. As a result, the solar wind slows down and is deflected around the planet in a bow shock, and the interplanetary magnetic field is draped back to form a magnetotail (Fig. 7.8). The solar wind also accelerates the charged, ionized particles in the upper atmosphere of Venus, giving some of them enough energy to escape the planet. So Venus slowly loses

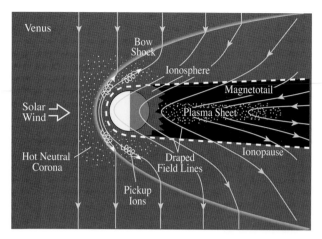

Fig. 7.8 Solar wind flows around Venus   Even though Venus has no appreciable magnetic field, the solar wind is prevented from reaching the surface by Venus' dense atmosphere and by electrical currents induced in its conducting ionosphere. The planet has a well-developed bow shock, but it does not have belts of trapped particles.

a small fraction of its atmosphere due to interaction with the solar wind (Fig. 7.9).

7.4 Unveiling Venus with radar

Venus rotates backwards, at an exceptionally slow rate

Although no human eye has ever seen the surface of Venus, radio waves can penetrate its obscuring veil of clouds and touch the landscape hidden beneath. By bouncing pulses of radio radiation off the surface, radar astronomers discovered in 1967 that Venus spins in the backward direction, opposite to that of its orbital motion. That is, unlike the other terrestrial planets, Venus does not rotate in the direction in which it orbits the Sun.

The radar observations also showed that Venus spins with a period longer than any other planet, at 243.018 Earth days. This rotation period is even longer than the planet's 224.7 Earth-day period of revolution around the Sun, so the

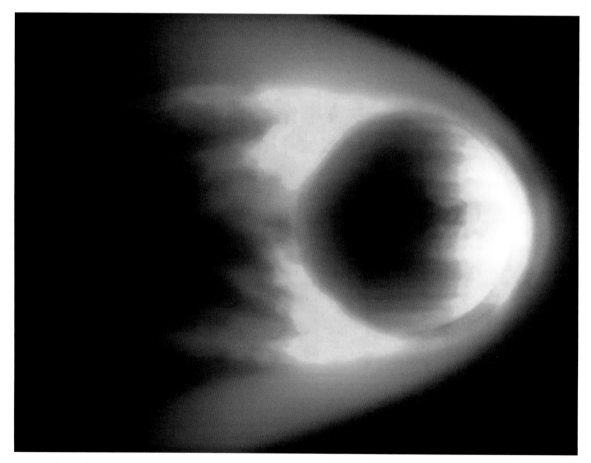

Fig. 7.9 Venus slowly loses some atmosphere   Venus is almost as large as Earth, and it is difficult for its atmosphere to escape from the planet's gravitational pull. However, the upper part of the atmosphere on Venus is ionized by solar radiation, and the solar wind accelerates some of these ions, giving them enough energy to escape and join the wind. In this way, the solar wind interacts with the thick atmosphere, and in effect pulls some of it into space. (Courtesy of ESA.)

day on Venus is longer than its year. The method of determining this slow rotation is the same Doppler technique used to discover Mercury's slow 58.6 Earth-day rotation period, and both planets have probably been slowed down by the Sun.

Tides raised by the Sun in the planet's thick atmosphere may explain why Venus turns very slowly and in the wrong way. The Sun's gravitational force produces two tidal bulges; the rotating planet drags these bulges along with it, causing them to twist out of alignment with the Sun. As a result, the Sun's gravitational attraction tends to oppose the rotation, slowing the planet down. Friction between the atmospheric tides and the planet's surface also helped to gradually slow Venus's rotation to the point where it stopped and reversed. The friction of the tides that our Moon produces in the Earth's oceans similarly slows down our planet's rotation.

Magellan and its predecessors reveal a volcanic surface on Venus

The surface of Venus cannot be seen at the visible wavelengths used by our eyes; it can only be sensed by radio transmissions. Radar, an acronym for radio detection and ranging, uses its own source of radio radiation, and does not need sunlight to probe the planet, gathering data day or night. Only radar is capable of piercing the thick clouds of sulfuric acid that blanket Venus.

The planet's topography is inferred from the length of time it takes for a radar pulse to reach a particular part of the surface and return an echo. The surface roughness is determined from the strength of the echo. Rough surfaces and slopes tilting toward the radar reflect more energy and return a stronger echo, thus appearing bright in a radar image. Surfaces that are smooth or tilt away from the incoming radio signal send less energy back, and appear dark in a radar image, somewhat like a wet road seen in the headlights of a car at night.

The surface features on Venus have been probed with radar systems of increasingly greater resolution over the decades. Blurred images were first obtained from powerful, ground-based radio telescopes, such as the one in Arecibo, Puerto Rico. The *Pioneer Venus Orbiter* then zoomed in to take a closer look in the 1970s, acquiring a global radar map of Venus at a coarse resolution of about 100 kilometers. This was followed by the twin spacecraft *Veneras 15* and *16*, whose radar instruments mapped much of the planet's northern hemisphere in the mid-1980s, resolving features as small as 1 kilometer across.

The radar system aboard the orbiting *Magellan* spacecraft next scanned Venus for more than four years in the early 1990s. It mapped details as small as 120 meters across, producing the most complete global view available for any planet, including Earth (Fig. 7.10). The spacecraft is named after the Portuguese explorer Ferdinand Magellan (1480–1521) whose expedition first circumnavigated Earth.

Magellan's radar images revealed a rich and varied landscape with stunning and unprecedented clarity, describing a surface whose nature and history turned out to be quite different from those of the Earth. The surface of Venus is shaped largely by volcanic activity. Its topography is dominated by massive, global outpourings of lava, punctuated by numerous shield volcanoes and unique volcanic constructs never seen before, scarred by sparse, pristine impact craters surrounded by beautiful outflows, and fractured, stretched, crumpled and split open by upwelling magma. Venus has more volcanoes than any other terrestrial planet. With the exception of Jupiter's moon Io, Venus is the most volcanic world in the solar system.

Perhaps the most stirring aspect of the *Magellan* images is the fresh, pristine nature of the features they reveal. Although no one expected to see significant evidence of erosion and weathering on the dry planet, observers were struck by the detailed sharpness of its craters, fractured plains, volcanoes and crumpled landmasses. Most of these surface features have been preserved for hundreds of millions of years. Everything is totally exposed and largely preserved — at least between periods of intense volcanic activity or internal upheaval.

As Aladdin said from his flying carpet in the Walt Disney movie *Aladdin*, it is:

A whole new world.
A dazzling place, I never knew.
But when I am way up here,
It's crystal clear.

Magellan had one instrument, the Synthetic Aperture Radar (SAR), capable of mapping its topography as an altimeter, and measuring electromagnetic radiation from the surface. Most typical radar systems send out one pulsed radio signal at a time, and process each echo by itself before sending out another pulse. In contrast, *Magellan* sent out several thousand radar pulses each second, and its SAR used computers to accumulate multiple echoes received from many locations simultaneously. The combined data simulated a large antenna and provided the superb resolution and fine detail.

As the spacecraft looped around the poles of Venus, the slowly rotating planet turned beneath it, exposing the entire globe to scrutiny during each 243-day rotation. Every second, the computers took in 36 million bits of data and

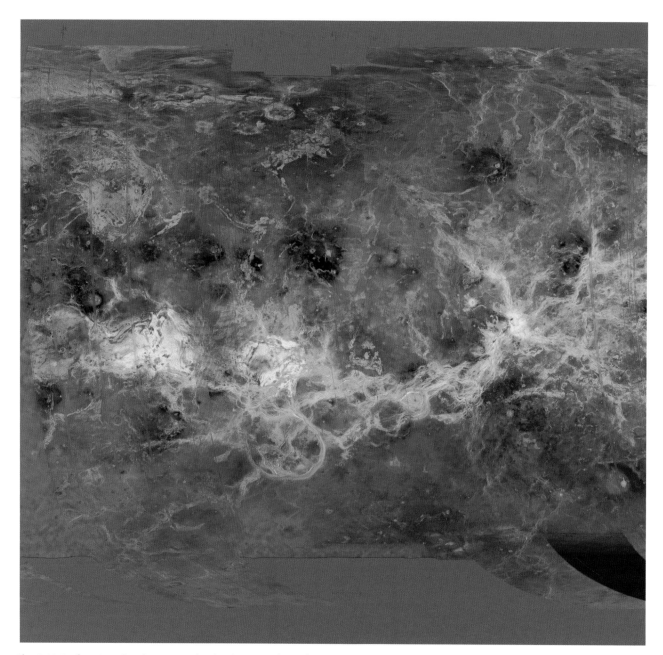

Fig. 7.10 Surface terrain of Venus   The cloud-penetrating radar system aboard the *Magellan* spacecraft lifted the perpetual veil of thick clouds on Venus, revealing its surface features in high resolution of 120 meters in the early 1990s. The eastern half of the planet's surface is portrayed in this cylindrical map, extending between 52 and 240 degrees east longitude (*left side to right side*) and from 90 degrees north latitude to 90 degrees south latitude (*top to bottom*). (Courtesy of NASA/JPL.)

relayed them back to Earth. Over four years, they obtained more than a million billion bits of information, exceeding the amount that had been obtained from all previous lunar and planetary spacecraft combined.

To complement the wide-angle, side-looking antenna, there was a second, narrow-beam antenna that looked straight down and used a pulsed signal and its echo to measure the topography. Elevations were measured with an accuracy of 30 meters with a resolution of about 5 kilometers.

Venus is a smoothed-out world

Radar data from the *Pioneer Venus Orbiter* and *Magellan* spacecraft showed that Venus is an extraordinarily smooth world, largely at one low level and quite different from the Earth (Fig. 7.11). Without its water, the topography of the Earth occurs at two distinct elevations, which correspond to the continents and ocean floors. In contrast, about 85 percent of the surface of Venus lies within one kilometer of the average planetary radius, 6051.8 kilometers.

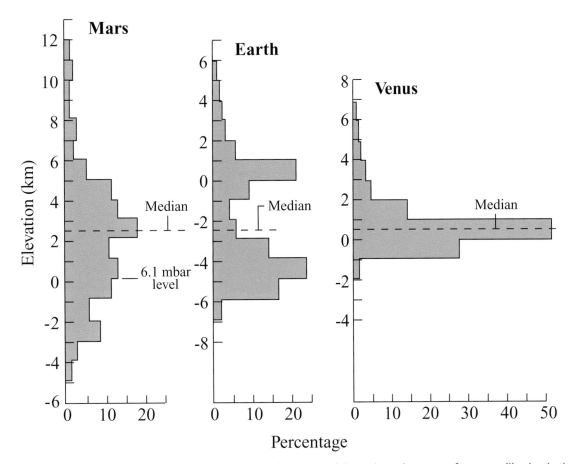

Fig. 7.11 Surface elevations for Earth, Venus and Mars Most places on Earth (*center*) stand near one of two prevailing levels, the high-standing continents or the low-lying ocean floors. In contrast, the surface of Venus (*right*) is unusually smooth and flat; about 60 percent of its terrain lies within 500 meters of the mean planetary radius of 6051.8 kilometers. A small percentage of its terrain consists of elevated highlands that are comparable in height to many continents on Earth. The surface features on Mars (*left*) are spread over a broader range of elevation than most of those on Venus; but the Martian surface elevations are not double-peaked as on Earth.

A coating of lava has smoothed these vast low-lying plains.

The lowest point of Venus is about 6048.0 kilometers and the highest point is at 6062.57 kilometers. Thus the variation of topography on Venus is almost 14.6 kilometers. The surface temperature and pressure varies with height, between 653 and 766 kelvin and between 45 and 119 bars at the highest and lowest elevations.

Although most of Venus's terrain consists of smooth, low-lying, volcanic plains, about 15 percent of the planet's surface consists of highlands that tower above the plains, rising an average 4 to 5 kilometers above the mean planetary radius. There are two large-scale elevated regions that punctuate the smoothed-out surface; they are Aphrodite Terra in the equatorial region and Ishtar Terra in the far north (Fig. 7.12).

Aphrodite Terra is over 10 000 kilometers long and covers a quarter of the planet's circumference at the equator. It contains tall volcanoes, long lava flows, and deep faults and fractures. Western Aphrodite is built from the massifs Ovda and Thetis Regiones; Atla Regio occupies the eastern part of Aphrodite. Ishtar Terra fills about half the planet's circumference at its high northern latitudes and is about the size of Australia. It consists of an elevated plateau encircled by narrow mountain belts (Fig. 7.12).

Each feature on Venus is described by two names — a woman's name that identifies it and a feature designation (Focus 7.1). As examples, Aphrodite and Ishtar are respectively the Greek and Babylonian goddesses of love, and a terra is an extensive landmass. The name of a goddess or a famous mortal woman identifies almost every feature on Venus. The only exceptions on Venus are Maxwell Montes, named in honor of the British physicist James Clerk Maxwell (1831–1879), whose 19th-century theories of electromagnetism made radar possible today, and Alpha Regio and Beta Regio, using the first two letters

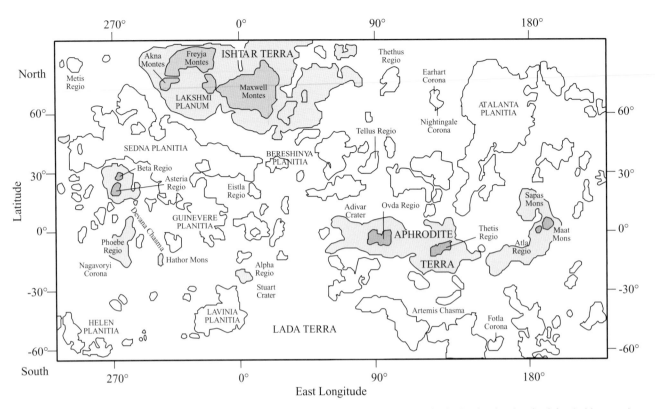

Fig. 7.12 Major surface features on Venus   Most of Venus lies at roughly the same radius, in the lowland, volcanic plains (*white areas*). The highland massifs (*dark areas*) include Aphrodite Terra, an elongated feature extending along the equator between 70 degrees and 210 degrees east longitude. Many of the elevated regions near the equator mark the sites of volcanism, such as Maat Mons and Sapas Mons (see Fig. 7.16). The other main elevated region is Ishtar Terra in the north at about 0 degrees longitude (*center top*). Ishtar Terra is roughly the size of Australia. The elevated plateau in the western part of Ishtar Terra is known as Lakshmi Planum, which is bounded on three sides by mountains (see Fig. 7.17), including Maxwell Montes. This mountain's 11-kilometer height above the average radius exceeds by two kilometers the height of Mount Everest above sea level.

Focus 7.1   Naming features on Venus

There are two names that describe every feature on Venus. They are a particular name that identifies it, plus a descriptive name that says what it looks like. As indicated in Table 7.3, each class of feature has a specific type of associated identifying female, either real or mythological.

The identifying names can be suggested by anyone, but eventually they have to be approved by the International Astronomical Union (IAU). The lists of approved names for the planets and their satellites are fascinating, and you can review them in the IAU's Gazetteer of Planetary Nomenclature located on the web at http://planetarynames.wr.usgs.gov/.

of the Greek alphabet. They were first seen in Arecibo radar images in the 1960s, before the female naming convention.

Craters larger than 20 kilometers in diameter are identified by the last name of a famous woman, who has to have been dead at least three years at the time of naming. Some of my favorites include the singer Billie Holiday, the anthropologist Margaret Mead, the writer Flannery O'Conner, the ballet dancer Anna Pavlova, the singer Edith Piaf, and the abolitionist Harriet Tubman. Smaller craters are given common female first names, as are hurricanes on Earth.

Some representative examples for the names of other topographic features include: Fotla Corona, named for a Celtic fertility goddess; Freyja Montes, with the name of a Nordic goddess who weeps gold tears for her lost husband; Maat Mons, with the namesake of an Egyptian goddess of truth and justice; and Ovda Regio, named after a female Finnish spirit who wanders the forest naked to find people to tickle to death.

Table 7.3 Common features on the surface of Venus and the category of women used to identify them

Descriptor term (singular, plural)	Feature type	Category of identifying female
Chasma, Chasmate	Canyon, steep-walled trough	Goddesses of hunt or the Moon
Corona, Coronae	Oval-shaped feature	Fertility and earth goddesses
Crater, Craters	Circular depression	Famous women
Dorsum, Dorsa	Ridge	Goddesses of the sky
Linea, Lineae	Elongate marking	Goddesses of war
Mons, Montes	Mountain	Miscellaneous goddesses[a]
Patera, Paterae	Irregular crater, shallow	Famous women
Planitia, Planitiae	Low plain, level surface	Mythological heroines
Planum, Plana	Plateau or high plain	Goddesses of prosperity
Regio, Regiones	Large area, broad region	Giantesses and titanesses[b]
Rupes, Rupes	Cliffs or scarps	Goddesses of hearth and home
Terra, Terrae	Highland, extensive landmass	Goddesses of love
Tessera, Tessarae	Tile-like, polygonal terrain	Goddesses of fate or fortune

[a] The one exception is Maxwell named after James Clerk Maxwell (1831–1879).

[b] The two exceptions are Alpha Regio and Beta Regio.

7.5 Volcanic plains on Venus

Planet-wide covering of lava on Venus

Venus has been extraordinarily volcanic. Extensive lava flows emerge from cracks in the crust and from towering volcanoes, and rivers of lava snake their way across the planet. Outpouring lava has covered much of the surface of the planet, and tens of thousands of small volcanoes are now found all across its face. Venus exhibits every type of volcanic edifice known on Earth, and some that have never been seen before.

The spreading lava has flooded and filled the low-lying regions of Venus, creating extensive smooth plains that cover about 85 percent of Venus's surface. The volcanic nature of these lowland plains, each designated by the term *planitia*, is demonstrated in the *Magellan* images. You can practically see the molten rock spreading like heavy cream across these plains, often running for hundreds of kilometers down gentle slopes. The magma has risen from within canyons as the crust split open, cooling and solidifying into lava flows that look like frozen river currents (Fig. 7.13).

In other places, the molten material has burned paths in the pre-existing lava deposits, following a narrow, sinuous smoothly curving course. They can meander for thousands of kilometers across the planet's surface. Many end in outflows that look like river deltas. These river-like channels were formed not by water, but by lava that was

Fig. 7.13 Lava flows This *Magellan* image shows a lava-filled canyon produced when Venus' crust was pulled apart and magma rose within the gap. The lowland plains have many similar canyon systems; typically about 10 kilometers wide and up to 1000 kilometers long, and apparently containing solidified lava flow. This region, located near the equator between Asteria Regio and Phoebe Regio, is about 10 kilometers wide and stretches 600 kilometers to the north (*top*). (Courtesy of NASA/JPL.)

hot enough to carve through solid rock, remaining hot and liquid over distances that are longer than the Nile, the longest river on Earth. The high surface temperature on Venus probably kept the lava liquid, and prevented the cooling flow front from damming up the molten rock behind it.

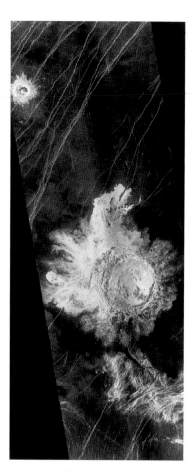

Fig. 7.14 Stuart Crater This beautiful crater exhibits asymmetric radar-bright ejected material attributed to an oblique impact and interaction with the dense, thick atmosphere of Venus. The crater's rim, which is 67 kilometers in diameter, encircles a bright floor that may have unique physical properties. This impact crater's name honors Mary Stuart, Queen of Scots (1542–1587). (Courtesy of NASA/JPL.)

A relatively young surface on Venus

Venus, like all planets, has been subjected to a continual rain of meteorite bombardment over the eons. The plains of Venus are uniformly peppered with impact craters (Fig. 7.14), the scars of this bombardment, though nowhere near as liberally as on the surfaces of the Moon, Mars, or Mercury. A comparison of craters with the same size on Venus and the Earth's Moon indicates that those on Venus are far fewer in number and more widely spaced than those on the Moon. At one time Venus was probably as heavily pockmarked with large craters as the Moon's ancient surface is, but the scarcity of the craters now on Venus indicates that the surface we now see is much younger than the lunar surface.

We can estimate when the lava flowed by counting the number of craters of a given size on the plains and comparing it to the number on the Moon — ignoring the smallest ones that are not found on Venus because the incoming projectiles burned up in the thick atmosphere. The Moon's crater record tells us the number of impact craters that should appear in a given time, and observations of the craters left on Venus tell us how many have been removed by burial under volcanic flows.

The *Magellan* spacecraft has logged about 1000 impact craters, which when compared to the lunar record indicates an average surface age of about 750 million years, with a large uncertainty of a few hundred million years. Estimates between 300 million and 1 billion years are possible for the global lava covering. At that time molten rock emerged from inside the planet, spreading across the surface, eradicating all previous craters, and wiping out all traces of the first 90 percent of Venus's history. That is a relatively young age, only about 10 percent of the age of the solar system and of the planet Venus itself. No matter how you look at it, the surface of Venus is practically newly born compared with the surfaces of Mars, Mercury and the Earth's Moon, which still bear the scars of a heavy bombardment about 3.9 billion years ago.

Theories for the volcanic makeover of Venus

Everyone agrees that the smooth plains covering most of Venus came from volcanic floods, emanating from the planet's interior, but the experts disagree over when and how it occurred. According to one "global catastrophe hypothesis", planet-wide volcanism wiped the face of Venus clean about 750 million years ago, resurfacing the entire globe and drowning any existing craters in a flood of lava.

There are two equally likely catastrophe interpretations that cannot be distinguished. One is a single resurfacing at about 750 million years ago. The second is that there was continuous resurfacing of the planet over most of its earlier history, and that this resurfacing slowed down sharply at about 750 million years ago. In either interpretation, the planet switched over to a low rate of localized volcanism about 750 million years ago, but it has not disappeared.

Most of the craters on Venus have lava on their floors, and some of them show exterior volcanism affecting their ejecta or breaking their rims, so there has been at least a modest level of ongoing volcanism.

A different model that has observational support is a series of discrete volcanic plains forming events extending over a considerable period of geologic time, with the volume of lava declining from one event to the next. In

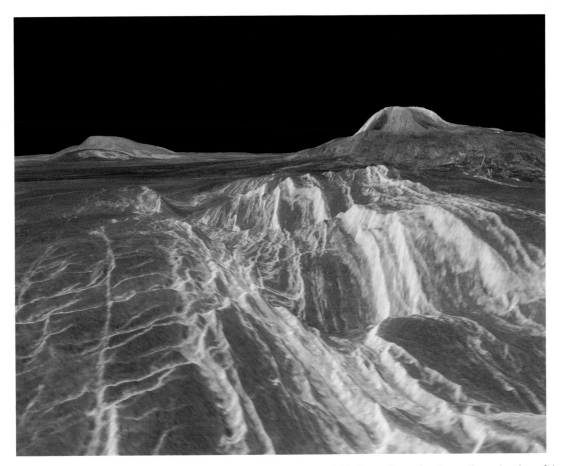

Fig. 7.15 Sif Mons and Gula Mons   Two volcanoes are located at the horizon of this three-dimensional *Magellan* radar view of the western Eistia Regio in the central part of Venus' surface, near the equator and zero degrees east longitude. In this computer-generated image, the vertical scale has been exaggerated to show the edge of a rift valley in the foreground. It extends all the way to Gula Mons, the volcano on the right horizon, which stands 3 kilometers high and is located 700 kilometers away. The volcano on the left horizon, called Sif Mons, has a height of about 2 kilometers and a diameter of about 300 kilometers. Eistia is a Norse giantess, Gula is the name of a Babylonian earth-mother and creative force, and Sif is the name of a Teutonic goddess and Thor's wife. The artificial tints are based on color images from the *Venera* 13 and 14 landers, simulating the color of sunlight that filters through the thick atmosphere. (Courtesy of NASA/JPL.)

this view, volcanism is an ongoing process, occurring at different places and times, further in the past and in the future from 750 million years ago. In fact, volcanic rises, coronae and rifting suggest that volcanism on Venus may extend up to the present.

The intense volcanism that repaved Venus may have transformed its climate and roasted its air. When massive volcanoes erupted roughly 750 million years ago, they should have ejected a lot of greenhouse gases into the atmosphere in a relatively short time. Calculations indicate that the volcanic release of carbon dioxide, water vapor and sulfur dioxide might have raised the temperatures well in excess of the "present inferno". Still, this volcanism is not the cause of the present intense greenhouse environment on Venus, which probably existed long before the volcanic resurfacing.

7.6 Highland massifs on Venus

Towering volcanoes and mountain ranges

Highland regions in the north and near the equator rise above the low terrain on Venus and account for about 15 percent of the planet's surface. Large-scale plumes of rising magma have probably pushed these elevated regions up from below. When the molten rock pierced the surface, lava flowed out to form towering volcanoes that are perched atop the raised highlands. Some of the volcanoes are found in Beta Regio on the western side of the equatorial highlands. Others are found in the western part of Eistia Regio (Fig. 7.15), located near the center of the equatorial highlands at about zero degrees longitude. Several rise out of Atla Regio in the eastern end of

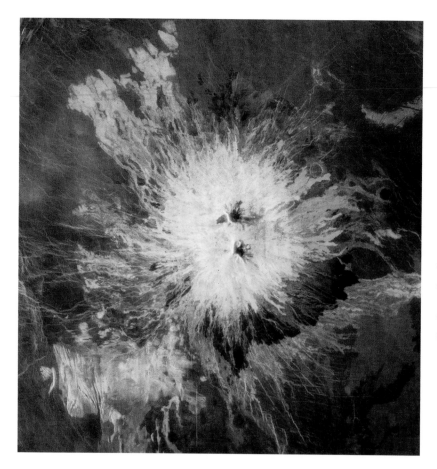

Fig. 7.16 Sapas Mons   Located in an equatorial highland called Atla Regio, the volcano Sapas Mons is about 400 kilometers in diameter and 1.5 kilometers high; it is named after a Phoenician goddess. The sides of the volcano are composed of numerous overlapping lava flows from flank eruptions similar to terrestrial volcanoes such as the Hawaiian shield volcanoes. The summit contains two smooth, radar-dark mesas, as well as groups of pits, some as large as 1 kilometer across. They may have formed when underground chambers of magma were drained, causing the surface to collapse above them. A 20-kilometer-diameter impact crater northeast (*upper right*) of the volcano is partly buried in lava flows. (Courtesy of NASA/JPL.)

Aphrodite Terra, including Maat Mons and Sapas Mons (Fig. 7.16).

One of the highest volcanoes, Maat Mons, rises 9 kilometers above the surface, and spreads 200 kilometers across it (see Fig. 2.30). Sapas Mons is shorter and broader (Fig. 7.16). Both peaks are known as shield volcanoes, because they have the shape of a shield or an inverted plate. Similar giant shield volcanoes are found in the Hawaiian Islands, each with a broad base and gentle slopes.

Scientists suspect that some volcanoes on Venus may still be erupting. Infrared observations from the *Venus Express* spacecraft have provided evidence of minerals which were formed in relatively recent volcanic flows that occurred a few hundred to 2.5 million years ago. They suggest that the planet may still be geologically active. The recurrent volcanic release of sulfur dioxide and water vapor could also help replenish sulfur dioxide in the atmosphere of Venus.

The region of Venus that most closely resembles terrestrial mountains is Ishtar Terra, located at far northern latitudes. It consists of an elevated plateau, Lakshmi Planum, which is bounded on three sides by mountain belts – the Danu, Akna and Freyja Montes (Fig. 7.17). Lakshmi just drops off into the surrounding plains on the fourth side, forming an immense cliff. The belts of mountains, with their banded ridges and narrow valleys, resemble mountain ranges on Earth, and the loftiest, Maxwell Montes, rises to Himalayan altitudes, standing over 11 kilometers above the surrounding terrain.

The raised plateau and surrounding mountains closely resemble the Tibetan Plateau and the Himalayan Mountains on Earth, which were produced by the collision of India with Asia. However, since Venus has no colliding continents, the intensely deformed mountains must have been pushed up into the sky by a different process. One possible explanation suggests that the material beneath the mountains was cooler than surrounding areas and sank. The crust would then be pulled together, bunching up into mountains and compressing the surface features. Or perhaps that part of the crust was compressed and folded after the surrounding plains were formed, like a carpet pushed against a wall. In either case, Ishtar Terra is a unique type of raised feature on Venus, and different processes hold up other highland regions on the planet.

Gravity's highs and lows

How can the high places on Venus stay up when its rock temperatures are halfway to their melting point? Something has to be holding the mountains up. So

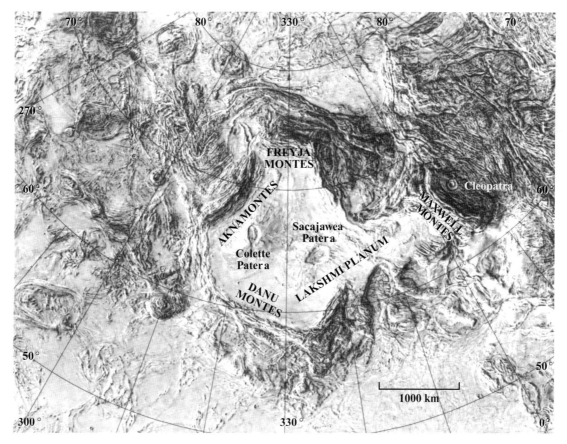

Fig. 7.17 Mountain ranges on Venus   The Earth and Venus are the only planets in our solar system that have mountain belts. The highland massif of Ishtar Terra in the northern hemisphere of Venus includes a huge plateau, named Lakshmi Planum. It is about the size of Africa, rises 3.5 kilometers above the surrounding terrain, and is bordered by the Akna, Danu, Freyja and Maxwell mountains, each about 1000 kilometers in extent. Maxwell Montes stands 11 kilometers above the mean radius. Akna and Freyja are the respective names of the Yucatan goddess of birth and a goddess in Norse mythology. Danu is the Celtic mother of god. Maxwell is named after the British physicist James Clerk Maxwell, while Cleopatra is the Egyptian queen who had affairs with Julius Caesar and Mark Anthony. The Blackfoot Indian woman Sacajawea guided the Lewis and Clark expedition to the Pacific Northwest, and Colette (1873–1959) was a French novelist. (Adapted from a USGS map using Arecibo, *Pioneer Venus* and *Veneras 15* and *16* radar data.)

spacecraft measured local variations in gravity and used them to look underneath the highlands and see what is there.

The movements of the orbiting *Magellan* spacecraft were tracked by recording small changes in the wavelength of its radio signal, and the corresponding small changes in the orbital speed, inferred from the Doppler effect, were used to specify local variations in the underlying mass and density.

It turned out that some of the highest places on Venus exert the strongest gravitational pull, a correlation first noticed by tracking *Pioneer Venus Orbiter* and confirmed by sending *Magellan* into a low-altitude orbit. These volcanic rises, such as Atla Regio and Beta Regio, are probably held up by active motions below, fed by rising plumes of sluggish, upwardly buoyant flow. The hot, low-density material wells up from deep within the planet, with glacial slowness in spite of its heat, eventually spreading out beneath the

surface, and pushing the volcanic rises up. Their low-density roots extend down under the high-standing regions and balance the mass excess. The combination is in "isostatic" equilibrium, like an iceberg floating on the ocean.

As the plumes near the surface, the lower pressures induce partial melting of the plume that sometimes punches holes in the crust, providing a conduit that permits lava to flow out from volcanoes and fissures in the planet. In this respect, the volcanic rises on Venus are thought to be very similar in origin to "hot spots" on Earth, such as the Hawaiian Islands.

7.7 Tectonics on Venus

Trapped heat deforms the surface

All planets have heat to get rid of. Gravity compressed and heated the planet's interiors when they formed, and

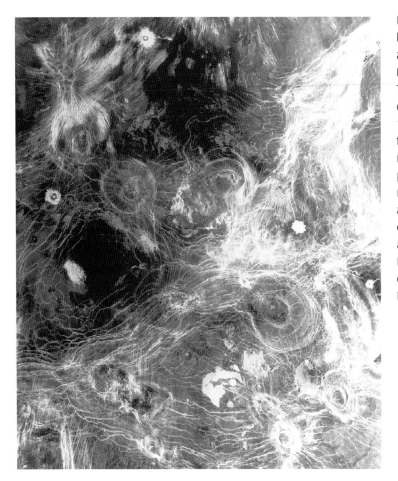

Fig. 7.18 Bereghinya Planitia   This high-resolution *Magellan* mosaic shows an area in the low-lying plain Bereghinya Planitia, named for the Slavic water spirit. The image is 1840 kilometers wide, and centered at 45 degrees north latitude and 11 degrees east longitude. The plain's fractured surface is attributed to plumes of magma that rise from inside Venus, pushing against the planet's crust. The most prominent features are the circular and oval structures that are sometimes called arachnoids for the spider-web-like appearance of their fractures (also see Fig. 7.20). Also visible are lava flows, impact craters and volcanic domes. (Courtesy of NASA.)

colliding bodies also brought heat to them in their early stages. Then the decay of radioactive elements added more internal heat. It is still being generated within Venus by radioactive decay, producing molten rock inside the planet.

Size is the main factor in determining how much heat remains inside a planet, for larger bodies lose heat more slowly and remain internally active for longer times. Thus, the Earth's Moon, which is relatively small when compared to a planet, has not been volcanically active for 2 or 3 billion years, even though lava flowed into its impact basins to form the lunar maria before then. The larger, rocky planet Venus must have had vast churning reservoirs of hot material beneath its crust for a much longer period, and probably still does.

The heat trapped inside a planet wants to get out, so it rises to crack and deform the planet's surface or spills out in volcanoes. In technical terms, the molten rocks become swollen by heat and lower in density, rising through the cooler, overlying high-density material and carrying heat upward, like the convective bubbles in a pot of boiling water. The upward-flowing material wants to crack open or puncture a hole in the overlying material, breaking on through to the other side to release all that heat. The crustal

deformations caused by the pent-up heat, and the internal changes affecting them, are known as *tectonics*, the Greek word for "building".

The hot rising material has buckled, fractured and stretched the crust on Venus, like a crumpled piece of paper or a face seamed and thickened by age (Fig. 7.18). It has split the crust open and spread it apart, forming rift valleys with steep sides and sunken floors. Some of them are found in Atla region of Venus alongside its volcanoes (Fig. 7.19). The linear rift zones in the equatorial highlands can extend for thousands of kilometers, but are cracked open by just a few kilometers. In contrast, rifts that split open the Earth's continents can, because of moving plates, widen up to make way for its biggest oceans.

When a bubble of hot material rises to just below the surface of Venus, it presses against the crust, causing the ground to bulge and crack. Circular and radial fractures are created around the edges of the rising dome, forming a network of radar-bright features that remind us of a spider's web (Fig. 7.20). Some of them have therefore been nicknamed *arachnoids*, from the Greek word for "spider". The term *coronae*, the Latin word for "crown", is used for the larger, elevated, circular structures that are also pushed up from below by rising molten rock trying to get out. Both

Fig. 7.19 Atla Regio Fractures, seen as bright, thin lines, criss-cross the volcanic deposits in part of the Atla region of Venus. The fractures are not buried by the lava, indicating that the convulsive tectonic activity post-dates most of the volcanic activity. This *Magellan* radar image is approximately 350 kilometers across, and centered at 9 degrees south latitude and 199 degrees east longitude. Several circular volcanoes, surrounded by radar-bright lobes, are also present. This region is named Atla after a Norse giantess, mother of Heimdall. (Courtesy of NASA/JPL.)

arachnoids and coronae are unique to Venus and have not been found on any other planet.

Coronae have concentric ridges and fractures that are hundreds of kilometers across, and large volcanic outpourings have occurred within them (Fig. 7.21). When enough lava spills out into a corona, the upwelling subsides and it is no longer supported from below. The bulge will deflate and buckle the surrounding terrain, producing an annulus of ridges and troughs that often surrounds coronae, like the moat around a castle. Or else, the magma cools and retreats as it ages and the molten rock drains back down the vent from whence it came. Then the dome will collapse like a giant fallen soufflé, creating ring-like fractures and a crumpled, cracked surface.

The increasing pressure of the upwelling magma can stretch the planet's skin until it bursts, like the broken cheese bubbles in a pizza or a split in an overcooked hotdog. Small volcanic domes, known as pancakes, are sometimes formed when pasty, sluggish lava breaks through and flows along the surface like toothpaste (Fig. 7.22). In other places the crust breaks and spreads open and lava flows into the gap like olive oil.

Just about everywhere that *Magellan*'s radar instrument looked, it found intersecting ridges, cracks and grooves, readily visible in the absence of overlying soil or vegetation and the absence of erosion. The intricate, tortured networks are found over the entire globe, ubiquitous throughout the volcanic plains and within the highlands. Internal forces associated with the pent-up heat have been pushing and pulling the crust, producing these patterns. They are known as *tesserae*, the Latin word for "tiles".

As the surface moves up in some locations and down in others, the associated stresses pull the surface apart or push it together. Over time, these stresses can be created in different directions, producing a regularly spaced grid pattern of fractured terrain that is only found on Venus at this scale (Fig. 7.23). Some of the cracked patterns of the tesserae have regular six-sided shapes that can be attributed to global heating and cooling of the surface. Repeated episodes of surface deformation in some highlands have additionally created a chaotic network of ridges, troughs and depressions with linear and curved structures.

Up, down and sideways on Earth and Venus

Radioactive elements decaying inside both the Earth and Venus generate enough heat to drive internal convection, in which hot material rises and cold material sinks on timescales of millions of years. The Earth's crust and uppermost mantle, its lithosphere, is broken into huge plates, each thousands of kilometers across, and convective

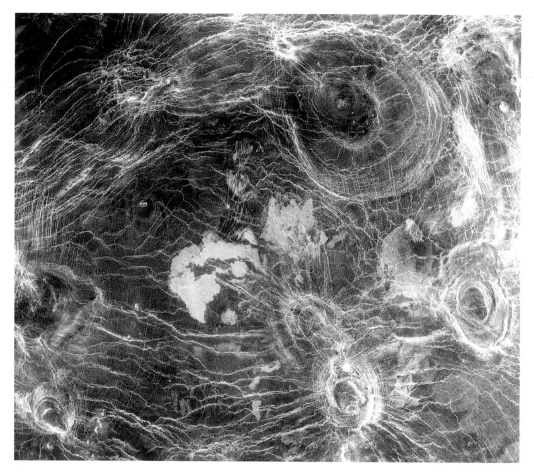

Fig. 7.20 Arachnoids　An enlarged view of a *Magellan* mosaic of Bereghinya Planitia (also see Fig. 7.18), showing circular structures of up to 230 kilometers in diameter. Their central domes are surrounded by concentric and radial fractures. Such features have been informally dubbed arachnoids for their spider-web-like appearance. They are similar in form, but usually smaller than, the circular volcanic structures known as coronae, shown in Fig. 7.21. (Courtesy of NASA.)

currents propel the plates laterally, or sideways, across the globe. On Venus there is no trace of such a process.

Plate tectonics dominate the Earth's geology. The plates separate at mid-ocean ridges, descend into deep-ocean trenches where long arcs of volcanoes are made, collide with each other to form the great mountain ranges, and grind horizontally together to set off earthquakes. The plates are rigid, so any movement on one side also happens at the other; in between the plates move sideways, carrying the continents with them.

The Earth's crust is also recycled laterally. New ocean floor rises out of the interior at the mid-ocean ridges, moves horizontally for great distances across the planet, and is destroyed by diving into the deep-ocean trenches.

Because Venus and Earth are about the same size, and have a similar bulk composition, the rates of internal heat generation by radioactive decay and the energy available to drive internal motions should also be similar. Thus, there ought to be enough heat trapped inside Venus to push plates around its surface.

Yet there is no convincing evidence for a system of plates that slide horizontally about the surface of Venus, as they do on Earth. On Venus there are no features comparable to the Earth's extensive mid-ocean ridges or to its deep-ocean trenches. Thus, it is unlikely that the crust of Venus is recycled in the same way as the Earth's is.

Perhaps the reason Venus has no sliding plates is related to the lack of water on the planet. There can be no liquid water on the torrid surface, and the planet may also be exceptionally dry inside. Water in the Earth's lithosphere is believed to be essential to making it weak enough to break into plates in response to the motions of convection in the interior. On Venus, the present-day lithosphere is too strong to allow plates to develop. The outer shell of the planet is probably seized up tight, like a car engine without oil. But this may not have been true in the past when greater amounts of internal heat might have overcome the lack of lubricating water.

Whatever the exact explanation, the outer solid shell of Venus seems to consist of one thick, rigid plate, not many

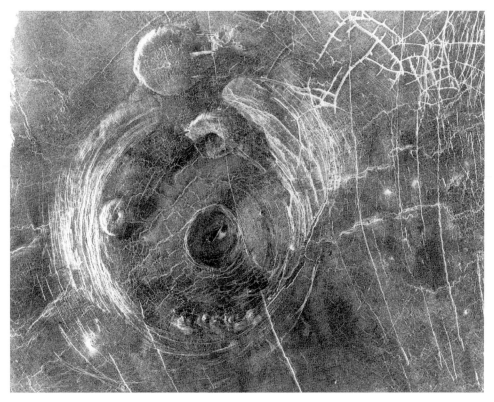

Fig. 7.21 Fotla Corona   Large circular and oval structures with diameters of 200 to 1000 kilometers are called coronae. The one shown at the center of this *Magellan* image is known as Fotla Corona, about 200 kilometers across and named after the Celtic fertility goddess. It is located in a vast plain to the south of Aphrodite Terra, and is centered at 59 degrees south latitude and 164 degrees east longitude. Molten rock rising from the interior of the planet most likely explains the corona's circular shape, raised topography, complex fractures and associated volcanism. Just north (*top*) of this corona is a flat-topped pancake dome, about 35 kilometers in diameter, thought to have formed by the eruption of sluggish, pasty lava. Another pancake dome is located inside the western (*left*) part of the corona. There is also a smooth, flat region in the center of the corona, probably a relatively young lava flow. Complex fracture patterns like the one in the northeast (*top right*) of the image are often observed in association with coronae. (Courtesy of NASA/JPL.)

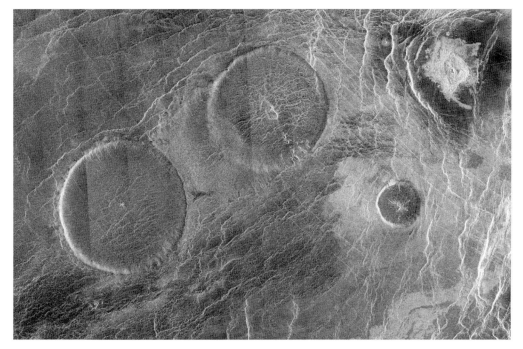

Fig. 7.22 Volcanic pancake domes   This *Magellan* radar image, centered in the Eistia Regio at 12.3 degrees north latitude and 8.6 degrees east longitude, shows an area 250 kilometers wide. The prominent circular features are volcanic pancake domes, each about 65 kilometers across, with broad, flat tops less than one kilometer in height. They are formed from viscous, or sticky, lava, and include cracks and pits that result from cooling and withdrawal of the lava. (Courtesy of NASA/JPL.)

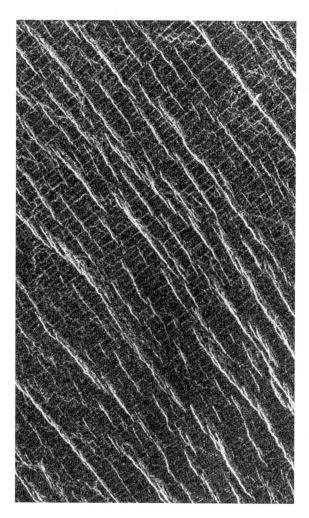

Fig. 7.23 Tesserae   Because there is little soil and no vegetation or erosion to confuse us, we can see tectonic patterns on Venus much more easily than on our own planet. This region, covering an area of 37 kilometers wide by 80 kilometers long, is located at 30 degrees north latitude and 333 degrees east longitude, on the low rise separating Sedna Planitia and Guinevere Planitia. It shows a criss-crossed pattern of radar-bright lines, which appear to be faults or fractures. The orthogonal system of ridges and grooves is formed in elevated terrain (by 1 or 2 kilometers), and spaced at regular intervals of 1 to 20 kilometers. Known as *tessarae*, from the Latin word for "tiles", the features suggest repeated episodes of intense surface fracturing that may be unrelated to volcanic activity. This type of terrain is not seen on any other planet. It covers about 8 percent of Venus's surface. Sedna is an Eskimo goddess whose fingers became seals and whales, and Guinevere is the legendary Queen of the British King Arthur. (Courtesy of NASA.)

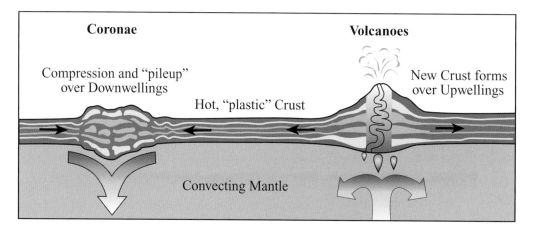

Fig. 7.24 Vertical motions on Venus   Some of the surface features on Venus have been formed by vertical, up-and-down motions driven by hot material welling up from the planet's interior. When rising bubbles of hot rock press against the crust, they can form domed, cracked features known as coronae. Larger plumes support volcanic rises on Venus, while sinking regions can lead to mountain formation.

shifting plates as on Earth. Such a strong and unbroken lithosphere would also help support the high plateaus on Venus over long timescales, and it might also explain why most of the surface is at one low level.

Given the absence of plate tectonics, there is only one main way for Venus to lose its heat. The hot mantle material pushes up and the dominant movement is often vertical, or up and down (Fig. 7.24). Upwelling material pushes

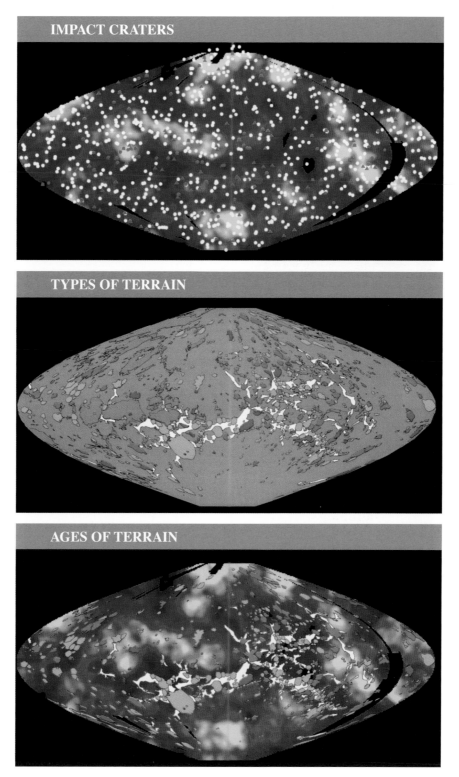

Fig. 7.25 Impact craters, terrain type, and terrain ages on Venus   Impact craters (*top diagram*) are randomly scattered all over Venus. Most are pristine (*white dots*). Those modified by lava (*red dots*) or by faults (*triangles*) are concentrated in places such as Aphrodite Terra. Areas with a low density of craters (*blue background*) are often located in highlands. Higher crater densities (*yellow background*) are usually found in the lowland plains.

The terrain type (*middle diagram*) is predominately volcanic plain (*blue*). Within the plains are deformed areas such as tessarae (*pink*) and rift zones (*white*), as well as volcanic features such as coronae (*peach*), lava floods (*red*) and volcanoes of various sizes (*orange*). Volcanoes are not concentrated in chains as they are on Earth, indicating that plate tectonics do not operate.

Terrain age data (*bottom diagram*) indicate that volcanoes and coronae tend to clump along equatorial rift zones, which are younger (*blue*) than the rest of the surface of Venus (*green*). Tesserae, ridges and plains are older (*yellow*). In general, however, the surface lacks the extreme variation in age that is found on Earth and Mars. (Courtesy of NASA/Mary Beth Price.)

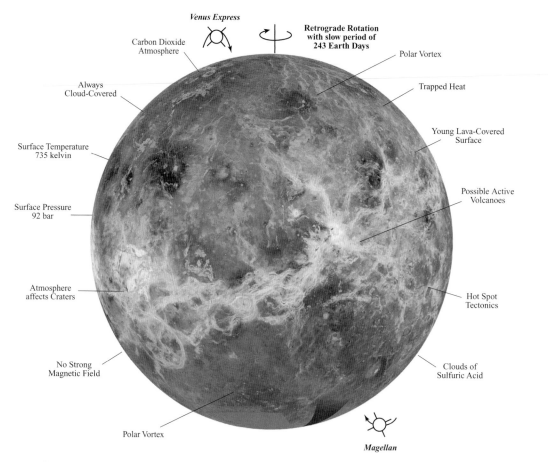

Fig. 7.26 Summary diagram

against the ground, creating arachnoids, coronae and tessarae, and punctures the surface to form volcanoes. Volcanic rises are held up by the hot rising material, and long mountain ranges may have been built during sinking, downward compression. Moreover, vast regions of the planet consist of flat, lowland plains with no substantial motion, either vertically or horizontally.

Although astronomers know virtually nothing about the first 4 billion years on Venus, they have been able to piece together a sequence of events during the last 750 million years (Fig. 7.25). At the beginning of the record, they see isolated surface deformation giving rise to a locally fractured crust. Widespread lava flooding created the flat lowland plains soon after this episode of tessera formation. After this brief but intense period of global volcanic floods, the style and rate of volcanism changed. Localized volcanoes grew on top of the vast plains and coronae were formed within them, but primarily in the equatorial regions where extensive rifts are also found.

8   Mars: the red planet

- Like the Earth, the planet Mars has an atmosphere, white clouds, polar caps and seasons.

- Mars has a partially liquid core, probably containing molten iron and perhaps surrounding a solid iron core, as within the Earth.

- Mars does not now have a global dipolar magnetic field to deflect lethal cosmic rays and energetic solar particles.

- The oldest terrain on Mars exhibits bands of magnetized material with alternating polarity, most likely originating about 4 billion years ago when the red planet might have had a global dipolar magnetic field.

- In the early 20th century it was thought that seasonal water melting from the polar caps in spring and summer produced a dark band of vegetation on Mars, and that intelligent Martian inhabitants had constructed canals to transport water across the planet.

- The seasonal dark regions on Mars are now attributed to winds, and the canals are now known to be an illusion caused when the eye arranges small, disconnected features into lines.

- Mars now has a thin, cold and dry atmosphere that is composed almost entirely of carbon dioxide, with a surface pressure of about a one-hundred-fiftieth of the Earth's atmosphere and a mean surface temperature that is well below the freezing temperature of water.

- Because of the low surface pressure and temperature of the Martian atmosphere, it cannot now rain on Mars. If any liquid water were now released on the planet's surface, it would survive for just a brief time before freezing or evaporating.

- The Martian atmosphere contains virtually no oxygen, so it has no ozone layer. The planet's surface is therefore exposed to the full intensity of the Sun's ultraviolet rays.

- Powerful and pervasive winds roar across Mars, sweeping up vast dunes of sand and fine-grained dust, creating tornado-like dust-devils, and occasionally producing global dust-storms that hide the entire planet from view.

- Seasonal polar caps of frozen carbon dioxide, or dry ice, wax and wane with the seasons on Mars. They lie on top of extensive caps of frozen water in both hemispheres. In the north, a residual or permanent cap of water ice remains in the summer heat, but carbon dioxide ice also remains in the southern winter.

- Layers in the polar caps of Mars suggest climate changes on timescales of 10 000 to 100 000 years, perhaps triggered by periodic variations in the planet's orbit and rotation axis.

- Mars is divided into two strikingly different hemispheres. In the south there are the older, elevated, heavily cratered highlands that resemble the lunar highlands. In the north there are the younger, lower-lying, smoother volcanic plains.

- Towering volcanoes and immense canyons are found on Mars.

- The dry tracks of past flowing water are etched into the surface of Mars, marking the site of ancient rivers and floods that occurred 3 to 4 billion years ago. Water networks are found in the heavily cratered southern highlands, and outflow channels are located in the equatorial regions running downhill from the highlands to the northern plains.

- Water might have once lapped the shores of long-vanished lakes and seas on Mars.

- Instruments aboard orbiting spacecraft have detected water-related minerals on the surface of Mars, especially in the ancient southern highlands.

- The *Mars Pathfinder* lander, with its *Sojourner Rover*, found that its landing site has been untouched by water since it flowed across the region more than 2 billion years ago. The two *Mars Exploration Rovers*, *Opportunity* and *Spirit*, have found evidence of ancient water flow on Mars, but no signs of recent running water.

- Huge amounts of water once flowed on the Martian surface, but exactly where all that water came from and where it all went is still uncertain. Colliding asteroids and comets could have deposited the water in the early history of the planet, or Mars might have been warmer long ago, with a thick, dense atmosphere, rain and flowing water. Most of the water that once flowed on Mars is now frozen into ice on or below the surface.

- Subsurface water ice is suggested by flow-like patterns of material ejected from impact craters on Mars, and substantial amounts of frozen water have been inferred from spectrometers aboard the *2001 Mars Odyssey* spacecraft.

- The *Phoenix* lander obtained evidence for subsurface water ice and past water-flow in the northern polar plains of Mars, and detected snow falling from Martian clouds of water ice.

- Liquid water may have been seeping out of the walls of canyons and craters on Mars in recent years, creating small gullies and depositing the debris in fanlike deltas. Other gullies have been attributed to landslides of loose dust or flows of sand related to carbon-dioxide frost.

- The lack of a global dipolar magnetic field and a thick atmosphere with an ozone layer, which would respectively divert cosmic rays or solar energetic particles and absorb ultraviolet rays, suggest that Mars' surface is now a sterile place where life cannot survive. If life did once exist there, it might have survived beneath the surface, within rocks or deep underground in the possibly wet and more temperate part of the planet's interior.

- Five spacecraft have successfully landed on the reddish-brown Martian surface and revealed no signs of life on Mars.

- There are no detectable organic molecules or cells in the Martian surface examined by the *Viking 1* and *2* landers. They found no unambiguous evidence for biological life at their landing sites. The chemically reactive, highly oxidized soil will destroy cells, while also rusting the Martian surface red.

- Cosmic impacts with Mars are capable of ejecting surface rocks into space, and some of them eventually arrive at the Earth. One such meteorite from Mars, named ALH 84001, was once thought to exhibit evidence for ancient, microscopic bacteria-like fossils, but non-biological explanations are now accepted for these features.

- The search for life on Mars is now focused on the hunt for liquid water, which might indicate that Mars is habitable and could have sustained life either in the remote past or recently.

- Methane has recently been found in the Martian atmosphere; it could be due to geochemical processes or to bacteria-like microorganisms.

- NASA has plans to launch a *Mars Science Laboratory* that would analyze the surface material near its landing site on Mars for proteins, amino acids and other molecules that are essential to life as we know it.

- Mars has two small moons, named Phobos and Deimos. Phobos is heading towards eventual collision with Mars.

8.1 Fundamentals

The red planet Mars is only half the size of Earth, and its mass is one-tenth of the Earth's mass. Mars has a day that is just 37 minutes longer than the Earth's day, and the length of the year on Mars is about twice that of a year on Earth. Mars exhibits clouds, variable polar caps and seasons, and its freezing surface lies beneath a cold, thin atmosphere that is composed mainly of carbon dioxide.

Table 8.1 Physical properties of Mars[a]

Mass	6.4169×10^{23} kilograms $= 0.1074\ M_E$
Mean radius	3.3895×10^6 meters $= 0.532\ R_E$
Mean mass density	3933.5 kilograms per cubic meter
Surface area	1.441×10^{14} square meters $= 0.2825\ A_E$
Rotation period or length of sidereal day	24 hours 37 minutes 22.663 seconds $= 8.86427 \times 10^4$ seconds $= 1.02596$ Earth days
Orbital period	1.8808476 Earth years $= 686.98$ Earth days $= 668.60$ Mars solar days
Mean distance from Sun	2.2794×10^{11} meters $= 1.52366$ AU
Orbital eccentricity	0.0934
Tilt of rotational axis, the obliquity	25.19 degrees
Distance from Earth	5.6×10^{10} meters to 3.99×10^{11} meters
Angular diameter at closest approach	14 to 24 seconds of arc
Age	4.6×10^9 years
Atmosphere	95.32 percent carbon dioxide, 2.7 percent nitrogen, 1.6 percent argon
Surface pressure	3 to 14 millibars $= 0.003$ to 0.014 bars
Surface temperature range	140 to 290 kelvin
Average surface temperature	210 to 220 kelvin
Magnetic field strength (remnant)	$\pm 1.5 \times 10^{-6}$ tesla $= \pm 0.05\ B_E$
Magnetic dipole moment	Less than 10^{-4} that of Earth

[a] Adapted from H. H. Kieffer, B. M. Jakosky, C. W. Snyder and M. S. Matthews (eds.), *Mars*, University of Arizona Press, Tucson, 1992. The symbols M_E, R_E, A_E, and B_E respectively denote the mass, radius, surface area and magnetic field strength of the Earth.

Fig. 8.1 Earth and Mars   This composite image demonstrates the relative size, similarities and main difference of Earth (*left*) and Mars (*right*). Mars is about half the size of the Earth. Both planets exhibit clouds and polar caps. Bluish-white clouds of water ice hang above volcanoes on Mars (*right – center left*) and large dark areas extend across its red surface (*right – top right*). The residual north polar cap on Mars (*right top*) is made of water ice, and is circled by dark dunes of sand and dust. The Earth also has clouds and polar caps composed of water ice. However, about 75 percent of the Earth is covered with oceans, while liquid water cannot now exist for long times on the surface of Mars. The Earth image was taken from the *Galileo* spacecraft in December 1990, and the *Mars Global Surveyor* acquired the Mars image in April 1999. (Courtesy of NASA/JPL/MSSS.)

8.2 Planet Mars

Mars is an Earth-like planet

Mars, fourth planet from the Sun, is our closest planetary neighbor after Venus, and it ranks third in brightness as seen from Earth – after Venus and Jupiter. The red planet is brighter than most stars, and it has intrigued humans since prehistoric times because of its reddish color and slow looping movement across the starry background. The ancients associated the blood-red color with destruction and warfare. The Babylonians knew it as Nergal, or Nirgal, the star of death, and the Greeks and Romans named the wanderer after their gods of war – Ares and Mars, respectively.

In several ways, Mars is similar to the Earth. The red planet spins on its axis with a rate and tilt that are almost identical to the Earth's. Surface markings have long allowed astronomers to determine the planet's rotation period. It rotates once on its axis every 24.62 hours, so the day on Mars is only 37 minutes longer than our own. Because Mars' rotational axis is inclined to its orbital plane, Mars has seasons. In fact, the present tilt of Mars' axis, at 25.2 degrees, is similar to that of Earth, at 23.5 degrees. Both planets therefore have four seasons – autumn, winter, spring and summer – although the Martian seasons last about twice as long since the Martian year is 687 Earth days long or 1.9 Earth years. But Mars is a considerably smaller planet, about half the size and one-tenth the mass of Earth (Fig. 8.1).

The eccentricity of Mars' orbit is larger than Earth's, 0.093 compared with 0.015, so the shape of the orbit of Mars is more elliptical, or out of round, than the orbit of Earth, which is closer to a circular shape. As a consequence, differences between the northern and southern seasons are much more pronounced on Mars than on Earth. The Martian pole that tilts toward the Sun at Mars' closest approach to the star, or at perihelion, has warmer summers than the other pole, and at present the south has the warmer summers. That is, the closest approach of Mars to the Sun currently occurs during the southern summer and northern winter. But because of slow changes in the direction of tilt of the rotational axis and the orientation of the perihelion, the hot and cold poles change on a 51 000-year cycle.

The Earth-like appearance of Mars in a telescope has intrigued humanity for centuries. Both planets have an atmosphere, clouds, polar caps and seasons. An atmosphere is required for the formation of the white clouds, which are composed of water-ice crystals that freeze out of the high cold air, like clouds on Earth. Clouds of carbon-dioxide ice also form above the Martian water-ice clouds. The polar caps on Mars grow in the local winter, when gases are extracted from the atmosphere, and recede with the coming of local spring and summer when gases are released into the Martian atmosphere.

Mars' kinship with the Earth has recently been extended to their interiors. After years of tracking the radio signals sent from spacecraft that have landed on Mars or are orbiting the planet, scientists have measured a small tidal bulge in the solid body of Mars, towards and away from the Sun, and determined the density at various depths in Mars. Like the Earth, the red planet is not merely a solid ball of rock, but instead contains a dense, metallic core. Early in Mars' history, the molten rock on Mars became differentiated, with heavy elements like iron sinking to the center and the light ones rising to the top. The size of this core is large enough to indicate that it must be partially liquid, most likely containing iron that is at least partly molten. But the absence of any detectable global dipolar magnetic field and the relatively high iron abundance at the surface of Mars suggest that Mars never melted as extensively as did Earth.

Right now Mars lacks an internally generated global dipolar magnetic field, which the Earth has, so the resemblance stops there. Nevertheless, *Mars Global Surveyor* detected bands of residual magnetism with alternating magnetic polarity, embedded in the ancient Martian surface (Fig. 8.2). Their presence indicates that Mars had a magnetic field in the past, but that it switched off about 4 billion years ago. The leftover magnetic bands may have been caused by a global magnetic field that repeatedly switched polarity as new crust was forming, somewhat like the alternating magnetic polarities found in the spreading ocean floor on Earth.

Early speculations about intelligent life on Mars

Over the past century, our fascination with Mars has been stimulated largely by the prospect that life may exist there, either in the past or the present. Large, varying dark regions form on the planet, seasonally distorting its ruddy face (Figs. 8.3, 8.4). The grayish-green regions flourish in the summer and become dormant in winter, as many plants do on Earth. Their seasonal growth on Mars has been called the "wave of darkening" since a dark band moves from the southern polar cap toward the equator as the cap shrinks. It was once thought that water melting from the polar cap might cause hypothetical vegetation to grow and progress from the southern polar region to the equator.

In the early 20th century, it was even widely believed that advanced civilizations had developed on Mars. These speculations resulted from ground-based telescopic observations apparently showing oases and canals stretching across the dusty plains of Mars, but often glimpsed at the limit of telescopic visibility.

In 1877, when Mars was exceptionally close to the Earth, the Italian astronomer Giovanni Schiaparelli (1835–1910), director of the Milan Observatory, reported that a maze of dark, narrow straight lines traverses the planet's surface (Fig. 8.5). He called them *canali*, the Italian word for "channels", assuming that they were natural features. Schiaparelli mapped them and gave the broadest ones the names of large terrestrial rivers, such as the Ganges and Indus.

The French astronomer Camille Flammarion (1842–1925) subsequently wrote in 1892 that the channels resemble man-made canals, redistributing scarce water across a dying Martian world. Flammarion was also convinced that the Martian inhabitants might be more advanced than terrestrial humans, describing them in popular books.

At about the same time, a wealthy Bostonian named Percival Lowell (1855–1916) convinced much of the American public that there was intelligent life on Mars. Rich enough to do as he pleased, Lowell built an observatory in the clear air of Flagstaff, Arizona, with the specific intention of observing and explaining the Martian canals. When Lowell turned his telescope toward Mars in 1894, he found what he expected to see – a vast network of canals bordered by vegetation. He also published popular books attributing the canals to a vast, planet-wide irrigation network, constructed by intelligent beings to transport water away from the melting polar caps to parched equatorial deserts.

Most astronomers, however, could not see the canals, concluding that they were an optical illusion if they existed at all. As it turned out, the "canals" have no objective reality, beyond the tendency of the human mind to seek order in chaos.

On close inspection from spacecraft, the so-called canals dissolve into swarms of elongated light and dark streaks, often tens of kilometers long, pointing in the

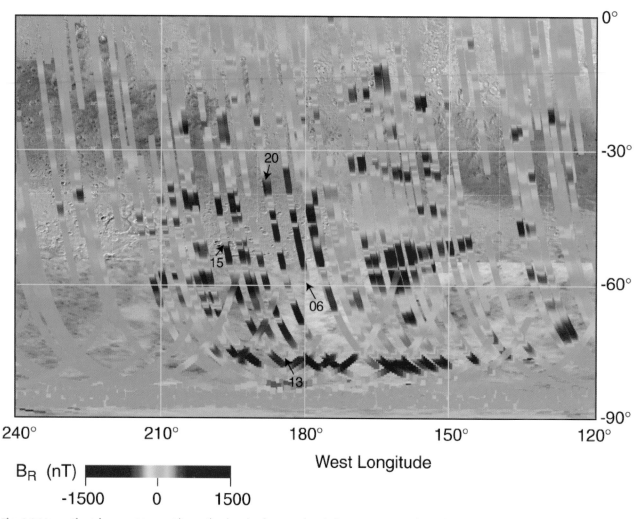

Fig. 8.2 Magnetic stripes on Mars   Alternating bands of magnetic polarity are most prominent in this part of the southern highlands, near Terra Cimmeria and Terra Sirenium. The magnetic data were obtained from an instrument aboard the *Mars Global Surveyor*. This map is color-coded red for a positive magnetic field pointing out of the planet and blue for a negative one pointing in, with a strength up to 1500 nanotesla, or 1.5×10^{-6} tesla. Stripes of alternating polarity, or direction, extend up to 2000 kilometers across the planet in the east-west direction. They are similar to the magnetic patterns seen in the Earth's crust at both sides of the mid-oceanic ridges, where the spreading crust has recorded flip-flop reversals in the Earth's dipolar magnetic field. (Courtesy of NASA/ JPL/GSFC.)

directions of strong prevailing winds. When all the streaks in a given area are integrated and superimposed by the human eye when looking through a telescope, they form canal-like features, in much the same way as dots in newsprint combine to make a picture. Moreover, the wave of darkening is not a sign of the seasonal revival of life, but instead develops when the surface rocks are scoured by powerful, seasonal winds that remove fine dust deposited on the darker terrain. Thus, the basis for early speculations about intelligent life on Mars was illusory.

8.3   The space-age odyssey to Mars

Detailed ground-based telescopic observations of Mars are difficult. One reason is that the planet is out of view for prolonged periods when it is on the other side of the Sun from the Earth. The other reason is that turbulence in the terrestrial atmosphere limits the angular resolution of even the most powerful telescope on Earth.

When Mars is closest to us and most easily observed with ground-based telescopes, the Earth moves between Mars and the Sun and the planet Mars is fully illuminated by sunlight, casting few shadows and preventing the detection of topographic detail. This alignment occurs every 780 days, or 26 months, and is known as an opposition, since Mars is then opposite the Sun in our sky (Table 8.2). Even when Mars is in opposition, the biggest telescope on Earth can separate or resolve details no smaller than about one-twentieth of the Martian diameter or 300 kilometers across, providing only a blurred vision

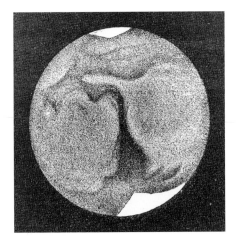

Fig. 8.3 Polar cap of Mars and its Syrtis Major   The English amateur astronomer Warren De La Rue (1815–1889) made this drawing of Mars on 20 April 1856, using a 0.33-meter (13-inch) reflector telescope. It shows bright polar caps and a dark, triangular feature now known as Syrtis Major Planitia (Gulf of Sirte Plains). The Dutch astronomer Christiaan Huygens (1629–1695) first sketched this feature in 1659. From his observations of Syrtis Major, Huygens concluded that the rotation period of Mars is about 24 hours. This drawing is reproduced from Camille Flammarion's (1842–1925) book *La Planete Mars et ses Conditions d'Habitabilité*, published in 1892.

Fig. 8.4 *Hubble Space Telescope* views Mars   This perspective of Mars was obtained from the *Hubble Space Telescope* (*HST*) on 10 March 1997 – the last day of spring in the Martian northern hemisphere. The red planet was near its closest approach to Earth and a single picture element of the *HST* spanned 22 kilometers on the Martian surface. The image shows bright and dark markings observed by astronomers for more than a century. The large dark feature seen just below the center of the disk is Syrtis Major Planitia, first seen telescopically by Christiaan Huygens (1629–1695) in the 17th century. To the south of Syrtis Major is the large circular impact basin Hellas (*center bottom*) filled with surface frost and shrouded in bright clouds of water ice. The seasonal north polar cap (*top center*) is rapidly sublimating, or evaporating from solid dry ice to carbon dioxide gas, revealing the smaller residual water ice cap with its collar of dark sand dunes. (Courtesy of NASA/JPL/STScI/David Crisp.)

of Mars. To see Mars in greater detail, one must use a spacecraft to approach it more closely.

The United States National Aeronautics and Space Administration (NASA) has dominated the modern exploration of Mars, sending numerous spacecraft to image the planet from orbit or to land on its surface. Systematic mapping of the Martian surface began in 1972, when NASA's *Mariner 9* orbiting spacecraft revealed the planet's volcanoes, deep canyons, huge outflow channels caused by ancient catastrophic floods, and extensive dune fields. A few years later, the two *Viking* orbiters fully mapped the surface with a resolution of about 250 meters, and the two *Viking* landers examined the chemistry of the local soil and carried out an unsuccessful search for signs of life on Mars.

In the late 20th and early 21st centuries, the NASA Mars Exploration Program has continued to seek evidence for whether Mars was, is, or can be a habitable world, using orbiters with high-resolution cameras, capable of resolving features a few meters across, and landers that directly sample the surface and its environment (Tables 8.3, 8.4). Since living things require water, the recent scientific strategy has also involved a search for water, either liquid or frozen and in the past or present (Sections 8.7, 8.8). The overall goals of the program are to determine if life ever arose on Mars, to characterize the climate and geology

of the planet, and to prepare for human exploration of Mars.

8.4 The atmosphere, surface conditions and winds of Mars

A carbon-dioxide atmosphere

During the first half of the 20th century, astronomers guessed that the Martian atmosphere was mostly nitrogen, like that of the Earth. Carbon dioxide had been identified in the planet's atmosphere by mid century, but it was thought to be a minor ingredient. Then instruments aboard *Mariner 4* showed in 1965 that carbon dioxide is the primary gas, and nitrogen is a minor constituent, while also revealing an ancient, cratered terrain that resembles the lunar highlands.

The *Viking 1* and *2* landers made detailed measurements of the composition of the Martian atmosphere in 1976 (Table 8.5). Carbon dioxide is indeed the principal constituent, amounting to 95.32 percent of the atmosphere at ground level, followed by nitrogen (2.7 percent) and

Table 8.2 Oppositions of Mars 2001 to 2035[a]

Opposition date	Right ascension (hours minutes)		Declination (degrees minutes)		Diameter (seconds of arc)	Distance[b] (10^7 km)
2001 June 13	17	28	−26	30	20.5	6.82
2003 Aug. 28	22	38	−15	48	25.1	5.58
2005 Nov. 7	02	51	+15	53	19.8	7.03
2007 Dec. 28	06	12	+26	46	15.5	8.97
2010 Jan. 29	08	54	+22	09	14.0	9.93
2012 Mar. 3	11	52	+10	17	14.0	10.08
2014 Apr. 8	13	14	−05	08	15.1	9.29
2016 May 22	15	58	−21	39	18.4	7.61
2018 July 27	20	33	−25	30	24.1	5.77
2020 Oct. 13	01	22	+05	26	22.3	6.27
2022 Dec. 8	04	59	+25	00	16.9	8.23
2025 Jan. 16	07	56	+25	07	14.4	9.62
2027 Feb. 19	10	18	+15	23	13.8	10.14
2029 Mar. 25	12	23	+01	04	14.4	9.71
2031 May 4	14	46	−15	29	16.9	8.36
2033 June 27	18	30	−27	50	22.0	6.39
2035 Sept. 15	23	43	−08	01	24.5	5.71

[a] An opposition occurs when the Earth moves between Mars and the Sun, and the two planets are closest. Adapted from William Sheehan, *The Planet Mars*, University of Arizona Press, Tucson, 1996.
[b] The distance between the Earth and Mars at opposition in units of 10 million kilometers (10^7 km). Because the Martian orbit is more elliptical than Earth's, the distance between the two planets at different oppositions varies as much as 50 million kilometers.

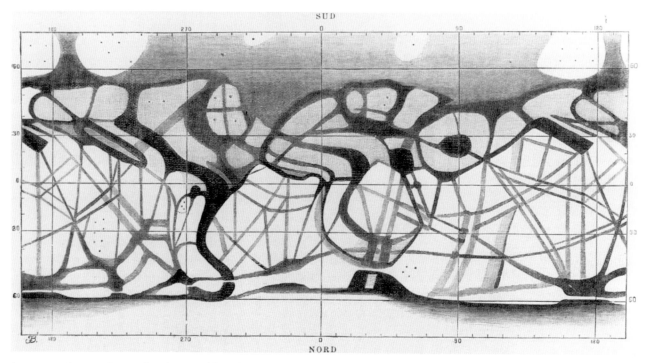

Fig. 8.5 Martian canals During the opposition of 1877, the Italian astronomer Giovanni Schiaparelli (1835–1910) mapped features he thought he saw on Mars, including a vast network of long, thin, straight lines criss-crossing the planet's surface. Some of the canals have apparently doubled, or divided in two, in this Mercator projection drawn by Schiaparelli during the opposition of 1881 using an 8.6-inch (22-centimeter) refractor. At this time, the apparent diameter of Mars was only 16 seconds of arc. Schiaparelli named these features *canali*, and they were likened to man-made water canals by subsequent observers, including Camille Flammarion (1842–1925) and Percival Lowell (1855–1916). Nevertheless, most astronomers failed to see the canals, and spacecraft have not detected them.

Table 8.3 Orbital missions to Mars[a]

Mission	Launch date	Orbit insertion	Discovery and/or accomplishments
Mariner 9	30 May 1971	13 Nov. 1971	Volcanoes, canyons, outflow channels
Viking 1	20 Aug. 1975	19 June 1976	Surface photographs, water ice in north polar cap
Viking 2	9 Sept. 1975	7 Aug. 1976	Same as *Viking 1*
Mars Global Surveyor	7 Nov. 1996	12 Sept. 1997	Laser altimeter, high-resolution images, gullies, ancient magnetism
2001 Mars Odyssey	7 Apr. 2001	24 Oct. 2001	Search for chemical, mineralogical and geological evidence for ancient water flow, vast amounts of subsurface water ice
Mars Reconnaisance Orbiter	12 Aug. 2005	1 July 2006	Water-related minerals, ancient water flow, gullies, high-resolution images

[a] Between 1971 and 1974 the Russians sent four *Mars* spacecraft into orbit around the red planet, with landers that crashed, missed the planet or failed within seconds of touchdown.

Table 8.4 Lander missions to Mars

Mission	Landing date	Site	Latitude (deg. N)	Longitude (deg. E)
Viking 1	20 July 1976	Chryse Planitia	22.27	311.81
Viking 2	3 Sept. 1976	Utopia Planitia	47.67	134.04
Mars Pathfinder	4 July 1997	Ares Vallis	19.09	326.51
MER Spirit[a]	4 Jan. 2004	Gusev crater	−14.57	175.47
MER Opportunity[a]	25 Jan. 2004	Meridiani Planum	−1.95	354.47
Phoenix	25 May 2008	North polar plains	68.22	234.25

[a] The acronym *MER* denotes the *Mars Exploration Rover*.

Table 8.5 Composition of the atmosphere at the surface of Mars[a]

Species	Abundance	Species	Abundance
Carbon dioxide (CO_2)	95.32 percent	Water vapor (H_2O)	0.016 percent[b]
Nitrogen (N_2)	2.7 percent	Neon (Ne)	2.5 ppm
Argon (Ar)	1.6 percent	Krypton (Kr)	0.3 ppm
Oxygen (O_2)	0.13 percent	Xeon (Xe)	0.08 ppm
Carbon monoxide (CO)	0.07 percent	Ozone	(0.04 to 0.2)[b] ppm

[a] Composition by volume in percent or in parts per million, denoted ppm. Because carbon dioxide varies seasonally due to condensation at the polar caps, all percentage abundances will vary seasonally. Adapted from H. H. Kieffer, B. M. Jakosky, C. W. Snyder, and M. S. Matthews (eds.), *Mars*, University of Arizona Press, Tucson, 1992.

[b] The abundance of water vapor and ozone vary with season and location. The annual global average of water vapor is 0.016 percent by volume.

Fig. 8.6 Frost on Mars   Atmospheric water vapor freezes onto the surface of Mars, producing a very thin coating of water ice on rocks and soil photographed from the *Viking 2* lander at its Utopia Planitia landing site on 18 May 1979. Scientists believe dust particles in the atmosphere pick up bits of solid water; carbon dioxide, which makes up 95 percent of the Martian atmosphere, freezes and adheres to the particles and they become heavy enough to sink. Warmed by the Sun, the surface evaporates the carbon dioxide and returns it to the atmosphere, leaving behind the water and dust in the white patches of frost shown here. The frost remained on the surface for about 100 Earth days. (Courtesy of NASA/JPL.)

argon (1.6 percent), while there is almost no oxygen (0.13 percent). In contrast the Earth's atmosphere is 77 percent nitrogen and 21 percent breathable oxygen.

The small amount of oxygen that is now present in the Martian atmosphere is the by-product of the destruction of carbon dioxide by energetic sunlight. This process also results in the production of exceedingly small amounts of ozone. Since there is so little ozone in the Martian atmosphere, it has no ozone layer, and the planet's surface is exposed to the full intensity of the Sun's lethal ultraviolet radiation. By way of comparison, the Earth has a thick ozone layer high in its atmosphere, which absorbs most of the dangerous ultraviolet sunlight and keeps it from reaching the ground.

There is now very little water vapor in the Martian atmosphere, about 0.016 percent near the surface. And that is about as much of the vapor that the tenuous atmosphere can hold. It is practically saturated with water vapor. When the temperature drops, water can condense and freeze out of the saturated air, forming low-lying mists or ground fogs in canyons and frosts on the surface (Fig. 8.6).

The concentrations of Martian water vapor vary with location and season, but they are always low. If all the water vapor above a given place on Mars could rain down to the surface as a liquid, the average depth would be 0.000 015 (1.5×10^{-5}) meters. By comparison, the Earth's

atmosphere normally contains 10 000 times as much water vapor, capable of raining onto the ground with depths of several centimeters. Moreover, the pressures and temperatures on Mars are such that it cannot now rain on Mars.

The thin, cold Martian atmosphere

Since Mars has only a tenth of the Earth's mass, the red planet has less gravitational pull. It might thus be expected to retain a less substantial atmosphere. And because Mars is 50 percent farther from the Sun than Earth, it receives about half as much sunlight and ought to be colder. The ground-level pressure and temperature on Mars are, in fact, lower than those outside the highest-flying jet airplane on Earth. The atmosphere near the surface of Mars is almost as thin as our best laboratory vacuums, and the temperatures are usually below the freezing point of water.

The surface pressure at the bottom of the thin Martian atmosphere was first accurately determined when *Mariner 4* passed behind the planet, and its radio signal penetrated the Martian atmosphere in order to reach Earth. The measurements indicated that the atmosphere at the surface of Mars is about 150 times thinner than the one on Earth, or that the average Martian surface pressure is about 0.007 bar or 7 millibar, at about one hundred-fiftieth that of Earth's atmosphere at sea level, pegged at 1.000 bars. As on the

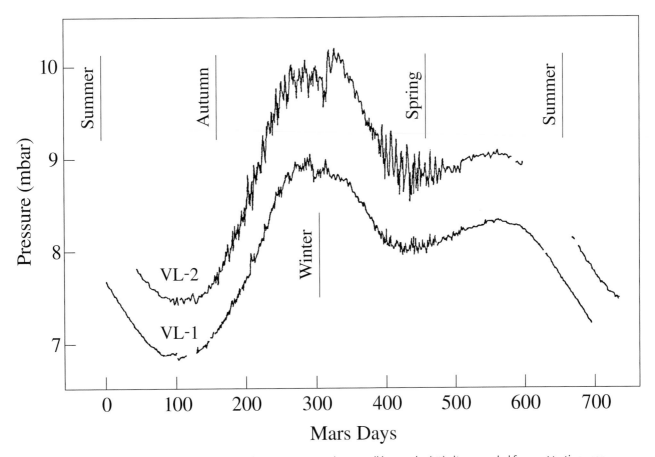

Fig. 8.7 Martian surface pressure Daily mean surface pressures at the two *Viking Lander* (VL) sites recorded for one Martian year, showing that the red planet periodically removes and replaces large amounts of carbon dioxide in its atmosphere. The seasons are for the northern hemisphere, and the pressure is given in millibars (mb), where 1 mb = 0.001 bar and 1 bar is the sea-level pressure of the Earth's atmosphere. At the *Viking 1* site (*bottom curve*), the pressure ranged from 6.7 mb during the northern summer to 8.89 mb at the commencement of northern winter. At the *Viking 2* site (*top curve*) the equivalent data were 7.4 mb and 10 mb. The higher values are probably due to the lower elevation; there was an approximate 1.1-kilometer difference in the elevation of the two landing sites. Dust storms are thought to produce some of the smaller non-seasonal pressure variations. The seasonal pressure differences seem to be dominated by the southern polar cap, which is larger than the northern one. In the southern winter, and northern summer, carbon dioxide is condensed out of the atmosphere, enlarging the southern polar cap and reducing the total surface pressure of the planet. In southern summer, and northern winter, the carbon dioxide has been released back into the atmosphere, with an increase in the total surface pressure.

Earth, the atmospheric pressure on Mars depends on the height of the surface, ranging from about 14 millibar in the bottom of the deep Hellas basin to about 3 millibar at the top of the tallest volcano.

The Martian surface pressure also varies with the season, as the polar caps grow or recede. The two *Viking* landers found, for example, that the surface pressure on Mars can change seasonally by 30 percent, while always remaining less than one-hundredth of that on Earth (Fig. 8.7). The frozen world becomes so cold during the southern winter that almost a third of its atmosphere freezes, dropping out of the sky and enlarging the southern polar cap. When the temperature rises in southern summer, the gas is released back into the atmosphere and

the surface pressure rises again. This enormous seasonal change in the mass of the atmosphere has no counterpart on Earth, where pressure changes of a couple of percentage points are cause for concern, often indicating a major storm.

Without a thick atmosphere to trap solar heat, there is no pronounced greenhouse effect on Mars and temperatures at the planet's surface are only a few kelvin warmer than that expected from direct sunlight. The surface temperatures vary from day to night, from summer to winter, and by proximity to the equator or poles. Even in the summer at latitudes below 60 degrees north or south, typical surface temperatures range from 180 kelvin at night to 290 kelvin at midday. Even though temperatures of

Fig. 8.8 Dunes in Endurance crater on Mars   Dust accumulates on the crests of dunes found on the floor of Endurance crater. This false-color image, taken from the *Mars Exploration Rover, Opportunity*, exhibits a "blue" tint in the flat surfaces between the dune flanks. This tint may be due to mineral-containing, sphere-like "blueberries," which accumulate on the flat surface and may be formed by ancient flows of liquid water. (Courtesy NASA/JPL/Cornell U.)

the immediate surface rise above the freezing point of water of 273 kelvin at low latitudes near midday, above freezing temperatures occur only within a thin upper crust; at a depth of just a few centimeters below the surface the mean temperature remains at 210 to 220 kelvin. At the poles in winter, temperatures drop to 150 kelvin, permitting carbon dioxide to condense out of the atmosphere.

Because the surface temperature and pressure are so low, liquid water cannot now stay on the surface of Mars. Over most of the surface, the temperature is usually below the freezing point of water, and when it warms above freezing, in the midday summer sunlight and near the equator, the water turns almost directly into vapor.

Water will stay liquid only if it is warm enough and at high enough pressure. Drop the temperature, and it will freeze; drop the pressure, and it will turn straight into water vapor, a process called *sublimation*. At the relatively low pressure that exists over large regions on Mars, any exposed liquid water would vaporize away. If liquid water were released onto the surface from a deep, possibly warmer interior, that water would survive for just a brief time before freezing into ice or evaporating and turning into water vapor.

The winds of Mars

Mars experiences substantial winds driven by temperature differences in the thin atmosphere. As on Earth, the Martian atmosphere is warmed by sunlight during the day and cools at night, but due to the absence of a thick moderating atmosphere or oceans, these daily temperature variations are extreme on Mars. Since the Sun also warms the summer hemisphere more than the winter one, Mars has global temperature differences that depend on the season, just as the Earth does.

The Martian atmosphere responds to both the daily and seasonal temperature differences by generating winds that blow from hot to cold regions, transporting heat and trying to equalize the temperatures. And since a rise in temperature is equivalent to an increase in pressure, it is high-pressure air that rushes toward low pressure, creating the wind.

The influence of the winds on Mars is pervasive. They stir up dense, billowing clouds of dust and scour the surface, creating time-varying light and dark patterns and carrying dust from one place to another. These are called *aeolian effects*, after Aeolus, the Greek god of the wind.

The tiny dust particles may be hazardous to future visitors to Mars. The dust will gum up spacesuits, scratch helmet visors, cause electrical shorts, sandblast instruments and clog motors. On the Moon, which is similarly dusty but has no substantive atmosphere or winds, spacesuits lasted only two days before they began to leak, and unlike the Moon, intense dust storms can arise on Mars, creating blizzards that will limit vision and drive dust into an astronaut's clothing and equipment.

The icy Martian winds have swept up vast dunes of sand and fine-grained dust. Rippled dunes have piled up in basins, chasms, and craters (Fig. 8.8), but the dunes cover only a small percentage of the land on Mars, probably less than 1 percent. Starkly beautiful patterns are created when the polar caps warm up in local spring and summer, exposing dark sand dunes.

As daily temperatures rise, winds stir up small, local dust storms, in much the same way that winds sometimes whip the terrestrial soils into towering columns called dust

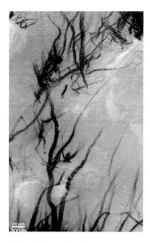

Fig. 8.9 Twisted paths of dust devils on Mars   Spinning columns of warm air, called dust devils, rise above the Sun-heated surface of Mars. Each tornado-like vortex picks up light-colored dust, exposing the darker surface underneath. Dust devils have created this wild pattern of criss-crossing dark streaks in the rippled flats of Argyre Planitia, covering an area 3 by 5 kilometers at latitude of 51 degrees south. This image was taken from the *Mars Global Surveyor* in March 2000. (Courtesy of NASA/JPL/MSSS.)

devils. These local dust storms form when the ground heats up during the day, warming the air immediately above the surface. The warm air rises in a spinning column that moves across the landscape like a miniature tornado, sweeping up dust that makes the vortex visible and leaving a dark streak behind. They have scratched tangled paths across some parts of Mars, often crossing hills and running across large sand dunes and through fields of house-sized boulders (Fig. 8.9).

Small dust storms can form simultaneously at several points on Mars, and then coalesce with each other, veiling the entire planet in a vast planet-wide dust storm, far larger than any seen on Earth. These awesome, globe-encircling storms occur during the hot summers in the southern hemisphere when Mars comes closest to the Sun and the surface temperatures are the highest. The rapid temperature increase generates hurricane-speed winds that can sweep fine dust particles high into the atmosphere. As more dust is carried aloft, it absorbs sunlight, further heating the atmosphere, strengthening the winds, and eventually covering the planet with a deep opaque cloud of dust (Fig. 8.10). But as the darkening cloud blots out the Sun, the lower layers of the atmosphere cool, the winds diminish, and the dust settles back down to the surface.

Such global dust storms are not generated during cool, long summers in the northern hemisphere, when Mars is 20 percent further away from the Sun than in southern summer, receiving less heat while also moving at a slower speed. It is the extra sunlight during the southern summer,

when Mars is close to the Sun, that provides the heat and winds that energize the most powerful storms.

8.5 The polar regions of Mars

Seasonal polar caps on Mars

As on Earth, there are large white caps at both poles of Mars, first observed with telescopes centuries ago. These seasonal deposits change in size during the Martian year, growing in the cold of local fall and winter and shrinking in the summer heat. But these varying caps are poles apart in size and seasonal change. The polar cap in the southern hemisphere is larger in the local winter than the seasonal northern cap ever gets, and the southern cap is smaller in the local summer than is the northern one.

This asymmetry is a direct consequence of the planet's eccentric orbit, which carries Mars closest to the Sun, at perihelion, in the southern summer and farthest away during the southern winter, which is longer and colder than the northern one. The relative warmth of southern summer near perihelion causes the south polar cap to almost disappear from sight while the relative cold of southern winter produces a larger polar cap. The north polar cap grows to a lesser extent in the shorter, colder northern winter, when the red planet is closest to the Sun, near perihelion, and moving at its fastest orbital speed.

During the fall and winter, when the temperature drops below 150 kelvin at either pole, carbon dioxide condenses to form a seasonal cap composed of frozen carbon dioxide, or dry ice. This is the same dry ice that is used on Earth to keep ice-cream, lobsters and other things cold for days at a time.

In the spring and summer, when the polar temperature rises above 150 kelvin, the carbon dioxide cap sublimates, or evaporates, from solid ice to gas, returning to the atmosphere. This condensation of carbon dioxide during winter and its subsequent sublimation in the spring is what gives rise to the familiar waxing and waning of the Martian polar caps. The process is entirely analogous to the snowfall that blankets the Earth's polar regions in the winter and evaporates in the summer, except the "snowfall" on Mars mainly consists of dry ice with just a few snowflakes of water ice. It also accounts for the enormous seasonal change in the surface pressure on Mars.

The residual, remnant, perennial or permanent caps on Mars

At both poles, the caps never completely disappear in the heat of the summer, when the temporary, seasonal

June 26, 2001   September 4, 2001

Fig. 8.10 Dust storm clouds out Mars   Nothing on our world matches the global dust storms on Mars, dramatically displayed in this pair of natural-color *Hubble Space Telescope* images. Surface features that were crisp and clear when the first picture was taken (*left*) were covered with blinding dust by the time of the second picture (*right*). (Courtesy of NASA/JPL/STScI/James Bell/Michael Wolf/and the Hubble Heritage Team.)

deposits of dry ice sublimate back into the atmosphere. Residual, or remnant, polar caps are left behind. Since they remain throughout the Martian year, these residual caps have also been called perennial or permanent caps.

In the northern spring and summer, the central portion of the seasonal carbon dioxide cap sublimates completely to expose a water-ice cap (Fig. 8.11). As the temperature rises, water vapor is released above the north pole of Mars, which was how the *Viking* spacecraft showed that the north polar cap contains residual water ice underneath its seasonal dry ice covering.

In the south, the carbon dioxide cap does not dissipate completely and never entirely disappears, but high-resolution thermal and radar images of the south polar cap, from the *Mars Global Surveyor, 2001 Mars Odyssey* and *Mars Express* spacecraft, indicate that water ice lies beneath the remaining dry ice. The south polar region apparently contains enough frozen water to cover the whole planet in a liquid layer approximately 11 meters deep – if it ever melted. As in the north, the seasonal carbon dioxide cap in the south freezes into a winter layer about one meter thick on top of permanent cap of water ice.

At both poles, layered deposits several kilometers thick extend out to roughly 80 degrees latitude (Fig. 8.12). Individual layers are best seen in the walls of valleys cut into the sediments, where up to 20 layers have been exposed, each a few tens of meters thick and with strata as thin as 10 centimeters. The extensive regular polar layers are interpreted as dark mixtures of ice and dust separated by bright layers of nearly pure ice. They trace the history of periodic climate change at least for the past few million years and perhaps longer.

The repetitive polar layering is attributed to cyclical changes in the planet's tilt and orbital orientation, similar to the Milankovitch cycles that regulate the distribution of solar radiation on Earth, producing its ice ages. These astronomical rhythms are much larger in amplitude on Mars than on Earth, and are believed responsible for layers deposited with different dust-to-ice ratios in both the northern and southern polar caps. Radar reflections from different north polar ice layers, observed from the *Mars Reconnaissance Orbiter*, are consistent with theoretical models of how changes in the tilt of Mars' rotational axis have produced changes in the planet's climate over the past 4 million years continuing to the present.

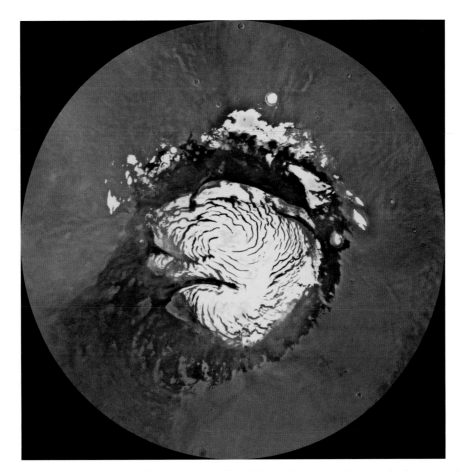

Fig. 8.11 Permanent, residual north polar cap of Mars The portion of the north polar cap that remains in summer is an imposing mountain of water ice, about 1200 kilometers across. The summit, which nearly corresponds with the planet's spin axis, stands about 3 kilometers above the flat surrounding plains. The north residual cap is surrounded by a nearly circular band of dark sand dunes formed and shaped by wind. This image, which was acquired by a *Viking* orbiter during the northern summer of 1994, strongly resembles that taken by the *Mars Global Surveyor* in the northern summer of 1999. Both the north and south residual caps contain deep valleys that curl outward in a swirled pattern that has been cut and eroded into the icy deposits, like a giant pinwheel. (Courtesy of NASA/JPL/USGS.)

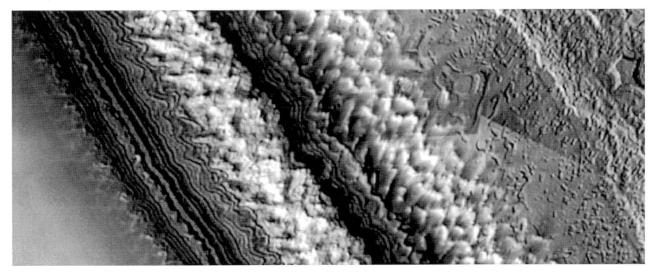

Fig. 8.12 Layered polar terrain on Mars These layers, exposed in the south polar residual cap of Mars, consist of bright ice and dark fine dust deposited over millions of years. The layered terrain in both the north and south residual caps is thought to contain detailed records of the climate history of Mars. This image covers an area of 10 by 4 kilometers. It was taken from the *Mars Global Surveyor* in October 1999 at 87 degrees south and 10 degrees west, near the central region of the residual south polar cap. (Courtesy of NASA/JPL/MSSS.)

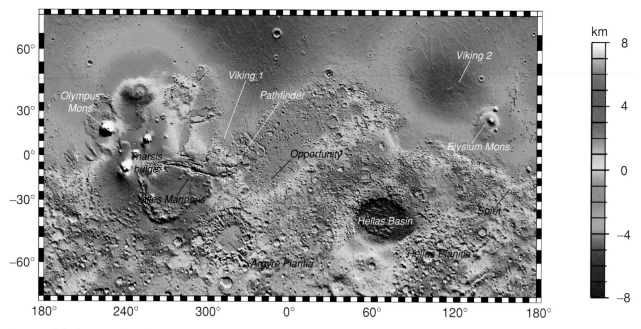

Fig. 8.13 Global topography of Mars By measuring the round-trip time of laser pulses bounced off the surface of Mars, the laser altimeter aboard the *Mars Global Surveyor* orbiter measured the altitude with a vertical accuracy of about 1 meter, providing this topographical map of Mars between latitudes of 65 degrees south (*bottom*) and 65 degrees north (*top*). The locations of five landing sites are labeled with the names of the landing spacecraft. This map portrays the great elevation difference between the northern, low-lying plains and the cratered southern highlands, and records the downhill direction that liquid water would flow. The red areas in the southern hemisphere are high regions, about 4 kilometers above the average surface height, and the blue regions of the northern hemisphere are low places, about 4 kilometers below the average height. The North Polar Basin, or Borealis Basin, is the large blue, low-lying area at the top of this topographical map; it covers about 40 percent of the planet. The map also shows the Tharsis bulge that lies near the Martian equator in the east longitude from 220 to 300 degrees. The bulge includes several major shield volcanoes, such as Olympus Mons. The huge Valles Marineris canyon system extends to the west of Tharsis. The giant Hellas impact basin, at 45 degrees south and 70 degrees east, is about 3000 kilometers across and lies about 9 kilometers deep. (Courtesy NASA/JPL/GSFC.)

8.6 Highs and lows on Mars

The crustal dichotomy of Mars

Mars is a divided world with a lopsided form, referred to as a great crustal dichotomy. The different age, height and crater density of the bottom, southern hemisphere and the top, northern hemisphere distinguishes the division. The planet's southern half is ancient, generally elevated, and rough, with a highly cratered surface that resembles the highlands on the Moon. The northern hemisphere, by contrast, consists mainly of young, lower-lying, smooth and flat plains with relatively few craters, not unlike the lunar maria. This hemispheric division has been most clearly revealed by a laser altimeter aboard the orbiting *Mars Global Surveyor* spacecraft, which analyzes pulses of laser light to measure the light time delay and distance between the spacecraft and the planet's surface (Fig. 8.13).

The cratered scars of impacting meteorites can be found all over Mars, but the craters are more densely concentrated in the southern hemisphere (Fig. 8.14). Like the lunar highlands, the extensive craters on Mars date back to an intense bombardment by meteorites early in its history, estimated at 3.8 to 4.0 billion years ago and known as the Noachian era on Mars.

The largest meteorites have gouged huge impact basins out of the Martian surface, throwing up mountains along their rims. They retain their classical designations made more than a century ago at the time of early ground-based telescopes – such as Argyre for the "silver" island at the mouth of the Ganges River, and Hellas, the Greek word for Greece. The giant Hellas basin, some 2300 kilometers across, is covered with white frost in southern winter, forming a brilliant white disk seen from Earth. The large craters, which are smaller than the impact basins, are named after astronomers and scientists who have studied the planet.

Most of the north is depressed by a few kilometers below the mean level on Mars, while the majority of the south is elevated by a few kilometers. The average elevation of the south is 5.5 kilometers above that of the north. But

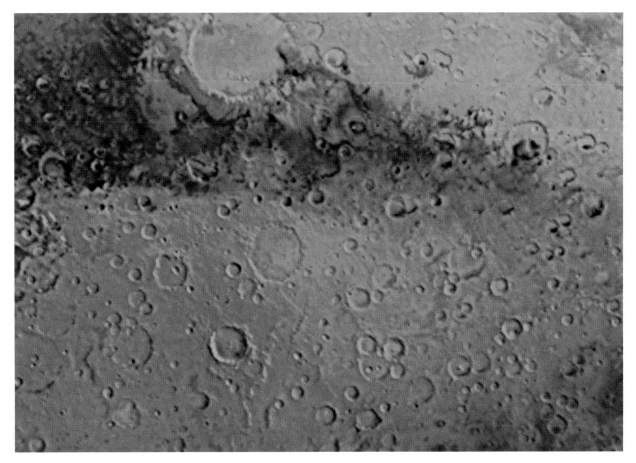

Fig. 8.14 Mars' Sinus Sabeus quadrangle   Heavily cratered highlands dominate the Sinus Sabeus region of Mars, located just south of the equator between 0 and –30 degrees latitude. A large impact crater named after the Italian astronomer Giovanni Schiaparelli (1835–1910) marks the northern part of this mosaic image, taken from the a *Viking* orbiter. (Courtesy of NASA/JPL/USGS.)

there are exceptions, the north includes lofty volcanoes that rise as much as 27 kilometers above the main surface level of Mars and the south includes giant impact basins, such as the Hellas basin whose floor marks the lowest point on Mars. The northern half is also distinguished by the presence of volcanoes, canyons, flood channels, and extensive lava flows of volcanic origin.

The plains of Mars are designated by the names of lands, followed by the Latin *planitia*, meaning "a level surface or plain". But they are not completely smooth. Volcanoes rise up in some of them, mesas and buttes in others; boulders or dunes give them a small-scale texture.

The Latin term *planum*, meaning "plateau or high plain" designates flat elevated regions, in contrast to the low-lying planitia. Most of the planitiae are located in the northern hemisphere, while the plana are found just south of the equator (Fig. 8.15). Another Latin name, *terra*, is used to designate an extensive landmass in the older, heavily cratered highlands.

The plains of Mars are sparsely cratered, and therefore post-date the period of heavy bombardment that gave rise to the profusely cratered Martian highlands. Different plains nevertheless exhibit varying crater densities, and this provides a method for tracing the planet's development. Such comparisons show that different regions on Mars span a large range in ages, from the ancient, heavily cratered highlands to very lightly cratered volcanoes that may be younger than a million years.

The formation of cratered highlands, in the Noachian era between 4.0 and 3.8 billion years ago, was followed by the Hesperian era from 3.80 to 3.55 billion years ago, when vast canyons were formed and catastrophic floods carved out huge channels. The extensive lowland volcanic plains were emplaced across the Martian surface during the subsequent Amazonian era, after 3.55 billion years ago, leading to the formation of the Lunae Planum, Chryse Planitia, Syrtis Major Planum, Amazonis Planitia, and Utopia Planitia, with estimated ages of 3.5, 3.0, 2.9, 2.8 and 1.8 billion years, respectively. Tall volcanoes were also formed throughout this era, from 3.5 billion years ago to relatively recently, perhaps even now.

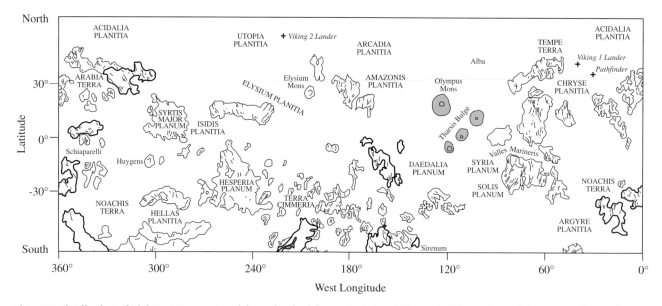

Fig. 8.15 Distribution of plains on Mars   Low-lying volcanic plains, each designated as a planitia, are located throughout the northern hemisphere of Mars. Other relatively smooth regions are found at the top of elevated plateaus, each called a planum; they are located in the southern side of the equator. Small solid dots denote volcanoes, including Olympus Mons that rises out of the Tharsis bulge, or uplift. The small crosses designate the landing sites of *Viking 1* and *Pathfinder* in Chryse Planitia (*upper right*) and *Viking 2* in Utopia Planitia (*top center*).

The topographic mismatch, with highs on the bottom and lows on the top, probably arose soon after Mars formed, and well before volcanic material began to "resurface" the northern hemisphere. According to one theory, the dichotomy may have been created within, by some unevenness in the young planet's molten subsurface layers. Internal convection somehow became lopsided, creating a global imbalance in the sweep of material and the release of heat.

In a competing theory, an ancient impact created the crustal dichotomy from outside, soon after the planet formed and when collisions with large objects were still common. A Pluto-sized body then struck the northern hemisphere at a shallow angle, ripping off the crust. This theory has received support from instruments on the *Mars Reconnaissance Orbiter* and *Mars Global Surveyor*, which have provided detailed information about the elevations and gravity of the red planet. They have discovered a large basin in the northern hemisphere, named the North Polar Basin or the Borealis Basin, which covers 40 percent of the planet and is between 8500 and 10 600 kilometers across. Its low, flat and relatively crater-free topography, previously shown in Fig. 8.13, suggest that it was created by a single large impact. Similar impacts with relatively large bodies about 4 billion years ago may account for the formation of our Moon, by a glancing blow of a Mars-sized object with the young Earth, and the removal of the missing low-density crust of Mercury.

Soaring volcanoes and immense canyons on Mars

Powerful forces have molded the face of Mars at an unexpected scale, including huge volcanoes and immense canyon-lands that dwarf their terrestrial counterparts. The colossal Martian volcano Olympus Mons has an elevation of 27 kilometers above the mean surface level, about three times that of Earth's Mount Everest above sea level and 2.6 times the height of Hawaii's Mauna Kea above its base. A gigantic canyon, dubbed Valles Marineris after *Mariner 9*, could stretch from New York City to San Francisco, putting our Grand Canyon to shame. This imposing system of interconnected canyons, or chasmata, extends along the Martian equator for 4000 kilometers, one-fourth the way around the planet. In places the chasms are as wide and deep as Mount Everest is high.

Unlike the Grand Canyon, which formed over eons of erosion by flowing water, the Valles Marineris was created when Mars cracked open about 3.8 billion years ago. All the chasmata found in this region are huge surface cracks, unlike canyons found on Earth that have been formed by running water. After the formation of the Martian canyons, erosive forces took over. Gigantic landslides widened the canyon walls, howling winds roared through the canyons, and liquid water apparently once existed in parts of the abyss.

Superposed on the global dichotomy is the Tharsis bulge, which straddles the equator. It is more than

Fig. 8.16 Ancient Martian volcano Ceraunius Tholus The Tharsis region of Mars includes both volcanic and tectonic features. This *Viking 1* image portrays the shield volcano, Ceraunius Tholus (*center right*), which is 115 kilometers in diameter. A two-kilometer wide channel extends from the summit caldera down the flanks of the volcano through a crater and into the adjacent plains. Smaller channels are just visible elsewhere on the flanks. The surrounding region (*left*) includes intensely fractured terrain. These structures are probably related to the uplift of the Tharsis region, causing fracturing and faulting. The term *tholus* means "small domical mountain or hill," and *ceraunius* means "thunderclap." Ceraunius Tholus is named for the Ceraunii Mountains on the coast of Epirus, Greece. (Courtesy of NASA/JPL.)

5000 kilometers across and 10 kilometers high, overlying the ancient highlands and lowland plains near the equator. It was formed roughly 2 billion years ago, after the creation of the crustal dichotomy. The tallest volcanoes on Mars are located on the bulge, including the Olympus, Arsia, Ascreus and Pavonis Montes, which are all much larger than any terrestrial volcano.

With their gently sloping flanks and roughly circular summit calderas, the Martian volcanoes resemble the shield volcanoes of Hawaii, such as Mauna Loa, which has a similar slope but one-third the height and one-twentieth the volume of Olympus Mons. Such volcanoes are formed by the repeated eruption of lava that cascades down the flanks in thousand of individual flows.

Why are some Martian volcanoes so much higher than their terrestrial counterparts? Perhaps it has to do with the planet's outer shell, or lithosphere. Because Mars is smaller than the Earth, it probably cooled faster, and its lithosphere became relatively stronger and thicker, not breaking up into moving plates as in the Earth's lithosphere. This gives Martian volcanoes a longer chance to grow in one spot.

Since the lithosphere on Mars is one thick, solid plate, the planet's crustal movements are mainly vertical rather than horizontal. In contrast, the Earth's plates slide horizontally over the deep-seated hot-spot sources of magma. This motion limits the growth of individual shield volcanoes on Earth, and produces chains of smaller volcanoes, such as the Hawaiian chain in the Pacific Ocean. Large-scale plate tectonics almost started on Mars, but the development was stifled by the planet's rapidly cooling outer layers. The Martian crust therefore does not move across the internal hot spots, so the lava can erupt from them for billions of years, building up volcanoes far larger than any volcano on the Earth.

The difference in gravity between Mars and the Earth also contributes to the size of Martian volcanoes. As a volcano grows in height, it eventually becomes too heavy for the underlying rock to support, and the added weight causes the entire mountain to spread outwards. Because the force of gravity on Mars is only about one-third as great as that on Earth, the Martian volcanoes can grow more than twice as tall as their terrestrial counterparts before reaching the limiting height caused by too much weight.

The lack of craters on some Martian volcanoes, such as the summit of Arsia Mons and Elysium Mons, indicate that lava flowed from them as recently as 10 million to 100 million years ago. Yet Martian volcanism dates back billions of years. Some volcanoes erupted 2 or 3 billion years ago; Ceraunius Tholus is an example (Fig. 8.16). The

Table 8.6 Common features on the surface of Mars

Descriptor term	Feature type	Example
Chasma, Chasmata	Canyon, steep-walled trough	Candor Chasma (6°S, 71°W)
Labyrinthus, Labyrinthi	Complex of intersecting valleys or canyons	Noctis Labyrinthus (7°S, 101°W)
Mons, Montes	Mountain, volcano	Olympus Mons (18°N, 133°W)
Planitia, Planitiae	Low plain	Elysium Planitia (20°N, 230°W)
Planum, Plana	Plateau or high plain	Sinai Planum (15°S, 87°W)
Terra, Terrae	Extensive land mass	Noachis Terra (35°S, 335°W)
Vallis, Valles	Valley	Ares Vallis (10.4°N, 25°W)

Focus 8.1   Naming features on Mars

There are two names that describe every feature on Mars. They are a particular name that identifies it, plus a descriptive name that says what it looks like. Some of the descriptor terms, or feature types, commonly used on Mars are given in Table 8.6.

high crater density on the outer flanks and outer edges of the Martian colossus, Olympus Mons, indicates a similar old age for the edifice, even though the paucity of impact craters at its summit implies that lava flows may have occurred less than 300 million years ago. By way of comparison, the oldest volcanoes now on Earth have ages of just a few million years or less.

Since it is bigger than Mars, the Earth has remained much more internally active, continually renewing the terrestrial surface and destroying most of the Earth's older terrain. And the internal heat of the Moon and Mercury, which are smaller than Mars, cooled off long ago; they have not been active for billions of years. Mars is between these extremes. It is just large enough to have remained active for most of the solar system's history, but not so active that all record of its early history has been erased.

Names of the surface features on Mars

The system for naming all the surface features on Mars, which have largely been discovered after close-up scrutiny by instruments aboard orbiting spacecraft, is described in Focus 8.1.

The particular names for large craters on Mars are those of deceased astronomers or physicists, ranging from Tycho Brahe (1546–1601) to Gerard Kuiper (1905–1973), as well as writers who have contributed to the lore of Mars, such as Isaac Asimov (1920–1992), Edgar Rice Burroughs (1875–1950) and Robert A. Heinlein (1907–1988). Small craters receive the names of small towns of the world with a population less than 100 000. Large valleys, described by the Latin term *vallis*, are given the name of Mars in various languages. Examples include the Ares, Kasei, Nirgal, Simud and Tiu Valles, named respectively for the word "Mars" in Greek, Japanese, Babylonian, Sumerian and Old English. Small valleys receive the classical or modern names of terrestrial rivers.

Other features on Mars are designated for the nearest named albedo (light and dark) feature on the maps of Giovanni Schiaparelli (1835–1910) or Eugene Antoniadi (1870–1944), whose appellations were drawn from classical literature and the Bible. The bright areas were named for continents or islands, such as Argyre, Arabia, Chryse, Elysium, Hellas and Tharsis. The main dark areas were given the names of bodies of water, such as Sinus Sabaeus, Solis Lacus, and Syrtis Major. The complete list is available in the International Astronomical Union's *Gazetteer of Planetary Nomenclature*, located on the Internet at http://planetarynames.wr.usgs.gov/.

8.7   Flowing water on Mars long ago

Cold, parched and wrapped in a thin, carbon dioxide atmosphere, Mars today is a frozen, desiccated and inhospitable world. It cannot now rain on Mars, and liquid water cannot now remain on its surface. Yet several lines of evidence point to running water on Mars in the distant past, 3 or 4 billion years ago, when huge amounts of water swept through outflow channels and emptied into russet flood plains, creating streamlined, washed-out landforms, and stately rivers slowly carved deep, winding valleys into the surface. There may even have been ancient lakes or shallow seas on early Mars.

Orbiting spacecraft have recently detected surface minerals created by water flowing on Mars more than 3.5 billion years ago, and roving spacecraft have found additional evidence for the planet's watery past.

The very old, water-cut features that indicate liquid water once flowed across the Martian surface take two main

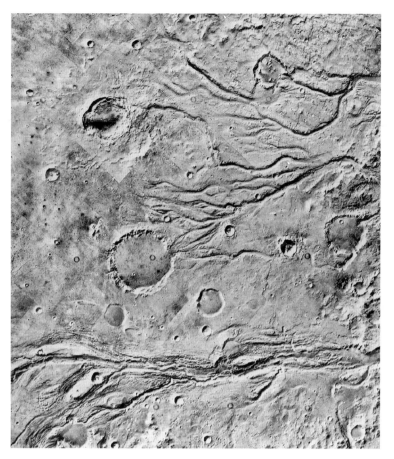

Fig. 8.17 Channels with tributaries on Mars Massive floods of water from the highlands into the Chryse basin in the lowlands may have carved these channels, located in the region of Mangala Vallis on Mars. The tributaries are rather shallow features, and join their main channels at quite acute angles. This image, taken from the *Viking 1* orbiter, has a width of 400 kilometers. (Courtesy of NASA.)

forms, known as the water networks and the outflow channels. The water networks, also known as valley networks and runoff channels, seem to have been derived from the gradual flow of liquid water. The immense outflow channels were gouged out of the surface by the powerful rush of short-duration floods.

Water networks on Mars

Networks of valleys cut across the oldest terrain on Mars, in the heavily cratered southern highlands, dating back to roughly 3.8 billion years ago. As the sinuous valley networks wind and meander downhill, they coalesce with several well-developed tributaries (Fig. 8.17), looking exactly like dry riverbeds on Earth. This suggests that the Martian water networks were not formed by rapid, surging floods.

The branching tributaries of some valley networks end abruptly in box canyons, suggesting that they formed by collapse into cavities formed by water running under the frozen, ice-rich surface. When viewed at high resolution, some individual valleys do not strongly suggest liquid flow, possibly due to subsequent modification of the surface. So some of the dry riverbeds, or water networks, might be best explained by collapse into features formed by underground rivers, and they may not require rain or running surface water. Instruments aboard the *Mars Reconnaissance Orbiter* and the *Mars Global Surveyor* have indeed revealed hundreds of small fractures in the equatorial regions of Mars, suggesting a network of ancient underground flows.

Ancient, water-charged torrents on Mars

Long, wide grooves have been gouged out of the equatorial regions of Mars, running downhill from the southern highlands into the northern lowland plains (Fig. 8.18). They tend to be narrow and deeply incised near their origins in the highlands and broad and shallow in the volcanic plains. Unlike the water networks, these enormous channels lack tributaries and are characterized by sculptured landforms such as scoured surface features, streamlined hills, and teardrop-shaped islands where the flowing water encountered an obstacle (Fig. 8.19).

Although some of the outflow channels date back almost to the end of the heavy bombardment 3.8 billion years ago, others were formed 2.5 to 3.5 billion years ago. That is relatively young on a cosmic scale, but still early in the planet's history and old in terms of geologic timescales.

The gigantic furrows bear all the marks of catastrophic outpourings of water, and are hence known as outflow

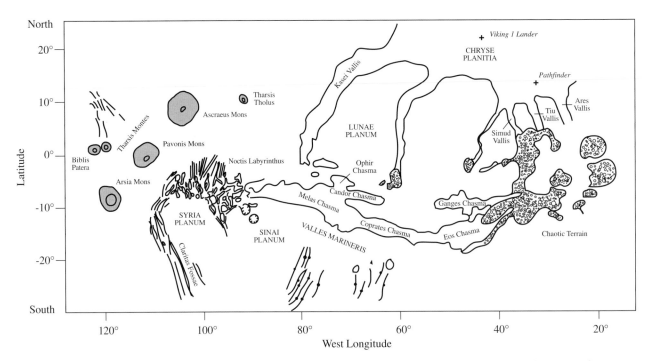

Fig. 8.18 Geological features of Mars   The uplift that created the Tharsis Montes and nearby shield volcanoes (*left*) seems to have fractured the terrain and opened up an enormous network of chasmata, or canyons, knows as Valles Marineris (*center*). Catastrophic floods originating in the vicinity of these canyons flowed north (*top*) into Chryse Planitia, which contains the site of the *Viking 1* lander and the *Mars Pathfinder* lander.

Fig. 8.19 Streamlined island in outflow channel on Mars   A raised crater rim acts as a barrier to the catastrophic floods that discharged from the outflow channel Ares Vallis. The water flowed from the southwest (*bottom left*) with a peak discharge more than 2000 times that of the Mississippi River. (Courtesy of NASA/Michael Carr.)

channels. They were formed by great, impulsive and short-lived floods of liquid water, somewhat like flash floods on Earth but on a more monumental scale.

A vast quantity of water was required to create the enormous outflow channels. They are sometimes more than 100 kilometers in width, up to 1000 kilometers in length, and as much as several kilometers deep. The discharge must have been enormous, flowing as rapidly as 75 meters per second and 1000 to 10 000 times faster than the Mississippi River on Earth. But we do not know how long the floods lasted, so we do not know the total volume of each flood.

Once released, the surging Martian torrents could not be stopped. Such discharges would not freeze even under present conditions on Mars. Large bodies of water must have been left in low-lying areas when the floods were over. Some scientists claim that Mars must have had oceans as extensive as those on Earth; others argue that seas larger than the Mediterranean were unlikely but that large lakes were possible. In either case, the lakes or seas would now be frozen.

Possible ancient lakes and seas on Mars

As we all know, water collects within holes in the ground, ranging in size from potholes in winter roads to stream-fed lakes and ocean basins. And if water once flowed across the surface of Mars, it would similarly pool in low-lying depressions, such as impact craters and basins, deep canyons and the northern lowland plains. At one time, they could all have been filled with water, forming ancient lakes and seas with perhaps a thin layer of ice on top, but all that now remains is their dried-out floors and sediment.

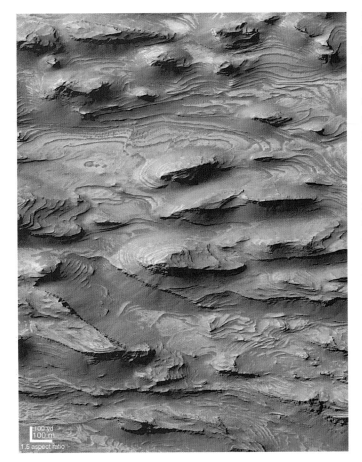

Fig. 8.20 Layered deposits in Mars' Candor Chasma The high-resolution camera aboard the *Mars Global Surveyor* orbiter revealed the presence of layered material in the floor of western Candor Chasma, at the far end of the main depressions within the extensive Valles Marineris. The numerous uniform deposits resemble regularly layered sedimentary rocks found on Earth. The Martian features could therefore be due to sediments that settled out of liquid water in ancient lakes or shallow seas a few billion years ago. Other layered deposits found in craters or basins may have existed before the canyons opened up in the surrounding terrain, and might be due to deposits of airborne dust settling out of the atmosphere, that were later buried and compacted, or to layers of volcanic material. (Courtesy NASA/JPL/MSSS.)

Widespread, stratified rock structures, found in topographical lows on Mars, could have been deposited by standing bodies of water. The layered material is located in impact craters, on parts of the Hellas impact basin, and on the floors of canyons in Valles Marineris, such as Candor Chasma (Fig. 8.20).

Thick sequences of hundreds of horizontal, regularly layered deposits are present in many places throughout the canyons, which seem to have contained lakes 3 to 4 billion years ago. The layered sediments were presumably deposited in these lakes, and then possibly compressed and cemented into rock. These lakes are supposed to have subsequently drained from the canyons into several large outflow channels, but the source of the sediment and the cause of the regular layering are still uncertain.

The low, flat northern regions of Mars could mark the dried-out bottom of a former ocean that once occupied up to one-third of the surface area of Mars, perhaps filling some or all of the North Polar Basin. Massive rivers that flowed through the outflow channels and into the northern plains might have fed the ocean. Possible ocean shorelines have been found, extending for thousands of kilometers and varying in elevation by several kilometers, perhaps due to the movement of Mars' rotation axis.

Orbiting spacecraft detect water-related minerals on Mars' surface

The infrared heat emission from the Martian surface has been imaged with spectrometers aboard orbiting spacecraft, revealing water-related deposits. The *Mars Global Surveyor* orbiter has, for example, mapped the presence of hematite, an iron oxide mineral that typically forms in the presence of water (Fig. 8.21). *Opportunity*, the *Mars Exploration Rover*, was subsequently directed to one of these locations of concentrated hematite, confirming the presence of flowing water in that location about 3.7 billion years ago.

The ancient southern highlands of Mars display a variety of mineral evidence for past water flow. An imaging spectrometer aboard the *2001 Mars Odyssey* orbiting spacecraft has found hundreds of deposits of chloride salts that typically lie in topographic depressions. They are most likely places where water was once abundant and then evaporated, leaving the salt deposits behind, perhaps 3.5 to 3.9 billion years ago.

More recently, the *Mars Reconnaissance Orbiter* has detected vast regions in the southern highlands that contain clay-like minerals, which can only form in the presence of liquid water. Volcanic lava buried the ancient regions

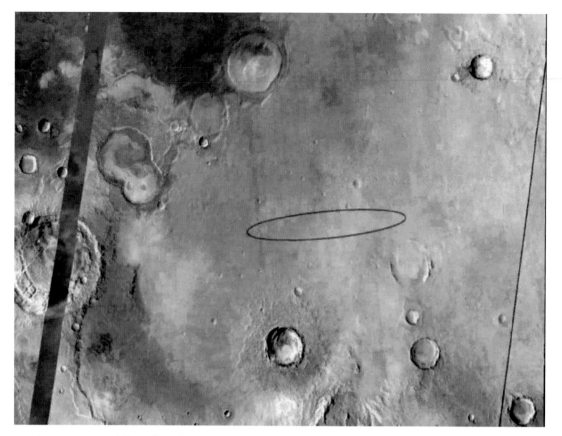

Fig. 8.21 Targeting water-containing minerals on Mars   This image shows the abundances and locations of the mineral gray hematite at the landing site of the *Mars Exploration Rover, Opportunity*, in Meridiani Planum. It was targeted to land within the oval, which is about 71 kilometers long. A colored map from an instrument aboard the *Mars Global Surveyor* orbiter displays high (*red and yellow*) and low (*green and blue*) concentrations of hematite, an iron-oxide mineral that typically forms in the presence of liquid water. The underlying surface image that includes the adjacent craters is from the *2001 Mars Odyssey* orbiter. The *Opportunity* rover found abundant evidence near its landing site for flowing water in the ancient past, about 3.7 billion years ago, including microscopic spherules dubbed "blueberries", which are rich in hematite (see Fig. 8.24). (Courtesy of NASA/JPL/ASU.)

during subsequent drier periods of the planet's history, but impact craters have exposed the clay minerals in thousands of locations across Mars. In at least one instance, rivers have apparently eroded the surface and formed deltas that indicate the sustained deposit of large amounts of clay in the vicinity of, and within, a pre-existing crater (Fig. 8.22). These delta clay deposits provide evidence for the long-term flow of liquid water in the Nili Fossae region of Mars. Minerals deposited on a small volcanic dome rising from a shallow bowl named Nili Patera in the Syrtis Major volcanic region of equatorial Mars suggest the presence of heated water, or hot springs, about 3 billion years ago.

Instruments aboard the *Mars Reconnaissance Orbiter* have detected other hydrated, or water-containing, mineral deposits spread over large regions of Mars between 2 and 3 billion years ago. They included hydrated silica, commonly known as opal, and hydrated sulfate, formed from the evaporation of salty and acidic water.

Carbonate minerals have also been detected. They can be created when water and atmospheric carbon dioxide react in the presence of dust or volcanic rock. The carbonates found so far are regional, not global deposits that might indicate an ancient Martian ocean.

2001 Mars Odyssey observations of ancient rocks containing olivine, which is easily destroyed by liquid water, nevertheless indicate that many areas of Mars have been dry for a very long time, and this has been confirmed by roving spacecraft on Mars.

Rovers obtain evidence for a watery past on Mars

The *Mars Pathfinder* lander was parachuted to Mars on 4 July 1997, in celebration of the United States Independence Day. It was directed to Chryse Planitia, the Plains of Gold, chosen because it lies at the mouth of the large outflow channel Ares Vallis and at low elevations where

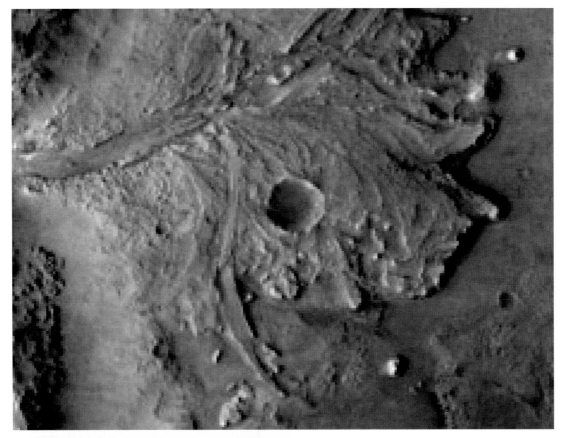

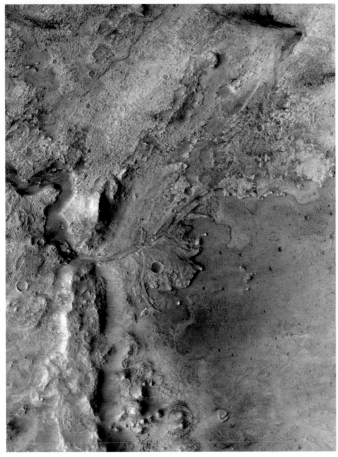

Fig. 8.22 Wet places on ancient Mars   A color-enhanced image of a delta in Jezero Crater obtained from an instrument aboard the *Mars Reconnaissance Orbiter* (*bottom*), and a close-up of the central area (*top*). A system of river channels apparently eroded clay minerals (*green*) out of the highlands and concentrated them into a crater lake, forming the delta. The distribution of clays inside the ancient lakebed suggests that standing water persisted for thousands of years in this location, and that liquid water may have persisted for thousands to millions of years as the clay formed, albeit more than 3.5 billion years ago. [Courtesy of NASA/JPL/JHUAPL/MSSS/Brown University; Bethany L. Ehlmann and colleagues, *Nature Geoscience* 1, 355–358 (2008).]

water might have accumulated. The spacecraft's *Sojourner Rover* therefore moved about the surrounding terrain in search for signs of water.

The size distribution and composition of the many rocks and boulders surrounding *Pathfinder* are consistent with their being deposited there by flowing water. In addition, the presence of numerous rounded pebbles implied the erosive action of running water in the past. The immediate vicinity of the landing site nevertheless appears to be dry and unaltered since catastrophic floods sent rocks tumbling across the plain more than 2 billion years ago. It has apparently been untouched by water ever since the ancient deluge.

The two *Mars Exploration Rovers*, *Spirit* and *Opportunity*, were deployed in January 2004, roaming the surface of Mars for more than five years. The *Spirit* rover was placed within Gusev Crater, an ancient impact crater whose smooth flat floor was interpreted as sediments deposited in a crater lake (Fig. 8.23). The water network Ma'adim Vallis cuts through the southern rim of Gusev Crater, apparently draining the ancient cratered highlands to the south. *Opportunity* was sent to Meridiani Planum, following the discovery of hematite there, an iron-oxide mineral that typically forms in the presence of liquid water. This mineral had previously been found in the region using a spectrometer aboard the *Mars Global Surveyor* orbiter.

As expected, *Opportunity* found that Meridiani has a water-rich history, gathering compelling chemical and mineral evidence that this region of Mars stayed wet for an extended period of time long ago. There were the "blueberries", formed by slow evaporation in mineral-rich liquid water (Fig. 8.24), and fossilized ripples in nearby sedimentary rock, attributed to the sloshing of shallow-water waves. The spherical blueberries are rich in the iron-bearing mineral hematite, explaining the mineral signatures seen by the *Mars Global Surveyor* orbiting spacecraft. When *Spirit* became stuck in 2009, its churning wheels broke through the crust, like the spinning wheels of a car stuck on an icy road, revealing former subsurface water flow in soil layers now covered by wind-blown sand and dust. It turned out, however, that the minerals at the landing site could only have precipitated from highly acidic water, more like battery acid than drinking water.

When the floor of Gusev Crater turned out to be just a rock-strewn plain with no water in sight, the disappointed scientists directed *Spirit* to the nearby Columbia Hills where they found what they were looking for. The hills have been extensively altered by flowing water.

So *Opportunity* and *Spirit* have provided evidence that opposite sides of Mars were once soaking wet. Nevertheless the Meridiani Planum evaporates and the Columbia Hills rocks only indicate a wet environment before

Fig. 8.23 Mars' Ma'adim Vallis and Gusev Crater   A mosaic of *Viking* images reveals the branching, winding valley network named Ma'adim Vallis, which is about 800 kilometers long. It drains from the heavily cratered terrain in the south (*bottom*) and breaches the southern rim of the Gusev Crater (*top*), which formed earlier and is about 150 kilometers in diameter. The smooth flat crater floor suggested it was filled with water and sediments, but the *Mars Exploration Rover*, *Spirit*, failed to find any water-related sediments in the hypothetical crater lake, perhaps because they have been covered by volcanic flows. (Courtesy of NASA/JPL/USGS.)

3.7 billion years ago when the conditions were also acidic, oxidizing and very salty. Since then the regions have dried out and frozen into ice, modified by winds, dust and impacts rather than flowing water.

Where did all the water on Mars come from and where did it all go?

There are several possible explanations for the origin of the water networks between 3.5 and 3.8 billion years ago. The water could have fallen as rain, during a sustained period of warm, wet climate, or it might have flowed just below the Martian surface, warmed by internal heat. According to another theory, the bombardment of comets and asteroids

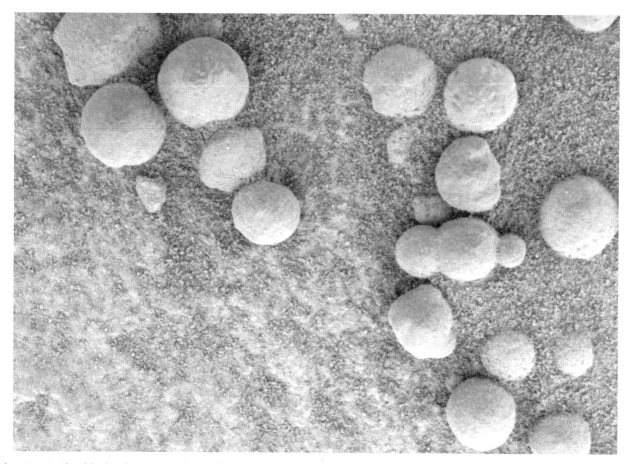

Fig. 8.24 Martian blueberries   This microscopic image taken from *Opportunity* shows sphere-like grains dubbed "blueberries" that fill a depression near the landing site of the *Mars Exploration Rover* in Meridiani Planitia. The spherical structures are concentrations of the mineral hematite formed in liquid water about 3.7 billion years ago. (Courtesy of NASA/JPL/Cornell U./USGS.)

on early Mars may have caused torrential rains and the formation of the ancient networks.

Where did the floods that formed the outflow channels come from? All of the channels emerge from discrete sources in areas that have undergone collapse, suggesting that the rapid melting of subsurface ice filled the outflow channels with raging floods. Three of the largest outflow channels, the Ares, Simud and Tiu Valles, originate in the chaotic terrain, regions of fractured, jumbled rocks that apparently collapsed when groundwater suddenly poured out.

What triggered the sudden release of such huge volumes of water in the outflow channels? Liquid water might have been trapped beneath a thick frozen expanse of permafrost. When volcanic eruptions or the formation of an impact crater breached the overlying frozen seal, the underground water would be suddenly released under great pressure. The rapid surge of water would create dramatic and sudden floods, each lasting only a few days, weeks or months. The overlying surface layer would then collapse, creating chaotic terrain. In an alternative

explanation for the outflow channels, Mars may have once been warmer and wetter, wrapped in a thicker atmosphere than it is now, with rain falling in torrents from the sky and coursing across the ground.

Everyone agrees that water once flowed on the surface of Mars in large quantities, and that much of that water has now left the surface and atmosphere. There is no liquid water residing on the surface of Mars today, and the amount of water vapor in its atmosphere is negligible. The Martian atmosphere might have once been warm and dense enough to allow water to remain liquid on or near the surface of Mars, possibly with rainfall. But then something changed. The atmosphere nearly disappeared and most of the water turned into ice.

8.8 Mars is an ice planet

Crater ejecta suggest subsurface ice on Mars

Although the round shapes of impact craters on Mars resemble those on the Moon, the ejecta blankets

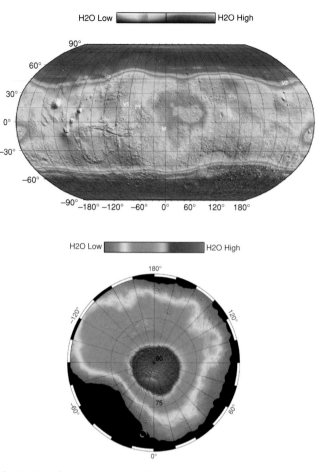

Fig. 8.25 Yuty Crater on Mars   The rounded, layered ejected material surrounding this crater may have been created when an impacting object melted the permafrost, or frozen ground, on Mars. Multiple layers of successive flows resemble the overlapping petals of a flower. The thin flow partly buries one crater, and is halted and deflected by the rim of another one. The muddy sludge seems to have sloshed across the surface, and was then refrozen. Such ejected features have not been found around craters on the Earth's Moon or the other planets. This crater, named Yuty after a town is Paraguay, is 19.9 kilometers in diameters and located in Chryse Planitia at 22.4 degrees north and 34.2 degrees west. (A *Viking* image courtesy of NASA.)

Fig. 8.26 Hydrogen on Mars   A gamma-ray spectrometer aboard the *2001 Mars Odyssey* orbiter has provided a global map of the element hydrogen (*top*). Regions of high hydrogen content at the north and south polar regions (*dark blue and violet*) are attributed to very high concentrations of buried water ice, at well over 50 percent by volume. The equatorial regions of Mars (*red and yellow*) are significantly drier than the polar regions, although they do exhibit low concentrations of hydrogen. The water-ice region near the north pole (*bottom*) is due to a permanent polar cap of water ice on the surface; elsewhere in the north polar region abundant buried subsurface ice is also found. (Courtesy of NASA/LPL/U. Arizona.)

surrounding many Martian craters look quite different from their lunar counterparts. Craters on the Moon are surrounded by secondary craters, which were formed by material thrown out by the initial impact, just as you would expect from an explosion on the dry surface. In contrast, the material around fresh-appearing craters on Mars, especially those between 5 and 100 kilometers across, flows out in rounded fronts, each outlined by a ridge, somewhat like the splashed pattern formed when a pebble is dropped in mud (Fig. 8.25).

The flowing pattern of the Mars ejecta can be explained by supposing that the Martian ground contained water ice when they were formed. The ice would have extended a few meters below the surface, somewhat like the layers of permafrost underlying the Arctic landscapes on Earth. The heat of the explosive impact would have melted the ice, and the resulting steam and liquid water would then act as a lubricant for the flowing debris. The muddy material would have sloshed outward like a wave until it dried and stiffened, or became cool and refroze.

An instrument aboard the *Mars Reconnaissance Orbiter* has identified several locations at northern mid-latitudes where bright material surrounds fresh, relatively recent craters. This material is attributed to water ice excavated from beneath the surface at depths of up to 10 meters.

Vast amounts of subsurface water ice on Mars

More direct evidence for subsurface ice has been provided by the gamma-ray and neutron spectrometers aboard the orbiting *2001 Mars Odyssey* spacecraft. They detected strong signals from hydrogen, which is attributed to extensive water-ice deposits up to a meter below the surface and all over the planet (Fig. 8.26). The amount of hydrogen in the upper meter rises to above 50 percent by mass in vast regions surrounding both poles; it makes up as much as 10 percent by mass in the top meter of material in some regions close to the equator. The concentration of hydrogen is so large that it can only be due to water ice.

The ground-penetrating radar aboard the *Mars Reconnaissance Orbiter* has additionally found evidence for thick masses of buried ice extending for hundreds of kilometers across the mid-latitude region of northern Mars, outside the polar regions and close to the equator. The hidden glaciers and ice-filled valleys are located beneath coverings of rubble and debris. They may be the protected remnants of a former ice sheet that retreated when the climate changed. Bright, exposed ice excavated from fresh craters provides additional evidence for water ice hidden just below the surface of mid-latitude Mars.

The various measurements of subsurface water ice cannot address how much water could be present below the one-meter depth to which the techniques penetrate. But the huge amounts of water ice detected in the outer parts of the planet are not inconsistent with a thick, cold reservoir of water that extends kilometers deep. Closer to the surface, a robotic spacecraft has dug a trench into the soil at its landing place in the north polar plains of Mars, detecting water ice.

Phoenix lands on Mars

The *Phoenix* lander parachuted down to a northern latitude of about 68 degrees on 25 May 2008 and verified the presence of subsurface water ice there. Its robotic arm dug into the Martian soil (Fig. 8.27), and returned it to a bake-and-sniff oven that identified water vapor produced by heating the soil. A shallow layer of water ice was uncovered at depths of 5 to 18 centimeters.

The *Phoenix* soil experiments also provided evidence that liquid water has interacted with the Martian surface throughout the planet's history into modern times, and that volcanic activity has persisted into geologically recent times, several million years ago. It is nevertheless freezing cold everywhere on Mars in modern times. An ingenious light detection and ranging instrument aboard *Phoenix* even detected snow falling from water-ice clouds just above the landing site. In other places, it looks as if liquid water might have been released from under the frozen surface of Mars, but most likely not recently.

Gullies on Mars

Close-up images taken from the *Mars Global Surveyor* have revealed gullies that suggest water might have flowed in brief spurts on Mars in the relatively recent past. Tens of thousands of the gullies are found on the steep inside walls of craters or the crests of sand dunes, with forms that resemble gully washes on Earth. The Martian flow features emerge high up on the wall, run downhill in deep, winding channels, and fan out with an abrupt ending in a fan-shaped apron of dirt and rock (Fig. 8.28).

Fig. 8.27 *Phoenix* trench in the northern plains of Mars   Bright, subsurface water ice has been exposed in this trench dug by the *Phoenix* lander in the northern Martian plains. The false-color image enhances the visibility of morning frost, especially around the edges of the trench, which are 4 to 5 centimeters deep and about 23 centimeters long. The details and patterns surrounding the *Phoenix* lander showed an ice-dominated terrain as far as it could view. (Courtesy of NASA/JPL/U. Arizona.)

The gully flows cut through terrain that is relatively young, including sand dunes and crater-free landscapes. Unlike the immense outflow channels gouged out of the terrain billions of years ago, the lack of small craters superimposed on the small, unassuming gullies and their debris indicates that they are no more than a few million years old, and possibly a lot younger; older regions are pocked with craters. Some of the gullies appear to be dust-free, which may imply that they happened within the last year or two. Otherwise, the planet's perennial dust storms would have partly covered or even buried them in dust.

A few fresh-looking gullies were found in *Mars Global Surveyor* images taken in 2004 and 2005, which were not present in images taken a few years previously, leading to the controversial speculation that the gullies might be forming today, from liquid water that is now present under the surface of Mars.

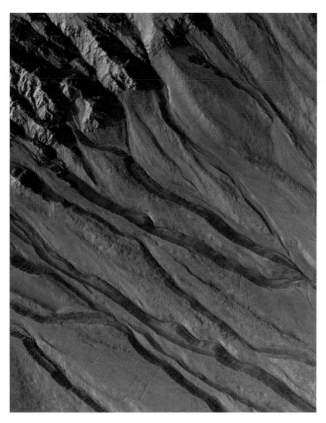

Fig. 8.28 Gullies with possible water-carved channels on Mars This false-color image, taken from a high-resolution camera aboard the *Mars Reconnaissance Orbiter*, shows gully channels in a crater in the southern highlands of Mars. The gullies emanating from the rocky cliffs near the crater rim (*upper left*) show meandering and braided patterns characteristic of water-carved channels. These gullies offer strong evidence of liquid water flowing on Mars, perhaps within the last few million years, whereas different, fresh gully deposits in other regions may have resulted from landslides of loose, dry material. North is approximately up and the image width is about 1 kilometer. (Courtesy of NASA/JPL/U. Arizona.)

When the Martian surface was scrutinized in greater detail, the planet nevertheless looked much drier than had been previously supposed. Features examined with high-resolution instruments aboard the *Mars Reconnaissance Orbiter* indicate that the bright-looking deposits of new gullies are most likely landslides of sand or loose dust, and that fresh flows on some sand dunes may be caused by carbon-dioxide frost rather than water.

Images of supposed ancient ocean floors and riverbeds also show no obvious signs that liquid water was ever present, suggesting that they are now buried beneath dust or lava. This might explain why spacecraft that landed on putative Martian lakebeds and outflow channels found no evidence for recent water flow.

There are some gullies, however, that still offer strong evidence of liquid water flowing on Mars within the last few million years. These gullies are on slopes too shallow for

dry flows and the high-resolution *Reconnaissance* images show clear indications of liquid flows, such as meandering and braided channels and terraces within the gullies.

A peculiar thing about both the dry and wet gullies is their location. Nearly all of them occur in places that are well below freezing all year round. Like moss on trees, the gullies form on the coldest slopes facing away from the Sun, where they are usually in shadow and rarely warmed by sunlight. Snow and ice would most likely now accumulate in these places, but when the polar regions tilted toward the Sun they would obtain enough sunlight to warm the topmost layer above freezing. This would explain why the Martian gullies are found at high latitudes and on poleward-facing slopes.

In this theory, gully formation isn't happening right now but when the swaying tilt of Mars' rotation axis is greatest. The poles then get more Sun exposure, the snow melts and gullies are formed. This tilt is measured by the planet's obliquity – the angle between the planet's axis of rotation and a line perpendicular to the plane of its orbit. Calculations indicate that the obliquity oscillates back and forth between 15 and 35 degrees every 100 000 years, with wider excursions occurring every 1 to 10 million years. The planet's rotation axis is currently tilted at 25.2 degrees, and when it becomes more than about 30 degrees the top layer of the Martian permafrost will thaw. Gullies might then be created by water mixed with ice.

The ongoing hunt for water on Mars is part of NASA's recent "follow-the-water" approach to the search for life on Mars. Liquid water is a key ingredient of life as we know it, so where there is liquid water there might be life. Once you find it on Mars you at least know that this region of the planet is habitable, either now or in the remote past depending on when the water flowed. Even if the water flowed millions or billions of years ago, hardy microbes might have survived until now, or perhaps there has never been a living thing on Mars. It's all speculation, until we land on the planet and find out. That is exactly what NASA is preparing to do, hoping to send robotic spacecraft to the sites of past water flow in the search for past or present life.

8.9 The search for life on Mars

Did any living things originate and survive on Mars?

Could life originate on Mars? No one knows for sure. There is abundant evidence for flowing water in the distant past, a few billion years ago, but not all water is fit to drink. The acidic, salty water that might have once flowed on Mars could have thwarted the chances of even microbes from developing or surviving.

Also, Mars subsequently lost its magnetic field and most of its atmosphere, making the Martian surface an extremely hostile place by Earth standards. Today Mars does not have enough oxygen, liquid water or heat for most, or possibly any, forms of terrestrial life. Its atmosphere is nearly all carbon dioxide, with very little oxygen or water vapor, and it is 150 times thinner than our air.

The lack of definite proof for liquid water on Mars today dims the chance for life there now. The red planet also now has no protective global dipolar magnetic field to keep potentially lethal solar energetic particles or cosmic rays from reaching the ground.

Moreover, without a protective ozone layer, the frozen surface is bathed in intense ultraviolet sunlight during the daytime, at levels that would quickly kill most organisms found on Earth today. The radiation damage to cells by the Sun's ultraviolet rays is indeed a serious risk to astronauts traveling to Mars. Any human visitor to the planet would have to wear a spacesuit to provide protection from the Sun's harmful radiation, as well as supplying oxygen to breathe.

In addition, the planet's soil now contains powerful oxidants that can break apart organic molecules, which provide living things with the capacity to evolve, adapt and replicate. Many scientists therefore remain skeptical about the chances of now locating living organisms in the dry, cold, hostile Martian world.

Optimists, on the other hand, have thought of plausible ways that primitive Martian life might survive when harsh conditions arrived. They could have taken refuge under the frozen Martian surface, for example, where it could be warmer and liquid water might be found. They might even be extracting water from the ice by chemical processes, developing internal antifreeze and hiding under or within rocks to avoid lethal ultraviolet sunlight and energetic solar and cosmic particles. After all, the optimists argue, single-celled life dominated the surface of the Earth for nearly 3 billion years, well before the rise of oxygen in its atmosphere, and microbial organisms now thrive on Earth under extreme conditions that were once thought to be lethal, including freezing temperatures underneath and within Antarctica sea-ice. All of these extreme life forms require water to survive, so the search for life on Mars ought to begin at places there might have been water. Even so, to send a spacecraft to Mars in search of life was an exciting long-shot gamble.

Viking 1 and 2 do not find evidence for life on Mars

One of humanity's most daring and imaginative experiments involved landing spacecraft on the surface of Mars,

Fig. 8.29 Apparently lifeless surface of Mars The Martian surface in western Chryse Planitia, as viewed from the *Viking 1* lander on 3 August 1976. Wind-blown dust clings against the eroded rocks, creates dust drifts and fills the sky. These drifts were little changed during the six years they were observed from *Viking 1*, and revealed no evidence for the movements of any living things. Mars is instead a cold and desolate world in which the silence is broken by the roar of winds, the hiss of dust, the rumble of mammoth landslides, and perhaps by outbursts of active volcanoes. (Courtesy of NASA/JPL.)

and searching for evidence of life there. The 4-billion-dollar gamble, in today's dollars, began on 20 July 1976, when the *Viking 1* lander came to rest on the western slopes of Chryse Planitia, the Plains of Gold, region of Mars. It appeared to have once been inundated by a great flood and was thus a promising place for life to have arisen. Six weeks later, the *Viking 2* lander settled down in the Utopia Planitia region on the opposite side of the planet, near the maximum extent of the north polar cap, again a favorable site for water and possible life.

How did the *Viking* landers test for life on Mars? The first, most obvious test consisted of looking to see if any living thing was moving about on the Martian surface. Pictures were taken of all the visible landscape, from the stubby lander-legs to the horizon, for two complete Martian years, but the view was always one of a desolate, rock-strewn, wind-swept terrain (Fig. 8.29). A careful inspection of all these pictures failed to reveal any motions or shapes that could suggest life, not a single wiggle or a twitch, or an insect or worm. So there are probably no forms of life on Mars larger than a few millimeters in size.

Of course, no one really expected that the *Viking* eyes would see living things frolicking about on the Martian surface, and each of the landers carried a biology laboratory designed to measure samples of the Martian soil for organic molecules and for signs of growth that might signal the presence of living microorganisms.

The presence of microbes might be inferred if the Martian soil contained organic molecules, which are formed by combining carbon atoms with other atoms such as hydrogen. All living things on Earth contain organic molecules, but non-biological organic molecules have already been found in comets, meteorites, Titan's atmosphere, and in abundant quantities in the space between the stars. It is not at all certain that living things could have originated from any of these organic molecules; so finding them on Mars would not be a conclusive indicator of life.

A test for biological organic molecules in the Martian soil might be called a dead-body test, for soil would be expected to contain a higher proportion of organic molecules derived from dead bodies than from living ones. But the *Viking* experiments did not detect even a single carbon compound, even though the instruments could have spotted organic molecules at a concentration of one in a billion. The tests were so sensitive that they would have easily detected organic molecules from the most barren and desolate environments on Earth.

The other experiments on board the *Viking* landers searched for the vital signs of living microbes. They did this by exposing the soil to various nutrients, and sniffing the atmosphere to see if any hypothetical microbes ate the food and released gas. Something did emit carbon dioxide and oxygen gas, but it wasn't alive.

When samples of the Martian soil were exposed to liquid food laced with radioactive carbon, large amounts of radioactive carbon dioxide poured out from the soil. This certainly suggested that animal-like microbes were digesting the food and exhaling carbon dioxide gas. But when additional nutrients were added to the soil, there was no additional increase in radioactive gas. Living creatures would have continued to ingest the food.

When the Martian soil was exposed to water, a burst of oxygen flowed from it. At first, the surprised scientists thought that plant-like microbes were emitting the oxygen, but they soon realized that the release was too fast and brief. Microbes would grow and produce more oxygen as time went on, releasing oxygen at a steady rate. Moreover, the oxygen was detected when the experiment was performed in the dark, and this behavior would not be expected from plant-like photosynthesis that depends on sunlight.

After further experiments, scientists concluded that the biological tests failed to detect any unambiguous evidence for life on Mars. Instead of being produced by organisms of any kind, all of the results were attributed to non-biological, chemical interactions. Highly oxidized minerals in the Martian soil were reacting with the nutrients, breaking them up and liberating some oxygen gas and even more carbon dioxide. Thus, at present we have no irrefutable evidence for even microbial life now on the surface of Mars.

The Sun's ultraviolet radiation has apparently turned the atmosphere and soil into an antiseptic, oxidized form.

The lethal soil would destroy cells, if there were any in the first place, living, dormant or dead, wiping them out with chemical reactions. The highly oxidizing conditions have even colored Mars red, turning iron in its surface material into rusted iron oxide.

The pioneering *Vikings* paved the way for the next spacecraft to land on the reddish-brown surface. They included the *Sojourner Rover*, deployed from the *Mars Pathfinder*, and the two *Mars Exploration Rovers*, *Opportunity* and *Spirit*. They moved across the Martian terrain, unlike the two *Viking* landers that were confined to one location, but their cameras have also never detected any signs of moving creatures. But these rovers did not carry out any chemical tests for living microbes. As we have seen, they instead searched for signs of liquid water, which might suggest that Mars could have supported life, either in the remote past or relatively recently. In the meantime, a dramatic, widely publicized frenzy resulted from speculative accounts of ancient microscopic life in a meteorite from Mars.

Possible life in a rock from Mars

Rocks that arrive from space and survive their fiery descent to the ground are given the name *meteorites*. Thousands of them have been recovered from the ice sheets of Antarctica, and most are chipped fragments of asteroids, a ring of rubble located between the orbits of Mars and Jupiter.

Only about a dozen meteorites have been identified as coming from Mars. They have been collectively named the SNC meteorites after the initials of the locations where they were first observed to fall from the sky – near Shergotty, India, in 1865, Nakhla, Egypt, in 1911, and Chassigny, France, in 1815. Pockets of gas trapped in the SNC meteorites have a unique composition that exactly matches that of the same gases in the Martian atmosphere, indicating that these rocks came from Mars and nowhere else.

The origin of each meteorite from Mars can be traced back to the jolt of a much larger object that struck the planet long ago. Most of the debris of this violent collision would fall back to the Martian surface, but some of it would be blasted off at a high enough speed to escape the weak tug of Martian gravity. That material would move in its own orbits around the Sun. These orbits would be gradually skewed by the gravitational pull of the distant planets, and very occasionally redirected on a collision course with Earth. Over an interval of 10 to 100 million years, a small fraction of the ejected debris would eventually strike the Earth.

Interest in the possibility of Martian life was heightened when scientists found possible signs of ancient, primitive bacteria-like structures inside one of these meteorites. It was recovered from the Allan Hills region of Antarctica, the first to be processed from the 1984 expedition; hence the designation ALH (for Allan Hills) 84001.

Everyone agrees that ALH 84001 originated on Mars soon after the planet formed, and was found in Antarctica 16 million years after being blasted away from the red planet. The controversy and excitement centers on suggestions that ancient microbial life took refuge in cracks within the meteorite a very long time ago.

In 1996, there was a great deal of public excitement over reports that organic molecules and possible microfossils of bacteria-like organisms had been found in ALH 84001, suggesting the existence of ancient primitive life on Mars. Newspaper headlines reported that the signs of life had been found on Mars. But the putative life forms had to be fossils of long-dead corpses, living several billion years ago, and they could only be seen with the most powerful electron microscopes on Earth.

The evidence for ancient microscopic life in the meteorite from Mars has not withstood the test of time. After more than a decade of study, all the excitement has died away. Even further, controversial evidence for structures with similar size, shape and arrangement in another Martian meteorite has not convinced the skeptics. Most scientists currently prefer non-biological explanations for all the evidence. The so-called fossils are, after all, smaller than bacteria on Earth, and much smaller than a cell; they are most likely just some kind of mineral formation.

So the meteorites do not provide convincing evidence that life once existed on Mars, but they do suggest that conditions suitable for life might have been present on the red planet early in its history. Many scientists subscribe to the belief that life will be discovered in some location other than Earth, and it might be on Mars or some other place.

The continuing hunt for signs of life on Mars

The apparently negative results of the *Viking* biology experiments, combined with the current hostile environment on Mars, strongly suggest that the planet is now inhospitable to any sort of existing life, and scientists have generally stopped looking for it. They are instead trying to determine if life ever arose on the planet in the past, focusing on the possibility that it once existed when running water and a warm dense atmosphere might have resulted in more favorable conditions for life.

There is even some lingering hope for detecting the chemical signatures of Martian life. In the early 21st century, astronomers detected the spectral signatures of methane molecules in the atmosphere of Mars. Methane, the simplest of hydrocarbon molecules, with one carbon and four hydrogen atoms, is easily broken apart by ultraviolet light from the Sun, so it must have been put into

the atmosphere within the past 300 years. There are two possible explanations for its existence. Geothermal chemical reactions, possibly involving underground water and heat, could be releasing the gas, or bacteria-like microorganisms currently living below the frozen Martian surface could produce the methane.

It seems that modern civilization has always anticipated finding life on Mars, perhaps because of its Earth-like seasons, clouds, past flowing water, ice caps and similar daily rhythm. And Mars remains the most likely, nearby place in the solar system to find extraterrestrial life. We have visited the Earth's Moon and concluded that there is no life there. The intense heat on the surface of Venus boiled away any water long ago; it would fry and vaporize all living things that we know of. Mercury has essentially no protective atmosphere, and its temperature extremes from day to night would alternately boil and freeze anything on its surface, except perhaps at the polar regions.

Thus, as dry and cold as it might be, Mars remains the most plausible nearby home for life in the solar system outside Earth itself, and future voyages to search for life on Mars are inevitable. NASA is, for example, now considering a *Mars Science Laboratory*, planned for launch as early as 2011, to search for life in the Martian soil. It will collect samples from the landing site and analyze them for organic compounds and environmental conditions that could have supported microbial life now or in the past. The onboard chemical analysis will include the possible identifications of proteins, amino acids and other complex molecules that are essential to life as we know it. The instruments will also be able to identify atmospheric gases associated with biological activity.

Robotic spacecraft, and eventually humans, will visit the most likely places to contain life on Mars, returning rocks and soil to be scrutinized in the terrestrial laboratory, and this brings up the possibility of exchanging microscopic life between the planets (Focus 8.2).

The discovery of life on Mars, even primitive life in the very distant past, would have profound implications. It would give us companionship in a vast and lonely Universe, and it would also be a little humbling. The discovery would raise the likelihood that life might be found elsewhere in the Universe as well, perhaps on one of the planets that surely exist in our Galaxy. Of course, the enduring idea of life on Mars could prove as illusory as the Martian canals, but even in that event many humans will still retain a passionate conviction that there must be life somewhere else in the Universe. So the quest will continue, and whatever happens the human spirit will remain as beautiful and glorious as ever.

Even if life within our solar system is only found on Earth, the discovery of planets around nearby stars other

Focus 8.2   Microbes from Earth and Mars

Terrestrial microbes will surely accompany astronauts to Mars, perhaps contaminating the planet and complicating the search for life. Conversely, rocks and soils brought back to Earth from Mars by a future space mission could be full of deadly microbes that might cause a global catastrophe on Earth. Visiting astronauts could conceivably get infected with an alien Martian plague, and be forced into quarantine if they make it back home. Precautions should also be taken against an unexpected crash landing of the returning spacecraft, which could release dangerous alien organisms. Since the surface of Mars is now sterilized by solar ultraviolet rays and the planet's oxidizing ground, these are unlikely possibilities, but precautions are being taken just in case the improbable occurs.

As a matter of fact, it is possible that Earth and Mars have regularly exchanged microbes over the years, without any modern spacecraft or astronauts being involved. Life forms may have arisen on Mars first and then migrated to Earth, hitching a ride on a meteorite; or it might have been the other way around, with life originating on Earth and traveling to Mars. Cosmic impacts with Mars have sent hundreds of tons of nomadic Martian rocks to Earth over recent centuries, and much more during the past eons. The rain of impacting debris was most intense in the early days of both planets, increasing the likelihood of biological exchange between them. And even now, two tons of Martian rocks are thought to rain down on Earth each year, and about the same amount of terrestrial rocks annually smashes into Mars.

So, life might not have emerged spontaneously on Earth. It could have come from Mars, and in that case we might all be Martians. Or life might have originated on Earth and was then delivered to Mars. Maybe life arose on the two planets independently and spontaneously, or perhaps impacting comets or asteroids pollinated both planets. And just maybe Earth is the only place to harbor life in the entire solar system. It's all a lot of fun to think about, but every one of these possibilities is mainly speculation with little hard scientific evidence.

than the Sun (Chapter 16) suggests that life might be common in the cosmos at large. These stars and their planets were formed in interstellar clouds that contain vast amounts of water, one of the key molecules of life, as well as all kinds of organic, carbon-bearing molecules, from formaldehyde to benzene. Many scientists think that somewhere out there, among the 100 billion stars in our Galaxy, there ought to be at least one habitable planet similar to Earth, swarming with organisms and perhaps with something more advanced than us.

8.10   The mysterious moons of Mars

Discovery and prediction of the moons of Mars

Mars has two little moons that are so dark and small that they remained unseen for centuries, even after the invention of the telescope. They were discovered in 1877 by Asaph Hall (1829–1907) using a 0.66-meter (26-inch) telescope at the United States Naval Observatory. He named the inner moon *Phobos*, the Greek word for "fear", and the outer one *Deimos*, Greek for "flight, panic or terror", after the attendants of the Greek god of war, Ares, in Homer's *Iliad*.

Johannes Kepler (1571–1630) suggested the possible existence of two Martian moons as early as 1610. Since Venus has no moons, the Earth has one, and Galileo had discovered four large moons orbiting Jupiter, it seemed logical to Kepler that Mars, with an intermediate orbit between Earth and Jupiter, would have two moons.

Then Jonathan Swift (1667–1745) endowed Mars with two fictional moons in his *Gulliver's Travels*, published in 1726. He placed them close to Mars, at 6 and 10 Martian radii, near to the 2.7 and 6.9 of Phobos and Deimos. Swift's prediction had to be a lucky guess, for there was no telescope at the time powerful enough to detect the two moons.

These are not the same kind of object as the Earth's large and spherical Moon. Both moons of Mars are tiny compared with their planet and they orbit very close to it. Phobos and Deimos are only a few tens of kilometers across and have insufficient gravity to mold them into a spherical shape (Table 8.7). They have an irregular shape and battered appearance, with craters large and small (Fig. 8.30), indicating that their surfaces are at least 3 billion years old. Both Martian moons move within the planet's equatorial plane, but they orbit so near to the surface of Mars that an observer at the poles of Mars could see neither moon. The two moons of Mars also have an unusually low mass density of less than 2000 kilograms per cubic meter.

A maverick moon of Mars

Phobos is the real maverick. It moves around Mars at 2.766 Martian radii, so close that it rises and sets three times in

Table 8.7 The satellites of Mars[a]

	Phobos	Deimos
Mass (kilograms)	1.065×10^{16}	1.47×10^{15}
Mean radius (kilometers)	11.1 ± 0.15	6.2 ± 0.18
Mean mass density (kilograms per cubic meter)	1872 ± 76	1471 ± 166
Mean distance from Mars[b]	$9378 \, km = 2.766 \, R_M$	$23\,479 \, km = 6.926 \, R_M$
Sidereal period (Martian days)[c]	0.3189	1.26244

[a] Adapted from JPL Planetary Satellite Physical Parameters.

[b] The mean radius of Mars is $R_M = 3389.5$ kilometers.

[c] One Martian day = 24 hours 37 minutes 22.663 seconds, so the orbital periods of Phobos and Deimos are 7 hours 39 minutes 13.84 seconds and 30 hours 17 minutes 54.87 seconds, respectively. Both satellites are locked into synchronous rotation with a rotation period equal to their orbital periods.

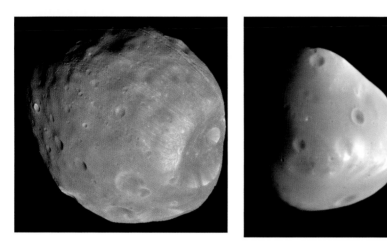

Fig. 8.30 Phobos and Deimos, moons of Mars A high-resolution camera on the *Mars Reconnaissance Orbiter* provided these close-up images of the Martian moons Phobos (*left*) and Deimos (*right*). The illuminated part of Phobos is about 21 kilometers across. The large crater Stickney, at the lower right, has a diameter of about 9 kilometers, nearly half the size of Phobos. This crater is named after Angeline Stickney (1830–1892), wife of Asaph Hall (1829–1907) who discovered the two small satellites of Mars. Deimos (*right*) is about 12 kilometers in diameter. (Courtesy of NASA/JPL-Caltech/U. Arizona.)

a single Martian day. That is, the orbital period of Phobos, of 7 hours and 39 minutes, is less than one-third of the planet's rotation period of 24 hours 37 minutes. From the surface of Mars, the small moon Phobos would be seen to move backwards across the sky, rising in the west and setting in the east.

Phobos is about as close to Mars as it can get. If it came much nearer to Mars, the planet's differential gravitational forces, between the sides nearest and furthest from Mars, would tear Phobos apart. In fact, the orbit of Phobos is steadily shrinking. If it continues to move toward Mars at the present rate, of about 1.8 meters closer to the planet every century, Phobos will either smash into the Martian surface or be ripped apart by the planet's gravity to make a ring around Mars in 50 million years. So astronomically speaking, we are now catching a fleeting glimpse of the last few moments of its life.

On the other hand, Deimos is near the outer limit for an object to be orbiting Mars. If it moved much further away, the Martian gravity would be too weak to hold on to the moon.

Phobos' suicidal motion has resulted in some fascinating, but untrue, speculations about the maverick moon. Two prominent scientists, the American astronomer Carl Sagan (1934–1996) and the Soviet astrophysicist Iosif Shklovsky (1916–1985), concluded in 1968, for example, that Phobos might be a hollow artificial satellite launched by a past Martian civilization; atmospheric friction, or air drag, would be causing the hypothetical satellite to move toward the planet.

We now know that tidal forces are pulling the small moon toward unavoidable destruction. Phobos produces two tidal bulges in the solid body of Mars, in much the same way that the Moon produces ocean tides on the Earth. As Phobos moves ahead of Mars, the closest tidal bulge pulls gravitationally on the moon, causing it to lose energy and move inexorably toward self-destruction. Because Phobos orbits so close and swiftly,

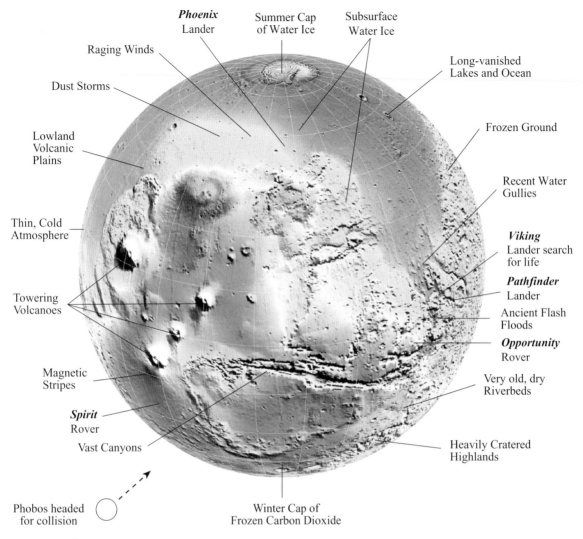

Phoenix Lander

Summer Cap of Water Ice

Subsurface Water Ice

Raging Winds

Long-vanished Lakes and Ocean

Dust Storms

Frozen Ground

Lowland Volcanic Plains

Recent Water Gullies

Thin, Cold Atmosphere

Viking Lander search for life

Pathfinder Lander

Ancient Flash Floods

Towering Volcanoes

Opportunity Rover

Very old, dry Riverbeds

Magnetic Stripes

Spirit Rover

Heavily Cratered Highlands

Vast Canyons

Phobos headed for collision

Winter Cap of Frozen Carbon Dioxide

Fig. 8.31 Summary diagram

this tidal action pulls the moon inward, instead of pushing it outward as tidal interaction does for the Earth's Moon.

Origin of the Martian moons

The irregular shapes, small sizes, and low mass densities of Phobos and Deimos closely resemble those of the numerous asteroids that orbit the Sun between the orbits of Mars and Jupiter. The two Martian moons and the asteroids have a similar battered appearance, with a profusion of craters, large and small. Moreover, the composition of the Martian moons seems to be unlike that of the planet they orbit. Their surfaces are as dark as some asteroids, known as the carbonaceous C-type, and nowhere near as lightly colored as the surface of Mars. Scientists therefore speculate that Phobos and Deimos were adopted from the asteroid belt.

One idea is that two asteroids wandered close by Mars and were pulled in by its gravity, perhaps at different times. In another variant on this theme, one larger asteroid was captured by Mars, and subsequently broke apart during a collision with another larger body, becoming Phobos and Deimos. In both explanations, the captured asteroids would have to lose energy during their encounter with Mars. Otherwise they would hurtle back into space rather than going into orbit around the planet.

They might be the last surviving remnants of a ring of debris blasted into orbit by a huge meteorite that collided with Mars during its formative years, about 4 billion years ago. Such a giant impact is now the favorite explanation for the origin of the Earth's Moon. It would provide a good explanation for the nearly circular orbits of Phobos and Deimos, which lie in the equatorial plane of Mars; such orbits are difficult to explain by the captured-asteroid hypothesis.

9 Jupiter: a giant primitive planet

- All we can see on Jupiter is clouds, swept into parallel bands of bright zones and dark belts by the planet's rapid rotation and counter-flowing, east–west winds.

- Jupiter turns to liquid under high pressures within its interior, so the cloudy atmosphere has no distinct bottom and Jupiter's weather pattern is free to flow in response to the giant planet's rapid spin.

- Jupiter's Great Red Spot and white ovals are huge shallow anticyclonic storms, which can have diameters larger than the Earth's and last for centuries.

- Large whirling storms on Jupiter gain energy by merging with, and engulfing, smaller eddies. The little storms obtain their energy from hotter, lower depths.

- White clouds of ammonia ice form in the coldest, outermost layers of Jupiter's atmosphere. Water clouds are expected to form at greater depths, and ammonium hydrosulfide clouds should condense between the water and ammonia clouds.

- All of the clouds on Jupiter ought to be white; their colors are attributed to an active chemistry that produces complex compounds in small amounts.

- Bolts of lightning illuminate deep, wet storm clouds on Jupiter.

- When the *Galileo* spacecraft parachuted a probe into Jupiter, the entry site, a region of downdraft, was missing the expected three layers of clouds and it was far drier and windier than anticipated.

- The fierce winds that give rise to Jupiter's banded appearance run deep, indicating that Jupiter's ever-changing weather patterns are driven mainly from within, by internal energy rather than by external sunlight.

- When compared to the outer layers of the Sun, the outermost atmosphere of Jupiter is slightly depleted in helium, and enriched in carbon, nitrogen and sulfur by a factor of about three.

- Jupiter is a primitive incandescent globe that radiates 1.67 times as much energy as it receives from the Sun, probably as heat left over from when the giant planet formed.

- Jupiter originated together with the Sun, and both the giant planet and the star are mainly composed of the lightest elements, hydrogen and helium.

- If Jupiter was about 80 times more massive, it could have become a star.

- Jupiter has a non-spherical shape with a perceptible bulge around its equatorial middle.

- The visible cloud tops and outer atmosphere of Jupiter form a very thin veneer that covers a vast global sea of liquid hydrogen.

- Most of Jupiter's interior consists of fluid metallic hydrogen formed under the extreme pressures that exist inside the planet.

- Jupiter probably has a dense, molten core with a mass that is less than or equal to 12 times that of the Earth.

- By re-creating extreme conditions like those inside Jupiter, modern laboratory experiments have compressed liquid hydrogen so that it becomes highly conductive like a metal.

- Jupiter's powerful magnetic field is generated by rotationally driven electrical currents inside its vast internal shell of liquid metallic hydrogen.

- The volcanoes on Jupiter's innermost large moon Io have turned the satellite inside out. It is the most volcanically active body in the solar system.

- Io's volcanoes emit plumes of sulfur dioxide gas that freeze onto the surface as a white frost.

- Volcanic vents on Io are filled with melted silicate rocks that are hotter than any place on any planet's surface, even Venus.

- Changing tidal forces squeeze Io's rocky interior in and out, making it molten inside and producing volcanoes.

- A vast current of 5 million amperes flows between the satellite Io and the poles of Jupiter, generating 2.5 trillion watts of power and producing aurora lights on both the satellite and the giant planet.

- Jupiter's magnetic field sweeps past Io, picking up a ton of sulfur and oxygen ions every second and directing them into a doughnut-shaped ring known as the plasma torus.

- There are no mountains or valleys on the bright, smooth, ice-covered surface of Jupiter's moon Europa; it has few impact craters, indicating a relatively young age.

- Long, deep fractures run like veins through Europa's icy covering, apparently filled by the upwelling of dirty liquid water or soft ice. Warmer, slushy material just beneath the crust also lubricates large blocks of ice that float like rafts across Europa's surface.

- An electrically conducting, subsurface sea within Europa may be responding to Jupiter's magnetic field, generating a time-varying magnetism in the satellite.

- Scientists speculate that subsurface liquid water in Europa may harbor alien life that thrives in the dark.

- Jupiter's moon Ganymede is bigger than the planet Mercury. The satellite's icy surface has been fractured and pulled apart, producing a grooved terrain, and surface depressions have been filled by eruptions from volcanoes of ice.

- Ganymede has an intrinsic magnetic field, and it is the only satellite that now generates its own magnetism.

- Jupiter's moon Callisto has one of the oldest, most heavily cratered surfaces in the solar system. Yet the satellite is covered by fine, dark, mobile material and it has a lack of small craters when compared to the surfaces of the Moon and Mercury.

- Like Europa, the outermost large moon Callisto has a borrowed magnetic field, apparently generated by electrical currents in a subsurface ocean as Jupiter's powerful field sweeps by. But Callisto has a largely homogeneous interior without any apparent dense iron core, and the buried sea has to lie deep enough to not affect its unaltered, cratered surface.

- Jupiter's faint, insubstantial ring system is made of dust. The ring particles might last for only a few thousand years, and they must be replenished if the ring system is a permanent feature.

- When interplanetary meteoroids, attracted by Jupiter's powerful gravity, pound into the small inner moons of Jupiter, they chip off dust fragments that go into orbit around the planet, forming its ring system.

9.1 Fundamentals

Jupiter's exceptional brightness and stately motion among the stars earned it the reputation as king of the planets to ancient astronomers. The giant planet outshines everything in the night sky except the Moon and Venus, and it revolves around the Sun at a leisurely pace with an orbital period of 11.86 Earth years.

Jupiter's complete orbital journey across the background stars is close enough to 12 years that the Chinese adopted it for their 12-year astrological cycle, using the giant planet's motion to mark out the years. Named *Sui Xing*, for the "Year Star", Jupiter passes through patterns of stars representing an ordered sequence of a dozen animals.

The arrival of the Chinese New Year at the end of January 2010 marked the beginning of the Year of the Tiger, and the succeeding years are designated as Rabbit, Dragon, Snake, Horse, Sheep, Monkey, Cock, Dog, Pig, Rat, and Ox. This system dates at least as far back as Marco Polo's (1254–1324) visit to the Mongol rulers of China. Jupiter also passes eastward through one of the twelve constellations of the Greek zodiac each year, but these stellar configurations are not related to the Chinese menagerie.

Jupiter's orbital radius is 5.2 times the radius of the Earth's orbit, so the planet's distance from Earth changes relatively little in the course of a year. As a consequence, its apparent size and brightness are fairly constant, unlike the behavior of Mars and Venus. When these nearby planets

Table 9.1 Physical properties of Jupiter[a]

Mass	1.8981×10^{27} kilograms $= 317.894\ M_E$
Equatorial radius at one bar	71 492 kilometers $= 11.19\ R_E$
Mean radius	69 911 kilometers
Bulk density	1326 kilograms per cubic meter
Rotation period	9.9249 hours $= 9$ hours 55 minutes 29.7 seconds
Orbital period	11.8626 Earth years
Mean distance from Sun	7.7833×10^{11} meters $= 5.203$ AU
Age	4.6×10^9 years
Atmosphere	86.4 percent molecular hydrogen, 13.6 percent helium atoms
Energy balance	1.67 ± 0.08
Effective temperature	124.4 kelvin
Temperature at one bar level	165 kelvin
Central temperature	17 000 kelvin
Magnetic dipole moment	20 000 D_E
Equatorial magnetic field strength	4.28×10^{-9} tesla or 14.03 B_E

[a] The symbols M_E, R_E, D_E and B_E denote respectively the mass, radius, magnetic dipole moment, and magnetic field strength of the Earth. One bar is equal to the atmospheric pressure at sea level on Earth. The energy balance is the ratio of total radiated energy to the total energy absorbed from sunlight, and the effective temperature is the temperature of a black body that would radiate the same amount of energy per unit area.

Fig. 9.1 Jupiter and its moon Ganymede · The icy surface of Ganymede, Jupiter's largest moon, is seen just before it passes behind the planet (*bottom*). The giant Jupiter is portrayed in close to natural color, including its Great Red Spot (*center*), the biggest and oldest known storm in the solar system. The spot's east–west diameter is more than twice that of Earth, and one-sixth the diameter of Jupiter itself, and the red swirling vortex has been observed for at least 300 years. By watching Jupiter eclipsing its moon Ganymede, astronomers have searched for haze that causes a slight dimming at different colors. This image was taken on 9 April 2007 from the *Hubble Space Telescope*. (Courtesy NASA/ESA/E. Karkoschka, U. Arizona.)

are on the same side of the Sun as the Earth, they appear much bigger and brighter than when they move to the opposite side of the Sun.

Jupiter is a true monarch of the planets, the largest planet in the solar system with a radius of about 11 times that of the Earth. The giant is so large that it could contain more than 1300 Earth-sized planets inside its volume. Yet Jupiter is only 318 times as massive as our planet. So Jupiter must be mainly composed of something lighter than the rock and iron that constitute the Earth.

If we divide the mass by the volume, we find that Jupiter has a bulk density of 1326 kilograms per cubic meter, only about one-quarter the mean mass density of the Earth. In fact, the mass density of Jupiter is only slightly greater than that of water, at 1000 kilograms per cubic meter, and this implies that Jupiter, like the Sun, is composed primarily of hydrogen. No other element is light enough to account for the low density of the planet.

Despite its great size, Jupiter rotates so fast that day and night each last about 5 hours and its full day is less than one-half of an Earth day. The rapid rotation stretches most of Jupiter's storm fronts into strips around the entire planet. The precise rotation rate is found by tracking radio bursts that are linked to the planet's spinning magnetic field, which emerges from deep within the planet. A rotation period of 9 hours 55 minutes 29.7 seconds = 9.9249 hours is obtained from the repeated passage of the radio storm centers. This rapid rotation can easily be detected in an hour or so with a small telescope if the planet's cloud markings are carefully watched, but not measured with precision.

9.2 Stormy weather on Jupiter

Zones and belts in Jupiter's atmosphere

The only features we can see on Jupiter are multicolored clouds, and occasionally a large moon that passes by (Fig. 9.1). The clouds circulate around the planet in dark

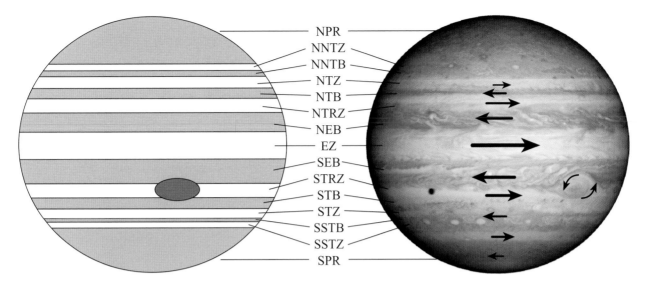

Fig. 9.2 Banded wind-blown clouds on Jupiter   The traditional nomenclature of Jupiter's light and dark bands of clouds (*left*) is given in abbreviated form (*center*). The dark bands are called belts, denoted by "B", the light bands are known as zones, or "Z", and the rest of each name is based on climatic regions at the corresponding latitudes on Earth. North, letter "N" is at the top, and south, denoted by "S", is at the bottom. The equatorial, or "E", bands are in the middle, the tropical, "TR", bands on each side of the equator, and the temperate, "T", ones at mid-latitudes. Far northern latitudes are denoted by "NN", far southern latitudes by "SS", and the polar regions by "P". The image of Jupiter (*right*) was taken from the *Cassini* spacecraft on 7 December 2000, when Jupiter's moon Europa cast a shadow on the planet. The arrows point in the direction of wind flow, and their length corresponds to the wind velocity, which can reach 180 meters per second near the equator. (*Cassini* image courtesy of NASA/JPL.)

belts and light zones. They are the sites of counter-flowing winds, or jets, that blow parallel to the equator at different speeds and in different directions (Fig. 9.2).

The zones and belts are thought to be associated with vertical, in-and-out motions called convection. Upwelling warm gas results in the light-colored zones, which are regions of high pressure. The darker belts overlie regions of lower pressure where cooler gas sinks back down into Jupiter's atmosphere. The zones and belts are therefore analogous to the high- and low-pressure systems that produce localized circulating storms on Earth, except Jupiter's rapid rotation has wrapped them all the way around the planet. Although this general convective pattern of the zones and belts was apparently supported by observations during the *Voyager* flybys of Jupiter, instruments aboard the *Cassini* spacecraft have also discovered large plumes of fast-rising gas scattered throughout the belts and absent in the zones.

Where the Earth has just one westward air current at low latitudes – a trade wind, and one nearly eastward current at mid-latitudes – a jet stream, at the cloud-top level Jupiter has five or six of these alternating jet streams in each hemisphere.

The cloud-top winds on Jupiter have higher speeds than the winds on Earth, and storms last longer on the giant planet. Its raging winds are powerful jets, moving at speeds of up to 180 meters per second, more than four times faster than any jet stream on Earth. And unlike most winds on our planet, the east–west winds on Jupiter are remarkably steady. They vary in strength and direction as a function of latitude with a regular pattern that apparently has remained unchanged for at least a century.

On Earth, heating by sunlight results in a large temperature difference between the poles and equator, which drives our winds and circulates the air. But Jupiter's pole and equator share about the same temperature, at least near the cloud tops, so its winds are not just due to solar heating. An internal heat source most likely drives Jupiter's turbulent weather system from below.

Jupiter's storm clouds

Small eddies and larger vortices interrupt the smooth profile of the belts and zones on Jupiter with turbulent red, white and brown spots. These whirlpools mark patterns of weather, as the clouds billow, churn and seethe. Huge storms larger than the Earth in size swirl across the planet, while smaller eddies chase each other, whirling and rolling about. Unlike storms on Earth, underlying landmasses do not break up the storms on Jupiter. The planet's deep, windy atmosphere overlies a liquid hydrogen ocean.

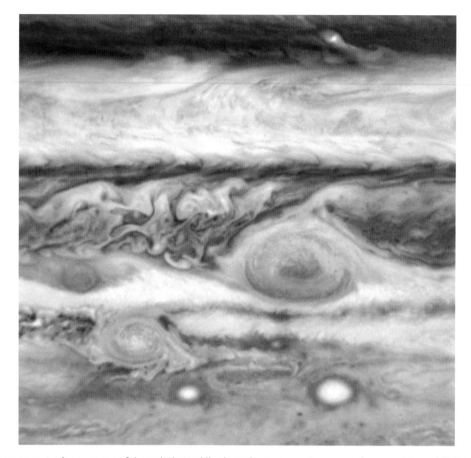

Fig. 9.3 New red spots on Jupiter   Some of the turbulent eddies in Jupiter's atmosphere move from west to east (*left to right*) above Jupiter's Great Red Spot (*center*), a giant high-pressure anticyclone that swirls in the opposite counter-clockwise direction. Some small eddies are sucked into the great red vortex, helping to sustain it, while other eddies roll around the perimeter, probably reinforcing the storm's circulation. Although the Great Red Spot has been observed for at least three centuries, a new red spot appeared in the spring of 2006, and in 2008 a white oval-shaped storm turned into a third, smaller red spot. The small red oval (*left*) will either be absorbed or repelled by the Great Red Spot when it moves into it, but the other, lower latitude red spot, which lies between the other two, will most likely pass the Great Red Spot unhindered. This visible-light image was taken on 9–10 May 2008 from the *Hubble Space Telescope*. (Courtesy of NASA/ESA/Mike Wong and Imke de Pater, U. C. Berkeley.)

The biggest whirlpools are visible from Earth using small telescopes of just 0.08-meters (3-inches) aperture, and have been observed for centuries. The earliest sightings of a large spot in the atmosphere of Jupiter have been credited to Robert Hooke (1635–1702) in 1664 and to Giovanni Domenico Cassini (1625–1712) the following year. The most prominent one – the Great Red Spot – appears in records and drawings dating back to 1831, and might have coincided with the earlier sightings.

Jupiter's famous Great Red Spot is essentially a huge weather system, with an east–west dimension greater than Earth's diameter. Because of rapidly increasing pressure with depth it cannot extend deeply into the planet. It is simply an enormous, shallow eddy trapped between counter-flowing jets, so large that the strong prevailing winds are forced to flow around it. The winds, in turn, funnel smaller eddies toward the Red Spot, helping to energize it.

Winds are swirling inside the awesome vortex in the counter-clockwise direction, at speeds up to 110 meters per second. Since it is in the southern hemisphere of Jupiter, this rotational direction indicates that the Red Spot is a high-pressure vortex, known as an anticyclone. A low-pressure cyclone would spin in the opposite direction.

The Great Red Spot is not unique, but just the largest of hundreds of different storms on Jupiter, including egg-shaped white ovals and smaller red spots that come and go or merge together (Fig. 9.3). At least at cloud-top levels, most of these long-lived vortices are high-pressure anticyclones that rotate counter-clockwise in the planet's southern hemisphere and clockwise in the northern hemisphere, with counter-flowing winds on their sides. Instead of wandering unpredictably like terrestrial hurricanes, the titanic whirlpools on Jupiter drift at a steady rate in either the eastward or westward direction, apparently rolling between and with the winds, like a giant ball-bearing. In contrast, storms that rotate in the other cyclonic direction on Jupiter are short-lived, lasting several days or less before being torn apart by the action of shearing winds. Lightning observed by the *Galileo* spacecraft was associated with these cloud systems.

One reason that the storms can last so long on Jupiter is that there is no solid surface directly below the clouds to interfere with the flow. The atmosphere has no distinct bottom, just shearing due to greater pressure. Thus, the weather pattern is free to flow in response to the giant planet's spin. A nearby solid surface would dissipate the energy of the storm clouds, as happens to hurricanes that make landfall on Earth.

But how can the biggest storms on Jupiter last for centuries within a constantly changing atmosphere? The large ovals and spots can survive by sucking in and engulfing smaller eddies that pass in their vicinity, like leaves in a whirlpool of water, consuming them and extracting their energy. The little short-lived storms feed their energy into the larger storms, just as a large fish eats smaller ones. And the food chain continues to the very top, with giant white ovals, each half the size of the Earth, occasionally merging together to become one.

The big storms engulf small ones, and the small storms probably pull up their energy from hotter, lower depths. As the moist, internal heat rises in the stormy updrafts, it can sustain the swirling clouds, supplying the energy that drives much of Jovian weather. Terrestrial weather systems can similarly include hot rising air as well as cool downdrafts, but the heat on Jupiter is generated from a completely different source – from deep in the planet and not from sunlight.

Cloud layers and colors on Jupiter

Radio signals can penetrate Jupiter's clouds and probe the planet's outer atmosphere, just as radio waves travel from a distant transmitter to your car radio on a cloudy day. Since the weight of overlying layers compresses the gas to greater density at lower depths, the radio signals experience more pronounced refractive alterations when passing through deeper regions of the Jovian atmosphere. These changes have been observed by monitoring homebound radio transmissions from *Voyager 1* and *2* when the spacecraft passed behind the planet, and they have been used to deduce the density, pressure and temperature as a function of altitude in and below the clouds.

If we could descend through Jupiter's thin cloud layer, we would find that the temperature and pressure increase with depth (Fig. 9.4). As in any planetary atmosphere, the atoms and molecules collide more frequently in the

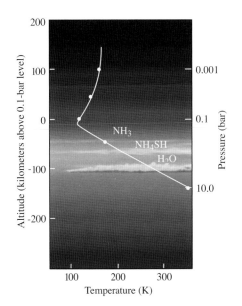

Fig. 9.4 Temperature and pressure at Jovian cloud levels   The fading radio signals when the *Voyager 1* and 2 spacecraft passed behind Jupiter in 1979 revealed the temperatures (*bottom axis*) and pressures (*right axis*) in its upper atmosphere. The temperature reaches a minimum of about 114 kelvin at a level called the tropopause where the atmospheric pressure is 0.1 bars, or 100 millibars. By way of comparison, the pressure of the Earth's atmosphere at sea level is 1.0 bar. The altitudes (*left axis*) are relative to the 0.1 bar level, and the dots are spaced to indicate tenfold changes in pressure. Solar radiation causes the temperature to increase with height just above the tropopause. At lower levels, the temperature and pressure increase systematically with depth. Three expected cloud layers of ammonia (NH_3), ammonium hydrosulfide (NH_4SH), and water ice (H_2O) are shown. The altitudes of the predicted cloud layers are based on a gaseous mixture that is of solar composition. An increase of the abundance of a condensable gas by a factor of 3 would lower the altitude of the cloud base by about 10 kilometers.

increasingly compressed, denser regions of Jupiter's atmosphere, so the pressure and temperature increase there. At the cloud-tops the temperature is a freezing 114 kelvin and the atmospheric pressure is about 0.1 bar, or one-tenth that of the Earth's air pressure at sea level. In slightly deeper layers, about 130 kilometers down, the temperature rises to a balmy 300 kelvin, well above the freezing point of water, at 273 kelvin. In these warmer regions, the pressure is comparable to the air pressure at the surface of the Earth, leading to speculations that living things might reside there (Focus 9.1). And above them it is cold enough to freeze various gases into ice to form the clouds.

Given the profile of temperature and pressure with altitude, it is possible to infer the altitude at which clouds of various types should form. The early calculations, initiated

Focus 9.1   Speculations about life in Jupiter's atmosphere

Once they established the temperatures and pressures in the outer atmosphere of Jupiter, scientists could speculate about the possibility of primitive life existing there. The outer atmosphere may be too cold for life to exist, for it would freeze to death. Deeper down inside the planet it is too hot to even allow organic molecules to exist; they would break apart into their constituent atoms. The molecular constituents of life might nevertheless survive in the region in between, where Earth-like temperatures and pressures exist, perhaps being synthesized from simpler molecules by the action of Jovian lightning.

But the warm part of Jupiter's extended atmosphere contains no solid surface on which primitive creatures could creep or crawl, and strong atmospheric currents would most likely either cycle them up into the frigid heights or drag them down to scalding depths. Heavy organisms would just sink down into the lethal heat. Some imaginative astronomers have therefore argued that buoyant inflated organisms could be floating in Jupiter's outer atmosphere, bobbing up and down like terrestrial jellyfish to seek more clement conditions without sinking too deeply into the planet.

Other researchers have argued that life on Jupiter is very implausible. They reason that biological compounds could not survive the harsh environment. This conclusion was reinforced when the *Galileo* spacecraft dropped a probe into the giant planet, showing that sophisticated organic molecules were not present at the entry site.

Nowadays, many scientists have forgotten about these early speculations about possible life inside Jupiter, and have turned their attention to Jupiter's moon Europa, which might have a life-sustaining ocean beneath its icy surface.

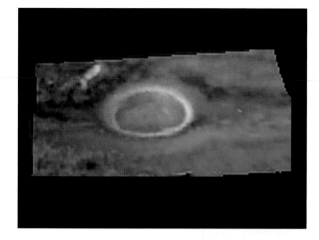

Fig. 9.5 Ammonia ice near Jupiter's Great Red Spot   A cloud of ammonia ice (*light blue*) is shown at the northwest (*upper left*) of the Great Red Spot (*middle*), inside its turbulent wake. The cloud was most likely produced by powerful updrafts of ammonia-laden air from deep within Jupiter's atmosphere. Reddish-orange areas show high-level clouds, yellow regions depict mid-level clouds, and green areas correspond to lower-level clouds. Darker areas are cloud-free regions. Light blue depicts regions of middle-to-high altitude ammonia ice clouds. This near-infrared image was taken on 26 June 1996 from the *Galileo* spacecraft. (Courtesy of NASA/JPL.)

by John S. Lewis (1941–) in 1969, assumed that the gas mixture is in chemical equilibrium and has a uniform composition like that of the Sun. And although there have been many refinements since then, three distinct cloud layers are always predicted. As one proceeds upward from the interior of Jupiter, the temperature and pressure fall to the point where the gases of water, ammonium hydrosulfide and ammonia are expected to condense to form clouds. Water clouds similarly form in the colder, higher parts of the Earth's atmosphere, which has only one layer of cloudy, stormy weather.

The visible cloud tops of Jupiter consist of ammonia ice crystals, which condense out of the atmosphere at the very low temperatures and pressures there (Fig. 9.5). They create graceful white clouds that probably make up the cold, light-colored zones observed from Earth. This is consistent with spectroscopic measurements of abundant ammonia at the cloud tops of Jupiter.

Below the ammonia layer, the models predict two other cloud layers. At a depth where the pressure is slightly higher, at about 2 bars, the ammonia combines with hydrogen sulfide to form clouds of ammonium hydrosulfide, which would smell something like rotten eggs if you could get near them. The lowest clouds are predicted to contain water ice, formed at a pressure of about 5 or 6 bars. The water clouds are obscured by the higher cloud cover of ammonia, and are almost never seen from outside the giant planet.

The layered cloud model does not explain Jupiter's highly colorful appearance. Ammonia, ammonium hydrosulfide and water form white ices, so all the expected clouds should be white. Scientists speculate that complex molecules formed by interaction of gases with solar ultraviolet radiation in the outer atmosphere might coalesce and grow to form brown and yellow smog particles. They could account for some of the colors found in the belts — and remind us of a bad smoggy day in Beijing, Los Angeles or Mexico City.

The origin of the red color in Jupiter's clouds is something of a mystery. It might arise when the chemical equilibrium is disturbed by something that energizes

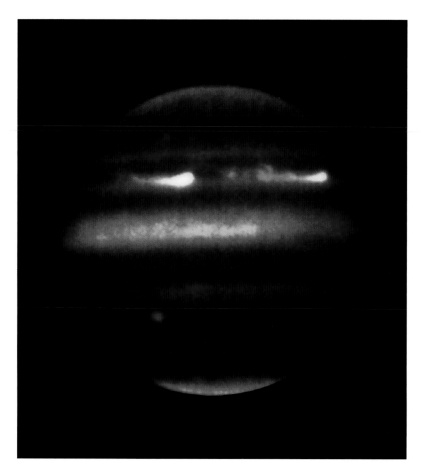

Fig. 9.6 Infrared storms on Jupiter The heat of two continent-sized storms (*top*) is detected in this infrared image. The bright plumes are attributed to storm systems triggered in Jupiter's deep clouds of water ice that move upward in the atmosphere and inject a fresh mixture of ammonia ice and water above the visible cloud tops. Models of the disturbance indicate that the associated jet stream extends deep in the atmosphere, at more than 100 kilometers below the cloud tops, and that the jet stream moves at a speed of about 600 kilometers per hour. This image was obtained from NASA's ground-based InfraRed Telescope Facility (IRTF) in Hawaii on 5 April 2007. (Courtesy of NASA/JPL/IRTF.)

an active chemistry. Lightning bolts, ultraviolet sunlight, high-speed particles, or extreme temperature variations might be responsible. One or more of these sources of energy probably breaks down molecules to produce coloring compounds of sulfur or phosphorus that are present in the atmosphere in only minute quantities. Alternatively, the red colors might be dredged up from greater depths by swirling storms, or perhaps attributed to rare organic, carbon-bearing compounds.

Lightning bolts in wet spots on Jupiter

Ancient mythology was close to the mark when it designated Jupiter as master of the rains, hurling thunderbolts at those who displeased him. Lightning flashes were discovered in *Voyager 1* and *2* images of the dark night-side of Jupiter, apparently illuminating the clouds in massive thunderstorms, and the lightning was confirmed by instruments aboard the *Galileo* spacecraft. Both missions showed that the lightning is concentrated near oppositely directed winds where storm clouds are found.

How deep the lightning occurs can be estimated from its size. The larger the flash, the deeper the lightning discharge. The observed sizes of Jupiter's lightning flashes suggest that they originate from layers in the atmosphere where water clouds are expected to form, at about

100 kilometers down. Only water could condense at these depths. When the *Galileo* cameras followed the night-side lightning sources into the dayside, they confirmed that the lightning originates in deep, moist clouds.

If the lightning bolts on Jupiter are similar to those on Earth, then they probably occur in water clouds where partially frozen water particles become electrically charged. The electrified rain or ice particles rise and fall in the turbulence, causing positive and negative charges to separate. A powerful discharge of current can then flow through the atmosphere, producing the lightning flashes.

So the lightning bolts on Jupiter most likely point to places where there are rapidly falling raindrops and quickly rising air currents. Infrared images of continent-size storms on Jupiter also indicate that they are triggered by upward-moving water clouds (Fig. 9.6). Such a moist convection might transport heat upward, carrying energy into the outer Jovian atmosphere. The circulating air seems to have been detected when *Galileo* sent a probe into the planet's clouds.

Plunging into a dry and windy spot on Jupiter

A pioneering descent into Jupiter's atmosphere took place on 7 December 1995 when an instrument-laden *Galileo Probe* was dropped from the *Galileo* spacecraft into the planet, taking the measure of its composition and winds to well below the visible clouds. The capsule returned data for just over an hour, down to the 20-bar pressure level, until the rising temperatures and crushing pressures wiped the probe out and it disappeared without a trace.

Scientists had expected that the *Galileo Probe* would pass through three cloud layers, composed of different chemicals that condense from tenuous gases at successively higher and colder levels, but contrary to expectations the clouds were not where everyone thought they would be. The probe's instruments saw almost no evidence for clouds. All of the expected cloud constituents were still in the gaseous state, and were found in increasing amounts through and well below the condensation levels where the cloud's ice particles should have been found.

Moreover, the planet was a lot drier than anticipated, at least in the vicinity of the probe-entry site, where the amount of water was five or ten times less than expected. Far less lightning was also detected during the probe's hour-long descent, supporting the conclusion that this part of the upper atmosphere contains little water. The missing clouds and water might be explained if the probe descended into an unusually clear hotspot of dry, down-welling air and reduced cloud cover (Fig. 9.7).

The previous models of Jupiter's clouds were probably too simplified, for they assumed a uniform layering with depth and ignored deep, vertical, up-and-down weather-related activity. Both wet and dry regions are found in the outer atmosphere of Jupiter, just as Earth has tropics and deserts, and they may be related to the circulation of rising and falling air. Winds that rise from the deep atmosphere could dredge up material that lacks water and other cloud-making ingredients. When these winds converge and drop back down, nothing is left to condense back into clouds and a dry clearing is created. Similar downdrafts occur over subtropical deserts on Earth, but, unlike our planet, Jupiter has no firm surface to quickly stop the air's fall.

One part of the weather forecast that proved correct below Jupiter's cloud tops was "windy". Instead of decreasing to a dead calm as the probe descended, the zonal winds stayed strong and even increased with depth. The zonal winds that create the planet's banded appearance continued to whip around the planet at speeds of up to 200 meters per second, until the capsule fell silent and stopped sending readings at about 600 kilometers down. Because little sunlight can penetrate to such depths, the winds must be driven mainly from below. Internal heat, probably left over from planetary formation or contraction of the planet as it slowly cools, is therefore the most likely driving force for Jupiter's powerful winds and ever-changing weather patterns.

Fig. 9.7 Hotspot on Jupiter   The dark region near the center of this image is an equatorial "hotspot", similar to the site where the *Galileo* spacecraft parachuted a probe into Jupiter's atmosphere in December 1995. Jupiter has many such regions, and they continually change, so the probe could not be targeted to either hit or avoid them. The dark hotspot is a clear gap in the clouds where infrared radiant energy from the planet's deep atmosphere shines through. Although hotter than the surrounding clouds, these so-called "hotspots" are still colder than the freezing temperature of water. Dry atmospheric gas may be converging and sinking in these regions, maintaining their cloud-free appearance. The bright ovals, shown in other parts of this image, may be examples of upwelling moist air. The images combined in this mosaic were taken on 17 December 1996 from the *Galileo* spacecraft. (Courtesy of NASA/JPL.)

Composition of Jupiter's upper atmosphere

When the *Galileo* spacecraft sent a probe into Jupiter, the relative amounts of several elements were measured in the planet's outermost atmosphere. The abundance of these ingredients has been compared to that of hydrogen, by far the most abundant element in Jupiter, and to the relative amounts found in the outer layers of the Sun (Table 9.2).

Helium, the second most abundant element in both Jupiter and the Sun, was found just a bit depleted from solar amounts, as had been suggested by previous measurements from the *Voyager 1* and *2* spacecraft. The *Galileo* results were more accurate, indicating somewhat more helium than the previous missions had, but still less than the outer layers of the Sun. Helium is apparently removed from Jupiter's outermost atmosphere, by helium raining into the interior of the planet. This slow helium-removal process operates on an awesome scale in neighboring Saturn, whose outer atmosphere is severely depleted in helium.

Heavier elements, such as carbon, nitrogen and sulfur, were enriched in the Jovian atmosphere by a factor of

Table 9.2 Element abundance in the outer layers of Jupiter and the Sun[a]

Element	Symbol	Chemical form	Jupiter	Sun	Jupiter/Sun
Helium	He	Helium	0.078	0.097	0.804
Carbon	C	Methane, CH_4	1.0×10^{-3}	3.6×10^{-4}	2.78
Nitrogen	N	Ammonia, NH_3	4.0×10^{-4}	1.1×10^{-4}	3.64
Oxygen	O	Water, H_2O	3.0×10^{-4}	8.5×10^{-4}	3.53
Sulfur	S	Hydrogen sulfide, H_2S	4.0×10^{-5}	1.6×10^{-5}	2.50
Deuterium	D	Deuterium	3.0×10^{-5}	3.0×10^{-5}	1.0
Neon	Ne	Neon	1.1×10^{-5}	1.1×10^{-4}	0.10
Argon	Ar	Argon	7.5×10^{-6}	3.0×10^{-6}	2.50
Krypton	Kr	Krypton	2.5×10^{-9}	9.2×10^{-10}	2.72
Xenon	Xe	Xenon	1.1×10^{-10}	4.4×10^{-11}	2.50

[a] Number of atoms per atom of hydrogen, designated by the symbol H.

about 3 when compared to a mixture of solar composition. This result had also been anticipated, for the carbon (C) and nitrogen (N) combine chemically with hydrogen (H) to form methane (CH_4) and ammonia (NH_3), and scientists have long known that Jupiter has about three times as much carbon and nitrogen in the form of these gases as the Sun. Still, even with this enrichment, hydrogen and helium comprise 99 percent of Jupiter's substance by volume.

The *Galileo Probe*'s instruments detected surprisingly high concentrations of argon, krypton and xenon. These three chemical elements are called noble gases, perhaps because they are very independent and do not combine with other chemical elements. They are enriched compared with solar composition by about the same factor as carbon, nitrogen and sulfur. In contrast, another noble element, neon, was starkly depleted with about ten times less than the solar amount.

These composition results are related to the formation and subsequent evolution of Jupiter. The apparent enrichment of elements like carbon, relative to light hydrogen, probably occurred when some of the light gas was blown out of the solar system by the active young Sun. In order to catch and retain the noble gases, Jupiter had to freeze them – which is not possible at the giant planet's current distance from the Sun. The noble gases might have been brought in from colder, more distant regions by comet-like bodies that helped build up young Jupiter, or the entire planet might have originated further away from the Sun and gradually migrated inward to where it is now.

The anomalous depletion of neon was even explained before its discovery. Scientists predicted that the neon would dissolve in helium, which rains down inside the planet and takes the neon with it.

9.3 Beneath Jupiter's clouds

Educated guesses about Jupiter's internal constitution

We cannot see beneath the clouds of Jupiter, but we can use external measurements to constrain its internal properties. As an example, we now know that the giant planet emits its own heat radiation, which means that it is hot inside. Since Jupiter and the Sun originated from similar material at the same time, a good initial assumption is that they have the same ingredients with similar proportions. The planet's low bulk density indicates that it is in fact composed largely of hydrogen and helium, just as the Sun is. Jupiter's oblate shape and rapid rotation also tell us something about the way it is constructed inside. Due to the enormous pressures inside Jupiter, most of the planet's hydrogen is compressed into a liquid metallic form, which has been created in the terrestrial laboratory and helps account for the giant's strong magnetic field. All of these constraints have been pieced together to make a picture of Jupiter's invisible interior.

Jupiter is an incandescent globe

With the advent of ground-based infrared measurements of the planets, pioneered by Frank J. Low (1933–2009) and his colleagues in the 1960s, astronomers were surprised to discover in 1969 that the giant planet is an incandescent globe with its own internal source of heat. This result was confirmed in greater detail with instruments aboard the *Voyager 1* and *2* spacecraft, which determined precisely how much infrared heat radiation was emerging from inside the planet. They showed that Jupiter is radiating

Focus 9.2   Stars that do not quite make it

Planets are supposed to shine only by reflected sunlight, while most stars generate their own radiation by thermonuclear reactions in their exceptionally hot cores. Jupiter has its own internal energy source, so it resembles a star in this respect, but unlike a star Jupiter has a relatively cold outer atmosphere. The planet shines by heat left over from its formation rather than nuclear fusion.

The process of gravitational contraction that warmed the inside of young Jupiter to its present central temperature of about 17 000 kelvin also heated the center of the Sun to about 15 million kelvin. That was hot enough to ignite the thermonuclear reactions that make our star shine. The Sun became much hotter inside because it is more massive, weighing in at 1000 times the mass of Jupiter. So Jupiter could have collapsed to form a star if it had more mass, and some astronomers therefore like to call Jupiter a star that didn't quite make it. In fact, calculations indicate that Jupiter might have become a star if it had been only 80 times more massive than it is now.

There are certain stellar objects, known as brown dwarfs, which also do not have enough mass to trigger nuclear fusion reactions in their core. They can shine faintly for about 100 million years as gravitational energy is converted into heat. Objects more massive than about 13 Jupiter masses, but not massive enough to ignite thermonuclear reactions, are thought to be brown dwarf stars, while those with less mass are classified as giant planets.

as much energy as it absorbs from the Sun. In contrast, a small planet like Earth would have radiated away the heat of its formation long ago. The giant planet's enormous size makes it a much better heat-trap than the terrestrial planets.

The Earth is now heated internally by the decay of radioactive elements contained in its rocks, but this has nothing to do with Jupiter's excess heat radiation. The giant planet contains relatively little rocky material, and the observed heat flow is 100 times too large to be explainable by radioactive heat production, even if Jupiter were composed almost entirely of terrestrial rocks rather than mostly hydrogen and helium.

Jupiter could also be shrinking slightly today, converting gravitational energy into heat and supplementing the leftover heat of its initial formation. But we cannot determine if that is happening. The giant planet needs to contract by only one meter per century to supply the observed amount of internal energy radiated by the planet, and astronomers could never measure its radius with that kind of precision.

Ingredients of Jupiter at formation

According to the widely accepted nebular hypothesis, the Sun and planets formed together during the collapse of a rotating interstellar cloud called the solar nebula. Most of it fell into the center, until it became hot enough to ignite the Sun's nuclear fires. Further out, the planets formed out of a whirling disk of the same material.

If the nebular hypothesis is correct, and the whole solar system originated at the same time, then you might expect Jupiter to have a similar chemical composition to the Sun. To a first approximation, the abundance of the elements in the giant planet does indeed mimic that of the Sun, with a predominance of the lightest element hydrogen. It is the most abundant element in most stars, in interstellar space, and in the entire Universe. The second most abundant element in both Jupiter and the Sun is helium, and hydrogen and helium together account for the low bulk density of both objects, at 1326 and 1409 kilograms per cubic meter respectively.

Spectroscopic observations indicate that the outer atmosphere of both Jupiter and the Sun also contain heavier elements, like carbon, nitrogen and oxygen. The giant planet has a bit more than the star, but even with this enrichment, heavy elements comprise less than 1 percent of the planet's composition by volume – all the rest is hydrogen and helium.

At the frigid temperatures where Jupiter and the other giant planets originated, the carbon (C), nitrogen (N), and

1.67 times as much energy as it receives from the Sun, and almost half of the total energy that Jupiter loses must come from its interior. That essentially meant that the planet had to be unexpectedly hot inside.

Jupiter must have been still hotter when it formed, thanks to the energy released during the gravitational collapse that accompanied its growth. When the newborn planet coalesced from a larger primordial cloud, gravitational energy was converted into heat as particles and small bodies fell inward and collided with each other. Such a process ignited the internal fires of the Sun and other stars, but Jupiter was not quite massive enough to become a star (Focus 9.2).

The compression inside Jupiter also excites the gas and leads to radiation that can carry off some of the energy. But so much heat is still left inside Jupiter since its time of formation that it is still cooling off, pumping out about twice

Table 9.3 Oblateness of the giant planets and the Earth[a]

Planet	Equatorial radius, R_e (km)	Polar radius, R_p (km)	Oblateness $(R_e - R_p)/R_e$
Earth	6 378.140	6 356.755	0.003 353
Jupiter	71 492	66 854	0.0649
Saturn	60 268	54 364	0.0980
Uranus	25 559	24 973	0.0229
Neptune	24 766	24 342	0.0171

[a] The radii are given in units of kilometers (km). The radii of the giant planets are those at the level where the atmospheric pressure is equal to one bar, the pressure of air at sea level on Earth.

Table 9.4 Range of pressures

Location	Relative pressure
Beneath the foot of a water-strider	0.000 01
Inside a light bulb	0.01
Earth's atmosphere at sea level	1.0
Inside a fully charged scuba tank	100.0
Deepest ocean trench	1 000.0
Pressure at which hydrogen becomes metallic	1 000 000.0
Center of Jupiter	70 000 000.0

oxygen (O) atoms would have bonded with the abundant hydrogen (H) to form methane (CH_4), ammonia (NH_3) and water (H_2O), respectively. These compounds are known as "ices" because they would have condensed into solid ice at the freezing temperatures far from the Sun. Rocky material, containing atoms of silicon and iron, would also have been present in lesser amounts. The ice and rock are now located at the center of Jupiter, probably because they coalesced to form a relatively massive nucleus that pulled in the surrounding hydrogen and helium, or perhaps because they settled gravitationally into the central core after the planet formed.

At the higher temperatures closer to the Sun, where the Earth and other terrestrial planets formed, the icy material would be vaporized and could not condense. That left only rocky substances to coalesce and merge together to form the terrestrial planets. Their modest mass and proximity to the Sun would not allow them to capture and retain the abundant lighter gases, hydrogen and helium, directly from the solar nebula.

An equatorial bulge

Observations with even a small telescope show that Jupiter is not a sphere. It has a perceptible bulge around its equatorial middle and is flattened at the poles. This elongated oblate shape is caused by Jupiter's rapid spin. The outward force of rotation opposes the inward gravitational force, and this reduces the pull of gravity in the direction of spin. Since this effect is most pronounced at the equator, and least at the poles, Jupiter has an oblate shape that is elongated along the equator. The same thing happens to all the giant planets, and even to the solid Earth (Table 9.3).

The amount of Jupiter's non-spherical extension depends on both its rate of rotation and the internal distribution of its material. The faster the spin, the more the outward push and the greater the elongation. And given its rotation, with the rapid period of 9.9249 hours, the size of the equatorial bulge depends on how Jupiter's mass is distributed inside. The more massive the planet's dense core, the smaller the equatorial bulge.

Scientists have measured the properties of Jupiter's equatorial bulge by accurately determining the motion of Jupiter's natural satellites, and closely tracking the *Pioneer 11* and *Voyager 1* and *2* spacecraft as they flew close to the giant. If Jupiter were a perfectly spherical planet, it would act as if all its mass was concentrated in a single central point and the motions of natural satellites or spacecraft would not depend on their orientation with respect to the planet's equator. In contrast, an oblate planet produces an extra force that tugs the moving object toward its equatorial bulge, and also toward any internal core. When combined with the known mass, volume and rotation rate of Jupiter, observations of these effects indicate that Jupiter has a dense core containing up to 12 Earth masses. Such a central object, presumably composed of non-gaseous rock and ice, was apparently required to initiate the accumulation of the giant planet's extensive hydrogen shell, which now compresses the core to high temperatures and pressures.

Enormous pressures and strange matter

The pressure inside Jupiter grows with depth, just as the pressure increases when you dive into the depths of an ocean. The particles at greater depths are compressed into smaller volumes by the weight of overlying material, so they collide with each other more frequently. This causes an increase in pressure at deeper levels inside the giant planet (Table 9.4).

Near Jupiter's cloud tops, the pressure is about the same as the air pressure at sea level on Earth, designated as 1 bar or 10^5 pascal, and the pressure increases to an

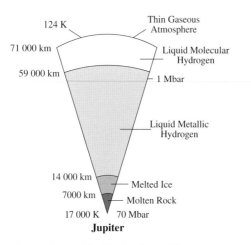

124 K — Thin Gaseous Atmosphere

71 000 km — Liquid Molecular Hydrogen

59 000 km — 1 Mbar

Liquid Metallic Hydrogen

14 000 km — Melted Ice

7000 km — Molten Rock

17 000 K — 70 Mbar

Jupiter

Fig. 9.8 Inside Jupiter   Giant Jupiter has a thin gaseous atmosphere covering a vast global ocean of liquid hydrogen. At the enormous pressures within Jupiter's interior, the abundant hydrogen is compressed into an outer shell of liquid molecular hydrogen and an inner shell of fluid metallic hydrogen. The giant planet probably has a relatively small core of melted ice and rock.

astonishing 70 million times that amount at the center of Jupiter. Higher pressures are associated with hotter temperatures, so the temperature also increases with depth inside Jupiter. Out at the one-bar level, the temperature is a freezing 165 kelvin. Deep down at the center, the temperature has risen to 17 000 kelvin, just over three times as hot as the visible disk of the Sun, at 5280 kelvin.

To understand the internal constitution of Jupiter, we need to know what happens to its most abundant ingredient, hydrogen, as the pressure increases. At the low pressure in the outer, visible parts of Jupiter, hydrogen forms a molecular gas, but this atmosphere is just a thin veneer. In proportion, the outer layer of gaseous hydrogen molecules resembles the skin of an apple. Deeper down, where the pressures and temperatures are higher, the hydrogen is liquefied. Indeed the planet is almost entirely liquid (Fig. 9.8). It is mostly just a vast, global drop of liquid hydrogen.

As suggested by Rupert Wildt (1905–1976) in 1938, the intense pressures and temperatures deep inside Jupiter will cause hydrogen molecules to break down, forming something he called "metallic hydrogen". Jupiter is indeed so massive that most of the fluid hydrogen inside is believed to be squeezed into metallic form. Below about one-seventh of Jupiter's radius down, the internal pressure exceeds 1 million bars (1 Mbar) and the liquid molecular hydrogen is transformed into liquid metallic hydrogen. It is said to be in a metallic state because, like a metal, it is an excellent conductor of electricity.

The hydrogen molecules are pushed so closely together that the electrons are squeezed free of any single atom or molecule. These mobile electrons can travel freely from one place to another, moving about and conducting electricity like the electrons in metals such as copper. But unlike a hard, shiny metal, the metallic hydrogen inside Jupiter has been melted at the high temperatures down there, and the molten metal behaves like a liquid rather than a solid.

Most of Jupiter is in the form of liquid metallic hydrogen. And underneath it all is a relatively small core of molten rock and ice. This relatively little core is up to 12 times heavier than the Earth.

The vast liquid shell of metallic hydrogen, which lies just outside Jupiter's core, is no longer just a theoretical conjecture. By re-creating extreme conditions similar to those inside Jupiter, modern laboratory experiments have turned liquid hydrogen into a metal. They have succeeded in pressurizing liquid hydrogen to 1.4 million bars and 3000 kelvin, squeezing the hydrogen into a liquid metallic state that conducts electricity just like any solid metal.

Electrical currents, driven by Jupiter's fast rotation within its liquid metallic shell, apparently generate the planet's strong magnetic field, in much the same way that currents in the Earth's molten metallic outer core produce our planet's magnetism. Jupiter's magnetic field is much more powerful than Earth's magnetism, with a magnetic moment that is 20 000 times as large and a cloud-top strength that is about 14 times Earth's surface magnetic field strength. The greater strength of Jupiter's magnetism could be attributed to the planet's faster rotation, its more extensive metallic region, and the relative proximity of the internal electrical currents to the cloud tops. By way of comparison, Earth's magnetic field is produced within a much smaller metallic core, which extends only halfway to the surface.

9.4 Introduction to the Galilean satellites

Galileo Galilei (1564–1643) discovered Jupiter's four largest moons in January 1610, using the newly invented telescope. They are bright enough to be seen in a pair of binoculars or a small telescope, and if it were not for the glare of Jupiter, these moons would be visible to the unaided eye.

These objects are now collectively called the Galilean satellites, even though Galileo wanted to name them after the Medici family of Firenze. They retain the individual names given to them by Simon Marius (1573–1624), also in 1610. In order of increasing distance from the giant planet, they are Io, Europa, Ganymede, and Callisto, all the names of mythological consorts of Zeus, the Greek equivalent of the Roman god Jupiter.

Table 9.5 Properties of the Galilean satellites[a]

Satellite	Distance from Jupiter center (Jovian radii)	Orbital period[b] (days)	Mean radius (km)	Mass (10^{22} kg)	Mean mass density (kg m^{-3})
Io	5.95 R_J	1.769	1821.61	8.930	3528
Europa	9.47 R_J	3.551	1560.8	4.799	3013
Ganymede	15.1 R_J	7.155	2631.2	14.815	1942
Callisto	26.6 R_J	16.69	2410.3	10.757	1834

[a] The mean distances from the center of Jupiter are in units of Jupiter's equatorial radius, $R_J = 71\,492$ kilometers. The radii are given in units of kilometers (km), the mass is given in kilograms (kg) and the mass density is in units of kilograms per cubic meter (kg/m^3). By way of comparison, the radius of our Moon is 1738 kilometers and the Moon's mean mass density is 3344 kg/m^3. The planet Mercury has a radius of 2439 kilometers and a mean density of 5430 kg/m^3.

[b] The orbital period of Europa is about twice that of Io, and the orbital period of Ganymede is nearly twice that of Europa.

Zeus changed the mortal Io, a beautiful river nymph, into a cow to hide her from his jealous wife. The Ionian Sea is named after the sea that Io the cow swam in during her wanderings. Europa, a Phoenician princess, bore Jupiter three sons, including Minos, the legendary ancestor of the Minoan civilization. Charlemagne subsequently named the continent which he had conquered "Europe", after the young lady. Ganymede was a beautiful Trojan boy, carried off by an eagle to be cupbearer of the gods. The nymph Callisto also conceived one of Zeus's sons, but Zeus's enraged wife turned her into a bear. Callisto and her son were placed in the heavens as the constellations Ursa Major and Ursa Minor, the big and little bears. Parts of these constellations are also known as the Big Dipper and the Little Dipper.

Many of the surface features on the Galilean satellites are named after persons or places in worldwide mythology, including those in the myths of Io, Europa, Ganymede and Callisto. Gods of fire and volcanoes from many cultures were also used for Io – this was changed after scientists realized the extent of the volcanism.

The Galilean satellites provided the first clear example of objects moving about a center other than the Earth, and for this reason they played an important role in the eventual acceptance of Copernicus' model of the solar system. They move in nearly circular orbits near Jupiter's equatorial plane with periods of days, revolving around the planet so quickly that their positions can be seen to change from hour to hour.

The first quantitative physical studies of these worlds became possible during the 19th century when Pierre Simon de Laplace (1749–1827) derived the satellite masses from their mutual gravitational perturbations. When large ground-based telescopes were constructed, astronomers could measure the sizes of these moons, and make approximate estimates of their mean mass densities accurate to roughly 10 percent. Precise determinations of these physical parameters became possible as the result of the *Voyager 1* and *2* flybys of Jupiter in 1979 and the *Galileo* orbital mission from 1995 to 2003 (Table 9.5).

When the two *Voyager* spacecraft flew past Jupiter in 1979, they got only a brief look at the Galilean satellites. However, it was time enough for their cameras to turn the four moons into astonishing places, including active volcanoes on Io, smooth ice plains on Europa, grooved terrain on Ganymede, and the crater-pocked surface of Callisto (Fig. 9.9). The incredible complexity and rich diversity of their surfaces, which rival those of the terrestrial planets, are only visible by close-up scrutiny from nearby spacecraft. Ground-based telescopes provide only a blurred view of the tiny, distant moons.

The volcano-ravaged surface of Io is being transformed before our very eyes, as spacecraft catch volcanoes in the act of erupting and watch lava flowing across its surface. Io must be at least partly molten inside to account for this rampant volcanism. The remarkably smooth surface of Europa, which is nearly devoid of large craters, also suggests a warm, active interior for this satellite. It has emitted material onto the surface that has erased crater-forming impacts in the recent past. An ocean of liquid water, or at least slushy ice, apparently exists at shallow depths beneath Europa's frozen surface, lubricating overlying ice rafts and oozing out into cracks in its icy covering.

The smaller and innermost of the Galilean satellites, Io and Europa, have the highest mass density, which is comparable to that of the rocks found on Earth. These

Fig. 9.9 The Galilean satellites   A composite of the four largest moons of Jupiter, which are known as the Galilean satellites. From left to right and increasing distance from Jupiter, the moons shown are Io, Europa, Ganymede, and Callisto. The images of Ganymede and Io are from the *Galileo* spacecraft, while those of Callisto and Europa are from the *Voyager 1* or *2* spacecraft. Io is subject to the strongest tidal stresses from the massive planet. These stresses generate internal heating, which is released at the surface and makes Io the most volcanically active body in our solar system. Europa appears to be strongly differentiated inside with a rock/iron core, a surface layer of bright water ice, and possibly local or global zones of liquid water between these layers. Tectonic resurfacing brightens terrain on the less active and partially differentiated moon Ganymede. Callisto, furthest from Jupiter, appears heavily cratered and shows no evidence of internal activity. (Courtesy of NASA/JPL/DLP.)

satellites are about the same size as our Moon, and have about the same mean mass density. In comparison, the larger, outer satellites Ganymede and Callisto are about the size of Mercury, but much less dense. The low mean mass densities indicate that they are composed in part of water ice. A mean mass density of 2000 kilograms per cubic meter would, for example, be explained if an object consists of half silicate rock and half water ice, with respective mass densities of 3000 and 1000 kilograms per cubic meter.

The compositions of the Galilean satellites were most likely affected by their relative proximity to Jupiter, in much the same way that the ingredients of the planets are related to their distances from the Sun. In both instances, the relatively small, dense objects are close to the center and larger, less dense objects are found farther out. It was probably too warm near the newborn Jupiter for water to condense, explaining why Io and Europa are largely composed of rock. Europa is now covered by ice, which may overlay an ocean of liquid water, but this blanket may be a relatively thin veneer. Recent models suggest that the Galilean moons grew gradually as material was added to a rotating disk surrounding young Jupiter, keeping the temperature low enough for ice in the region of Ganymede and Callisto.

In contrast, the relative cold of regions farther from Jupiter permitted Ganymede and Callisto to retain their ice and become mixtures of ice and rock. This would explain their low mass density, as well as the fact that they are more massive than Io and Europa.

The high reflectivity of the Galilean satellites, combined with the very cold temperatures at their remote distances from the Sun, has long suggested that ice might be present on their surfaces. In fact, Europa is so bright, and reflects so much incident sunlight, that it ought to be covered by pure water ice. Spectroscopic observations in the 1970s, using Earth-based telescopes at infrared wavelengths, indeed identified the expected water ice on the surfaces of Europa, Ganymede and Callisto. Although Io has the high reflectivity one might expect from an ice-covered sphere, the infrared observations failed to detect any signs of water ice. Instead, sulfur dioxide frost was identified on Io's surface by its spectroscopic signature.

Between 1995 and 2003, the orbiting *Galileo* spacecraft sharpened our view of Jupiter and its largest moons, providing captivating images of their diverse surfaces. The spacecraft also effectively looked beneath the surfaces of the four largest moons, inferring their interior structure from their gravitational forces and magnetic fields. It flew close enough to each satellite to measure small changes in the spacecraft's trajectory produced by each moon's gravitational forces, and this provided information about how the satellite's mass is distributed inside (Fig. 9.10). Io, Europa and Ganymede all have a large metallic core, a rocky silicate mantle, and an outer layer of either water ice, for Europa and Ganymede, or rock, for Io. In contrast, Callisto is a relatively uniform mixture of ice and rock.

Additional evidence for a metallic core inside Ganymede was provided by the *Galileo* spacecraft's

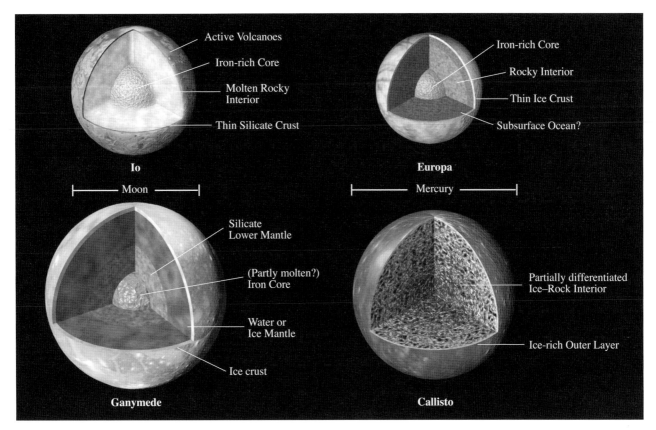

Fig. 9.10 Inside the Galilean satellites   Cutaway views of the possible internal structures of the Galilean satellites. Ganymede is at the lower left, Callisto at the lower right, Io on the upper left, and Europa on the upper right. The surfaces of the satellites are mosaics of images obtained in 1979 by the *Voyager* 1 or 2 spacecraft, and the interior characteristics are inferred from gravity field and magnetic field measurements from the *Galileo* spacecraft. With the exception of Callisto, all the satellites have iron-nickel cores surrounded by rock shells. Io's rock or silicate shell extends to the surface, while the rock layers of Ganymede and Europa are surrounded by shells of water ice or liquid water. Callisto seems to be a relatively uniform mixture of comparable amounts of ice and rock. *Galileo* images of Europa suggest that a liquid water ocean might now underlie a surface layer of water ice. (Courtesy of NASA/JPL.)

discovery of its intrinsic magnetic field, which ought to be generated within a massive molten core. The metallic cores of Io, Europa and Ganymede probably originated when the satellites were molten inside, the heavier material sinking toward the center and the lighter material rising toward the surface.

9.5 Jupiter's volcanic moon Io

Io: a world turned inside out

The innermost Galilean satellite, Io, has a radius and density that are nearly identical to those of our Moon, but contrary to expectation there are no impact craters on Io. The dramatic landscape is instead richly colored by hot, flowing lava and littered with the deposits of volcanic eruptions (Fig. 9.11). The active volcanoes emit a steady flow of lava that fills in and erases impact craters so fast that not a single one is left.

Io's volcanoes are literally turning the satellite inside out. Each volcano can churn out 100 cubic meters of lava every second, fast enough to fill an Olympic-sized swimming pool every minute, and the active volcanoes collectively provide 45 000 tons (45 million kilograms) of lava every second. They eject enough material to cover the satellite's surface to a depth of 100 meters in a short span of a million years or less. So all of the material that we now see on Io was probably deposited there less than a million years ago. Evidently Io's mantle and crust have been recycled many times over the span of Io's history.

Sulfur and sulfur dioxide give rise to Io's colorful appearance. Its red and yellow hues are attributed to different forms of sulfur, probably formed at different temperatures. Volcanic plumes of sulfur dioxide gas fall and freeze onto the surface, forming white deposits that were first detected by ground-based infrared spectroscopy in the 1970s.

Fig. 9.11 Jupiter's volcanically active moon Io   This composite mosaic shows Jupiter's moon Io in true colors, at approximately what the human eye would see. Black, brown, green, orange and red spots mark active volcanic centers. These volcanoes are turning the satellite inside out, continuously forming and re-forming its surface by lava flows. The absence of impact craters suggests that the entire surface is covered with new volcanic deposits much more rapidly than craters are created. This image was taken from the *Galileo* spacecraft on 3 July 1999. (Courtesy of NASA/JPL/U. Arizona.)

Whereas our Moon has been geologically inactive for eons, Io is the most volcanically active body in the solar system (Fig. 9.12). It exhibits gigantic lava flows, fuming lava lakes, and high-temperature eruptions that make Dante's *Inferno* seem like another day in paradise. Scientists estimate that Io has about 300 active volcanoes, and the hotspots of at least 100 of them have been observed.

The cameras aboard the *Voyager 1* spacecraft discovered nine active volcanoes during its flyby in 1979, and the most active volcanoes, such as Prometheus, Loki and Pele, were observed from the *Galileo* spacecraft two decades later (Fig. 9.13). Prometheus is the "Old Faithful" of Io's many volcanoes, remaining active every time it has been observed. Loki is the most powerful volcano in the solar

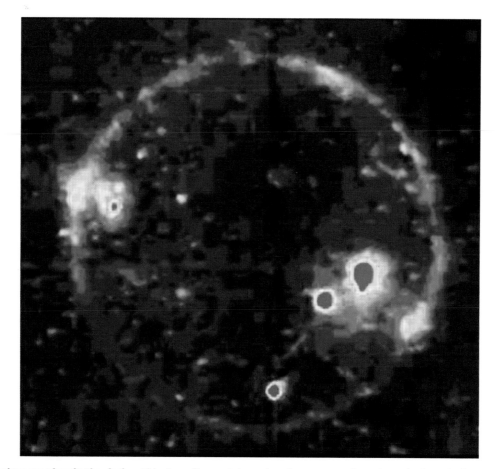

Fig. 9.12 Io's volcanoes glow in the dark This view of Io was taken when the moon was in Jupiter's shadow. The image is color coded so blue to yellow to red represents increasing brightness. The bright spots indicate the locations of volcanic vents on Io, which are spewing hot lava. The brightest spot (*right*) is from the volcanic caldera Pillan Patera with a temperature that exceeds 1700 kelvin, at least 200 kelvin hotter than the temperatures of the hottest volcanic eruptions on Earth today. The volcano Pele is just below and to the left of Pillan, Acaia is at the far left, and Svarog is near the bottom. This image was taken from the *Galileo* spacecraft in 1998. (Courtesy of NASA/JPL/U. Arizona.)

system, consistently putting out more heat than all of Earth's active volcanoes combined. And Pele (Fig. 9.14), the first volcano to be seen in eruption on Io, has repeated the performance for *Galileo* and the *Hubble Space Telescope*. More than 100 active volcanoes have been discovered on Io; many of them have been named for gods of fire, the Sun, thunder and lightning (Table 9.6).

Plumes of volcanic gas erupt from Io's active volcanic vents, rising up to 500 kilometers above the surface. They spread out in graceful, fountain-like trajectories, depositing circular rings of material about 1000 kilometers in diameter. Instruments aboard *Galileo* have practically smelled the hot, sulfurous breath of the eruptions, monitoring the sulfur dioxide gas as it rises, cools and falls. Diatomic sulfur, consisting of two sulfur atoms joined in pairs, has also been detected gushing out of the active volcanoes by instruments on the *Hubble Space Telescope*.

Although some of the kaleidoscopic colors on Io's surface are attributed to sulfur, the bulk of the lava is melted silicate rock. Instruments on the ground and in space have taken the temperature of the searing lava, showing that it sizzles at temperatures of 1700 to 2000 kelvin. That is hotter than any surface temperature of any planetary body in the solar system, even Venus at 735 kelvin. These temperatures rule out substances that melt at lower temperatures, such as liquid sulfur.

So the volcanic vents on Io must be spewing out melted rock, somewhat like terrestrial volcanoes whose lava is rich in iron, magnesium and calcium silicates, but much hotter than lava from Earth's volcanoes. The ubiquitous high-temperature volcanism on Io also has nothing to do with water, a common propulsive agent for some terrestrial volcanoes. There is no evidence that any water now exists on Io, and there may never have been any there. Io is close enough to Jupiter that heat received from the young planet during the satellite's formation may have kept water from condensing. And if any water managed to collect in Io, the interior heat would have probably boiled it away.

Fig. 9.13 Volcanic eruptions on Jupiter's moon Io A bluish plume rises about 140 kilometers over the bright limb, or edge, of Io (*left*), above the volcanic depression, or caldera, named Pillan Patera. In the middle of the image, near the night/day shadow line or terminator (*right*), the ring-shaped Prometheus plume is seen rising 72 kilometers above Io while casting a shadow to the right of the volcanic vent. Named after the Greek god who gave mortals fire, the Prometheus plume is visible in every image of Io from the *Galileo* spacecraft, including this one acquired in 1997, as well as every *Voyager* image of Io acquired 18 years before in 1979, suggesting continued activity from the same volcano for decades. (Courtesy of NASA/JPL/U. Arizona.)

Table 9.6 Major eruptive volcanic centers on Io

Name	Latitude (degrees)	Longitude (degrees)	Origin of name
Amirani	24.5N	114.7W	Georgian god of fire
Kanehekili	18.2S	33.6W	Hawaiian thunder god
Loki	18.2N	302.6W	Norse blacksmith, trickster god
Marduk	29.3S	209.7W	Sumero-Akkadian fire god
Masubi	49.6S	56.2W	Japanese fire god
Maui	19.5N	122.3W	Hawaiian demigod who sought fire
Pele	18.7S	255.3W	Hawaiian goddess of the volcano
Prometheus	1.3S	153.9W	Greek fire god
Surt	45.2N	336.5W	Icelandic volcano god (Surter)
Thor	39.2N	133.1W	Norse god of thunder
Volund	28.6N	172.5W	Germanic supreme smith of the gods
Zamana	18.4N	172.6W	Babylonian Sun, corn, and war god

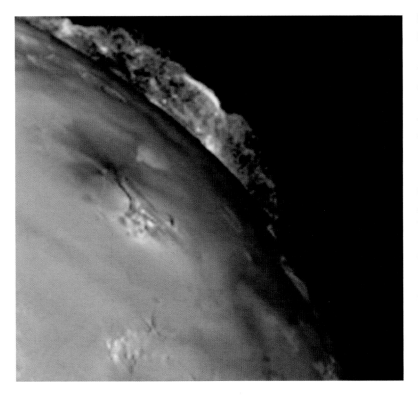

Fig. 9.14 Volcano Pele erupts on Jupiter's moon Io During its flyby on 4–5 March 1979, the *Voyager 1* spacecraft captured this image of an active volcano on Jupiter's energetic moon Io. The volcano has been named Pele, after the Hawaiian goddess of the volcano. Its erupting plume is visible at the upper right, rising to a height of about 300 kilometers above the surface in an umbrella-like shape. The plume has been ejected from the triangular-shaped blue and white complex of hills (*right center*). In this enhanced color image, we see the plume fallout as concentric brown and yellow rings, the largest stretching across 1400 kilometers and covering an area the size of Alaska. Pele remained active for at least two decades, when the *Galileo* spacecraft imaged new deposits from its plumes in the later 1990s. Large tidal distortions raised in Io by Jupiter heat the moon's interior, and the hot magma then expands, rises and forces its way out through volcanoes. (Courtesy of NASA/JPL/USGS.)

Io's tides of rock

What is keeping Io hot inside, warming up its interior, melting its rocks, and energizing its volcanoes? The heat released during the moon's formation and subsequent radioactive heating of its interior should have been lost to space long ago, just as our Moon has lost the internal heat of its youth and become an inert ball of rock. Unlike the Earth, whose volcanoes are energized by heat from radioactivity and friction due to mass motion, it is tidal distortions, created by Jupiter and its other moons, which sustain Io's molten state.

Just as the gravitational force of the Moon pulls on the Earth's oceans, raising tides of water, the gravitational force of massive Jupiter creates tides in the rocks of Io. Since the pull of gravity is greatest on the closest side to Jupiter, and least on the farthest side, Io's solid rocks are drawn into an elongated shape. But this tidal distortion does not melt the rocks by itself. If Io remained in a circular orbit, one side of the moon would always face Jupiter, its tidal bulges would not change in height, and no heat would be generated.

Shortly before the *Voyager* spacecraft encountered Io in 1979, Stanton Peale (1937–), Patrick Cassen (1940–), and Raymond Reynolds noticed that Io's orbit is slightly out of round, and predicted that the resultant tidal flexing of Io would cause "widespread and recurrent volcanism". The three Galilean satellites Io, Europa and Ganymede resonate with each other in a unique orbital dance, known as the Laplace resonance, in which Io moves four times around Jupiter for each time Europa completes two circuits and Ganymede one. This congruence allows small forces to accumulate into larger ones.

Although firmly gripped in Jupiter's gravity Io is also yanked in the opposite direction by the repetitive gravitational pulls of Europa and Ganymede. The resultant gravitational tug-of-war between Jupiter and the satellites distorts the circular orbits of all three moons into more elliptical ones. The effect is greatest for Io, which revolves nearest to Jupiter, but there is a noticeable consequence for Europa and perhaps even Ganymede.

During each lap around its slightly eccentric orbit, Io moves closer to Jupiter and further away, wobbling back and forth slightly as seen from Jupiter (Fig. 9.15). The strong gravitational forces of the planet squeeze and stretch Io rhythmically, as the solid body tides rise and fall. Friction caused by this flexing action heats the material in much the same way that a paper-clip heats up when rapidly bent back and forth. This tidal heating melts Io's interior rocks and produces volcanoes at its surface.

Magnetic connections between Jupiter and Io

Earth-based observations in the 1970s revealed a vast cloud of sodium atoms that envelops Io, forming an extended atmosphere that is nearly as big as Jupiter. The sodium cloud stretches backward and forward along Io's orbit, until the sodium atoms become ionized and no longer emit the light that makes them visible. The neutral, or un-ionized, sodium atoms have probably been chipped off the surface of Io by the persistent hail of high-energy particles found near the giant planet.

The volcanoes on Io provide the raw material for the satellite's tenuous atmosphere of sulfur dioxide (SO_2) that gathers above the erupting vents like localized umbrellas. The volcanic plumes are like fountains, with eruptions

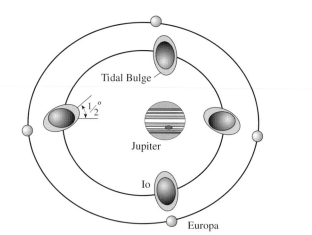

Fig. 9.15 Tidal flexing of Io   Due to an orbital resonance with nearby Europa, Jupiter's satellite Io has a non-circular orbit. The forced eccentricity makes Io travel at different speeds along its orbit, and the side facing Jupiter nods back and forth slightly, as seen from the planet. Although only half a degree in extent, this movement causes varying tidal forces inside the satellite, flexing it in and out like squeezing an exercise ball with your hand. This, in turn, generates internal friction and heat, leading to the active volcanoes seen on Io with instruments aboard the *Voyager 1* and *2* and *Galileo* spacecraft. In this drawing, Io's size and the eccentricity of its orbit are exaggerated when compared with Jupiter.

that arch gracefully back to Io's surface, and the gas is not propelled with sufficient velocity to escape the satellite's gravitational pull. Nevertheless, atoms of sulfur (S) and oxygen (O) can escape from Io once they are ionized by exposure to radiation from the Sun or from the hail of energetic particles in Io's vicinity. These ions have been detected from the *Voyager* and *Galileo* spacecraft by their ultraviolet glow.

Since charged particles cannot cross magnetic field lines, Jupiter's spinning magnetic field confines and directs the sulfur and oxygen ions into a doughnut-shaped ring known as the plasma torus (Fig. 9.16). As the giant planet rotates, it sweeps its magnetic field past Io, stripping off about a ton, or 1000 kilograms, of sulfur and oxygen ions every second. This material is lost from Io forever, and is continuously replenished by its volcanic activity, albeit indirectly through subsequent ionization.

Once coupled to the Jovian magnetic field, the sulfur and oxygen ions are accelerated to high velocity. Carried by the field, which is anchored inside Jupiter, the ions revolve around the planet once every 9.9249 hours, while Io orbits Jupiter in a more leisurely period of 42.48 hours. So the ions are always catching up with the satellite, and some of them slam into its surface, dislodging and energizing material and lifting it into the thin atmosphere.

As Jupiter's magnetic field sweeps past Io, it generates an enormous electrical potential of 400 000 volts, allowing

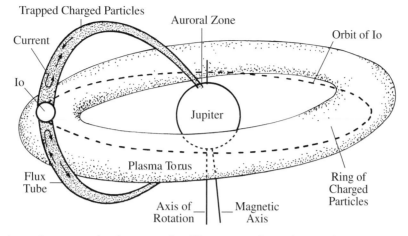

Fig. 9.16 Flux tube and plasma torus   An electric current of 5 million amperes flows along Io's flux tube. It connects Io to the upper atmosphere of Jupiter, like a giant umbilical cord. The plasma torus is centered near Io's orbit, and it is about as thick as Jupiter is wide. The torus is filled with energetic sulfur and oxygen ions that have a temperature of about 100 000 kelvin. Because the planet's rotational axis is tilted with respect to the magnetic axis, the orbit of the satellite Io (*dashed line*) is inclined to the plasma torus. Currents are generated as the plasma from the Io torus spreads into the vast, rotating magnetosphere of Jupiter, and these currents couple the moon to Jupiter's atmosphere where they stimulate a ring, or oval, of aurora emissions.

a powerful electric current of 5 million amperes to flow from Io to the poles of Jupiter and back again. The electrons move along Jupiter's magnetic field lines, within a magnetic flux tube that attaches the moon to its planet like a giant electromagnetic umbilical cord. Instruments aboard the *Galileo* spacecraft have detected beams of electrons flowing along the flux tube. They generate an awesome natural power of 2.5 trillion watts, vastly exceeding that of any terrestrial energy-generation plant.

When the electrons in this huge electrical circuit collide with the atoms in Io's tenuous atmosphere, they generate a dazzling light show of red, green and blue emissions (Fig. 9.17). And when the electrons are directed into the atmosphere of Jupiter, at the opposite end of the circuit, they trigger its bright aurora emissions, marking the glowing foot of the flux tube. Currents in this cosmic power-station also generate powerful bursts of radio noise, noticed since 1964, which are strongly controlled by Io's orbital position.

9.6 Jupiter's water moon Europa

Europa's bright, smooth icy complexion and young face

The smallest and yet brightest of the Galilean satellites, Europa, has a density comparable to that of rock, but its surface is as bright and white as ice. In fact, it is water ice! With surface temperatures of 110 kelvin or less, the water ice on Europa is frozen as hard and solid as granite.

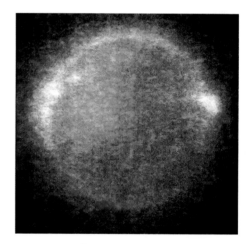

Fig. 9.17 Aurora on Jupiter's moon Io   The ghostly glow of aurora are detected in the atmosphere of Io when the satellite is in Jupiter's shadow. The aurora is produced by energetic particles, which are trapped in Jupiter's magnetic field and collide with Io's atmospheric gases, creating a red and green glow analogous to the aurora of Earth. Blue light is caused by dense volcanic plumes and may indicate regions that are electrically connected to Jupiter itself. (Courtesy of NASA/JPL/U. Arizona.)

Sunlight and charged particles cause some of the water ice to vaporize, and ultraviolet sunlight splits the molecules of water vapor into hydrogen and oxygen atoms. The hydrogen escapes into space, leaving behind a very tenuous atmosphere of oxygen.

Europa's surface is nearly devoid of impact craters, and there are no mountains or valleys on its bright smooth surface. No features extend as high as 100 meters, making Europa the smoothest moon or planet in the solar system.

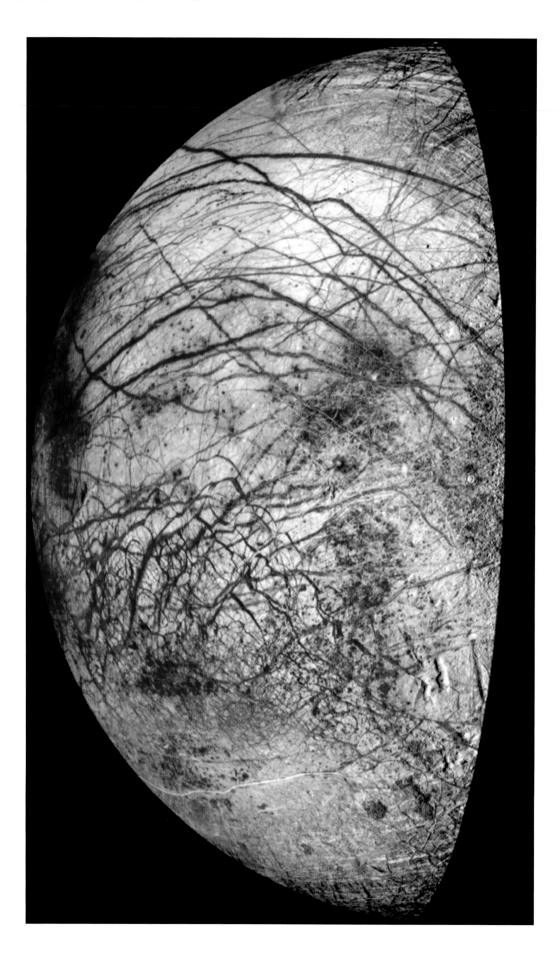

The paucity of cratered impact scars indicates that Europa has a comparatively young surface, showing few signs of age. Since its surface must have been accumulating impact craters as time goes on, Europa must have been resurfaced in recent times, geologically speaking, probably in the past few hundred million years. Whatever is keeping Europa smooth is doing it from beneath the frozen crust, as eruptions of liquid water to the surface or flows of soft water ice. But it is unknown if the eruptions or flows are still occurring.

Long cracks, ice rafts and dark places on Europa

A veined, spidery network of long dark streaks marks Europa's young face, suggesting great inner turmoil. The fine lines run for thousands of kilometers, intersecting in spider-web patterns (Fig. 9.18). They give Europa a broken appearance that resembles a cracked mirror or an automobile window that has been shattered in some colossal accident. The dark lines are most likely deep fractures formed when that part of the ice cracked open, separated, and filled with darker, warm material seeping and oozing up from below. Dirty liquid water or warm dark ice has apparently welled up and frozen in the long cracks, producing the lacework of dark streaks (Fig. 9.19). They provide evidence for a young and thin, cracked and ruptured ice shell, probably moving slowly over the top of an ocean that is 100 kilometers or more deep.

The surface of Europa is fragmented everywhere, as if pieces of ice have broken apart, drifted away and then frozen again in slightly different places (Fig. 9.20). Large blocks of ice have floated like rafts across the moon's surface, shifting away from one another like moving pieces of a jigsaw puzzle. Some of them are tilted; others rotated out of place, like plastic toys bobbing in a bathtub. This shows that the ice-rich crust has been or still is lubricated from below by either slushy ice or maybe even liquid water.

The size and geometry of the ice floes on Europa suggest that internal heat has melted the ice just a few kilometers below the surface, producing an immense ocean that is hidden beneath the moon's frozen crust. The warmth and currents have broken the thin crustal ice into pieces that slide over the underlying watery slush. They resemble disrupted pack ice seen on Earth's polar seas during springtime thaws. But the thaw on Europa is coming from heat below, not from sunlight above.

Explosive ice-spewing volcanoes and geysers may erupt from the buried seas, reshaping the chaotic surface of the frozen moon and leaving dark scars behind. Extended dark regions may, for example, have formed when the subsurface ocean melted through Europa's icy shell, exposing darker material underneath, or when upwelling blobs of dark, warm ice broke through the colder near-surface ice.

Europa's subsurface sea of melted ice

Tidal distortions could explain how water ice has melted in the frigid environment near Europa. The satellite has a slightly eccentric orbit due to gravitational interactions with Io and Ganymede, which respectively revolve closer and farther away from Jupiter than Europa. Over the course of one trip around Europa's elongated path, Jupiter's strong gravity stretches and compresses the satellite in a process called tidal flexing. Frictional heat associated with similar tidal flexing melted the rocks inside Io, and it operates on Europa as well – to a smaller extent since Europa is further from Jupiter. But the warmth generated by tidal heating may have been, and may still be, enough to soften or liquefy some portion of Europa's icy covering, perhaps sustaining a subsurface ocean of liquid water.

The tidal flexing that warms Europa's interior may also crack the blanket of ice that traps the liquid water below. The varying distance of Europa from Jupiter causes the tides in the underground sea to rise and fall as much as 30 meters. The pressure of this continual, rhythmic in-and-out motion probably cracks the brittle crust apart.

Fig. 9.18 Broken ice on Jupiter's moon Europa This enhanced-color image shows subtle differences in the materials that cover the icy surface of Europa. Reddish linear crack-like features (*top*) extend for a thousand kilometers. They are caused by the tides raised in Europa by the gravitational pull of Jupiter. As the moon travels along its eccentric orbit the tides vary and fracture the thin, icy crust. The fractures probably open and become filled with a dirty slush from a possible ocean below. Mottled, reddish "chaotic terrain" exists where the surface has been disrupted and ice blocks have moved around. The red material at the fractures and chaotic terrain is a non-ice contaminant and could be salts brought up from a possible ocean beneath Europa's frozen surface. Also visible are a few circular features, which are small impact craters. Europa's surface has very few craters, indicating that recent or current geologic activity has removed the traces of older impacts. The paucity of craters, coupled with other evidence such as the red material, has led astronomers to propose that there might be an ocean of liquid water beneath Europa's surface. This view combines images from the *Galileo* spacecraft taken in violet, green and near-infrared wavelengths in 1995 and 1998. (Courtesy of NASA/JPL/U. Arizona.)

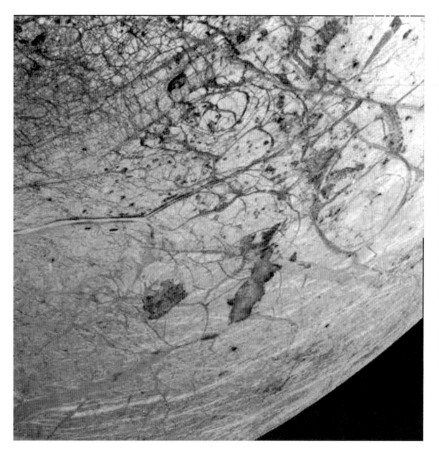

Fig. 9.19 Europa's frozen, disrupted surface   Old impact craters are not visible on Jupiter's moon Europa. They must have been erased, perhaps by fresh ice produced along cracks in the thin crust or by cold glacier-like flows. The number of impact craters found on the bright, smooth surface indicates an age of approximately 100 million years. The thin, water-ice crust has undergone extensive disruption from below (*upper left*). Two irregular, chaotic dark features (*just below center*) were most likely formed when liquid water or warm ice welled up from underneath Europa's icy shell. These dark spots, technically called macula, are named Thera and Thrace after two places in Greece where Cadmus stopped in his search for Europa. This image, approximately 675 kilometers across, was taken from the *Galileo* spacecraft on 20 February 1997. (Courtesy of NASA/JPL.)

As Europa moves along its elliptical orbit and approaches Jupiter, the planet pulls more strongly on the moon, stretching its surface and cracking it open. Then as Europa recedes from Jupiter the gravitational stresses stop. By the time they begin again, Europa has a slightly different orientation, and the cracks start in a different direction, resulting in long, looping cracks that snake across the moon's scalloped surface.

Magnetic measurements from the *Galileo* spacecraft provide more evidence for an other-worldly ocean inside Europa. The satellite's magnetism changes direction as Jupiter's magnetic field sweeps by in different orientations to the satellite, owing to the tilt between the planet's rotation axis and magnetic axis. This means that the magnetic field at Europa is not generated in a core, but is instead induced by the passage of Jupiter's field through an electrically conducting liquid, such as salt water, beneath the ice. Although this evidence for a subsurface liquid ocean is indirect, it is the only indication that buried water is there now, rather than in the geological past.

So it is highly likely that Europa had liquid water near its surface at one time, and it might still be there. Gravity data tell us that the water moved to the top long ago, within an outer layer about 100 kilometers thick, and the cracked surface, floating icebergs, and changing magnetic field provide strong circumstantial evidence for internal seas just below the crust of ice. If the liquid water is still there, we can stretch our imagination and speculate that life might reside within its lightless depths (Focus 9.3).

Focus 9.3   Life in Europa's ice-covered ocean

The possibility of liquid water just below Europa's surface has led to speculation that life could have gained a foothold there. Tidal flexing might make the internal seas warm enough, and they would be wet enough; there might even be organic molecules down there. A global sea of liquid water could seethe with alien microbes hidden beneath Europa's gleaming ice-covered surface. After all, we know that the heat, minerals and chemical energy of underwater volcanoes on Earth's sea-floors sustain life in the dark without sunlight. Foot-long tubeworms and giant white clams thrive near the hot vents beneath the Earth's oceans, and microorganisms even live inside these volcanoes. This does not mean that there *is* life inside Europa, and there is no direct evidence for it, but it is an interesting speculation.

Fig. 9.20 Jupiter's moon Europa under stress   This mosaic of the Conamara Chaos region on Europa indicates relatively recent resurfacing of the moon. Many sets of parallel and crosscutting ridges and fractures are detected. They are the frozen remnants of surface tension and compression, probably produced by heating and upwelling from below. The break-up and movement of the existing crust have formed irregularly shaped blocks of water ice. The blocks have been shifted, rotated, and even tipped and partially submerged within some lubricating material at the time of disruption, most likely soft ice or liquid water, or an ice-and-water slush below the surface. Some of the blocks have acted like rafts, separating and moving into new positions, somewhat like pack ice in the Earth's polar seas. The presence of young fractures cutting though this region indicates that the surface froze again after the resurfacing, forming solid, brittle ice. This image was taken from the *Galileo* spacecraft in February 1997. (Courtesy of NASA/JPL.)

9.7 Jupiter's battered moons, Ganymede and Callisto

Cratered, wrinkled Ganymede

Ganymede, the largest moon in the solar system, has a radius that exceeds that of the planet Mercury, although Ganymede is less dense and less massive than the planet. The satellite's mean mass density is so low that it must contain substantial quantities of liquid water or water ice. Like Europa, it is covered with ice. Ganymede's icy surface has experienced a violent history involving crustal fractures, mountain building and volcanoes of ice (Fig. 9.21).

Bright regions on Ganymede's surface contain sets of parallel ridges and valleys, termed grooved terrain, which looks like the swath of a giant's rake. The grooved terrain was most likely formed when the moon's water-ice crust expanded and stretched, cracking and rifting open as it was pulled apart. The crustal expansion might have happened when the satellite's rocks melted and moved into its interior while its water migrated to the top where it froze.

Sets of intersecting mountain ridges overlap and twist into each other. Some of the ridges cut across craters, while craters appear on other ridges (Fig. 9.22). Ganymede evidently experienced several epochs of mountain building. These crustal deformations may have continued for a billion years.

Water-ice volcanism played a role in creating the bright terrain on Ganymede. Prominent depressions were apparently flooded with liquid water or icy slush, and then froze

Fig. 9.21 Jupiter's large ice moon Ganymede   Dark patches of ancient terrain are apparently frozen in swaths of bright, younger and translucent ice on Jupiter's satellite Ganymede, the largest moon in the solar system. More recent impact craters have splashed fresh, bright ice across the surface of the ice-covered moon. This image was taken on 27 February 2007 from the *New Horizons* spacecraft on its way to an expected encounter with Pluto in July 2015. (Courtesy of NASA/JHUAPL/SRI.)

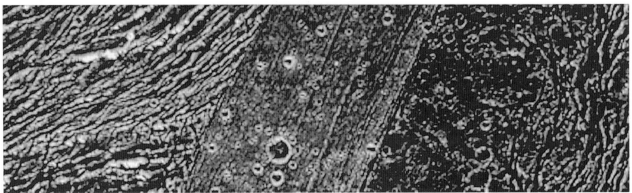

Fig. 9.22 Ganymede close up   The bright icy crust on Jupiter's moon Ganymede contains both young and old terrain with bright grooves, caused by internal stress, and craters due to external impact. The youngest terrain (*center*) is finely striated and relatively lightly cratered. The oldest terrain (*right*) is rolling and relatively heavily cratered. The highly deformed grooved terrain (*left*) is of intermediate age. This image, approximately 89 kilometers across, was taken from the *Galileo* spacecraft on 20 May 2000. (Courtesy of NASA/JPL.)

into bright smooth bands that now cover much of the moon. Craters found in these areas indicate that this also happened early in the satellite's history, at least a billion years ago.

Darker regions on Ganymede are older and more heavily cratered. Ancient, densely packed impact craters on parts of Ganymede testify to the great age of the terrain, dating back 3 or 4 billion years. Some of the large polygonal blocks rise about a kilometer above the bright, grooved terrain, and look as if they have moved sideways for tens of kilometers along the moon's surface.

Ganymede: a moon with its own magnetic field

One of the major surprises of the *Galileo* mission was the discovery that Ganymede has its own intrinsic magnetic field. The moon is generating a magnetic dipole similar to those of many of the major planets, and roughly a thousandth of the strength of Earth's magnetic field. No other satellite now has such a magnetic field, but our Moon might have had one in the distant past.

Currents stirred inside Ganymede's large, dense molten iron core may have produced its self-generated magnetic field, but a molten core would cool down and the inner flows might last for only a million years or so. Scientists have therefore speculated that the satellite's orbit has shifted over time, and that strong tidal flexing once heated its interior more than it does now. If the moon moved in a closer or more eccentric orbit in the past, Jupiter's gravity would have squeezed it in and out by greater amounts, heating it up inside and generating a strong magnetic field that lingers today. The wrinkled, grooved terrain on Ganymede's icy surface might record this earlier period of intense heating. The hot core might still be cooling off, with internal currents that generate the magnetism seen today.

Since Ganymede's magnetic field is nestled within Jupiter's stronger and more extensive one, the giant planet's magnetic field sweeps past the moon. Magnetic readings taken from *Galileo* record the shifting magnetism that is thereby induced in Ganymede. Although this induced magnetic signature is much weaker than the satellite's intrinsic dipolar field, it suggests that a thick internal layer of salty water cause the induced field. Electric currents coursing through Ganymede's internal shell of salt-water can also contribute to its intrinsic magnetic field.

Callisto: an ancient, battered world

Remotest of the Galilean moons, Callisto is the third biggest moon in the solar system, after Ganymede and Saturn's moon Titan. Due to its larger distance from Jupiter, Callisto has had a much more sedate and peaceful history than the other Galilean satellites, with little sign of internal activity. It is a primitive world whose surface of ice and rock is the most heavily cratered in the solar system (Figs. 9.23, 9.24). Unlike nearby Ganymede, the moon Callisto has no grooved terrain or lanes of bright material, and it exhibits no signs of icy volcanism. So, Callisto is a long dead world unaltered by resurfacing since it formed and ancient impacts molded its face, a fossil remnant of the origin of the planets and their moons. In fact, with a surface age of about 4 billion years, Callisto has the oldest landscape in the solar system.

Yet, when seen close up by the *Galileo* spacecraft, there are indications of subdued, youthful activity on Callisto's surface. It is blanketed nearly everywhere by fine, mobile dark material, interrupted only where bright crater rims poke up through it (Fig. 9.25). Small impact craters are mostly absent, and those that are found sometimes appear worn down and eroded (Fig. 9.26). Thus, the smaller craters seem to have been filled in and degraded over time, perhaps by the dark blanket of debris that might have been thrown out by the larger impacts. Ice flows may have alternatively deformed and leveled many craters, because ice, which is rigid to sharp impact, can flow gradually over long periods of time, as glaciers do on Earth. The lack of small craters on Callisto might also be explained if the ancient population of impacting objects near the remote satellite had relatively few small objects when compared to the population near the Moon and Mercury.

Perhaps because Callisto is farther from Jupiter than the other Galilean moons, its ingredients are somewhat separated but still largely mixed, like a half-baked potato that is hard on the inside and soft on the outside. Unlike the other three Galilean satellites, Callisto has a homogeneous interior, without a dense metallic core, and it has no magnetic field of its own. But Callisto does have a crust of ice that may cover a subsurface ocean of liquid water.

Like Europa and Ganymede, the battered Callisto has a variable magnetic field, apparently generated by electrical currents as Jupiter's powerful field sweeps by. A shell of liquid water can explain the internal conductivity if it has the salinity of terrestrial seawater, but it would have to be deep enough inside the moon that the water could not rise to the surface, keeping it unaltered.

Since it does not participate in the orbital push and pull of Io, Europa and Ganymede, tidal flexing by Jupiter has not kneaded or heated Callisto inside. So Callisto's internal ocean can only be heated by radioactive elements. The lack of tidal flexing may also help explain the unwrinkled nature of Callisto's pockmarked face.

9.8 Jupiter's mere wisp of a ring

The rings of Saturn were discovered in the 17th century. About three centuries later, in 1977, several faint and unsuspected narrow rings were discovered about the planet Uranus. Jupiter was next to join the group of ringed planets, but this time the discovery was not a complete surprise. In 1974, the *Pioneer 11* spacecraft had encountered an unexpected reduction in the amount of high-energy charged particles when the spacecraft passed near Jupiter. Not much was made of this anomaly, although some

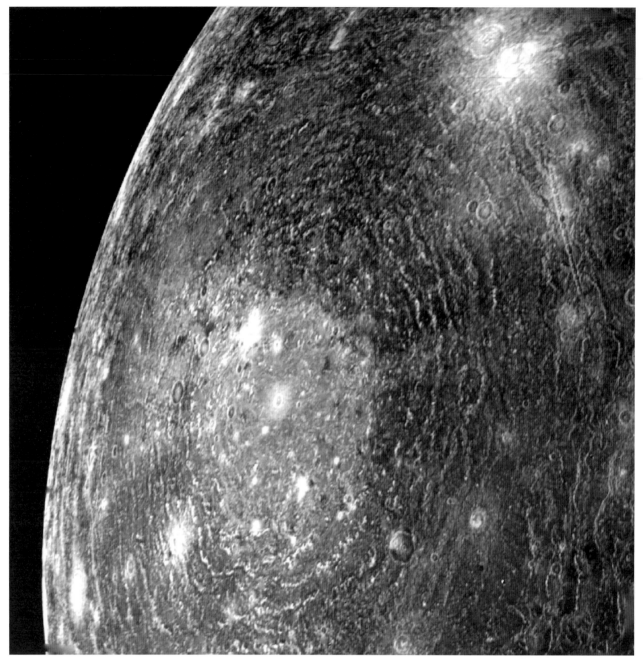

Fig. 9.23 Valhalla multi-ring basin on Jupiter's moon Callisto   The icy crust of Callisto is as rigid as steel, and it therefore retains the scars of an ancient bombardment by impacting meteorites. An exceptionally powerful impact produced a multi-ring basin, named Valhalla after the home of the Norse gods. The extensive system consists of a light-floored central basin some 300 kilometers in diameter, surrounded by at least eight concentric mountainous ridges, which resemble ripples produced on a pond by a rock striking the water. The impacting object on Callisto apparently punctured the surface and disappeared. Today only the frozen, ghostlike ripples remain. The great number of rings observed around this basin on Callisto is consistent with the moon's low density and probably low internal strength. Although there are very few large craters, the rest of Callisto is pockmarked with smaller impact craters that are flat for their size, and many of them have bright rims that resemble clean water-ice splashed upon the dirtier surface ice. This image was taken from the *Voyager 1* spacecraft on 6 March 1979. (Courtesy of NASA/JPL.)

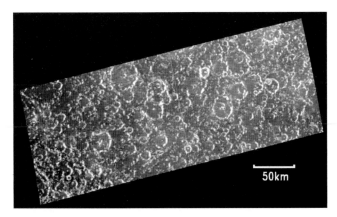

Fig. 9.24 Opposite the Valhalla impact on Callisto This close-up image was purposely taken on the opposite point, or antipode, of Callisto's huge Valhalla impact basin, created when an exceptionally large meteorite must have sent shocks into the moon's interior and waves rippling across the surface. The internal shocks ought to focus at the antipode, as they did for the Caloris impact on Mercury, creating a grooved and hilly terrain (also see Chapter 2, Fig. 2.23). The absence of such terrain at the Valhalla antipode suggests that Callisto has liquid water in its interior, which dispersed the seismic shocks. Magnetic field measurements also suggest that Callisto has a layer of liquid water deep below its surface. This image was taken from the *Galileo* spacecraft on 25 May 2001. (Courtesy of NASA/JPL/U. Arizona.)

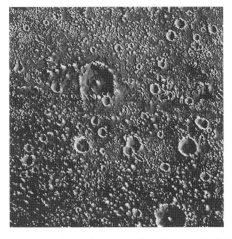

Fig. 9.25 Dark material and few small craters on Callisto A dark, mobile blanket of fine material covers Callisto's surface, sometimes collecting within crater walls. While Jupiter's moon Callisto is saturated with large impact craters, it has fewer very small craters when compared with the Earth's Moon and Mercury. One explanation is that the smaller craters have been filled by dark material that has moved down surface slopes. An alternative explanation for the paucity of little craters on Callisto is that there were fewer small impacting objects in its vicinity when compared with the amount within the inner solar system. This image, about 74 kilometers across, was taken from the *Galileo* spacecraft on 17 September 1997. (Courtesy of NASA/JPL.)

scientists thought that the falloff could be due to a previously unknown satellite or ring that blocked the energetic particles.

Finally, in 1979 – after much debate about the likelihood of finding a ring – a search was carried out with a camera on *Voyager 1*, and a narrow faint belt of material was found encircling the planet in its equatorial plane near the same distance that the energetic particles had disappeared. The ring was not previously observed from Earth because it was too faint and close to the bright planet. Since its discovery, Jupiter's main ring has been detected by Earth-based telescopes sensing infrared radiation, from the *Hubble Space Telescope*, and more fully explored by the inquisitive eyes of the *Galileo* spacecraft (Figs. 9.27, 9.28).

When the retreating *Voyager 1* camera looked back at the shadowed side of Jupiter, the main ring became brighter. The Sun was backlighting the ring's tiny particles, making them shine brightly. This behavior is typical of very small particles that scatter light in the forward direction, like tiny salt grains on the windshield of an automobile, the smoky haze in some movie theaters, or the condensation trails of airplanes.

The size of the ring particles can be inferred from the way they scatter light, and the conclusion is that they are a few millionths of a meter across or about the same size as flour dust or the grains of pollen. The particles that make up cigarette smoke and the hazes in the Earth's atmosphere have similar sizes. Numerous, larger ring particles would have reflected sunlight, making the rings appear brightest when approaching them, which did not happen.

Voyager 1 and *2* viewed a three-ring system around Jupiter, consisting of a flattened main ring, an inner extended cloud-like ring, called the halo, and a third outer ring, known as the gossamer ring because of its transparency. Observations from the *Galileo* spacecraft in 1996 and 1997 showed that the gossamer ring consists of two parts, one embedded in the other.

Jupiter's insubstantial ring system is practically made of nothing at all, no dustier than a typical living room. And the individual dust particles only reside temporarily in Jupiter's rings, just as the dust in the air of your room settles onto the room's furniture and bookshelves.

The dust particles in Jupiter's rings don't stay in the rings forever. They can last no more than a few thousand years before being tossed out of the ring plane or spiraling down into Jupiter's upper atmosphere and eventually disappearing. Given this short lifetime, the fine particles are probably replenished if the Jovian rings are permanent features.

Fig. 9.26 Structures within the Valhalla impact basin on Jupiter's moon Callisto This mosaic of two images shows an area within the Valhalla region on Jupiter's outermost large moon, Callisto. Numerous impact craters, ranging in size from 155 meters to 2.5 kilometers across, are seen in the mosaic, which is approximately 33 kilometers wide. There is an unexpected absence of small craters and the ragged rims of the existing craters suggest an erosion process, which is believed to be due to sublimation of volatiles from the surface ice. A prominent dark fault scarp, or cliff, and smaller parallel ridges are also pictured (*top right*). These images were obtained on 4 November 1996 from the *Galileo* spacecraft. (Courtesy of NASA/JPL/ASU.)

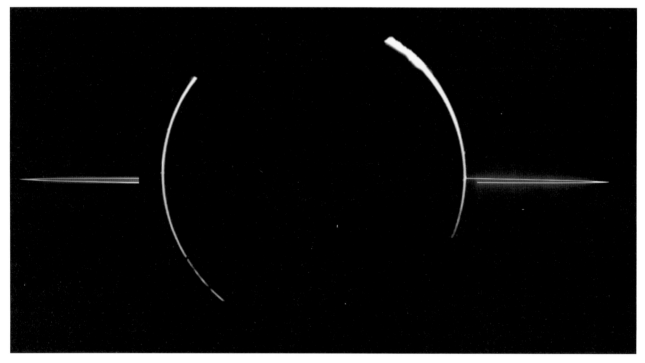

Fig. 9.27 Jupiter's rings The upper atmosphere of Jupiter and the planet's main ring can be seen when the Sun is behind the planet, and an imaging spacecraft is in Jupiter's shadow peering back toward the Sun. In such a configuration, very small dust-sized particles are accentuated so both the ring particles and the smallest particles in the upper atmosphere of Jupiter are highlighted. It is somewhat like looking back at a movie projector in a dusty theater or at a bright light in a smoky room, which permits you to see the dust or smoke in the air. The small particles in Jupiter's rings are believed to have human-scale lifetimes, and must be continuously replenished if the ring persists. In addition to the flat, main ring, Jupiter's ring system includes a toroidal halo interior to the main ring and a gossamer ring, which only become visible when images are overexposed with respect to the main ring. This image was taken on 9 November 1996 from the *Galileo* spacecraft. (Courtesy of NASA/JPL/Cornell U.)

Fig. 9.28 Jupiter's main ring and halo   Jupiter's bright, flat main ring (*bottom*) is a thin strand of material encircling the planet with an outer radius of 128 940 kilometers, or about 1.8 times Jupiter's radius, located very close to the orbit of the giant planet's small moon Adrastea, at 128 980 kilometers. The brightness of the main ring drops markedly very near the orbit of another moon, Metis. A faint mist of particles, known as the ring halo, surrounds the main ring and lies above and below it (*top*). The vertically extended halo is unusual for planetary rings, which are normally flattened into a thin plane by gravity and motion. The halo probably results from the "levitation" of small charged particles that are pushed out of the main ring plane by electromagnetic forces. These images were obtained from the *Galileo* spacecraft on 9 November 1996 when it was in Jupiter's shadow, looking back toward the Sun. The rings of Jupiter proved to be unexpectedly bright when seen with the Sun behind them, just as motes of dust or cigarette smoke brighten when they float in front of a light. A third gossamer ring, which consists of two components, is not shown here; it lies beyond the main ring, at greater distances from Jupiter. (Courtesy of NASA/JPL.)

They might be supplied by dust that is blasted off Jupiter's four small, innermost moons by interplanetary meteoroids, the fragments of comets and asteroids. The meteoroids are drawn in by Jupiter's very strong gravity, which also greatly increases their speed. When the high-velocity cosmic meteorites slam into one of the small inner moons, they create a dust cloud. The dust is thrown off at such high velocity that it escapes the moon's relatively small gravitational pull, orbiting Jupiter and contributing to one of its rings.

The outer edge of the main ring lies just inside the orbit of the tiny moon Adrastea, just 15 kilometers in size and too small to be seen from Earth. It was discovered by the *Voyager* spacecraft, as was another tiny moon, named Metis, which is embedded near the bright midpoint of the main ring. The dust generated by meteorite impact on Adastea and Metis can easily escape the small gravity of these moons, accounting for the dense accumulation of particles in the main ring. Some of the microscopic particles are small enough that if they are slightly charged then electromagnetic forces can overpower the effects of Jupiter's gravity, pumping them into the inner halo that is seen above and below the main ring.

The two much fainter, and wider, gossamer rings, which are more distant from Jupiter than the main ring, lie just inside the orbits of the small moons Amalthea and Thebe. Detailed observations from the *Galileo* spacecraft indicate that dust particles knocked off these two satellites feed the two gossamer rings, with a thickness that corresponds to each satellite's elevation above the planet's equatorial plane.

Jupiter has more than 63 moons, and most of them have been discovered using ground-based telescopes instead of by close-up spacecraft observations. Large charge-coupled device cameras and computers have been attached to the world's largest telescopes and programmed to detect anything that moves near Jupiter. The new moons are all small, between 1 and 10 kilometers across, and most of them move along retrograde orbits, in the opposite direction to the major satellites of Jupiter. The numerous small moons were probably captured by Jupiter, perhaps when its former, more extended atmosphere slowed the passing moons down enough to be held in the giant's gravitational embrace.

Altogether, Jupiter's dark rings are as wide as Saturn's yet invisible from terrestrial telescopes. Unlike Saturn's rings, which are made of bright, highly reflective

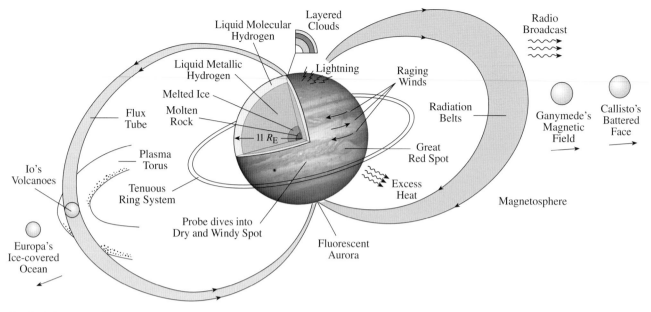

Fig. 9.29 Summary diagram

water ice, in pieces as large as houses, Jupiter's rings consist of fine dust grains, so dark that they reflect barely 5 percent of the sunlight that hits them. They are also spread so thin that Jupiter's rings are almost transparent. So this naturally brings us to Saturn, lord of the rings.

10 Saturn: lord of the rings

- Saturn has the lowest mass density of any planet in the solar system, low enough for the planet to float on water, and this means that Saturn is primarily composed of the lightest element, hydrogen.

- Saturn's rapid rotation has pushed its lightweight material out in the planet's middle, creating the most pronounced equatorial bulge of any planet.

- Saturn is just a great big liquid drop, covered by a thin atmosphere of gas, slightly smaller than Jupiter and less than a third of its mass.

- Liquid hydrogen is compressed inside Saturn's depths to form an electrically conducting, liquefied metal.

- There is no solid surface anywhere inside Saturn, though it might have a core of melted ice and molten rock that is about ten times as massive as the Earth.

- Saturn radiates almost twice as much energy as it receives from the Sun, and most of the planet's excess heat is generated by helium raining down into its liquid metallic hydrogen core.

- Saturn's rings are completely detached from the planet and separated from each other.

- The rings of Saturn are not solid, but instead composed of innumerable small water-ice particles and larger chunks of water ice.

- The icy constituents of Saturn's main A, B and C rings are as big as hailstones, snowballs and even icebergs; there are more smaller ones, but the big kind supply most of the ring mass.

- The total mass of the rings of Saturn is comparable to that of its medium-sized satellite Mimas, which is 396 kilometers across.

- Saturn has a retinue of diffuse, tenuous, and nearly transparent rings, designated the D, E, F and G rings, that are most likely composed of microscopic ice crystals, smaller than snowflakes and about the size of the dust in your room.

- Two small moons confine the edges of Saturn's narrow F ring and shepherd the ring particles between them.

- The icy material in the prominent rings of Saturn has been marshaled into thousands of individual ringlets, resembling ripples on a pond, but with circular, oval and even spiral shapes.

- Gravitational interaction with nearby external satellites can sculpt Saturn's ring material into numerous ringlets and produce waves in it.

- Small moons embedded within Saturn's rings can sweep out gaps, keeping them open and also sharpening their edges.

- The gaps within Saturn's rings are not completely empty; the Cassini Division contains about 100 ringlets.

- Enigmatic dark spokes stretch in the radial direction across Saturn's rings, moving at constant speed regardless of distance from the planet, in apparent violation of the laws of gravity.

- Saturn's dark ring spokes consist of microscopic dust-sized particles that may become electrically charged and levitated above the larger ring particles. They might then be swept around Saturn by its rotating magnetic fields.

- Planetary rings lie closer to a planet than its large satellites, within the Roche limit where the planet's tidal forces will rip a large satellite to pieces and prevent small bodies from coalescing to form a larger moon.

- The rings of Saturn could have formed when a moon was pulled toward the planet by tidal forces and eventually ripped apart.

- Small moons embedded in the rings of Saturn might sustain them.

- Saturn's relatively small moon Enceladus emits jets of ice particles, powdery snow, water vapor and organic compounds, which vent from warm fissures, known as tiger stripes, in the moon's south polar crust.

- The active jets on Enceladus suggest that tidal effects may make the moon hot inside. The tiger-stripe fractures could rub against each other, creating heat, or open to expose explosive ice to the vacuum of space. The interior ice might be melted into subsurface seas of liquid water containing organic chemical elements.

- Saturn's largest moon, Titan, is a planet-sized world with a substantial atmosphere whose surface pressure is about 1.5 times the air pressure at sea level on Earth.

- Titan's atmosphere is composed of 98.4 percent molecular nitrogen, nearly 1.6 percent methane, and trace amounts of other hydrocarbons; so nitrogen molecules are the main constituents of Titan's atmosphere, as they are in the Earth's air.

- Hazy smog, composed of complex organic molecules, envelops Titan's atmosphere and hides its surface from view.

- The *Huygens Probe* touched down on the surface of Titan on 14 January 2005, detecting methane rainfall and dark narrow riverbeds on the way down. The probe landed at equatorial latitudes, on a damp, moist riverbed littered with pebbles that were apparently rounded by flowing liquid.

- Radar pulses from the *Cassini* spacecraft in orbit about Saturn have seen through the haze that shrouds Titan, revealing long, deep, meandering channels on Titan's surface, which resemble terrestrial rivers but are attributed to flowing methane or ethane rather than water.

- The *Cassini* radar instrument has imaged dark, flat, smooth places with shore-like boundaries. They have been attributed to large lakes of liquid methane and ethane, and the spectral signatures of liquid ethane have been detected in at least one of them.

- Seasonal variations might account for the fact that about 20 more lakes were found at high northern latitudes on Titan than high southern ones. Clouds may rain methane during winter in the north, when southern lakes are evaporating in the local summer.

- Vast dunes accumulate near Titan's equator, shaped by strong winds blowing east to west. Unlike Earth's sand dunes, Titan's dunes are thought to be composed of organic material that has rained down from its smoggy skies.

- The dunes and lakes on Titan may contain hundreds of times more hydrocarbons than all the oil and gas reserves on Earth.

- Saturn has six mid-sized icy moons that retain impact craters dating back to their early history; some of them exhibit signs of internal activity and ice volcanism. Impacting objects almost broke the moons Mimas and Tethys apart.

- A number of unique small, irregularly shaped moons revolve around Saturn with remarkable orbits. The co-orbital moons have almost identical orbits, the Lagrangian moons share their orbit with a larger satellite, and the shepherd moons confine the edges of rings.

- Saturn's mid-sized moon Hyperion is so light that it must be about half-filled with empty spaces, and it tumbles chaotically along its orbit with no definite rotation period or orientation in space.

- The enigmatic moon Phoebe moves around Saturn in the opposite, retrograde direction to the planet's other mid-sized satellites. Phoebe has sharp-edged craters and a varying brightness that suggest thin, dark, surface deposits overlying bright water ice.

- An enormous, exceptionally distant, wide and diffuse ring of Saturn is apparently replenished by micrometeorite impacts with Phoebe; the ejected material might also move in toward Saturn, striking the next innermost moon Iapetus and accounting for the dark side of its two-faced surface.

10.1 Fundamentals

Majestic Saturn, the sixth planet from the Sun, was the most distant world known to the ancients, and it moved least rapidly around the zodiac. The Greeks identified the planet with Kronus, the father of Zeus, while the Romans named the planet Saturn after their god of sowing. Both the Greeks and the Romans associated Saturn with the ancient god of time, who later became Father Time.

You can see Saturn's oblong, golden disk with a small telescope, girdled by its beautiful rings, unattached to the globe. They set Saturn apart from all the other planets. Even though we now know that all four of the giant planets possess ring systems of some kind, Saturn's rings easily outclass the others.

Saturn's realm has been scrutinized in close-up detail from the *Voyager 1* and *2* flyby spacecraft, in 1980 and 1981

respectively, and from the orbiting *Cassini* spacecraft from 2004. The two *Voyagers* sent back marvelous images of Saturn's yellow-brown clouds, magnificent rings and large icy satellites (Fig. 10.1). The instruments aboard *Cassini* have provided captivating images as beautiful as a work of art (Fig. 10.2). The *Hubble Space Telescope* has also been returning high-resolution images of Saturn's rings and banded atmosphere, and ground-based telescopes have been monitoring their infrared heat radiation.

Saturn's orbital radius is 9.5 times the radius of the Earth's orbit, and it takes slightly more than 29 Earth years for Saturn to complete one revolution around the Sun. Perhaps because of its remote orbit and slow motion, the planet's name has been adopted for the word "saturnine", to describe a cool and distant temperament.

At its large distance from the Sun, the ringed planet and its satellites receive only about 1 percent as much sunlight

Fig. 10.1 Saturn's realm   The magnificent rings of Saturn encircle the planet, never touching its cloud tops. From the outside in, there are the bright A and B rings separated by the Cassini Division. The narrow Encke Gap in the outer A ring is also visible, as is the dark C ring nearest to the planet. The yellow-brown atmosphere of Saturn, shown here in enhanced color, has a banded structure, but it lacks Jupiter's bright zones and belts. Three icy satellites (Tethys, Dione and Rhea) are visible as small white spots against the darkness of space (*bottom left*), and another smaller satellite (Mimas) is visible just above them on Saturn's bright edge or limb. Small black round shadows cast by Mimas and Tethys are visible on Saturn's disk, and the planet blocks light from getting to the rings at the right just outside the planet. Because of its rapid spin, Saturn has an oblong, egg-like shape, flattened at the poles and extended at the equator. This *Voyager 2* image was acquired on 4 August 1981. (Courtesy of NASA/JPL.)

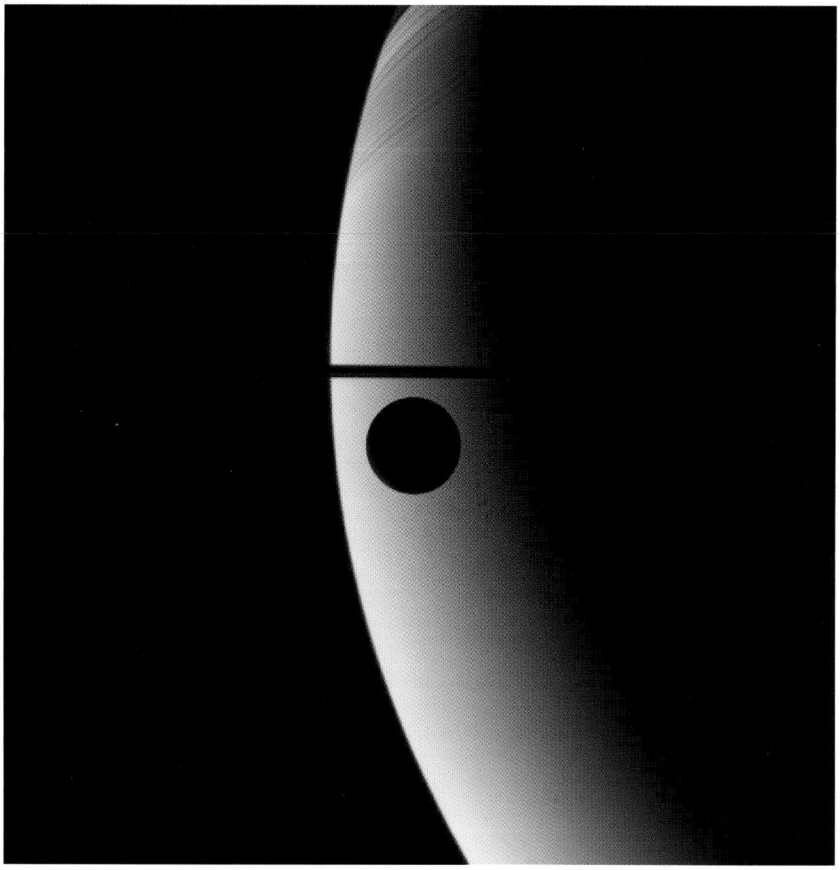

Fig. 10.2 Rhea passes in front of Saturn
The moon Rhea glides above the featureless, golden face of Saturn. This view looks down onto the unlit side of Saturn's normally impressive rings, which are visible here only as a thin line. This image was acquired from the *Cassini* spacecraft on 21 March 2006. (Courtesy of NASA/JPL/SSI.)

and solar heat as the Earth does. The surfaces of many of Saturn's satellites are therefore covered with water ice. And even though Saturn generates some of its own heat, its cloud tops have a temperature of only 95.0 kelvin.

Saturn is the second largest planet in the solar system, overshadowed only by Jupiter. The radius of Saturn, without the rings, is about four-fifths the radius of Jupiter and slightly more than 9 times the radius of the Earth.

The volume of Saturn is great enough to encompass 764 Earth-sized planets. But Saturn's mass is only 95 times greater than the Earth's mass, so the giant planet must be composed of material that is much lighter than rock and iron, the primary ingredients of the Earth.

From Saturn's mass and volume, we calculate its average mass density, or bulk density, to be only 687 kilograms per cubic meter, the lowest of any planet and less than that of liquid water. If Saturn were placed in a large enough ocean of water, it could float. It has a low average density because it is mainly composed of the lightest elements, hydrogen and helium, in the gaseous and liquid states.

Saturn rotates with a day of only 10.6562 hours – only 44 minutes longer than Jupiter's rotation period of 9.9249

hours. This is the rotation period of Saturn's magnetic field that is anchored inside the planet, and it is inferred from the observed periodic modulation in Saturn's radio emission, generated in the spinning magnetic fields. The visible clouds spin at different speeds, faster at the equator and slower at the poles. Like Jupiter, the rapid rotation has stretched Saturn's clouds around the planet in bands, but with a buttery color and weather patterns that are harder to see.

10.2 Winds and clouds on Saturn

Ferocious winds circle Saturn in jet streams that reach 500 meters per second, almost four times the speed of Jupiter's fastest winds and ten times hurricane force on the Earth. The dominant winds on Saturn blow eastward, in the same direction as the planetary rotation, at almost all latitudes, with the most powerful nearest to the equator. Reversals in wind direction are only found near Saturn's poles, where the clouds counter-flow in the eastward and westward direction. They form banded belts and zones similar to those observed almost everywhere on Jupiter. At each pole, the winds spiral downward in a high-speed vortex larger than the Earth, like the

Table 10.1 Physical properties of Saturn*a*

Mass	$5.683\,19 \times 10^{26}$ kilograms $= 95.184\,M_E$
Equatorial radius at one bar	$60\,268$ kilometers $= 9.46\,R_E$
Mean radius at one bar	$58\,232$ kilometers
Mean mass density	687.1 kilograms per cubic meter
Sidereal rotation period	10 hours 39 minutes 22.3 seconds $= 10.6562$ hours
Sidereal orbital period	29.4475 Earth years
Mean distance from Sun	1.4294×10^{12} meters $= 9.539$ AU
Age	4.6×10^9 years
Atmosphere	97 percent molecular hydrogen, 3 percent helium
Energy balance	1.79 ± 0.10
Effective temperature	95.0 kelvin
Temperature at one-bar level	134 kelvin
Central temperature	$13\,000$ kelvin
Magnetic dipole moment	$600\,D_E$
Equatorial magnetic field strength	0.22×10^{-9} tesla or $0.72\,B_E$

a The symbols M_E, R_E, D_P, B_E denote respectively the mass, radius, magnetic dipole moment and magnetic field strength of the Earth. One bar is equivalent to the atmospheric pressure at sea level on Earth. The energy balance is the ratio of total radiated energy to the total energy absorbed from sunlight, and the effective temperature is the temperature of a black body that would radiate the same amount of energy per unit area.

deep eye of a cosmic hurricane complete with multiple thunderstorms.

Despite its raging winds, Saturn lacks the dynamic and colorful storm clouds of Jupiter. Stormy weather on Saturn is apparently masked by an upper deck of dirty, smog-coated particles that give the planet a pastel, butterscotch hue at the visible wavelengths we see with our eyes. Jupiter, being warmer than Saturn, has less of this smoggy haze, and its cloud features are more distinct at visible wavelengths.

The clouds and hazes in Saturn's atmosphere are perceived in greater detail at ultraviolet and infrared wavelengths (Figs. 10.3, 10.4). Infrared images can also detect the heat radiated by the warmer parts of Saturn's atmosphere.

On rare occasions, a gigantic storm cloud of fresh, clean, white ammonia ice warms up, rises and punches through the opaque upper cloud deck, somewhat like the upwelling of warmer air in terrestrial thunderheads. The swirling, white equatorial ovals have been recorded by ground-based telescopic observers, but only a few times in the past two centuries. High-resolution images from the *Hubble Space Telescope* are clear enough to pick out the details of large storm systems and to record minor storms that look like bright cloud features. Observations from the *Cassini* spacecraft suggest that the rotating storms power the jet streams.

Instruments aboard *Cassini* have also detected intense ammonia-ice blizzards, apparently energized from below, where lightning has been observed and the clouds include water. The powerful storms produce lightning flashes that are as bright as the brightest ones seen on Earth, and the radio noise produced by Saturn's lightning bolts produce as much static at the spacecraft as terrestrial lightning causes in an AM radio on Earth.

When *Voyager 2* passed behind Saturn, its homebound radio signals penetrated the upper atmosphere, and alterations in these transmissions have been used to deduce the pressure and temperature below the clouds (Fig. 10.5). Because there is no solid surface directly below the clouds, altitudes are referred to the level in the atmosphere where the pressure is equal to 0.1 bars, or one-tenth the sea-level pressure on Earth. This is the approximate level where the temperature bottoms out, at about 82 kelvin, and the obscuring veil of haze may be formed.

Under the assumption that Saturn's gas mixture is in chemical equilibrium, with a uniform composition like that of the Sun, ammonia is expected to condense and form clouds at about 100 kilometers below the reference level, where the pressure has risen to about 1 bar. These clouds of ammonia ice presumably rise to form the bright, white storms that are occasionally seen above the global haze. Water clouds may form much lower in the atmosphere,

Fig. 10.3 Saturn at ultraviolet wavelengths   The southern hemisphere of Saturn reaches its maximum 27-degree tilt toward Earth, revealing the planet's banded clouds as well as its rings. Particles in Saturn's atmosphere and rings reflect different wavelengths of radiation in different ways, as indicated in this ultraviolet image that reveals smaller particles that do not absorb or scatter radiation at the longer visible or infrared wavelengths. Saturn's counter-flowing east-west winds have aligned the cloud particles, believed to be ammonia ice crystals, within fixed latitude bands. The rings are made up of particles of water-ice. This view of Saturn was recorded from the *Hubble Space Telescope* on 7 March 2003. (Courtesy of NASA/E. Karkoschka, U. Arizona.)

Fig. 10.4 Saturn at infrared wavelengths   This image was taken at infrared wavelengths that are sensitive to temperatures in Saturn's upper troposphere. The prominent "hotspot" at the bottom of the image is located at the planet's south pole, which is in summer, but at the temperature of only 91 kelvin it is still freezing cold. Warm regions at 88 to 89 kelvin are found above 70 degrees southern latitude. Ring particles are not at uniform temperature everywhere in their orbit around Saturn. They are cold just after having cooled down in Saturn's shadow (*lower left*). As they orbit Saturn, the particles increase in temperature to a maximum just before passing behind Saturn again in shadow. This image is a mosaic of 35 individual exposures made at the Keck I Observatory in Mauna Kea, Hawaii, on 4 February 2004; the small missing section is due to incomplete mosaic coverage during the observing sequence. (Courtesy of NASA/JPL.)

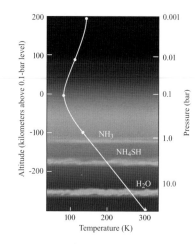

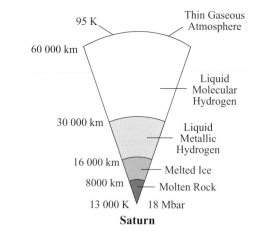

Fig. 10.5 Temperature and pressure at Saturn's cloud levels
The fading radio signals when the *Voyager* 1 and 2 spacecraft passed behind Saturn in 1980 and 1981, respectively, revealed the temperatures (*bottom axis*) and pressures (*right axis*) in its upper atmosphere. The temperature reaches a minimum of about 80 kelvin at a level called the tropopause where the atmospheric pressure is 0.1 bars, or 100 millibars. By way of comparison, the pressure of the Earth's atmosphere at sea level is 1.0 bar. The altitudes (*left axis*) are relative to the 0.1 bar level, and the dots are spaced to indicate tenfold changes in pressure. Solar radiation causes the temperature to increase with height just above the tropopause. At lower levels, the temperature and pressure increase systematically with depth. Three possible cloud layers of ammonia (NH_3), ammonium hydrosulfide (NH_4SH) and water-ice (H_2O) are shown. The altitudes of the predicted cloud layers are based on an equilibrium gaseous mixture that is of solar composition. An increase in abundance of a condensable gas by a factor of 3 would lower the altitude of the cloud base by about 10 kilometers.

Fig. 10.6 Inside Saturn   Giant Saturn has a thin gaseous atmosphere covering a vast global ocean of liquid hydrogen. At the enormous pressures within Saturn's interior, the abundant hydrogen is compressed into an outer shell of liquid molecular hydrogen and an inner shell of fluid metallic hydrogen. The giant planet may have a relatively small melted rock-ice core.

where the pressure rises to almost 10 bars, but no one has ever seen them.

10.3 Beneath Saturn's clouds

The internal constitution of Saturn

Saturn's low mass density indicates that the lightest element, hydrogen, is the main ingredient inside the planet, just as it is for Jupiter and the Sun. The lightweight material, just 68.7 percent as dense as water, is hurled outward in its equatorial regions by the planet's rapid 10.6562-hour rotation, making Saturn the most oblate planet in the solar system. Its equatorial bulge amounts to about 10 percent of the radius, and is about as big in extent as the Earth. Or, as some view it, the polar regions of Saturn are squashed and flattened by this amount.

The oblong shape of Saturn can be seen with a small telescope, and measured precisely from its satellite orbits and ring positions, as well as by the trajectories of the passing *Voyager* 1 and 2 spacecraft and the orbiting *Cassini* spacecraft. When these measurements are combined with Saturn's known mass, volume and rotation rate, scientists can obtain information about its internal distribution of mass.

The model the experts come up with is just a scaled-down version of Jupiter, with a small, dense core of melted ice and molten rock surrounded by a vast globe of liquid hydrogen and topped by a thin gaseous atmosphere (Fig. 10.6). Like Jupiter, giant Saturn is not a solid world, and is essentially a large, round drop of liquid.

Deep down inside, the liquid hydrogen is compressed to such high pressures that it conducts electricity like a metal. But since Saturn's mass is less than a third of the mass of Jupiter, and only slightly smaller, the internal pressure at a given depth is less, and the liquid hydrogen turns into a metal further down in the ringed planet. Saturn therefore has a smaller shell of liquid metallic hydrogen.

Also like Jupiter, the giant planet Saturn has a magnetic field generated by rotationally driven electric currents in the planet's liquid metallic shell. But since Saturn has a thinner shell, the strength of its magnetic field is about one-twentieth of Jupiter's, despite the fact that both planets rotate with about the same period. Auroras are produced in Saturn's polar regions when charged particles spiral down along the magnetic field lines and collide with gases in the atmosphere.

One unusual characteristic of Saturn's magnetism is that its magnetic axis is almost precisely aligned with its

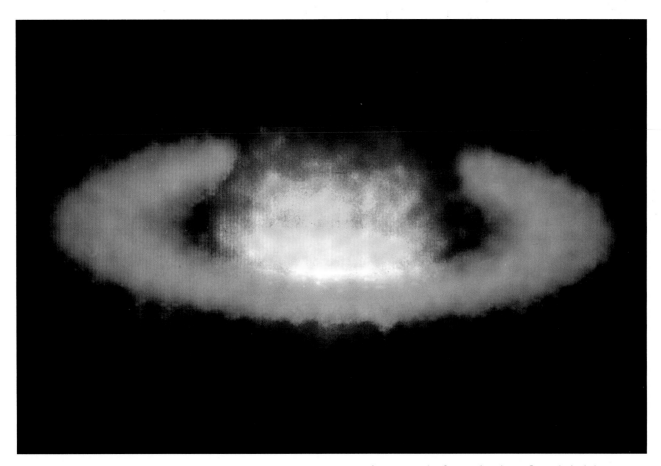

Fig. 10.7 Saturn and its rings at infrared wavelengths   Because Saturn's rings are made of water-ice, they reflect relatively large amounts of sunlight at an infrared wavelength of 3.8 micrometers (*blue*). Methane in Saturn's outer atmosphere absorbs radiation at this wavelength, but the incandescent globe has its own internal source of heat that makes it shine brightly at the longer infrared wavelength of 4.8 micrometers (*orange*). The heat welling up from within Saturn is partly due to helium raining inside of the planet. (Courtesy of David Allen, Anglo-Australian Telescope Board © 1983.)

rotation axis. No other planet has such an alignment of the two axes, and it is difficult to explain how the magnetic field can be maintained in such a way.

Interior heat and helium rain

Precise measurements from the *Voyager 1* and *2* spacecraft indicate that Saturn is radiating 1.79 times more energy in visible and infrared light than it absorbs from incoming sunlight. This excess energy must be coming from within the planet. It implies that Saturn, like Jupiter, is an incandescent globe with an internal source of heat (Fig. 10.7).

Both Jupiter and Saturn radiate almost twice as much energy as they receive from the Sun, but the dominant source of internal heat is different for the two giant planets. Jupiter's internal heat is primarily primordial heat liberated during the gravitational collapse when it was formed, and Saturn must have also started out hot inside as the result of its similar formation. But being somewhat smaller and less massive than Jupiter, the planet Saturn was not as hot in its beginning and has had time to cool. As a result, Saturn lost most of its primordial heat, and there must be another source for most of its internal heat.

Saturn's excess heat is generated by the precipitation of helium into its metallic hydrogen core. The heavier helium separates from the lighter hydrogen and drops toward the center, somewhat like the heavier ingredients of a salad dressing that hasn't been shaken for a while. Small helium droplets form where it is cool enough, precipitate or rain down, like water vapor condensing out of the Earth's atmosphere, and then dissolve at hotter deeper levels. As the helium at a higher level drizzles down through the surrounding hydrogen, the helium converts some of its energy to heat. In much the same way, raindrops on Earth become slightly warmer when they fall and strike the ground; their energy of motion – acquired from gravity – is converted to heat.

Fig. 10.8 Saturn's rings open up   These images were taken from the *Hubble Space Telescope* during a four-year period, from 1996 to 2000 (*left to right*), as Saturn moved along one-seventh of its 29-year journey around the Sun. As viewed from near the Earth, Saturn's rings open up from just past edge-on to nearly fully open as it moves through its seasons, from autumn towards winter in its northern hemisphere. (Courtesy of NASA/STScI.)

The helium-rain theory has apparently been confirmed by *Voyager* measurements of a lower abundance of helium in the outer atmosphere of Saturn than in Jupiter or the Sun. The number of helium molecules in Saturn is only 3 percent (and hydrogen 97 percent), while the number is 16 percent in the Sun and 13.6 percent for Jupiter. Since Jupiter is just slightly depleted in helium when compared to the Sun, helium rain is also probably operating inside this giant planet, but in more modest amounts because of Jupiter's greater mass and internal temperature.

At the higher temperatures and pressures found in Jupiter's interior, most of the liquid helium dissolves in liquid hydrogen, like cooking ingredients which dissolve more easily in hot liquids than cold ones. Inside Saturn, which has a lower internal temperature, some of the helium forms droplets instead of dissolving.

10.4 The remarkable rings of Saturn

Billions of whirling particles of water ice

The austerely beautiful rings of Saturn are so large and bright that we can see them with a small telescope. Because the glittering rings are tipped with respect to the ecliptic, the plane of the Earth's orbit about the Sun, they change their shape when viewed from the Earth (Fig. 10.8). The rings are successively seen edge on, when they can briefly vanish from sight in a small telescope, from below, when they are wide open, edge-on again and then from above. The complete cycle requires 29.4475 Earth years, the orbital period of Saturn, so the rings nearly vanish from sight every 15 years or so.

The three main rings of Saturn have been observed for centuries. They are the outer A ring and the central

Fig. 10.9 Saturn's rings of ice   The narrow-angle camera aboard the *Cassini* spacecraft took this image from beneath Saturn's ring plane on 21 June 2004. The brightest part of the ring system, extending from the upper right to the lower left, is the central B ring. It is separated from the outermost A ring by the wide, dark Cassini Division, discovered in 1675 by the Italian-born French astronomer Giovanni (Gian) Domenico (Jean Dominique) Cassini (1625–1712). Below the B ring, closer to the planet, is the C ring. All three rings are composed of innumerable particles of water-ice. The different shades of the rings are attributed to different amounts of contamination by other materials such as rock or carbon compounds. When viewed close up, the broad icy rings break up into thousands of individual wave-like ringlets. (Courtesy of NASA/JPL/SSI.)

B ring, separated by the dark Cassini Division, and an inner C, or crepe, ring that is more transparent than the other two. They remain suspended in space, unattached to Saturn, because they move around the planet at speeds that depend on their distance, opposing the pull of gravity.

The motions of Saturn's rings can be measured using spectral features in their reflected sunlight. When part of a ring moves toward or away from an observer, the spectral features are displaced in wavelength by an amount that depends on the velocity of motion. There is a shift toward shorter wavelengths when the motion is toward the observer, while motion away produces a shift toward longer wavelengths. Observations of this Doppler effect, by the American astronomer James Edward Keeler (1857–1900) in 1895, showed that the inner parts of the rings move around Saturn faster than the outer parts, all in accordance with Kepler's third law for small objects revolving about a massive, larger one. They orbit the planet with periods ranging from 5.8 hours for the inner edge of the C ring, to 14.3 hours for the outer edge of the more distant A ring. Since Saturn spins about its axis with a period of 10.6562 hours, the inner parts of the main rings orbit at a faster speed than the planet rotates, and the outer parts at a slower speed.

The difference in orbital motion between the inner and outer parts of the rings means that they are not a solid sheet of matter, for they would be torn apart by the differential motion. As demonstrated by James Clerk Maxwell (1831–1879) in 1867, the rings are instead made up of vast numbers of particles, each one in its own orbit around Saturn, like a tiny moon. Billions of ring particles revolve about the planet. They have been flattened and spread out to a thin, wide disk as the result of collisions between particles.

The rings of Saturn are flat, wide and incredibly thin. Measured from edge to edge, the three main rings span a total width of 62 200 kilometers, so they are a little wider than the planet's mean radius, at 60 228 kilometers (Fig. 10.9; Table 10.2). When observed edge on, from on or near the Earth, the rings practically disappear from view (Fig. 10.10). They look about a kilometer thick, but this thickness has been attributed to warping, ripples, embedded satellites and a thin, inclined outer ring. When instruments on *Voyager 2* monitored starlight passing through the rings, they found that the ring edges extend only about 10 meters from top to bottom. If a sheet of paper represents the thickness of Saturn's rings, then a scale model would be two kilometers across.

When sunlight hit the rings exactly edge-on, instruments on the *Cassini* spacecraft showed that there is a lot

Table 10.2 Saturn's rings[a]

Ring	Width (km)	Closest distance (km)	Distance range (R_S)	Particle size (m)	Optical depth	Mass (kg)
D	7540	66 970	1.11 to 1.235	$<10^{-6}$	0.0001	
C	17 490	74 510	1.235 to 1.525	0.01 to 3.0	0.05 to 0.35	1×10^{17}
B	25 580	92 000	1.525 to 1.949	0.01 to 5.0	0.4 to 2.5	2×10^{19}
A	14 610	122 170	1.949 to 2.025	0.01 to 7.5	0.05 to 0.15	4×10^{17}
F	50	14 180	2.324	10^{-7} to 10^{-5}	0.1	
G	500 to 3000	170 180	2.82	3×10^{-8}	2×10^{-6}	
E	302 000	181 000	3 to 8	1×10^{-6}	1.5×10^{-5}	7×10^{8}

[a] The ring widths and closest distances are given in kilometers (km), the particle size in meters (m), and the mass in kilograms (kg). The distance range is given in units of Saturn's apparent equatorial radius, $R_S = 60\,330$ kilometers. At the one-bar pressure level, the equatorial radius is 60 268 kilometers.

Fig. 10.10 Edge-on view of Saturn's rings   When the Earth is in the plane of Saturn's rings, an observer on the Earth views the rings edge on. Because the rings are so thin, they are then barely visible. Saturn's largest satellite, Titan, is seen just above the rings (*left*); it is enveloped in a dark brown haze and casts a dark shadow on Saturn's clouds. Four other moons are clustered near the other edge of Saturn's rings (*right*), appearing bright white because their surfaces are covered with water-ice. From left to right, these icy satellites are named Mimas, Tethys, Janus and Enceladus. This image was taken on 6 August 1995 from the *Hubble Space Telescope*. (Courtesy of NASA/STScI.)

of vertical relief above and below the paper-thin rings. Rippling corrugations undulate though the innermost D ring and the neighboring C ring, even into the B ring. Bumps and ridges pile up to unexpected heights. One ridge of icy ring particles loomed as high as 4 kilometers, apparently pulled up by a moon as it traveled through the plane of the rings. So the rings aren't perfectly flat, and depart from thinness in places.

What are the ring particles made out of? At visible wavelengths, the rings are bright and reflective, but at infrared wavelengths they are dark and less reflective. This suggests that the particles are cold and made of ice. In fact, they are composed largely, and almost exclusively, of water ice.

Detailed Earth-based infrared spectroscopy of the main rings in the 1970s showed that incident sunlight is absorbed by water ice at the surfaces of the particles. Subsequent spectral investigations indicated that the frozen water is exceptionally clean and pure, with negligible amounts of dust or rock. They are also poor absorbers and emitters of microwaves, which implies that more than 99 percent of the mass of the rings is water ice, and that less than 1 percent consists of dirty contaminants.

The total mass of the prominent A, B and C rings is about equal to that of Saturn's medium-sized satellite Mimas, which has a mean radius of 198 kilometers and weighs in at 3.749×10^{19} kilograms. Such a mass is consistent with particles composed of water ice. To check that, just multiply the mass density of water, at 1000 kilograms per cubic meter, by the total volume of the main rings – 10 meters thick, 60 000 kilometers wide, and a

Fig. 10.11 Particles in Saturn's rings   When the *Cassini* spacecraft passed behind the rings of Saturn, it sent three simultaneous radio signals, at 0.94, 3.6 and 13 centimeter wavelength, through the rings to Earth. The observed changes of each signal as the spacecraft moved behind the rings provided a profile of the distribution of ring material as a function of distance from Saturn, or an optical-depth profile. The image shown here was constructed from these profiles, depicting the observed ring structure at about 10 kilometers in resolution. Color is used to present information about the presence or absence of small ring particles in different regions based on the measured effects of the three radio signals. Purple color indicates regions where there is a lack of particles of size less than 5 centimeters. Green and blue shades indicate regions where there are particles smaller than 5 centimeters and 1 centimeter, respectively. The saturated broad white band is the densest region of the B ring, which blocked two of the three radio signals. From other evidence in the radio observations, all ring regions appear to be populated by a broad range of particle sizes that extend to several meters across. (Courtesy of NASA/JPL.)

circumference of about 600 000 kilometers. Since the particles are not jammed tightly together, and probably separated by 5 to 10 times their size, the resulting mass has to be diluted by a corresponding factor.

Typical chunks of ice in the main rings vary in size from hailstones to fist-sized chunks of ice; some are as small as snowflakes, and a few of the icebergs are as large as a house. In other words, the ring particles range in size from a hundredth of a meter to ten meters across (Table 10.2). There are more and more particles of smaller and smaller size within this range, so the main rings consist primarily of the smaller particles about 0.01 meters in size. Though far less numerous, the larger particles greater than 1 meter across contain most of the ring mass.

The ring particles are too small for spacecraft cameras to see individually, but scientists can infer their size from radio measurements. Since the rings are very reflective to ground-based radar transmissions, we know that their particles are comparable to, or larger than, the radar wavelength of about 0.1 meters. The particle-size distribution has been determined from the way the rings blocked the radio signals from *Cassini* when the spacecraft passed behind the rings (Fig. 10.11). This method showed that there are remarkably few particles either larger than 5 to 10 meters in size or smaller than 0.01 meters. Within these bounds, the number of particles in the main rings decreases with increasing size, in proportion to the inverse square of their radius.

However, four additional rings, designated the D, E, F and G rings (Fig. 10.12), consist of much smaller, microscopic ice crystals. These rings, discovered using ground-based or spacecraft observations, are all very

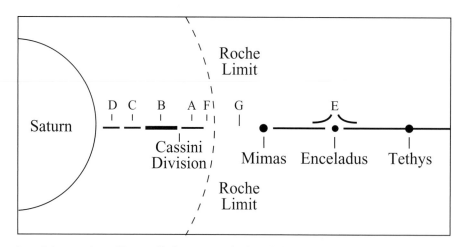

Fig. 10.12 Cross-section of rings and satellites   All of Saturn's main rings lie inside the Roche limit (*dashed curve*) within which the planet's gravity will tear a large satellite apart. The A and B rings have been observed for centuries. The more tenuous C ring was discovered in the 19th century, and definite observations of the transparent D ring awaited the arrival of the *Voyager 1* spacecraft on 12 November 1980. The icy satellite Enceladus feeds the tenuous E ring, also revealed from *Voyager 1*, as well as from the *Cassini* spacecraft. For clarity, the thickness of the rings has been exaggerated.

diffuse, tenuous and nearly transparent. The way that their particles scatter light indicates that they are the smallest of all, roughly a micron in size – a micron is a millionth of a meter in size, or 10^{-6} meters.

Lying between the C ring and the planet is the D ring. Although terrestrial observers had reported a faint ring between C and the globe as early as 1969, these reports remained controversial until the *Voyager* spacecraft definitely verified the existence of the D ring. It is so tenuous and transparent that it is probably impossible to see from the Earth using the best telescopes. According to one hypothesis, splintered chips from colliding ice particles drift down into Saturn's atmosphere and form the D ring.

Outside the traditional system lies the huge, sparse E ring, a broad tenuous band of small particles that is five times the combined breadth of rings A, B and C, and roughly centered on the orbit of Enceladus. Discovered with ground-based telescopes in 1966, the E ring becomes visible when the ring system is viewed approximately edge-on. It is composed almost exclusively of small grains just one micron in size.

Because they have relatively short lifetimes of several thousand years, these tiny particles must be continually replenished if the E ring is a permanent feature. As discussed in greater detail later in this chapter, watery eruptions from ice jets and geysers on Enceladus feed small bits of ice into Saturn's E ring. Once lofted into space, the pressure of sunlight, the gravitational tugs of Saturn, and possibly electromagnetic effects spread the particles out.

Initially discovered by instruments on the *Pioneer 11* spacecraft and verified in *Voyager* images, the tenuous G ring lies between the A and E rings. The G ring consists of similar micron-sized particles to the E ring, and may be renewed from small moons embedded within it.

The *Pioneer 11* instruments also discovered the incredibly narrow F ring, that lies just outside the A ring, by its absorption of energetic particles; while images from the *Voyager* spacecraft showed the F ring in great detail, demonstrating that its width varies from a few to tens of kilometers. Moreover, it is not just a single ring; *Voyager 1* spotted a contorted tangle of narrow strands that had smoothed out by the time *Voyager 2* arrived about 9 months later. Because the F ring particles are brighter when backlit by the Sun, and fainter in reflected sunlight, we know that the particles are also micron-sized, much smaller than snowflakes and comparable in size to the dust in your room.

But how can this ring retain such narrow boundaries? In the absence of other forces, collisions between ring particles should spread them out, causing the particles to fall inward toward Saturn and expand outward from it, thus creating a broader and more diffuse ring. Two tiny moons, named Pandora and Prometheus, flank the F ring and confine it between them, thereby keeping the particles of the F ring from straying beyond the ring's narrow confines (Fig. 10.13).

These shepherd satellites, discovered by the *Voyager* spacecraft, chase each other around the narrow F ring and keep it from spreading, as though they were two gravitational sheepdogs herding sheep into a narrow path. Each shepherd tends one edge of the ring. The moon outside the ring moves more slowly than the ring particles, which in turn are outpaced by the fast inner moon. The gravity of

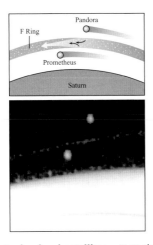

Fig. 10.13 Saturn's shepherd satellites Two shepherd satellites confine Saturn's narrow F ring. The outer shepherd gravitationally deflects ring particles inward, and the inner shepherd deflects ring particles outward. (Courtesy of NASA/JPL.)

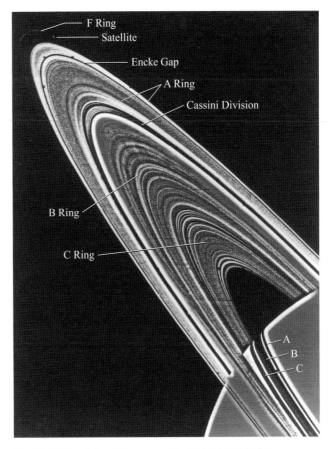

Fig. 10.14 Ringlets When viewed with high resolution, approximately 100 concentric features are seen within Saturn's rings, including some in the Cassini Division. The ring system would probably separate into countless ringlets if we could detect fine enough detail. A small satellite, discovered by *Voyager 1*, is seen (*upper left*) just outside the narrow F ring. The *Voyager 1* spacecraft took this mosaic of Saturn's rings on 6 November 1980. (Courtesy of NASA/JPL.)

the faster-moving inside satellite pulls the inner F ring particles forward as it passes, causing them to accelerate and spiral outward. The slower-moving outer shepherd exerts a net backward force on the outer ring particles, causing them to move inward. The result is a very narrow ring. Such confining moons were originally proposed to account for the narrow rings surrounding Uranus, constraining their edges from the otherwise inevitable spreading, and such a pair of satellites was eventually found astride one of its rings.

The shepherd satellites that flank the F ring, and possibly one or more embedded satellites as well, gravitationally interact with the ring material and distort its normal, circular ring shape, producing temporary kinks and twists in the clumpy, braided ring. Instruments aboard the *Cassini* spacecraft have revealed details of the interaction, showing that Prometheus can draw out material when passing closest to the F ring. Because their orbits are slightly eccentric, these two satellites produce a varying perturbation of ring particles, perhaps accounting for the changing smooth and contorted appearance of the F ring. Through similar gravitational interactions, small moons can produce ripples and waves on the surface of Saturn's main rings, and clear gaps within them.

Ringlets, waves, gaps and spokes

From a distance, the principal rings of Saturn look like smooth, continuous structures. Up close, however, from the views provided by the *Voyager 1* and *2* and *Cassini* spacecraft, the icy material is marshaled into thousands of individual ringlets (Fig. 10.14). Some of the ringlets are perfectly circular, others are oval-shaped and a few seem

to spiral in towards the planet like the grooves on an old-fashioned record. In some places, the flat plane of the rings is slightly corrugated, and ringlets are seen at the crests and dips of the corrugations, like ripples running across the surface of a pond.

An outside hand is at work sculpting at least some of the intricate ring structures through the force of gravity. The combined gravitational pull of Saturn and the accumulated pull of nearby moons can redistribute the ring particles, concentrating them into many of the observed shapes. Although small nearby moons have only a weak gravitational pull on the particles in the rings, the pull becomes pronounced when the particle orbital period and the satellite orbital period are exact fractions or whole-number ratios, such as $\frac{1}{2}$ for the 2:1 resonance, and the gravitational perturbations are repeated over and over again in the same resonant location. The interplay of this

Fig. 10.15 Two waves in Saturn's rings   Gaps and wave-like concentrations in ring particles are due to the gravitational influence of Saturn's moons. A small, nearby moon orbiting at varying distance from Saturn's rings is thought to produce waves of density, causing the ring particles to bunch together and disperse like the crests and troughs of ocean waves (*bottom left*). The gravity of a nearby moon can also produce a vertical wave in the ring particles, known as a bending wave (*upper right*). This image, which spans about 220 kilometers, was taken form the *Cassini* spacecraft on 29 October 2004. (Courtesy of NASA/JPL/SSI.)

effect and Saturn's inward gravitational pull can repel and attract the ring particles, pushing and pulling them into localized alternating high and low concentrations such as ringlets.

The additive gravitational perturbations between a nearby moon and the ring particles can also make waves of density propagate radially through the rings (Fig. 10.15). The particles tend to congregate at the crests and troughs of these density waves, like automobiles in traffic intersections or a crowd starting a "wave" in a stadium. Similar spiral density waves on a vastly greater scale are thought to create the stellar arms of spiral galaxies.

Resonance can also confine ring edges and clear gaps. Particles straying into a gap at a resonant location are removed by repetitive interaction with a particular moon. The ring particles at the outer edge of the B ring and the inner edge of the Cassini Division, for example, are traveling almost twice as fast as Saturn's largest inner satellite, Mimas, with a period one-half as large and occupying the 2 : 1 resonance with this moon. The razor-sharp outer edge and scalloped hem of the A ring similarly result from a 7 : 6 resonance with the external co-orbital satellites Janus and Epimetheus. In addition, many low-density regions in Saturn's A ring are located at positions resonant with small moons that lie just exterior to the ring.

Tiny satellites embedded within the rings can also sweep out a tidy gap with neat edges. The small moon Pan, embedded within Saturn's A ring, apparently plows its way through Encke's gap, keeping it open and creating wavy radial oscillations around the inner and outer edges of the gap.

But simple interactions with known moons have not been completely successful in accounting for all of the intricate detail found in Saturn's rings. The apparent gaps in the system are not completely empty. The Cassini Division, for example, contains perhaps 100 ringlets (Fig. 10.16), with particles just as large as those in the neighboring ring. Some gaps do not even occur at known resonant positions or contain detected moons embedded within them. Unseen moons might influence the clumping and removal of material in these locations.

Backlit images also reveal several faint rings of microscopic particles within the Cassini Division, overlying the orbits of small inner moons such as Janus and Epimetheus and Pallene, and ring arcs extending from other small moons along their orbits, like Anthe and Methone.

Perhaps the most bizarre *Voyager* discovery was the long, dark streaks, dubbed spokes, which stretch radially across the rings, keeping their shape like the spokes of a wheel. These ephemeral features are short-lived, but regenerated frequently. They are found near the densest part of the B ring that co-rotates with the planet at a period of 10.6562 hours. But the inner and outer parts of Saturn's dark spokes also whirl around the planet with this period, at constant speed in apparent violation of Kepler's third law and Newton's theory of gravity. If the spokes consisted of dark particles embedded in the rings, the particles would move with speeds that decrease with increasing distance from Saturn, and the spokes would quickly stretch out and disappear.

The exact mechanism for generating and sustaining the mysterious spokes remains obscure, but according to one hypothesis, the small dust particles may become charged, perhaps as the result of collisions with energetic electrons. Electromagnetic forces then raise or levitate the tiny, charged particles off the larger ring bodies, and the spokes are swept around Saturn by its rotating magnetic field. It sounds bizarre, but subtle forces are required to overcome gravity.

Why do planets have rings?

One might expect the particles of a ring to have accumulated long ago into larger satellites. But the interesting feature of rings – and a clue to their origin – is that they do not coexist with large moons. Planetary rings are usually closer to the planets than their large satellites.

The rings are normally confined to an inner zone where the planet's tidal forces would stretch a large satellite until

Fig. 10.16 Beneath the rings of Saturn   When the *Voyager 1* spacecraft dove beneath Saturn's rings, it could view sunlight transmitted through the rings, presenting a reversed image of the sunlit side. Both the C ring and Cassini Division appear bright because they are sparsely populated with small particles that efficiently scatter light in the forward direction, whereas the B ring appears dark because its densely packed particles absorb all the incident sunlight. This perspective is not available from Earth, where we always see the sunlit side of the rings. Two bright, icy satellites are seen at the upper center and upper left. (Courtesy of NASA/JPL.)

it fractured and split, while also preventing small bodies from coalescing to form a larger moon. The outer radius of this zone in which rings are found is called the Roche limit after the French mathematician Eduoard A. Roche (1820–1883), who described it in 1848 (Fig. 10.17).

For a rigid satellite with the same mass density as its planet, the Roche limit is 1.26 times the planet's radius (Focus 10.1). The Roche limit for a solid body is 1.38 times that radius, and Roche's initial calculation, for fluid objects, was 2.446 times the planetary radius. Anywhere inside this distance a large satellite can no longer remain intact, but instead gets torn apart by planetary tides. Nevertheless, because of their material strength and great internal cohesion, small moons less than 100 kilometers across can exist inside the Roche limit without being tidally disrupted, just as the ring particles can.

For a satellite with no internal strength and whose density is the same as the planet, the Roche limit is 2.446 times the planetary radius, or about 175 000 kilometers for Jupiter, 147 000 kilometers for Saturn, 62 000 kilometers for Uranus, and 59 000 kilometers for Neptune. Jupiter's

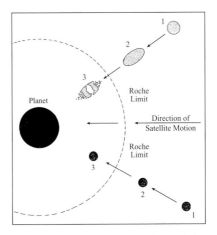

Fig. 10.17 The Roche limit   A large satellite (*top*) that moves well within a planet's Roche limit (*dashed curve*) will be torn apart by the tidal force of the planet's gravity. The side of the satellite closer to the planet feels a stronger gravitational pull than the side farther away, and this difference works against the self-gravitation that holds the body together. A small solid satellite (*bottom*) can resist tidal disruption because it has significant internal cohesion in addition to self-gravitation.

Focus 10.1 The Roche limit

To visualize the significance of the Roche limit, consider two particles of mass m, separated by a distance R, and located at a distance D from a planet of mass M_P. The gravitational pull of the planet on the particle closest to it will be greater than the pull on the more distant particle. If the difference in pull on the near and far particles, the tidal force, exceeds the mutual gravitational attraction between the two particles, they cannot remain close to each other and will disperse. The outcome of the tug-of-war between the tidal force and the mutual attraction is primarily decided by the particles' distance from the planet. At distances less than the Roche limit, D_{Roche}, particles are pulled apart, and this prevents the accumulation of larger moons. The tidal force will also tear apart any large moon-like object that ventures within the Roche limit.

The gravitational force F_P of a planet of mass M_P on a smaller mass m, whose center is located at a distance D from the center of the planet, is

$$F_P = \frac{G M_P m}{D^2}$$

where G is the Newtonian constant of gravitation. The planet will pull harder on the side of the object that is closer to it and less hard on the side that is further away. The difference ΔF between the force felt by one side and the center of the mass m is

$$\Delta F = \frac{G M_P m}{2} \left[\frac{1}{(D + R_m)^2} - \frac{1}{(D - R_m)^2} \right]$$

$$= \frac{2 G M_P m}{D^3} R_m$$

where R_m is the radius of the smaller object. If it approaches the planet, D becomes smaller and this tidal force will increase, eventually pulling the object apart at a critical distance D_{Roche} from the center of the planet.

The gravitational binding force F_B, which attracts the extreme sides of the object and holds it together, is $G m / R_m^2$ per unit mass, or for the total mass

$$F_B = \frac{G m^2}{R_m^2}$$

The Roche limit is reached when the tidal disruptive force, ΔF, equals the binding force, F_B, and when we set these two expressions equal and collect terms we obtain

$$D_{\text{Roche}} = \left(\frac{2 M_P}{m} \right)^{1/3} R_m$$

This result is expressed in terms of R_m, the radius of the small object, but by using the mass densities ρ_P and ρ_m, with $M_P = 4\pi \rho_P R_P^3 / 3$ and $m = 4\pi \rho R_m^3 / 3$, we obtain the Roche limit in terms of the planet radius, R_P:

$$D_{\text{Roche}} = \left(\frac{2 \rho_P}{\rho_m} \right)^{1/3} R_P$$

which for a planet and smaller object of the same mass density becomes

$$D_{\text{Roche}} = 1.26 \, R_P$$

The calculation by Roche used liquid objects whose shapes can distort continuously, and his result is

$$D_{\text{Roche}} = 2.446 \left(\frac{\rho_P}{\rho_m} \right)^{1/3} R_P \approx 2.446 \, R_P$$

mere wisp of a ring, the icy snowballs of Saturn's rings, and the dark boulders in the narrow rings encircling Uranus and Neptune all lie within the Roche limit for the relevant planet. The Earth's Roche limit is 18 470 kilometers, and if our Moon ever ventured within this distance from the Earth's center, it would be pulled apart by tidal forces and our planet would have rings.

And where did Saturn's rings come from? There are two possible explanations for their origin. In the first explanation, the rings consist of material left over from Saturn's birth about 4.6 billion years ago. This hypothesis assumes that the rings and moons originated at the same time in a flattened disk of gas and dust with large, newborn Saturn at the center. According to the second explanation, a

former moon or some other body moved too close to Saturn and was torn into shreds by the giant planet's tidal forces, making the rings. In this case, the rings could have formed after Saturn, its satellites and much of the rest of the solar system.

It has long been thought that Saturn's rings and satellites are both primordial leftovers of the planet formation process. Any disk material initially within the Roche limit soon after Saturn formed would have been prevented by the giant planet's tidal stresses from gathering or accreting into a large satellite. But outside the Roche limit, satellites could have coalesced from smaller bodies. Thus, any primordial, circumplanetary disk should develop into rings near the planet and exterior satellites, as we see today around Saturn. This could also help explain why only the giant planets have rings and a retinue of satellites, while the rocky terrestrial planets have no rings and either no satellite or just one or two of them. But there is a recent controversy over this long-standing explanation, centered on the fact that the planetary rings might be relatively young and therefore cannot be as ancient and enduring as the planets themselves.

Data obtained from the *Voyager* missions suggested that the shimmering rings of Saturn are ephemeral, created about 100 million years ago when a huge moon or comet came too close to the planet and shattered into pieces. The dazzling, sparkling brightness of Saturn's rings suggested such relatively young rings. They glisten with clean particles of pure water ice, unsullied by the constant pelting of cosmic dust. The rings could look much darker if they were very old, just as new-fallen snow becomes dirty over time.

The gravitational tugs of Saturn's moons on the rings will shorten the lives of the rings, providing another argument for their youth. When setting up density waves in the rings, nearby moons extract momentum from the ring particles, causing them to slowly spiral toward Saturn. To conserve momentum in the overall system, the moons gradually move away from the planet.

But subsequent data obtained from the *Cassini* spacecraft provided a different perspective, suggesting that some of the material in the rings is very long-lived, staying near the planet for billions of years. Ultraviolet light reflected off and passing through the water-ice particles indicated three times the mass assumed from the *Voyager* observations. This provides more raw materials for continuously recycling the ring material.

In this picture, small moons may be endlessly shattered into ring particles and gathered together, reforming. A larger ring mass indicates that the rings clump, rather than being uniformly distributed. There is a competition between the clumping of the ring particles due to gravitational interaction and collisions by micrometeorites, which shatter and disperse these clumps. The two processes can go on for several billion years.

So, it was once imagined that Saturn's resplendent rings formed with the planet about 4.6 billion years ago and the rings were next thought to be no older than 100 million years or they would not be there now. If we lived at the time of the dinosaurs, we might not have seen any rings around Saturn, and the rings we see today were once thought to be just temporary embellishments destined to disappear from sight in 100 million years or so. But now some of the very scientists who were once arguing for the rings' youth now say that they may have been created roughly 4.6 billion years ago when the solar system was formed, which means that Saturn may have always had rings.

10.5 Introduction to Saturn's moons

Saturn's medium-sized icy moons

Saturn has only one large satellite, Titan, comparable in size to Jupiter's four Galilean satellites, but it has an extensive family of smaller moons, including six mid-sized icy bodies that range from 98 to 765 kilometers in radius. In order of increasing orbital distance from Saturn, and also of increasing size, they are Mimas, Enceladus, Tethys, Dione, Rhea and Iapetus (Table 10.3). They were all discovered long ago – one by Christiaan Huygens (1629–1695), two by William Herschel (1738–1822) and four by Giovanni Domenico Cassini (1625–1712). Saturn's smaller retrograde satellite Phoebe wasn't discovered until 1899, by William H. Pickering (1858–1938).

The mean mass densities of these moons are often low, between 1100 and 1300 kilograms per cubic meter, which suggests that they are mainly composed of pure water ice. A few of them, such as Enceladus, Titan and Phoebe, have higher mass densities, indicating a composition of rock and ice. This is consistent with their highly reflective surfaces. With the exception of Iapetus and Phoebe, they all reflect more than 50 percent of the incident sunlight, and one of them, Enceladus, reflects almost 100 percent of the sunlight that strikes it. Surface water ice was also identified by infrared spectroscopy in the years prior to the *Voyager* missions.

The cold surfaces of all of Saturn's mid-sized icy satellites preserve ancient impact craters dating back to their formative years. Several of them also exhibit evidence for melting ice and surface activity after their formation.

There are several ways to group these objects. Mimas, Rhea and Iapetus, for example, have surfaces dominated by craters with no signs of internal activity. Or the moons can be paired by approximately the same size – Mimas and Enceladus, Tethys and Dione, and Rhea and Iapetus.

Dione provides a representative example of the icy cratered surfaces of these moons (Fig. 10.18). It has a mean mass density of 1476 kilograms per cubic meter, with perhaps enough rock in its makeup to produce internal heat from radioactivity. Most of the surface is heavily cratered, with varying numbers suggesting resurfacing by flows of erupting material.

Small moons of Saturn

Instruments on the *Voyager 1* and *2* and *Cassini* spacecraft have discovered a host of small moons that reside within

Table 10.3 Properties of Saturn's largest moons[a]

Name	Mean distance from Saturn (radii)	Orbital period (days)	Mean radius (km)	Mass (10^{19} kg)	Mass density (kg m^{-3})
Mimas	3.08	0.942	198.2	3.75	1150 ± 4
Enceladus	3.95	1.370	252.1	10.79	1608 ± 3
Tethys	4.88	1.888	533.0	61.75	973 ± 4
Dione	6.26	2.737	561.7	109.57	1476 ± 4
Rhea	8.73	4.518	764.3	230.70	1233 ± 5
Titan	20.22	15.945	2575.5	13454.43	1880 ± 4
Hyperion	24.53	21.277	135.0	0.558	542 ± 48
Iapetus	59.01	79.331	735.6	180.58	1083 ± 7
Phoebe	214.97	550.565R	106.6	0.829	1634 ± 46

[a] The orbital distances are given in units of Saturn's equatorial radius, which is 60 268 kilometers and nearly 10 Earth radii. The satellite radii are given in units of kilometers (km). By way of comparison, the mean radius of the Earth's Moon is 1738 kilometers. The mass is given in units of 10^{19} kilograms (kg) – our Moon's mass is 7348 in these units. The mass density is given in units of kilograms per cubic meter (kg m^{-3}).

Fig. 10.18 Saturn's moon Dione   The surface of pale, icy Dione appears to hover above Saturn's rings seen edge-on (*horizontal stripes near the bottom*), with the tranquil gold and blue hues of the planet in the distant background. Observations at blue, green and infrared wavelengths have been combined to produce this color view, which approximates the scene as it would appear to the human eye. This image was taken on 11 October 2005 from the *Cassini* spacecraft. (Courtesy of NASA/JPL/SSI.)

Fig. 10.19 Saturn's bright moon Enceladus   Ice-covered Enceladus, about 504 kilometers across, looks suspended in front of rings darkened by Saturn's shadow. From the outer ring edge inward, the observed ring features include the A ring, the Cassini Division and the B ring. The C ring is the dominant, darker region. This image was taken in visible light on 7 March 2005 from the *Cassini* spacecraft. (Courtesy of NASA/JPL/SSI.)

the inner parts of Saturn's satellite system. They are all bright objects; probably composed of ice, and many of them have orbits that are remarkable in one way or another.

At least six of these tiny moons are associated with the rings: Pan, Atlas, Pandora, Prometheus, Janus and Epimetheus. Pan disturbs particles in the A ring to form the Encke Division. Pandora and Prometheus shepherd the F ring; Atlas shepherds the outer margin of the A ring.

Saturn's two co-orbital satellites, Janus and Epimetheus, are even more bizarre. Janus and Epimetheus move in almost identical orbits. The satellite on the inner orbit that is closest to Saturn moves slightly faster, overtaking the outer satellite every four years. But the bodies' diameters are greater than the distance between their orbital paths, so they cannot pass without some fancy pirouetting. They avoid a collision at the last moment by gravitationally exchanging energy and switching orbits. The inner one is pulled by the outer one and raised into the outer orbit, and vice versa. They then move apart, only to repeat this *pas de deux* four years later, and exchange again.

Three so-called Lagrangian satellites move along the orbits of Saturn's larger satellites Tethys and Dione. The satellite Tethys shares its orbit with two small companions, Telesto and Calypso, one about 60 degrees ahead and the other about 60 degrees behind. These two positions, first specified by the Italian-born French mathematician Joseph Louis Lagrange (1736–1813) in the 19th century, are places where the gravitational pull of Saturn and the larger moon are equal. One additional miniature moon, named Helene, shares Dione's orbit, leading it by 60 degrees.

Like Jupiter, the planet Saturn also has a host of small moons discovered from ground-based telescopes.

But there are four fascinating moons of Saturn that have been most extensively studied. They are bright Enceladus, hazy Titan, tumbling Hyperion, and remote, renegade Phoebe, and we now turn to these four, starting with Enceladus.

10.6 Saturn's active water moon Enceladus

Spacecraft view Enceladus close up

Saturn's enigmatic moon Enceladus is small, white and bright, as fresh as new-fallen snow (Fig. 10.19). The diminutive satellite, a mere 504 kilometers across, reflects nearly 100 percent of the incident sunlight, making it one of the brightest known moons at visible wavelengths. Telescopic infrared spectra also indicated a surface of almost pure water ice, which might explain the high reflectivity.

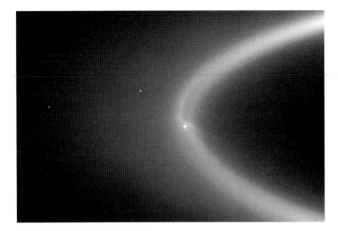

Fig. 10.20 Active Enceladus feeds Saturn's E ring   Geysers near the south pole of Saturn's moon Enceladus (*middle of ring*) send water ice and water vapor tens of thousands of kilometers into space, where they are trapped by Saturn's gravity into orbit around the planet, forming the E ring. Saturn's moon Tethys is at the far left, and a background star is seen between Tethys and the ring. The Sun was almost directly behind Saturn when this image was taken, and this backlighting makes the small particles in the E ring appear brighter than they normally are. The dark region extending up from Enceladus follows the moon in its orbit; this hollowed-out core is probably caused by the sweeping action of Enceladus as it moves in the center of the E ring. This image was taken in visible light from the *Cassini* spacecraft on 15 September 2006. (Courtesy of NASA/JPL/SSI.)

When viewed close up from *Voyager 2* in 1981, just a few craters were found on the face of Enceladus, and large, smooth, crater-free regions were discovered, apparently coated with water ice. Parts of the surface contained long cracks and groves, where liquid water might have risen from a warm interior, freezing into smooth ice on the surface. The moon's bright surface suggested that this resurfacing occurred in the fairly recent past, since the ice had not existed long enough to become cratered or darkened.

At the time, it was known that Enceladus orbits Saturn in the middle of the wide, diffuse E ring, discovered in 1966, which is composed of tiny water-ice particles. Theoretical calculations indicated that such a ring will spread out into invisibility over cosmic time intervals, and ought to be replenished in order to survive, so it was hypothesized that Enceladus is feeding the E ring (Fig. 10.20).

Thus, after the *Voyager 2* encounter, scientists postulated that ice, or even liquid water, might be expelled from a currently active Enceladus, accounting for its smooth, highly reflective surface and location near the core of the E ring. But this was an educated guess, an unproven speculation, until the *Cassini* spacecraft passed close enough to test the idea.

Between 2005 and 2010, the *Cassini* spacecraft swooped down as low as 25 kilometers above Enceladus, showing that its south polar landscape exhibits sustained, continuous activity, spewing out ice particles, water vapor and trace amounts of organic chemicals into space. Most of the material does not rise fast enough to escape the moon's gravitational pull, and falls back to cover it with freezing water, but some of the ejected ice has sufficient velocity to escape and go into orbit around Saturn, populating the E ring.

Although the north polar region of Enceladus is pitted with craters, the southern surface is completely without craters and has been twisted and buckled, producing long, parallel, linear troughs, dubbed "tiger stripes", which are stained with dark organic material, run north–south, and are surrounded by roughly circular, circumpolar mountain ridges and valleys (Fig. 10.21).

Active, geyser-like jets arise from the warm tiger-stripe fractures

It is the tiger stripes that mark the location of Enceladus' activity, the places where jets blast, vent and spray out water, lofting it into space somewhat like the Old Faithful Geyser in Yellowstone National Park. These nearly evenly spaced fractures are deep, narrow and long. The V-shaped cracks are about 500 meters deep, with slopes that seem to be coated with ice particles, the fallout of active jets. The fractures are less than 1 kilometer wide, and can extend up to 175 kilometers across the south polar terrain, terminating in hook-shaped bends.

Under the low gravity on Enceladus, the geyser-like jets spew out for thousands of kilometers into space, bending into giant flame-shaped plumes, creating a halo of ice, dust and gas around the moon, and pumping icy material into Saturn's E ring. Water vapor, water ice, powdery snow, carbon dioxide, nitrogen, methane, ammonia, and small amounts of carbon-bearing organic molecules, including acetylene and propane, have been observed in the jets.

Numerous large, small and variable jets have been detected all along the tiger stripes (Fig. 10.22). An individual vent probably stays active until condensing vapor and falling ice particles plug it up and close it off, while the pressure of underground heat forces new vents to open somewhere else along the fracture. The cracks could also open and close under the tidal pull of Saturn's gravity, further controlling the timing of the eruptions. Altogether an endless sequence of jets is produced, varying in space and time with the opening up of new vents and the closing off of old ones.

Infrared detectors aboard the *Cassini* spacecraft have shown that the tiger-stripe sources of the jets glow all along

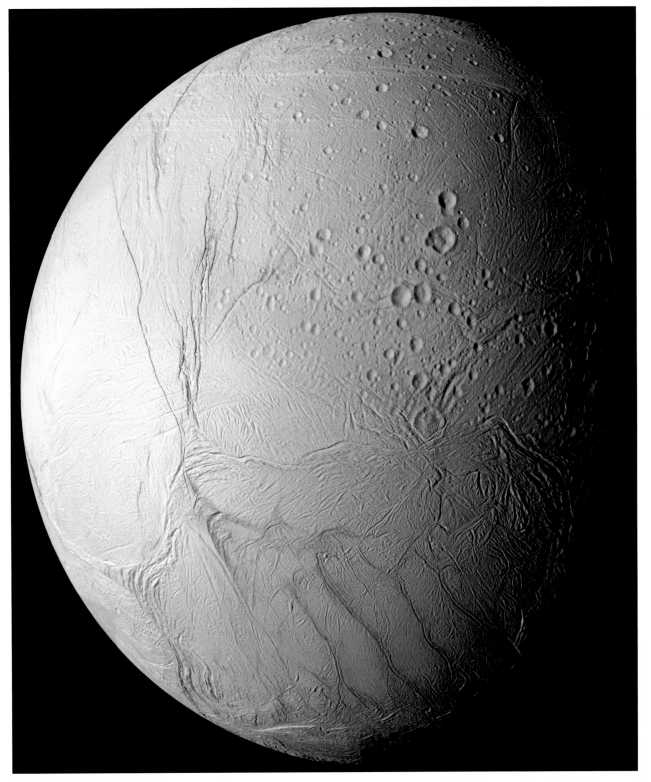

Fig. 10.21 Tiger stripes on Enceladus   Saturn's enigmatic moon Enceladus is a jumbled world of fresh snow plains (*middle*), old cratered terrains (*top*), and prominent tiger-stripe fractures (*bottom, false color blue*). The fissures spray ice particles, water vapor and organic compounds outward, some of them forming Saturn's E ring and others falling back on the moon. In the mosaic shown here, three prominent tiger stripes extend from the bottom center upwards toward the center. From left to right, they are named Alexandria Sulcus, Cairo Sulcus, and Baghdad Sulcus, the longest tiger stripe. Across the middle of the image, near the northern end of the tiger stripes, 90-degree bends curve along similar paths, starting in a direction parallel to each tiger stripe and then turning perpendicular. Changes in tectonic stresses most likely cause the bends and narrow ridges. This mosaic of images was obtained in visible light from the *Cassini* spacecraft on 21 November 2009. (Courtesy of NASA/JPL/SSI.)

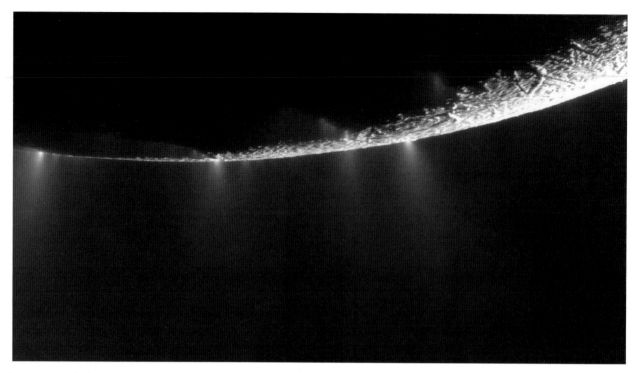

Fig. 10.22 Enceladus vents water jets   Dramatic plumes, both large and small, spray out water-ice particles, water vapor and organic compounds from many locations along tiger-stripe fractures near the south pole of Saturn's moon Enceladus. More than 30 individual jets of different sizes can be seen in this image, and more than 20 of them had not been identified before. The south pole of Enceladus lies near the limb in the top left quadrant of the mosaic, near the large jet that is second from left. (Courtesy of NASA/JPL/SSI.)

the fractured lengths. It can remind one of William Blake's (1757–1827) poem, *The Tyger*, with the famous opening line "Tyger! Tyger! Burning bright". Unexpected warmth is found within the giant fractures, with temperatures that can exceed 180 kelvin, more than twice the average temperature of the moon's surface and well above the 70 kelvin that would be expected by heating from sunlight alone. The excess heat appears to be confined within narrow, elongated regions, no more than a kilometer wide and stretching along the fractures.

Why is Enceladus hot inside?

So where does the unexpected warmth in the tiger stripes come from? Enceladus is too small to generate much heat on its own. With a bulk density of 1600 kilograms per cubic meter, it might have a rock core surrounded by a mantle of water ice, but the radioactivity of those rocks cannot produce the observed heat.

There are three possible explanations for the heat, all related to the tidal effects of Saturn's varying gravitational pull on the small moon as it moves along its non-circular orbit. The moon's stretched-out, elliptical trajectory, with an orbital eccentricity of 0.047, is retained by an orbital resonance with the bigger moon Dione, which makes one circuit around Saturn during every two

orbits of Enceladus and repeatedly pulls on the smaller satellite.

The varying tidal forces on Enceladus will cause the tiger-stripe fractures to move back and forth, generating heat on their sides as they rub against each other, somewhat like rubbing your hands together to keep warm on a cold winter day. The frictional heat might cause nearby ice to sublimate, or evaporate, into water vapor, which might drag ice particles into space, without the need of any liquid water inside Enceladus.

The tidal flexure of Enceladus might also open and close its fractures over the course of an eccentric orbit, uncovering a buried form of ice called clathrate, which will explosively decompose when exposed to the vacuum of space. Carbon dioxide, methane and nitrogen ought to be released during the explosion, which might propel the water-rich jets outward.

In a third possibility, the tidal squeezing of Enceladus as it moves around its non-circular orbit could make it hot enough inside to melt interior ice. Water vapor produced by this internal heating source could well up in the tiger-stripe fractures, propelling ice particles further out. The temperatures might become high enough at lower depths to produce liquid water inside Enceladus.

The moon is freezing cold on the outside, and under normal circumstances water ice will not melt until it is hotter than 273 kelvin. However, some ammonia has been detected in the jets, and if present in large enough quantities the melting temperature of water might be lowered to about 170 kelvin. Ammonia dissolves in water and acts like the antifreeze in your car, keeping water liquid at a lower temperature than would otherwise be possible.

Although ammonia has not yet been observed in great abundance in the jets, unusually high levels of salt have been detected in ice grains expelled during the eruptions. Scientists conclude that liquid water must be present within Enceladus in order to dissolve enough minerals to account for the levels of salt detected.

A hypothetical sea under the ice-covered south polar region of Enceladus would contain liquid water, organic chemical elements, and an internal source of heat energy, making it a habitable place where living things could survive. It might be analogous to the mid-ocean ridges on Earth's dark sea-floor where exotic creatures thrive on volcanic heat and chemicals in the water, in the complete absence of sunlight. Such a possibility had already been suggested for Jupiter's moon Europa. But of course the possibility of similar things living inside of either Enceladus or Europa is pure unproven, though informed, speculation.

This brings us to Saturn's biggest satellite Titan, 10 times larger than Enceladus. Although liquid water cannot now lap the shores of Titan, it contains seas of liquid methane and ethane.

10.7 Hidden methane lakes and organic dunes on Saturn's moon Titan

Titan's thick, hazy atmosphere

The Dutch astronomer Christiaan Huygens (1629–1695) discovered Titan on 25 March 1655; he was also the first to notice that Saturn has rings. Titan is the largest of Saturn's satellites, the second largest satellite in the solar system, and the only satellite possessing an extensive, dense atmosphere.

The sharp-eyed Catalan astronomer Josep Comas Solà (1868–1937) noticed that Titan's tiny disk is dark at the edges, suggesting that it has an atmosphere. It was confirmed by the Dutch-American astronomer Gerard Kuiper (1905–1973), by spectroscopic observations of methane in Titan's atmosphere.

As the result of space-age investigations, we now know that the surface pressure of Titan's atmosphere is an ear-popping 1.467 bars. That is one and a half times the 1-bar air pressure at sea level on Earth, and equivalent to the pressure experienced by a deep-sea diver at about 6 meters under the ocean's surface.

But it is very cold out there. Besides being about a billion kilometers from the Sun's heat, the surface of Titan lies below a haze which blocks out about 90 percent of the incident light. As a result, the surface temperature is 93.5 kelvin.

When the *Voyager 1* and *2* spacecraft passed behind Titan, as seen from Earth, the homebound radio signal penetrated the atmosphere, permitting an accurate determination of the surface radius from the time the signal disappeared. Under its thick atmosphere, the solid surface of Titan has a mean radius of 2576 kilometers, just slightly smaller than Jupiter's satellite Ganymede, at 2631 kilometers, and larger than the Earth's Moon, with a radius of 1737 kilometers. Titan is also a little larger than the planet Mercury, whose mean radius is 2440 kilometers. Titan is big enough to be a small planet except it is in orbit around Saturn, so Titan is a moon, not a planet.

The trajectories of the *Voyager* and *Cassini* spacecraft have been deflected by a small amount due to Titan's gravitational pull. The size of the deflection permitted an improved determination of Titan's mass. From the mass and radius we can determine the mean mass density, or bulk density, of Titan – 1880 kilograms per cubic meter. That is almost twice the mass density of water ice. If Titan were solid rock, like the Earth's Moon, its average density would be about three times that of water ice. So Titan must be about half rock and half ice.

Precise tracking of *Cassini* during its close encounters with Titan has been used to determine the distribution of materials in the moon's interior, from their gravitational tugs on the spacecraft. The ice and rock are mixed in roughly equal proportions inside Titan, with no separation into distinctive layers other than the outermost 500 kilometers of relatively pure ice.

By way of comparison, Mercury has an average mass density greater than five times that of water. It has a dense iron core in addition to a rocky mantle, and a magnetic field generated by currents in the core. Titan does not have such a core or any detectable magnetic field.

Visible light cannot penetrate the veil of red-orange smoggy haze that covers Titan's surface (Fig. 10.23). In the satellite's dry, cold atmosphere, the smog builds up to an impenetrable haze that extends above the bulk of the atmosphere. On Earth, smog similarly forms by the action of sunlight on hydrocarbon molecules in the air. The urban smog usually forms within a kilometer of the Earth's surface. Titan's atmosphere extends far above its surface because of the high atmospheric pressure and the relatively low mass and gravitational pull of Titan.

Fig. 10.23 Saturn's hazy moon Titan
The murky orange disk of Titan passes in front of Saturn and just below the planet's rings. Titan's photochemical smog completely obscures the moon's surface. High-altitude hazes are visible against the disk of Saturn, attenuating light reflected by the planet. Images taken at red, green and blue wavelengths have been combined to create this natural-color view, taken on 1 August 2007 from the *Cassini* spacecraft. (Courtesy of NASA/JPL/SSI.)

Titan's atmosphere extends at least 10 times further than does our own. Moreover, the high-flying stratospheric haze material, named tholin for the Greek word for "muddy", lies at altitudes above the surface of greater than 1000 kilometers. The reddish tholins are very large, complex organic molecules thought, by some, to be the chemical precursors to life.

Instruments aboard the *Voyager 1* spacecraft revealed several fascinating things about Titan, though its surface remained largely hidden. They found that the dominant constituent of the thick, heavy atmosphere is molecular nitrogen, now pegged at 98.4 percent of the atmosphere, with the remaining 1.6 percent composed of methane and trace amounts of hydrocarbons.

So, nitrogen molecules account for the bulk of Titan's atmosphere as they do on Earth – 77 percent of our air is molecular nitrogen. But, unlike Earth, the atmosphere of Titan contains no molecular oxygen, which accounts for 21 percent of the Earth's atmosphere. The freezing temperature on Titan is way too low for any living things, such as plants, to supply oxygen. Titan therefore has a hydrogen-rich atmosphere, rather than an oxygen-rich one.

High above Titan's surface, abundant nitrogen and methane molecules are being broken apart continuously by the Sun's energetic ultraviolet light and by the bombardment of electrons trapped in Saturn's magnetic environment. Some of the fragments then recombine to form more complex molecules that have been detected in small amounts by the *Voyager 1* infrared spectrometers (Fig. 10.24). Almost 20 organic gases were identified in Titan's atmosphere. In addition to methane (CH_4), the list includes hydrocarbons like ethane (C_2H_6), acetylene (C_2H_2), and propane (C_3H_8), and nitrogen compounds such as hydrogen cyanide (HCN). Many of these molecules can join together in chainlike polymer structures that contribute to Titan's dark, smoggy haze.

Although it could not see beneath the smog surrounding Titan, the radio signals sent home from *Voyager 1* have been used to infer the vertical structure of the pressure and temperature down to the surface (Fig. 10.25). Titan's temperature profile is very similar to that of the Earth's atmosphere, where the temperature initially drops with increasing altitude above the surface and then rises again, so Titan's atmosphere has the equivalent of our ground hugging troposphere and higher stratosphere.

On Earth, it is ozone that absorbs ultraviolet sunlight, making the stratosphere. On Titan, it is a photochemical haze, produced by sunlight's destruction of methane. Also unlike Earth, the temperatures in Titan's atmosphere are everywhere below the freezing point of water, and peak at 175 kelvin at about 40 kilometers above the surface.

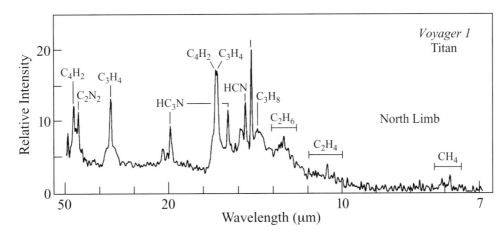

Fig. 10.24 Molecules in Titan's atmosphere   Emission features in the infrared spectrum of Titan's reflected sunlight identify the molecular constituents of its atmosphere. Sharp peaks in this spectrum, acquired from the *Voyager 1* spacecraft in 1980, are attributed to methane (CH_4), acetylene (C_2H_2), ethane (C_2H_6), and more complex hydrocarbon molecules, as well as nitrogen compounds. The wavelength is given in units of microns (μm), or 10^{-6} meters.

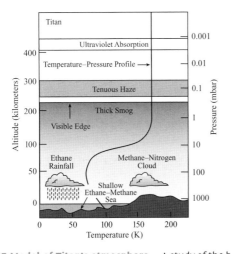

Fig. 10.25 Model of Titan's atmosphere   A study of the bending and fading of homebound radio signals when the *Voyager 1* spacecraft passed behind Titan led to this plot of the temperature and pressure in the satellite's atmosphere. The temperature (*bottom axis*) decreases with height until about 40 kilometers altitude, and then increases rapidly at higher altitudes (*left axis*). The entire atmosphere is well below the freezing temperature of water, at 273 kelvin, but the lower atmosphere is just warm enough to allow the condensation of liquid nitrogen. The high-altitude smog on Titan covers clouds of methane, and liquid ethane and methane probably rain down to the surface. The pressure (*right axis*) is given in units of millibars or 0.001 bars, and it reaches 1500 millibars or 1.5 bar near the surface of Titan. The air pressure at sea level on Earth is 1.0 bar. The pressure axis is logarithmic, which means that the numbers on the altitude axis are not uniformly spaced.

Focus 10.2   Titan could be an early Earth in a deep freeze

The chemistry in Titan's hydrogen-rich atmosphere may be similar to that in Earth's atmosphere several billion years ago, before living things released molecular oxygen into the air and modified it. So Titan could serve as a time machine, taking us back to a simpler era on Earth before life began to contaminate the planet. Titan could even provide clues to how life got started when the Earth was young. Nevertheless, the exceedingly low surface temperature rules out current life on Titan.

All the life that we know about depends on molecules that contain carbon and hydrogen atoms, and such hydrocarbon molecules have been found in Titan's atmosphere. The chemical study of these compounds is known as organic chemistry – but it has nothing to due with organic foods grown without artificial fertilizers. And on Titan the organic chemistry is going on without concurrent life.

So Titan is a frozen moon that resembles the early Earth in a deep freeze. About seven billion years from now, the Sun will near the end of its life and swell up to become a bright giant star. The intense heat from the aged and swollen Sun will warm Titan's surface and may bring it to life. The moon will become an oasis of liquid water and organic chemicals, ready to initiate life.

The surface temperature on Titan, of 93.5 kelvin, is far too cold to permit life, but its atmosphere may nurture chemical reactions similar to those at work on Earth before life began there (Focus 10.2). But Titan's temperature is close to the triple point of methane, which would permit it to exist in solid, liquid and gaseous form, just as Earth's temperature is near the triple point of water. The methane can condense in Titan's cold atmosphere to produce thick clouds that lie beneath the haze. Infrared observations that penetrate the smog suggest the presence of short-lived methane clouds in Titan's lower atmosphere, which form briefly and irregularly. Since the atmosphere is not fully saturated with methane, there cannot be extensive oceans of pure methane on the surface, but both ethane and methane can rain out of the atmosphere. They can exist as a liquid rather than a solid at the surface temperature of 93.5 kelvin. Evaporation of the liquid seas can re-supply the hydrocarbons to the atmosphere, completing the cycle.

Thus, we might expect seas, lakes and ponds of liquid hydrocarbons on Titan, consisting of ethane and methane, and the *Cassini* spacecraft and its companion *Huygens Probe* have now confirmed their existence.

Huygens Probe lands on Titan

On 14 January 2005, the NASA-built *Cassini* spacecraft dropped the European-built *Huygens Probe* on a 2.5-hour parachute descent through Titan's atmosphere to its long-hidden surface. This was the first time that a spacecraft had touched down on another planet's moon. The probe was designed to survive the impact and splash down on a liquid surface, sending back data for no more than three hours, but no ocean was in sight.

Instruments on the descending probe obtained images of dark, snaking and branching features that merge into a large channel, suggesting methane flows that may have once been fed by methane rainfall. Methane rain and drizzle were, in fact, detected on the way down, but the probe landed in a damp riverbed at equatorial latitudes, rather than a lake or sea. It found only small rocks and pebbles, perhaps rounded by tumbling methane rivers in the past.

The *Huygens Probe* could see only a small section of Titan, but radar images from the *Cassini* spacecraft, still in orbit around Saturn, revealed long, deep channels extending over great distances in other locations, meandering along with few tributaries like some lazy terrestrial rivers of water.

Scientists had perhaps been misled by images taken from afar, which showed dark regions straddling Titan's equator. They were once thought to be possible methane seas, but they contain no more liquid than water on the maria of Earth's Moon. Nevertheless, the long-anticipated liquid lakes were spotted in the north polar regions of Titan.

Lakes of liquid methane and ethane

Radar observations from the *Cassini* spacecraft in 2005 to 2010 have revealed numerous flat, dark features on Titan, which were first reported, with coarser resolution, in 2003 by radio astronomers using the ground-based Arecibo telescope. The radio waves can penetrate the haze and smog to obtain images of the moon's surface. Dark radar images correspond to weaker radar echoes and a smoother terrain, while bright regions correspond to stronger echoes and a rougher surface.

Radar images of parts of the northern regions of Titan are completely black, reflecting essentially no radar signal, and are therefore extremely smooth. They have been attributed to lakes of liquid methane, possibly with ethane dissolved in them. The numerous large lakes discovered by *Cassini* radar are found mainly at northern latitudes, which was then in winter and 2 to 3 kelvin colder than the equator. Since ethane and methane are both highly flammable, much of the satellite could go up in flames, but it won't ignite because of the lack of molecular oxygen in the atmosphere.

Some of the lakes are many tens of kilometers in length, and their perimeters are reminiscent of the shores of terrestrial lakes (Fig. 10.26). Clouds that rain methane have been observed above some of the lakes, plausibly filling them in season. These liquids are also thought to carve the meandering rivers and channels spotted elsewhere on the moon's surface.

Though less common, the lakes also form in Titan's south polar regions. Low-altitude passes over these regions by *Cassini* have confirmed the presence of liquid ethane in one of the largest southern lakes, Ontario Lactus, by the way it absorbs and reflects infrared radiation. Observations of this lake's sloping, receding shoreline suggest that the lake is evaporating in the summer warmth. Seasonal variations of the precipitation and evaporation of methane and ethane, similar to those of water lakes on Earth, might explain why the lakes in Titan's northern high latitudes were 20 times more common than lakes in the southern high latitudes, with rainfall filling the northern lakes in winter.

Organic dunes

Titan's entire surface is not wet. At low latitudes near the moon's equator, the *Cassini* radar instrument detected vast dune fields instead of lakes. The giant, rippled dunes are aligned in the longitudinal direction, suggesting that they are shaped by strong winds blowing mainly from east to west. The linear dark streaks extend hundreds of kilometers and rise to a few hundred meters, with strips between the dunes that are nearly free of dark material. They bear a striking resemblance to sand dunes on Earth, but instead of sand Titan's dunes are made out of solid organic grains. The tiny ice particles are coated with organic substances, and likely derive from organic chemicals in Titan's smoggy skies.

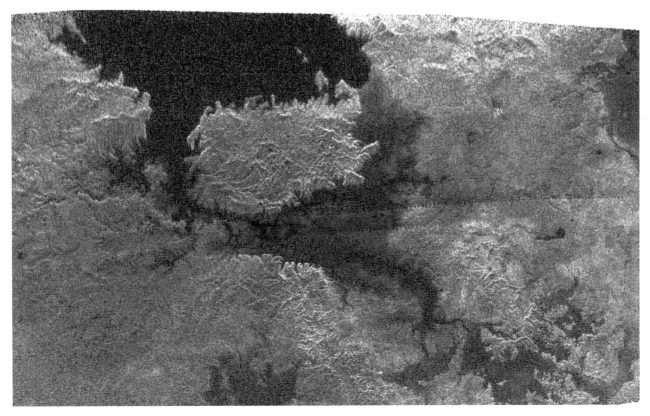

Fig. 10.26 Lakes of methane and ethane on Saturn's moon Titan   A radar image shows a bright, central island in a dark, smooth lake, surrounded by a bright shoreline with numerous inlets. The island is about 90 kilometers across. This is one of many large lakes formed at high latitudes on Titan, with more in the northern polar regions than the southern ones. Seasonal rains of liquid methane and ethane probably fill these lakes. The radar instrument aboard *Cassini* obtained this image during a close south polar flyby on 22 February 2007. (Courtesy of NASA/JPL.)

10.8 Alien worlds, distant ring

The *Cassini* spacecraft has also zeroed in on one of Saturn's most unusual, mid-sized moons, named after the god Hyperion. Attention was drawn to this alien world in the 1980s after *Voyager 2* images indicated that it is one of the most irregularly shaped, non-spherical bodies in the solar system, and it was realized that Hyperion's rotational period is not constant. The strange moon also has a very low mean mass density of about 540 kilograms per cubic meter, indicating that it is probably composed of water ice and filled with empty spaces that occupy about half its volume.

Many natural satellites are locked into synchronous rotation through tidal interaction with their planet, including the Earth's moon, the four Galilean satellites, and most of the large or mid-sized moons of Saturn. The rotation period of these satellites is equal to their orbital period, so the same side of the moon always faces their planet. The exception is Hyperion, whose rotation is chaotic. As shown by Stanton J. Peale (1937–) and Jack Wisdom (1953–) in the mid-1980s, strong gravitational torques on the asymmetric satellite, coupled with its large orbital eccentricity and resonance with Titan, cause Hyperion to tumble in a random manner. Its axis of rotation wobbles so much that its orientation in space is unpredictable.

Even more fantastic aspects of Hyperion were revealed when *Cassini* was targeted to pass within 500 kilometers of the moon, on 26 September 2005. A combination of images taken at different wavelengths reveals a surface that is covered with deep, sharp-edged craters with dark material at their bottoms (Fig. 10.27). The largest crater on Hyperion is about 122 kilometers across, or nearly half the moon's average diameter of 270 kilometers, perhaps lending support to the idea that Hyperion is an impact fragment of a former larger body.

Phoebe is another irregular satellite, for it moves around Saturn in the backward retrograde direction and is inclined by 152 degrees to Saturn's equator. For more than 100 years Phoebe was Saturn's outermost known moon, about four times more distant from the planet than Iapetus, the next closest moon, which orbits very nearly in the plane of Saturn's equator.

Fig. 10.27 Saturn's odd moon Hyperion   This stunning view of Saturn's asymmetric, tumbling, impact-cratered moon Hyperion reveals numerous sharp-edged craters, which make the moon look like a giant sponge. Hyperion's unusual appearance can be attributed to the fact that it has an exceptionally low mean mass density, of about 540 kilograms per cubic meter, with a large porosity of 0.46 and a weak surface gravity. In other words, the moon seems to contain a lot of empty holes, somewhat like a sponge. This processed, false-color view combines images taken at infrared, green and ultraviolet wavelengths from the *Cassini* spacecraft during a close flyby on 26 September 2005. (Courtesy of NASA/JPL/SS.)

When approaching Saturn in 2004, the *Cassini–Huygens* spacecraft first encountered outlying Phoebe, providing an opportunity to test the mission's imaging equipment before traveling on to encounter the planet. The high-resolution images reveal Phoebe to be a heavily cratered body with large variations in brightness, suggesting that bright water ice lies beneath a thin blanket of dark surface deposits (Figs. 10.28, 10.29). Although once believed to be a captured asteroid, scientists now speculate that Phoebe is a former Centaur, one of a number of icy objects originally residing in the distant Kuiper belt and now orbiting the Sun between Jupiter and Neptune.

Phoebe circles Saturn within an enormous ring in the far reaches of the planet's realm and way outside its Roche limit. The infrared emission of this ring was discovered in 2009 using the *Spitzer Space Telescope* (Fig. 10.30). The ring is tilted 27 degrees from Saturn's equatorial plane, and all the other rings, extends from 128 to 207 times the radius of Saturn, and is about 20 times as wide as the diameter of the planet. The diffuse ring is thought to originate from micrometeorite impacts of the surface of Phoebe.

Fig. 10.28 Saturn's retrograde moon Phoebe   This image of Phoebe suggests that the moon may be an ice-rich body coated with a thin layer of dark material, resembling the nucleus of comets. Small bright craters in the image are probably young features. When impacting projectiles slammed into the surface of the moon, the collisions excavated fresh, bright material, probably water ice, underlying the surface area. Dark material on some crater walls appears to have slid downwards, exposing more light-colored material. Phoebe orbits Saturn in the backward, retrograde direction to all of the other mid-sized satellites of the planet, and the moon's dark and irregular cratered surface, retrograde orbit, and low mean mass density suggest that Phoebe was once part of the Kuiper belt of icy comets beyond Neptune before passing near Saturn and being captured by its gravity. This mosaic of two images was acquired from the *Cassini* spacecraft during its Phoebe flyby on 11 June 2004. (Courtesy of NASA/JPL/SSI.)

Fig. 10.29 Bright and dark layers on Phoebe   Saturn's mid-sized, retrograde moon, Phoebe, is most likely an ice-rich body overlain with a thin layer of dark material. The sharply defined crater at *above center* exhibits two or more layers of alternating bright and dark material. The layering might have occurred during the crater formation, when material ejected from the crater buried the pre-existing surface that was itself covered by a relatively thin, dark deposit over an icy mantle. This image was taken from the *Cassini* spacecraft during its Phoebe flyby on 11 June 2004. (Courtesy of NASA/JPL/SSI.)

The new ring lends support to the idea there is a connection between Phoebe and the dark side of its neighboring moon Iapetus. *Voyager 2* images indicated that two-faced Iapetus is a divided world, with a heavily cratered side as bright as ice and another side as dark as asphalt or coal. The side of Iapetus that faces forward in its orbit is apparently being darkened in some mysterious way, and the darkening particles may be coming from Phoebe's ring. Once deposited, the dark material absorbs sunlight and warms the surface of Iapetus, perhaps vaporizing and releasing water ice that moves away from the dark side.

This concludes our survey of Saturn, the most distant planet known to the ancients. We will now travel out beyond this enchanting world to the next wanderers, Uranus and Neptune.

Fig. 10.30 Saturn's enormous infrared ring This artist's conception illustrates the infrared glow of cold dust particles in Saturn's largest ring, discovered using the *Spitzer Space Telescope* in 2009. The very tenuous collection of ice and dust particles is spread out in an enormous belt at the far reaches of Saturn's system, with an orbit tilted 27 degrees from the main ring plane. The bulk of its material starts about 6 million kilometers away from the planet, and extends outward another 12 million kilometers. The ring's diameter is equivalent to roughly 300 times the diameter of Saturn. The planet appears as just a small dot in the middle of this portrayal. The inset shows an enlarged image of Saturn, as seen by the W. M. Keck Observatory at Mauna Kea, Hawaii, in infrared light. Saturn's retrograde moon Phoebe circles within the newfound ring, and is likely to be the source of its material. Dark material dislodged from Phoebe may also be the source of the dark side of Saturn's next innermost moon, two-faced Iapetus. (Courtesy of NASA/JPL-Caltech/Keck.)

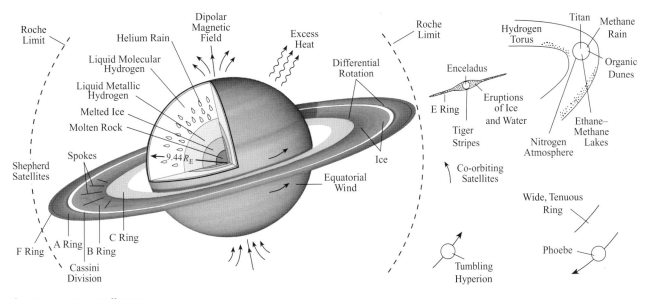

Fig. 10.31 Summary diagram

11   Uranus and Neptune

- Uranus and Neptune were unknown to ancient astronomers, and were not discovered until after the invention of the telescope.

- Uranus is just barely visible to the unaided eye, and Neptune requires a telescope to be seen.

- The blue-green, turquoise color of Uranus and the indigo blue color of Neptune come from methane in their clouds.

- Uranus and Neptune have a similar size, mass, and bulk composition, and they are both much smaller, less massive, and denser than the other two giant planets, Jupiter and Saturn.

- In contrast to all the other planets in the solar system, Uranus is tipped sideways so its rotation axis lies nearly within the planet's orbital plane, leading to extreme seasonal variations in solar heating of the planet's polar regions.

- Although Uranus apparently has no strong internal source of heat, Neptune radiates 2.7 times the energy it absorbs from the Sun. The source of Neptune's excess energy is most likely heat left over from the planet's formation.

- The cloud bands and winds on Uranus blow parallel to the planet's equator, apparently controlled by the planet's rapid spin rather than by direct heating from the Sun.

- Despite receiving a relatively small amount of sunlight compared to the other major planets, Neptune's atmosphere is surprisingly active and dynamic, with large storm systems and high-speed winds that may be driven by internal heat.

- As the southern hemisphere of Neptune turned slowly toward the Sun's faint heat, the temperature of the planet's south polar region increased by just 10 kelvin and methane storm clouds became more frequent in the southern hemisphere.

- Unlike Jupiter and Saturn, there is no liquid metallic hydrogen inside Uranus and Neptune, but they both contain deep atmospheres of molecular hydrogen.

- The internal structure of Uranus and Neptune includes vast internal oceans of water, methane and ammonia "ices", melted at the high temperatures inside.

- The magnetic fields of Uranus and Neptune are askew, tilted from their rotation axes; rotation-driven currents in internal shells of ionized water could generate these magnetic fields.

- The austere, skeletal rings of Uranus are very narrow and widely spaced from each other, and made of very dark material.

- Shepherd satellites are most likely responsible for the narrowness of the rings of Uranus and Neptune.

- The rings around Uranus are not quite circular, do not lie exactly in Uranus' equatorial plane, and vary in width; these irregularities are attributed to the gravitational interaction of ring particles with small nearby moons.

- The material in one narrow ring around Neptune has been concentrated into three clumps, probably by the gravity of a nearby moon or moons.

- The sparse rings around Neptune contain no more material than that found in a single small moon only a kilometer across.

- Most of Neptune's rings that we see now will probably be ground into dust by collisions and meteoritic bombardment in a few hundred million years, eventually being consumed by their central planet and vanishing from sight. But the rings can easily be replaced by debris blasted off small moons already embedded in them.

- The five major moons of Uranus are dark and dense, made up predominantly of rock and water ice.

- Miranda, the innermost mid-sized satellite of Uranus, exhibits a bizarre variety of surface features that suggest repeated violent impacts in the past. It may have been shattered by catastrophic collisions and reassembled, or it became frozen in an embryonic stage of differentiation.

- Uranus' large moons all revolve in the planet's equatorial plane, almost perpendicular to the planet's orbital plane, in circular synchronous satellite orbits with the same side always facing Uranus.

- Neptune's largest satellite, Triton, revolves about the planet in a direction opposite to that in which Neptune rotates.

- Triton has a very tenuous, nitrogen-rich atmosphere, bright polar caps of nitrogen and methane ice, frozen lakes of water flooded by past volcanoes of water ice, and tall geysers that may now be erupting on its surface.

- Triton may have formed in orbit around the Sun and was subsequently captured by Neptune, whose tidal forces kept Triton molten for much of its early history. These tides are now pulling the satellite toward a future collision with the planet.

11.1 Fundamentals

Planetary twins

Saturn was the most distant planet known to the ancients. Uranus and Neptune are both so far away, and so faintly illuminated by the Sun, that telescopes were required to discover them. One can just barely discern Uranus with the unaided eye, but few astronomers have seen it without a telescope. Neptune cannot be seen without the aid of a telescope. William Herschel (1738–1822) discovered Uranus in 1781 during his telescopic survey of the faint stars located near bright ones, and Neptune was found in 1846, as the result of a mathematical

Table 11.1 Some comparisons of Uranus and Neptune[a]

	Uranus	Neptune
Mass (Earth mass)	14.53	17.14
Equatorial radius (Earth radius)	3.98	3.91
Bulk density (kilograms per cubic meter)	1270	1638
Sidereal rotation period (hours)	17.24	16.11
Sidereal orbital period (Earth years)	84	165
Mean distance from Sun (AU)	19.19	30.06
Outer atmosphere	82.5 percent hydrogen	80.0 percent hydrogen
	15.2 percent helium	18.5 percent helium
	2.3 percent methane	1.5 percent methane
Energy balance	Less than 1.4	2.7 ± 0.3
Effective temperature	59.3 kelvin	59.3 kelvin
Temperature at one-bar level	76 kelvin	73 kelvin
Central temperature	5000 kelvin	5000 kelvin
Magnetic dipole moment[b]	50 D_E	25 D_E
Equatorial magnetic field strength	0.23×10^{-4} tesla	0.14×10^{-4} tesla

[a] The Earth's mass is 5.9722×10^{24} kilograms, the Earth's equatorial radius is 6378 kilometers, the astronomical unit (AU) is the mean distance between the Earth and the Sun with a value of 1.496×10^{11} meters. The energy balance is the ratio of total radiated energy to the total energy absorbed from sunlight, the effective temperature is the temperature of a black body that would radiate the same amount of energy per unit area, and a pressure of one bar is equal to the atmospheric pressure at sea level on Earth.

[b] The magnetic dipole moment is the product of the equatorial magnetic field strength and the cube of the planet's radius, given here in units of the Earth's magnetic dipole moment $D_E = 7.91 \times 10^{15}$ tesla meters cubed.

prediction based on its gravitational effect on the motion of Uranus.

These two outermost major planets are considered giant planets, but they are much smaller, less massive and denser than the other two giants, Jupiter and Saturn. Uranus and Neptune are just four times bigger than the Earth, only about one-third the size of Jupiter, and less than half the size of Saturn. The mass of Uranus and Neptune has been inferred from the orbital periods and distances of their largest moons. They have just 14.53 and 17.14 times the Earth's mass, for Uranus and Neptune respectively, which is about the same mass as the ice–rock cores of Jupiter or Saturn.

Uranus is about 19 times as far away from the Sun as the Earth is, and Neptune is about 30 times as distant. As a result, it takes 84.0 Earth years for Uranus to complete one revolution about the Sun and nearly twice that for Neptune.

These two distant planets remain little more than dim, fuzzy spots of light in even the most powerful telescope. From the Earth, the planet Uranus subtends an angle of just 3.5 seconds of arc, and Neptune 2.0. Since the Earth's atmosphere blurs features smaller than about 1.0 seconds of arc, ground-based observers cannot distinguish features in the outer atmospheres of Uranus and Neptune.

One can still infer enough about Uranus and Neptune from telescopic observations to know that they have similar bulk physical properties. The size, mass, composition and rotation of Uranus and Neptune are in fact so similar that they are often called planetary twins (Table 11.1). From each planet's mass and size, we calculate its bulk density, which is intermediate between the low-density giants, Jupiter and Saturn, and the dense rocky Earth or Mars. Most of Uranus and Neptune must therefore be composed of something less dense than rock, but more substantial than the hydrogen and helium that dominate the composition of Jupiter and Saturn. The main ingredients of Uranus and Neptune are probably liquid water and other melted ices.

The blue-green, turquoise color of Uranus and the deeper indigo blue of Neptune are attributed to methane. At the low temperatures prevailing at their cloud tops, the methane condenses to form a top layer of clouds made of methane ice crystals. And since methane absorbs red light quite strongly, the sunlight reflected off the thick, deep clouds of Uranus and Neptune has a blue color.

The blue cloud deck of methane ice forms where the atmospheric pressure is about one bar, or roughly the same as the air pressure at sea level on Earth. At this level, the equatorial radius of Uranus is 25 559 kilometers while that of Neptune is 24 764 kilometers. Rapid internal rotation produces an equatorial bulge on both planets; the polar radius is 586 kilometers shorter on Uranus and 424 kilometers on Neptune.

The sidereal rotation periods of Uranus and Neptune lie between the roughly 10-hour day of Jupiter and Saturn and the 24-hour rotation period of Earth and Mars. Periodic variations in the radio emission of Uranus and Neptune, detected from the *Voyager 2* spacecraft, indicate that their magnetic fields, which are anchored deep inside the planets, rotate once with respect to the stars every 17.24 hours and 16.11 hours, respectively.

Thus, Uranus and Neptune bear an uncanny resemblance to each other, but there are two remarkable differences. Uranus is tipped sideways and has no significant heat inside, while Neptune is more upright and has a lot of internal energy.

Uranus is tipped on its side and has no strong source of internal heat

Blue-green Uranus is tipped sideways, with its poles where its equator should be. Unlike all the other planets, whose rotation axes are roughly perpendicular to their orbital plane, the ecliptic, the rotation axis and poles of Uranus lie almost within the ecliptic. The equatorial plane of Uranus is inclined 97.9 degrees from its orbital plane, with a tilt that is just a bit more than a right angle.

This knowledge comes not from watching Uranus rotate, but instead from observing the orbits of its major moons. The orbits are all circular, and they lie in one plane, which is turned at right angles to the plane of Uranus' orbital motion about the Sun. As a result, the moons form a bullseye pattern, revolving around Uranus like a Ferris wheel. Since these satellites should be orbiting within the plane of Uranus' equator, the entire planet has to be tipped on its side (Fig. 11.1). One speculation is that Uranus was knocked sideways during a glancing collision, between the planet and another massive body, perhaps when the planet was still forming.

Because the rotational axis of Uranus lies near its orbital plane, first one pole and then the other points toward the Sun as the planet slowly progresses around its orbit. Each pole faces the Sun for about 21 Earth years, with a corresponding period of darkness at the opposite pole. When either pole points at the Sun, an imaginary observer near that pole would never see the Sun set. Terrestrial astronomers see the moons of Uranus moving in circles around the planet. Between the long summer and winter at each pole, the equator turns toward the Sun, and

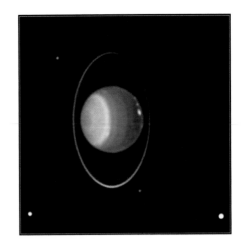

Fig. 11.1 The tilted planet   Thin, spidery rings and several small moons encircle Uranus. The planet is tipped on its side, so its equator, rings and direction of rotation are almost perpendicular to the plane of the planet's orbital motion around the Sun. Storm clouds, found in the northern upper atmosphere (*right*) circle the planet at more than 500 kilometers per hour. The clouds are tinted pink in this false-color image to show that it was taken at infrared wavelengths, but the clouds are white in visible light. This image was taken on 8 August 1998 from the *Hubble Space Telescope*. (Courtesy of NASA/JPL/STSCI.)

we observe the moons traveling vertically up and down as they move around the equator.

When *Voyager 2* arrived at Uranus, on 24 January 1986, the spacecraft's infrared detectors found that the planet is radiating about as much energy as it receives from the Sun. This means that Uranus lacks a strong internal heat source, in contrast to Jupiter and Saturn that produce heat in their centers. These two giants each radiate away about twice as much energy as they absorb from the Sun. But like Jupiter and Saturn, and unlike Uranus, the planet Neptune glows in the infrared with its own internal heat discovered when *Voyager 2* flew past Neptune in August 1989, twelve years after launch.

The temperature at Neptune's cloud tops is 59.3 kelvin, about the same as that of Uranus. But since Neptune is 50 percent farther from the Sun, it should have been a lot colder than Uranus; the temperature that would result from sunlight alone is 46 kelvin at the cloud tops of Neptune. The hotter measured temperature implies that Neptune radiates 2.7 times as much energy as it absorbs from the Sun, and its outer atmosphere must therefore receive energy from the interior.

11.2 Storm clouds on the outer giants

The outer atmospheres of Uranus and Neptune are quite similar in composition to those of Jupiter and Saturn.

Molecular hydrogen accounts for 82.5 percent of the outer atmosphere of Uranus, followed by 15.2 percent helium atoms and 2.3 percent methane. Neptune has roughly comparable amounts of these elements in its outer atmosphere. Although they are mostly composed of the light gases, hydrogen and helium, it is the methane in the colder cloud levels that accounts for their colors.

Mild weather on Uranus

At the low temperatures prevailing near the top of Uranus' atmosphere, methane gas freezes and forms a methane cloud deck; haze particles are also formed there due to the action of ultraviolet sunlight on methane. The haze and methane clouds hide the lower atmosphere from view. Ammonia and water clouds probably form deeper in the atmosphere and are difficult to see. On warmer Jupiter and Saturn the topmost light-colored clouds that we see are composed of ammonia ice crystals rather than frozen methane.

With no appreciable heat rising from the interior to drive the weather system, Uranus seemed to present a dull and placid face to the world. Looking at Uranus was something like gazing down into a bottomless ocean. Yet some zonal banding was extracted from the *Voyager 2* images and subsequently confirmed using the Earth-orbiting *Hubble Space Telescope* (Fig. 11.2) and the ground-based Keck II telescope. So storms and clouds have been detected on Uranus, perhaps related to seasonal changes in solar heating. The clouds are arranged in bands that circle the planet's rotation axis, running at constant latitudes parallel to the equator like the more vivid bands seen at Jupiter.

The features at different latitudes on Uranus move in the same direction as the planet rotates, but at faster speeds. The difference is greatest at high latitudes, where the clouds circle the poles in 14 hours, and it gets progressively smaller toward the equator, closer to the internal rotation period of 17.24 hours.

Since the high-latitude clouds are rotating faster than the interior of Uranus, the clouds cannot be simply carried by the planet's rotation. They are being blown by winds in the same direction as the planet rotates, just as clouds on Earth, Jupiter and Saturn. But unlike these planets, the winds on Uranus blow fastest near the poles of the planet.

The long, alternating periods of sunlight and darkness have little effect on the winds of Uranus. During the *Voyager 2* encounter, the south pole of Uranus was facing the Sun almost directly. The equator was in constant twilight, and the north pole had been in darkness for 20 years. So you might expect the south pole to be the warmest

Fig. 11.2 Bands and vortex in the atmosphere of Uranus Although Uranus is similar in size and atmosphere composition to Neptune, it does not appear to have as active an atmosphere. A bright band at about 40 degrees southern latitude (*left*) nevertheless stretches out in the direction of rotation, and a dark, elongated whirling vortex, up to 3000 kilometers across, formed at about 27 degrees latitude in the northern hemisphere (*right*), which is just beginning to be exposed to sunlight after many years of being in shadow. This three-wavelength composite image was taken from the *Hubble Space Telescope* on 23 August 2006. (Courtesy of NASA/ESA/L. Sromovsky and P. Fry, U. Wisconsin, H. Hammel, STSCI, and K. Rages, SETI Institute.)

place on the planet and the north pole the coldest, with a temperature difference that would drive winds from pole to pole. But the thin clouds on Uranus move parallel to the equator, orthogonal to the direction of the pole.

The *Voyager* thermometer showed that the atmospheric temperatures are much the same everywhere on the globe. The temperature at the equator and the pole facing the Sun were about the same, and the dark winter side may have even been a few degrees hotter. So something has to be redistributing the solar heat, exchanging it between warm and cold places.

Stormy weather on Neptune

The *Voyager 2* flyby in 1989 forever changed our view of Neptune's weather. Despite its great distance from the Sun, the dimly lit atmosphere of Neptune is one of the most turbulent in the solar system, with violent winds, large dark storms and high-altitude white clouds that come and go at different places and times (Fig. 11.3).

Neptune has strong zonal winds driven and defined by the planet's rotation. The clouds near the equator circulate slower than Neptune's interior rotates so the prevailing winds blow in the westward direction, opposite to the

Fig. 11.3 Neptune's dynamic atmosphere   White clouds lie above the swirling Great Dark Spot (*right center*), and raging winds reach speeds of 300 meters per second, creating a global banding in the atmosphere of Neptune. The faint sunlight at Neptune's great distance cannot provide the energy of such winds; they are probably energized by heat from the interior of the planet. This image was sent from the *Voyager 2* spacecraft on 14 August 1989, after a 12-year journey to the planet, using a 20-watt transmitter with less power than an ordinary incandescent light bulb. Traveling at the speed of light, the signals took more than 4 hours to reach Earth. (Courtesy of NASA/JPL.)

There are some important differences between the two Great Spots. Jupiter's red spot has survived for centuries, while Neptune's dark one disappeared from view within a few years of its sighting from *Voyager 2*. And Jupiter's storm lies above the clouds while Neptune's seems to form a deep well in the atmosphere, providing a window-like opening to the deeper, darker clouds below.

White, fleecy cirrus-like clouds cast shadows on the blue cloud deck below, indicating that they are high-altitude condensation clouds that rise about 100 kilometers above the surrounding ones. They form as atmospheric gas flows up, over and around the storm center, without being consumed by it. When the rising methane gas cools, it forms white clouds, fashioned from crystals of frozen methane. Water in the Earth's atmosphere freezes in a similar way into ice crystals that form white cirrus clouds. When strong upwelling carries the wispy white methane clouds to great heights in Neptune's atmosphere, they are sheared out like anvils of terrestrial thunderstorms.

Although the global wind pattern on Neptune resembles the Earth's trade winds and jet streams, they cannot be energized in the same way. Solar heating of the atmosphere and oceans drives the terrestrial winds. At Neptune's distance the Sun is 900 times dimmer and the winds should be correspondingly weaker if they are driven by the feeble sunlight. The fast winds on Neptune and the planet's complex stormy weather must instead be energized by heat generated in the planet's interior. The internal heat warms Neptune from the inside out, producing convection currents of rising and falling material, somewhat like a pot of boiling water on a stove. Uranus, on the other hand, shows no signs of substantial internal heating, and this may explain why its atmosphere is relatively benign and inactive.

You wouldn't want to forecast the variable and unpredictable weather on Neptune. When the *Hubble Space Telescope* (*HST*) took a look at the planet, the violent storms seen by the *Voyager 2* cameras had vanished without a trace, and other storms had appeared. The infrared detector on *HST* showed that Neptune's dynamic weather changes from day to day and over the years as well. White methane storm clouds as large as the Earth became more frequent in the southern hemisphere as it turned slowly toward the Sun's faint heat from local winter to spring (Fig. 11.4). In 2007, a team of astronomers using the Very Large Telescope in Chile reported that the temperature at Neptune's south pole had become 10 kelvin hotter than the rest of the planet, whose average temperature was about 73 kelvin. The extra heat in the warm south polar region provided an avenue for previously frozen methane to turn to gas and escape outward to colder cloud heights.

rotation of the planet. The equatorial wind speed on Neptune is about 325 meters per second relative to the core, almost as fast as Saturn's equatorial wind and faster than those on Jupiter or Uranus.

The wind pattern on Neptune lacks Jupiter's multiple zonal winds that flow in opposite directions. Neptune has just one westward air current at low latitudes, like the Earth's trade winds, and one meandering eastward current at mid-latitudes in each hemisphere, resembling the Earth's jet streams. And like Uranus, the polar and equatorial temperatures on Neptune are nearly equal.

The largest dark storms on Neptune are probably high-pressure systems that come and go with atmospheric circulation. The most prominent one was the Great Dark Spot, a vast, circulating storm almost as large as Earth. It is called the Great Dark Spot because it resembled the Great Red Spot of Jupiter and is dark rather than red. Both storms are found in the planetary tropics — at about one-quarter of the way from the equator to the south pole, both rotate counter-clockwise, in the direction of high-pressure anticyclones, and both are about the same size relative to their planet. As on Jupiter, some of the small dark spots on Neptune may be whirling in the opposite direction to the bigger one, perhaps indicating that they are little cyclones with descending material at their centers.

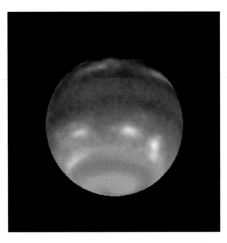

Fig. 11.4 Neptune's stormy disposition   The weather on Neptune is recorded in this image, which combines simultaneous observations on 11 August 1998 with the *Hubble Space Telescope* and NASA's InfraRed Telescope Facility on Mauna Kea, Hawaii. The predominant blue color of the planet is a result of absorption of red and infrared light by Neptune's methane atmosphere. Clouds elevated above most of the methane absorption appear white. Neptune's powerful jet stream, where winds blow at nearly 400 meters per second, is centered at the dark blue belt near Neptune's equator. The Great Dark Spot detected when the *Voyager* 2 spacecraft visited Neptune in August 1989 has completely disappeared in this image, taken nine years later. (Courtesy of NASA/STScI/IRTF.)

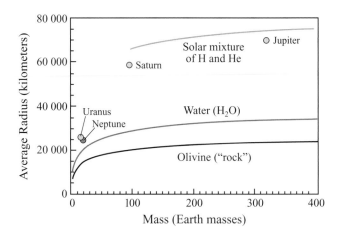

Fig. 11.5 Radius–mass relations   A liquid body of solar composition describes a radius–mass relation (*top*) that approximates the mass and radius of Jupiter and Saturn. They consist mainly of the lightest element hydrogen (H) and next lightest abundant element helium (He). Uranus and Neptune contain little hydrogen or helium, for their radii are much too small to be consistent with a solar composition. Instead, they lie only slightly above the mass-radius relation for liquid water, so they probably contain large quantities of water. For comparison purposes, the bottom curve describes a giant planet composed of pure rock, as the Earth is but with a much smaller size.

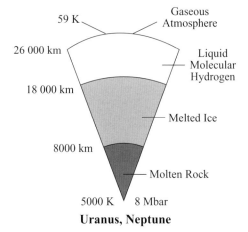

Fig. 11.6 Inside Uranus and Neptune   An outer shell of liquid molecular hydrogen covers a thick inner shell of melted ice with liquid water, methane, and ammonia within the interior of both Uranus and Neptune. Because of their relatively low mass and hydrogen abundance, neither planet contains an inner shell of liquid metallic hydrogen, as Jupiter and Saturn do.

11.3 Interiors and magnetic fields of Uranus and Neptune

Uranus and Neptune are water worlds

Although Uranus and Neptune contain deep outer atmospheres of molecular hydrogen, their relatively small size, low mass, and high bulk mass density imply that hydrogen cannot be their main ingredient, unlike Jupiter and Saturn (Fig. 11.5). The hydrogen in Uranus and Neptune is confined within thin gaseous atmospheres and liquid molecular shells that do not extend to great depths and contribute only about 15 percent of the planetary mass (Fig. 11.6). These two planets do not have enough hydrogen, or sufficient mass and internal pressure, to squeeze the hydrogen into a metallic state. So there is no internal shell of liquid metallic hydrogen inside Uranus and Neptune.

The relatively large bulk density of Uranus and Neptune, when compared to Jupiter and Saturn, implies that heavy material contributes a much greater fraction of the mass for the two outermost major planets. Most of the interiors of Uranus and Neptune probably consist of a vast internal ocean of water, mixed with liquid methane and liquid ammonia (Fig. 11.6). So these planets are sometimes called water worlds, or alternatively ice giants since the liquids can all be frozen into ice at the colder cloud tops. These substances are nevertheless kept liquid inside the planets due to the higher temperatures there. Uranus and Neptune most likely also have a central molten rocky core beneath their oceans of melted ice.

So most of the interiors of Uranus and Neptune are not unlike the cores of Jupiter and Saturn, which are thought to contain 10 to 20 Earth masses of melted ice and molten rock.

The differences between the four giant planets apparently derive primarily from the amounts of hydrogen and helium that they were able to attract and hold on to as they formed. According to one explanation, the cores of Jupiter and Saturn accreted, or gravitationally gathered in, the surrounding hydrogen before it dissipated. The hydrogen dispersed into a larger volume and lower density at the more distant orbits of Uranus and Neptune. So their cores of ice and rock accumulated slowly and took a longer time to grow. Little hydrogen or helium was left to capture by the time they had grown large enough to start gravitationally collecting the surrounding gas. In another scenario, a blast of radiation from a luminous nearby star, other than the Sun, boiled away any hydrogen or helium that may have collected around Uranus and Neptune, while Jupiter and Saturn were protected by the greater density of the nearby material.

Tilted magnetic fields

Like the Earth, Jupiter and Saturn, both Uranus and Neptune have strong internal magnetic fields. But the resemblance ends there. Here on Earth our magnetic poles are very near our geographic poles, which is very useful for navigation with a compass. The magnetic and rotational axes of Jupiter and Saturn are also closely aligned. But they are way off kilter on both Uranus and Neptune.

The magnetic axis is tipped by 58.6 degrees to the rotation axis of Uranus, and the two are separated by 46.8 degrees for Neptune. By way of comparison, the displacement is 11.7 degrees on Earth, 9.6 degrees on Jupiter, and less than 1 degree on Saturn. If the disparity on Uranus and Neptune existed on Earth, a compass needle would point somewhere near Cairo, Egypt, instead of the North Pole. Theoreticians expected a closer alignment between rotation and magnetism on Uranus and Neptune.

It is almost certain that the same dynamo process as that responsible for Earth's magnetic field sustains the magnetic fields of Uranus and Neptune. This happens in Earth's molten metallic core, and it occurs within the liquid metallic hydrogen inside Jupiter and Saturn. Unlike these two giants, there is no shell of metallic hydrogen inside Uranus and Neptune. It is probable that the electrical conductivity within Uranus and Neptune is provided by water-rich material that has a conductivity that is about two orders of magnitude less than that of metals. It is also likely that this conductivity comes from protons, not electrons, within the ionized waters.

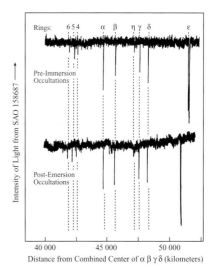

Fig. 11.7 Discovery of the thin rings of Uranus Astronomers recording the light of a star that was expected to disappear behind Uranus, on 10 March 1977, unexpectedly recorded short dips in the starlight before the star passed behind the planet (*top*). The same pattern was repeated when the star reappeared (*bottom*), indicating that narrow rings briefly block out the starlight at the same distance on opposite sides of the planet. The strong and abrupt absorption of starlight indicates that the narrow rings are quite opaque and have well-defined edges. These observations were taken from high above the Indian Ocean aboard the Kuiper Airborne Observatory. (Courtesy of James L. Elliot.)

11.4 Rings of Uranus and Neptune

Narrow, widely spaced rings around Uranus

Astronomers have had a history of happy accidents concerning Uranus, starting with William Herschel's (1738–1822) serendipitous discovery of the planet in 1781. Another lucky incident occurred on 10 March 1977, when the planet was scheduled to pass in front of a faint star. By observing such a stellar occultation, astronomers hoped to determine properties of the planet's atmosphere, and to accurately establish its size from the duration of the star's disappearance behind it.

Because of uncertainties in the predicted time of the star's disappearance, one telescope was set into action about 45 minutes early. Soon after the recording began, the starlight abruptly dimmed but then it almost immediately returned to normal, producing a brief dip in the recorded signal. At first, the dip was attributed to a wisp of cloud on Earth or to an unexpected change in the telescope's orientation. But the star blinked on and off several times before and after the planet covered it. Moreover, each dip on one side of Uranus was matched by another one on the other side, at the same distance from the planet (Fig. 11.7). The brief dips were due to something very

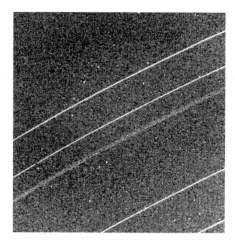

Fig. 11.8 The rings of Uranus   This *Voyager 2* image, taken on 23 January 1986, shows five of the nine rings of Uranus that had been previously inferred from Earth-based observations of their brief occultation of a star's light. In this view, sunlight striking the rings' particles was reflected back toward the camera, showing that the dense parts of the ring system consist of narrow rings with wide gaps. In contrast, Saturn's main rings are broad with narrow gaps. From bottom to top, the rings are designated by the Greek letters α, β, η, γ and δ. (Courtesy of NASA/JPL.)

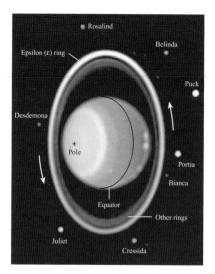

Fig. 11.9 Rings and small satellites of Uranus   Eight of Uranus' small satellites circle the planet just outside its bright epsilon ring. This image, taken with the *Hubble Space Telescope* on 28 July 1997, is a false-color composite of three images taken at different infrared wavelengths in which Uranus appears relatively dim but the rings and moons do not. The satellites range in size from 40 kilometers across for Bianca, to 150 kilometers for Puck. The arrows denote their direction of revolution about Uranus. White clouds are seen just above the planet's blue-green methane atmosphere. (Courtesy of NASA/STScI.)

narrow and very close to Uranus, and their mirror-like symmetry indicated that a family of narrow rings, which blocked out the star's light but could not be seen directly from the Earth, surrounds the planet.

During the next few years, observations of more than 200 stellar occultations by Uranus revealed the details of nine narrow rings. In order of increasing distance from Uranus, the rings are named 6, 5, 4, α, β, η, γ, δ and ε, following the differing notation of the discoverers. From the brief duration of the dips of blocked starlight, astronomers concluded that all but one of the individual rings could be no wider than 10 kilometers. The relatively long time between the dips indicated that the threadlike rings are separated by hundreds of kilometers of nearly empty space. These skeletal, web-like rings are unlike any seen before, all exceptionally narrow and widely spaced from each other. Since the rings are so narrow, and separated by wide spaces of almost nothing, they are extremely difficult to see with even a large telescope on Earth.

When *Voyager 2* arrived at Uranus in 1986, nearly a decade after the discovery of its narrow rings, instruments on the spacecraft confirmed all the known rings, and added at least two (Fig. 11.8). They found the λ ring, a narrow strand between the δ and ε rings, and another one interior to ring 6. The spacecraft also discovered at least 10 small moons that are located just outside the ring system.

The irregular orientation, narrowness, and shapes of the rings of Uranus are attributed to nearby small moons

(Fig. 11.9). The repeated gravitational tugs of two of them, Cordelia and Ophelia, pull the ε ring into its oval shape and restrain its edges. These tiny moons flank the ring, controlling its shape in much the same way that the shepherd satellites, Pandora and Prometheus, constrain Saturn's F ring. Nearby moons probably sharpen the edges of the other rings, keeping them from spreading out as the result of particle collisions.

The *Hubble Space Telescope* was used to discover two fainter outer rings of Uranus, which were not fully recognized until 2005. That brought the total of known dusty rings of Uranus to 13. The space telescope has also been used to image the rings as they slowly closed up, as viewed from near Earth, snagging a rare edge-on view in August 2007 (Fig. 11.10). Astronomers only see the ring edges every 42 years as the planet follows its leisurely 84-year orbit about the Sun.

Both the ring particles and the inner moons of Uranus are very dark, quite unlike the bright particles and tiny moons found in Saturn's wide rings. In fact, the particles of Uranus's rings reflect only about 2 percent of the sunlight falling on them, making them as dark as charcoal. Most investigators agree that the material is dark because it is rich in carbon, either derived from a methane coating or primordial in origin.

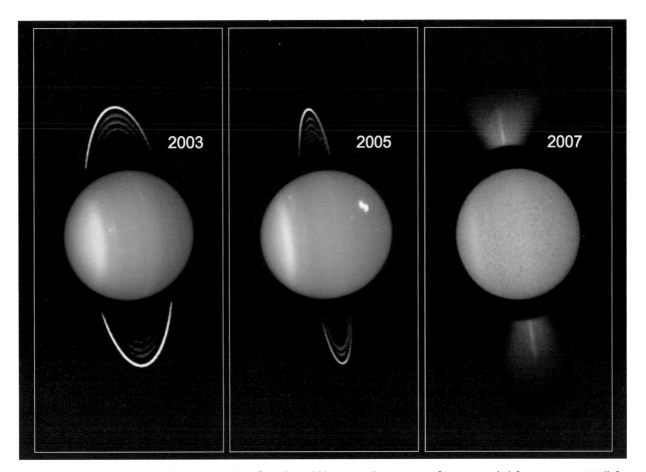

Fig. 11.10 Rings on edge   These images were taken from the *Hubble Space Telescope* over a four-year period, from 2003 to 2007 (*left to right*), as Uranus moved along one-twentieth of its 84-year journey around the Sun. As viewed from the Earth, the rings are always nearly perpendicular to the planet's orbital plane, but they gradually close up from partly open to edge on. When tiled edge-on to Earth, on 14 August 2007, the rings appeared as spikes above and below the planet, within a fan-shaped glare – an image artifact. The fainter outer rings appear in the 2003 image, but were not noticed until they were seen in the 2005 image. Uranus has a total of 13 dusty rings. (Courtesy of NASA/ESA/M. Showalter, SETI Institute.)

Neptune's sparse thin rings and arcs

After the discovery of the rings of Uranus by watching a distant star pass behind the planet, astronomers hoped to repeat the achievement by observing stellar occultations by Neptune, but the results were inconclusive. Sometimes the starlight would remain unchanged before and after the planet directly occulted the star. At other times the star would blink on and off, but always on just one side of the planet. Because the brief dimming of starlight was not symmetrical about the planet, and not all stellar occultations produced a blinking signal, the hypothetical rings became shortened, in the minds of the astronomers, to ring-arcs that only reached partway around the planet. Chance might then dictate which astronomers would detect the obscuration.

Voyager 2 clarified the problem. Neptune's ring-arcs turned out not to be isolated segments, but rather three thicker portions of one very thin ring (Fig. 11.11). The ring is continuous, stretching all the way round the planet just like any well-behaved ring, but it does not have an even width or density all around. Its material is generally spread so thinly that it does not noticeably dim a star's light, and the ring is only wide and dense enough to hide a star in three arc-like concentrations. They have been named Liberté, Egalité and Fraternité after the French revolutionary slogan. It was these wide high-density clumps that had been detected from Earth, blocking starlight and giving the impression of disconnected arcs. The rest of the ring couldn't be seen from Earth because it is so transparent, and hence below the threshold of detection.

The existence of such clumps, or concentrations, in the rings was an enigma. Every time that the ring particles collide and bump together, they must change speed, and gradually spread around the ring away from the clumps.

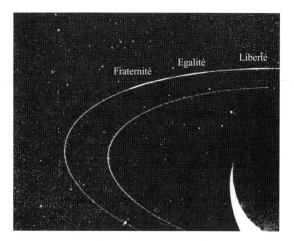

Fig. 11.11 Neptune's rings As *Voyager 2* left Neptune in August 1989, the planet's narrow rings were backlit by the Sun, enhancing the visibility of the rings' dusty particles. The outer ring consists of at least three dense clumps of orbiting debris, named Liberté, Egalité and Fraternité, which stand out from the thinner remainder of the ring. Astronomers on the ground had only detected the clumps during some stellar occultations, and assumed that the ring was incomplete. The outermost ring is named Adams, and the innermost is designated Le Verrier. They are named for John Couch Adams (1819–1892) and Urbain Jean Joseph Le Verrier (1811–1877) who independently predicted the existence of the then unknown planet Neptune. A third, fainter ring is located closer to Neptune than the two main rings; it has been named Galle after Johann Gottfried Galle (1812–1910) who found the planet close to both of Adams' and Le Verrier's predicted positions using a 0.23-meter (9-inch) refractor. (Courtesy of NASA/JPL.)

Unless something is confining the material in the arc-like concentrations, it should spread uniformly around the entire ring in just a few years. Some external force must therefore be holding the material in place, and keeping it within the three arcs. A small moon might hold the clumps together by its gravity, and astronomers think they have found it – the satellite Galatea orbits Neptune just inside the ring.

Altogether, *Voyager 2*'s cameras found five rings around Neptune, named in honor of astronomers. From the innermost to outermost, they are the Galle, Le Verrier, Lassell, Arago and Adams rings. The faint, innermost Galle ring is broad, roughly 1700 kilometers wide, and may extend all the way down to the top of the planet. The outermost Adams ring is bright and narrow, and contains the three bulging clumps.

Unlike the main narrow rings of Uranus, the brightest Adams and Le Verrier rings of Neptune contain a vast amount of microscopic dust, apparently produced by grinding down or eroding larger particles. In fact, the sparse, dusty rings of Neptune contain only about a thousandth as much matter as the rings around Uranus, and a million times less matter than Saturn's rings. Put together, all the particles in Neptune's rings would make a body only a few kilometers across. So small nearby satellites could have been the source of Neptune's rings. Meteoritic bombardment over the eons could, for example, have pulverized such a moon into rubble, producing all that dust. Even now, three small moons are located deep within Neptune's ring system, between the Galle and Le Verrier rings.

Formation and evolution of the rings of Uranus and Neptune

The austere rings that now circle Uranus and Neptune may have had a violent and chaotic past, arising from catastrophic collisions of moons or when one larger satellite moved inward by tidal interaction with the planet until it was close enough to be ripped into pieces. The inner moons and larger particles in the rings were then probably gradually broken up by collisions into smaller ones.

All the ring particles we see today might eventually be eroded away by meteoritic bombardment, ground into fine dust by particle collisions, or displaced by gravitational interaction with neighboring satellites. Once the rings have been turned into dust, as they have in parts of Uranus's ring system and most of Neptune's, they must eventually spiral inward into the planet's atmosphere where they will burn up. Thus an entire dusty ring system will inevitably decay and disappear over astronomical times.

This need not imply that rings will vanish from Uranus and Neptune. The rings could easily be replaced by the collisional breakup or meteoritic erosion of tiny moons already embedded in the rings. If you broke up all the satellites now within Neptune's rings, you might produce rings as magnificent as Saturn's present ones. This now brings us to some of the larger satellites lying outside the rings of these two planets.

11.5 The large moons of Uranus and Neptune

Five major moons of Uranus

Uranus possesses five major satellites discovered telescopically from Earth before the space age, and named Miranda, Ariel, Umbriel, Titania and Oberon (Table 11.2), and as the result of modern telescopes and the *Voyager 2* spacecraft we now know that the planet possesses at least 27 moons. All of the large satellites orbit Uranus near its equatorial plane, perpendicular to its orbital plane, and they are

Table 11.2 The five large moons of Uranus[a]

Name	Mean distance from Uranus (radii)	Orbital period (days)	Radius (kilometers)	Mass (10^{21} kilograms)	Mass density (kg m^{-3})
Miranda	5.08	1.41	236	0.065	1214
Ariel	7.47	2.52	579	1.29	1592
Umbriel	10.41	4.15	585	1.22	1459
Titania	17.07	8.70	789	3.42	1662
Oberon	22.82	13.46	761	2.88	1559

[a] The orbital distances are given in units of the radius of Uranus, which is 25 559 kilometers, or about four Earth radii. The satellite radii are given in units of kilometers. By way of comparison, the mean radius of the Earth's Moon is 1738 kilometers. The mass is given in units of 10^{21} kilograms – our Moon's mass is 73.5 in these units. The mass density is given in units of kilograms per cubic meter (kg m^{-3}).

all tidally locked into synchronous orbits with one side perpetually facing the planet. As a group, they are similar in size to the mid-sized satellites of Saturn. The two largest and outermost moons, Titania and Oberon, are roughly half the size of the Earth's Moon; the smallest and innermost, Miranda, is about one-seventh the lunar size. Infrared spectroscopy from Earth indicated that they all have water ice on their surfaces, but their icy surfaces are dark looking. Their bulk mass densities (Table 11.2) imply that the four largest moons of Uranus are about half rock and half water ice, so they are rocky on the inside, as well as dirty on the outside.

Because they are relatively small and very cold outside, these moons were expected to be heavily cratered ice-balls, devoid of any signs of internal activity. When *Voyager 2* photographed their icy surfaces at close range, the expected craters were seen, but the surfaces also displayed surprising diversity. They have been fractured and cracked open, probably as the result of surface expansion, and covered by flowing ice or perhaps even liquid water that subsequently froze. Internal heat supplied by radioactive elements may have produced the surface expansion, melted the ice, and generated icy volcanic flows. Because radioactive elements are embedded in rock, the relatively large amount of rock in the large moons of Uranus, as compared to Saturn's satellites of about the same size, produces greater internal heat.

It was Prospero's daughter Miranda who, in Shakespeare's *The Tempest*, proclaimed:

O, Wonder!
How many goodly creatures are there here!
How beauteous mankind is! O, brave new world,
That has such people in't!

The Tempest, Act V, Scene i, lines 182–5

and her counterpart at Uranus certainly has all the earmarks of a "brave new world".

The landscape on Miranda is one of the most amazing yet observed in the solar system. It includes deep canyons, old cratered plains, bright younger terrain, and an eclectic mixture of ridges, grooves, mountains, valleys, fractures and faults (Fig. 11.12).

There are two possible explanations for Miranda's jumbled surface. One explanation supposes that the satellite was once blasted apart by one or more catastrophic collisions and then pulled together again under the combined gravitation of its pieces, reforming into a born-again moon. It may even have been broken up and reassembled several times. The alternative explanation holds that Miranda is a half-grown world whose development was interrupted. It supposes that the moon heated up inside, soon after formation, and that the dense rocky material began to sink toward the center while the lighter icy substances started rising to the surface. But as Miranda cooled the internal heat dwindled, and the differentiation process could not be finished.

Neptune's Triton, a large moon with a retrograde orbit

Jupiter, Saturn and Uranus have a flock of satellites whose orbits mimic those of the planets around the Sun. Their larger moons revolve in regularly spaced, circular orbits in the same direction as the rotation of the planet and close to the planet's equatorial plane, presumably because they share the rotation of the nebular disk from which the planet and its satellites formed. The radii, orbital distances and other characteristics of these regular satellites also tend to differ in smooth progression. In sharp contrast to Jupiter,

Fig. 11.12 Miranda   The innermost large moon of Uranus, Miranda, is seen at close range in this *Voyager 2* image, which was taken on 24 January 1986. The great variety and directions of the fractures and troughs, and the different densities of impact craters on them, signify a long, violent history for Miranda. Two distinct types of terrain are visible. They are rugged, higher-elevation terrain, disappearing into the shadows (*upper right*), and the lower-elevation, striated terrain (*upper left*). The light gray cliff (*bottom right*) is 20 kilometers high; the larger crater (*lower middle*) is about 25 kilometers across. (Courtesy of NASA/JPL.)

Saturn and Uranus, the planet Neptune lacks a system of regular larger satellites.

Neptune's largest moon, Triton, is the only large satellite in the solar system to circle a planet in the retrograde direction, opposite to the planet's direction of rotation (Fig. 11.13). This oddity is compounded by the high orbital inclination. The satellite's orbital plane is tilted at an enormous 157 degrees from the planet's equator. Nereid, the outermost moon of Neptune, discovered by Gerard Kuiper (1905–1973) in 1949, adds to the mayhem. It has the most elongated orbit of any planetary satellite, seven times as distant from the planet at its farthest compared with its closest approach.

Triton's unusual retrograde orbit was quickly established, soon after its discovery – by William Lassell (1799–1880) in 1846, just three weeks after the discovery of Neptune. But due to its great distance, the radius, mass and reflectivity of Triton remained uncertain until 25 August 1989 when instruments on the *Voyager 2* spacecraft measured them. With a radius of about 1353 kilometers, Triton is about three-quarters of the radius of the Earth's Moon, at 1738 kilometers. The slight gravitational tug exerted on

Voyager 2 by Triton yielded a mass of 2.141×10^{22} kilograms for the satellite. The size and mass are combined to give a mean mass density of about 2061 kilograms per cubic meter. This is significantly more dense and rock-rich than the large moons of Uranus, between 1400 and 1700 kilograms per cubic meter.

Triton's frozen surface, thin atmosphere and geyser-like eruptions

Measurements from *Voyager 2* indicated that Triton is the ultimate icebox, with a daytime surface temperature of 37.5 kelvin. This deep freeze is only about three dozen degrees above absolute zero, when nothing can move, not even an atom. In fact, Triton has the coldest measured surface of any natural body in the solar system! It is so cold because it is very far away from the Sun, therefore receiving little sunlight, and also because Triton reflects more of the incident sunlight than most satellites — only Enceladus and Europa are comparable. As a result, the total amount of sunlight absorbed by Triton's surface is less than that of any other planet or satellite. A frosty ice coating overlies

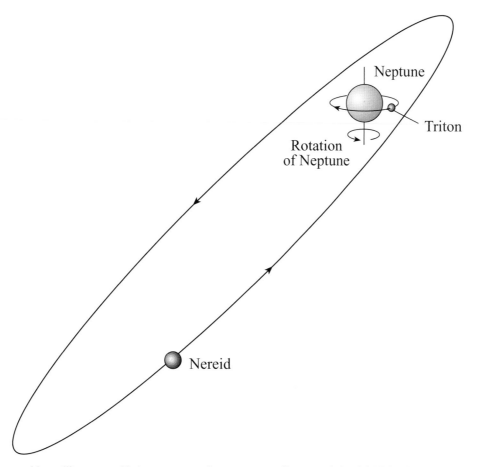

Fig. 11.13 Neptune's odd satellites   Nereid, the outermost of Neptune's satellites, travels in a highly inclined, eccentric orbit, in the same direction as that of the planet's rotation. Triton, the largest satellite of Neptune, travels around Neptune in a circular obit, but, unlike any other largest satellite of the giant planets, it travels in the opposite retrograde direction to the rotation of Neptune. In addition, careful analysis of Triton's motions shows that the satellite is in a decaying orbit and is slowly being pulled toward Neptune.

all of the surface features, reflecting the incident sunlight (Fig. 11.14). The brilliant ice has a salmon-pink tint with peach hues, possibly due to organic compounds derived from methane by the bombardment of energetic particles from the solar wind and Neptune's radiation belts.

Although nitrogen and methane frosts are apparently the dominant constituents of Triton's visible disk, water ice is needed to support and preserve the observed topography, including cliffs and ridges that exceed one kilometer in height. At the frigid temperature of Triton, water ice is as strong as steel, and behaves like hard rock on Earth; but the methane and nitrogen ice do not have sufficient strength to support the elevated features, which would deform and collapse under their own weight. Thus, thin, brilliant veneers of nitrogen and methane ice apparently overlie a rigid crust of water ice.

Despite its frigid surface conditions, an exceedingly tenuous atmosphere envelops Triton. It consists mainly of nitrogen molecules, the same gas that makes up most of the atmospheres of Earth and Saturn's moon Titan.

But the cold, thin atmosphere on Triton has a surface pressure of only 15 microbars, or 15×10^{-6} bars, which is ten-millionths the surface pressure of the atmosphere on Titan and fifteen-millionths of the air pressure at sea level on Earth.

Voyager 2's cameras showed that Triton's southern polar cap was then in the process of dissipating. Nitrogen ice was sublimating, or changing directly from the solid to vapor form, and supplying the atmosphere with gas. The vaporized nitrogen gas is probably carried by atmospheric winds to the dark northern hemisphere.

Even in the middle of southern summer, a bright ice cap extends from the south pole three-quarters of the way to Triton's equator (Fig. 11.14). It is so cold that some of Triton's air freezes out at its poles, coating them with a huge ice cap of frozen nitrogen. It is also too cold for water ice to vaporize from Triton's surface and enter its atmosphere. By way of comparison, the Earth's polar caps contain frozen water ice, for it is too warm for nitrogen to freeze at our planet's poles.

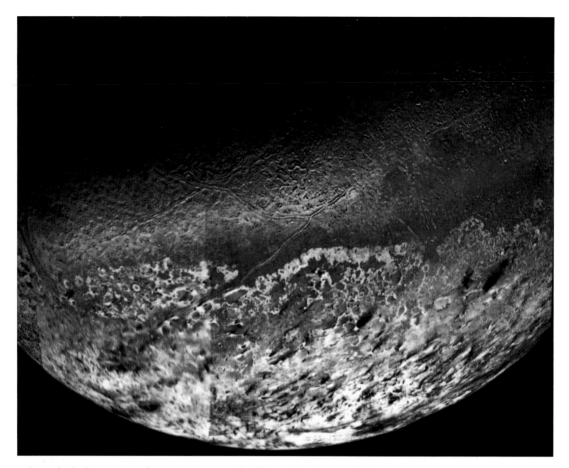

Fig. 11.14 Triton, the largest moon of Neptune   A mosaic of the south polar cap of Triton, the largest satellite of Neptune, taken in August 1989 from the *Voyager 2* spacecraft. At the time of this flyby, Triton was the coldest measured object in the solar system with a surface temperate of 37.5 kelvin. It is so cold that most of Triton's thin nitrogen atmosphere is condensed as frost, making it the only satellite in the solar system known to have a surface coated with deposits of solid nitrogen ice. Small admixtures of highly reflective methane ice are also present on the surface; they have been reddened and darkened by the action of energetic radiation. Dark plumes and streaks probably mark vents of ice volcanoes, where nitrogen gas has been driven outward in geyser-like eruptions from beneath the surface and carried by the winds into elongated streaks. (Courtesy of NASA/JPL.)

Numerous parallel dark streaks in the bright southern cap are apparently blown by the prevailing winds in Triton's tenuous atmosphere and strewn over the ice. The dark streaks emanate abruptly from specific points, and thin away toward their ends. Some are only a few kilometers in length, while others extend more than 100 kilometers. They must have formed relatively recently, for they seem to overlie deeper ice deposits, and it is unlikely that they could survive sublimation of the polar ice.

The dark streaks are probably related to jets and geysers of nitrogen seen erupting from the sunlit polar ice. These plumes rise in straight columns to an altitude of about 8 kilometers, where dark clouds of material are left suspended and carried downwind for over 100 kilometers, like smoke wafted away from the top of a chimney. Since the active plumes occur where the Sun is directly overhead, the solar heat might energize them.

Scientists have not reached a consensus about what produces the plumes, but one likely explanation is that geysers are sending up plumes of nitrogen gas laced with extremely fine dark particles. Triton is far too cold to spout steam and water, like geysers on Earth, but Sun-powered geysers might expel dark material when pent-up nitrogen gas becomes warm and breaks through an overlying seal of ice.

On Triton, the subterranean heat might be accumulated from sunlight, which passes through the translucent ice and is absorbed by darker methane or other carbon-rich material encased beneath. The overlying nitrogen ice would trap the solar heat, for it is opaque to thermal infrared radiation, producing a solid-state greenhouse effect. Nitrogen gas, pressurized by the subsurface heat, then explosively blasts off the iced-over vents or lids, launching volcanic plumes of gaseous nitrogen and

Fig. 11.15 Smooth and bumpy terrain on Triton Smooth volcanic plains, created by ice volcanoes on Triton, form the flat, frozen surfaces of ice lakes (*left*), most likely filled with water ice and perhaps coated with deposits of nitrogen ice. The absence of any impact craters indicates that the surface is relatively young. The rugged terrrain (*right foreground*) is Triton's cantaloupe terrain, a network of interfering, closely spaced dimples or depressions termed cavi, each 25 to 35 kilometers across. They are also of internal origin, but not due to volcanic flow. The cantaloupe terrain may be explained by a gravitational instability in which less dense material rises through overlying dense material, overturning the icy crust. The rising blobs of ice are known as diapirs. (Courtesy of NSA/JPL.)

ice-entrained darker material into the tenuous atmosphere, just as the water in an overheated car radiator is explosively released when the radiator cap is removed.

Astronomers expected to find a surface covered by craters, but there are almost no craters in sight on Triton. Much of the visible surface outside the polar cap resembles the skin of a cantaloupe, containing numerous pits or dimples with low raised rims and ridges that snake their way through them (Fig. 11.15). But the dimpled pits are too similar in size and too regularly spaced to be impact craters, and appear to have an internal origin.

Long cracks or faults on Triton seem to have been partially filled with oozing water ice. Vast frozen basins found within Triton's equatorial regions have apparently been filled by icy extrusions flowing out from the warm interior, like a squeezed slush cone. These frozen lakes of water ice look like inactive volcanic calderas, complete with smooth filled centers, successive terraced flows and vents (Fig. 11.15).

Origin and evolution of Triton

Triton's retrograde and inclined orbit suggests it may not date back to the planet's formation, and that Triton was once a separate world which was captured into an eccentric, backwards and tipped orbit in the remote past. Neptune's lack of an ordered family of large satellites might also be explained if Triton was born in its own independent orbit around the Sun, and was subsequently captured by Neptune, destroying any regular satellite system the planet may have had in the process.

Neptune could only pull Triton into its gravitational sphere of influence if the passing body lost some energy as it went by. Otherwise, Neptune and Triton would just continue to go on their separate ways. One possibility is that Triton smashed into one of the planet's more substantial satellites, losing enough energy that it could not escape Neptune's gravitational pull and went into orbit around it. Or Triton might have grazed the fringes of Neptune's

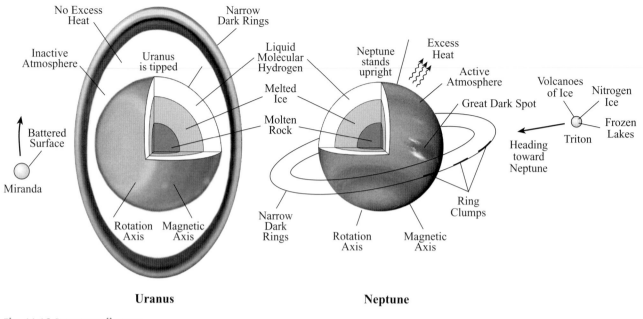

Fig. 11.16 Summary diagram

atmosphere where friction slowed it down enough to be captured.

When Triton began revolving around Neptune, its orbit would have most likely been very elliptical. But tides raised on Triton by Neptune could have circularized the satellite's orbit. While its orbit was evolving, Triton could have cannibalized or ejected satellites it encountered, thereby removing any large regular satellites Neptune may have once had, while also knocking Nereid into its unusual, highly eccentric orbit.

This explanation for Triton's origin is considered likely but still speculative, for we do not know for certain that the moon formed separately from Neptune. But we are fairly certain about Triton's suicidal future. Tidal interaction between Neptune and Triton is now gradually drawing the moon inward. Some time in the very distant future, Triton will be pulled close enough to Neptune for the planet's tidal forces to rip the moon apart, perhaps forming rings as bright and magnificent as Saturn's are today.

12   **Asteroids and meteorites**

- There are billions of asteroids in the main asteroid belt, located between the orbits of Mars and Jupiter.

- The asteroid belt is largely empty space, and a spacecraft may safely travel through it.

- Hundreds of Trojan asteroids circle the Sun in the same orbit as Jupiter. These asteroids are located near the two Lagrangian points where the gravity of the Sun balances that of Jupiter.

- The Earth resides in a swarm of asteroids. Many of these near-Earth asteroids travel on orbits that intersect the Earth's orbit, with the possibility of an eventual devastating collision with our planet.

- Asteroids can be chaotically shuffled out of certain orbits in the main belt, and redirected into the inner solar system.

- The asteroids are the pulverized remnants of former worlds that failed to coalesce into a single planet.

- Groups of asteroids, known as families, have very similar orbits. The members of each family are the collision fragments of a larger object, which was itself much smaller than a major planet.

- The combined mass of billions of asteroids is less than 5 percent of the mass of the Earth's Moon.

- The largest body in the main asteroid belt, 1 Ceres, and the first to be discovered there, is about 950 kilometers across and contains about one-third of the total mass of all the asteroids.

- Unlike most, if not all, of the other asteroids, 1 Ceres has a round shape, suggesting a differentiated interior with a rocky core, thick mantle of water ice, and a dusty outer crust. 1 Ceres has been designated a dwarf planet.

- The gravity of the large majority of asteroids is too weak, and the mass too low, to pull them into a round shape. Their irregular, chipped shapes have been formed by eons of collisions.

- The colors of sunlight reflected from asteroids indicate that they formed under different conditions prevailing at varying distances from the Sun.

- Roughly 75 percent of the main-belt asteroids are the dark, black carbonaceous C-type orbiting the Sun in the outer half of the belt; about 15 percent of asteroids are bright, red, silicate S-type, residing on the sunward side of the main belt. Metallic, M, asteroids account for some of the remaining ones, and a few V asteroids are covered with volcanic basalt.

- Asteroids may be mined for minerals or water.

- Periodic brightness variations tell us that most asteroids are non-spherical objects spinning with periods of a few hours.

- Radar images of asteroids reveal diverse shapes and cratered surfaces, ranging from solid rock to loose rubble, as well as binary asteroids and even a triple one.

- The close-up view obtained by passing spacecraft indicates that asteroids have been battered and broken apart during catastrophic collisions in years gone by.

- Some asteroids are thought to be rubble piles, the collected fragments of past collisions held together by gravity; other asteroids are solid rocks of uniform internal composition. Asteroids 253 Mathilde and 25143 Itokawa are rubble piles, whereas asteroid 433 Eros is a battered but solid rock.

- The *Near Earth Asteroid Rendezvous* (*NEAR*) spacecraft was the first to orbit an asteroid and the first to land on one. *NEAR* circled the near-Earth asteroid 433 Eros for a year, examining its dusty, boulder-strewn landscape in great detail, obtaining an accurate mass for the asteroid, and showing that much of it is solid throughout.

- The Japanese *Hayabusa* spacecraft has orbited the small, near-Earth asteroid 25143 Itokawa, whose orbit crosses that of the Earth. This asteroid, just 535 meters in its longest dimension, has a lumpy, oblong shape composed of two parts with rough terrain joined by a smooth region. It is a loose collection of rubble held together by its weak gravity, with surface rocks sorted by shaking and jostling. *Hayabusa* landed on the asteroid and may return a sample of it to Earth.

- The *Dawn* spacecraft, launched in 2007, is expected to orbit the asteroid 4 Vesta from September 2011 to April 2012 and to orbit the dwarf planet–asteroid 1 Ceres in 2015. Astronauts may land on an asteroid by 2025.

- Asteroid rotation periods range from a couple of minutes to a few months, and some of them wobble instead of rotating uniformly.

- Meteorites are rocks from space that survive their descent to the ground.

- The number of recovered meteorites more than doubled when scientists discovered a large number of them on the blue ice of Antarctica.

- A few meteorites may have been blasted off the Moon or Mars, but most of them are chips off asteroids.

- The organic matter found in meteorites predates the origin of life on Earth by a billion years; but the meteoritic hydrocarbons are not of biological origin.

12.1 The orbits of asteroids

The main belt of asteroids

Millions of asteroids are confined within a wasteland between the orbits of Mars and Jupiter, like so much rubble left over from the creation of the solar system. Most of them occupy a great ring, known as the asteroid belt, at mean distances of 2.2 to 3.3 AU from the Sun and with orbital periods of 3 to 6 years. For comparison, the mean distance between the Earth and the Sun is 1 AU, about 150 million kilometers. Not all asteroids lie in this belt, but those that do are said to belong to the main belt.

The Sicilian astronomer Giuseppe Piazzi (1746–1826) discovered the first asteroid accidentally, on 1 January 1801, the first day of the 19th century. He named the tiny object Ceres, in honor of the patron goddess of Sicily. Another asteroid, named Pallas, was located the following year, and the third and fourth, designated Juno and Vesta, were found in 1804 and 1807, respectively.

Ceres is the largest body in the main belt. The mass of this asteroid comprises about one-third the total mass of all the asteroids. It is about 950 kilometers across, and has a water-rich surface. Smaller Pallas and Vesta are about half the size of Ceres and a quarter of the mass. Vesta has experienced significant heating and differentiation, showing signs of a metallic core and a dry surface covered by basaltic lava flows.

We now know that there are very many smaller asteroids. After a gap of nearly a century, astronomical photography enabled them to be found by the hundreds. During a long-exposure photograph made through a telescope tracking the stars, the faster-moving asteroids make short trails while the stars look like dots.

When an asteroid is discovered, it receives a temporary designation, consisting of the year of discovery followed by two letters. The first letter indicates the half-month of the asteroid discovery – the letters "I" and "Z" are not used – to make a total of 24 half-months. The second letter shows the order of discovery within that half-month, but the letter "I" is not used for this second letter. Thus, the asteroid 1998 KY was the 24th (letter Y) asteroid found during the second half of May (letter K) in 1998.

After an asteroid has been observed often enough for an accurate orbit to be established, it receives a number corresponding to the chronological order of reliable orbit determination – but not the order of discovery. The number is often followed by a name provided by the discoverer. As of January 2010, there were 231 665 asteroids with accurate orbits and official numbers, and about 200 000 more have been observed well enough to obtain preliminary orbits.

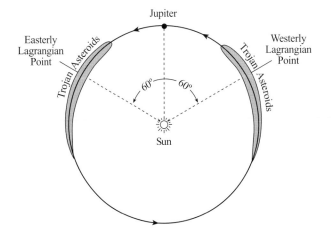

Fig. 12.1 Trojan asteroids Asteroids that share Jupiter's orbit are known as the Trojan asteroids. They are located near the point where the gravitational force of Jupiter and the Sun are equal. The gravitational perturbations of the inner planets produce slight swinging motions, so the Trojan asteroids oscillate within the two shaded regions. Some of the Trojan asteroids may occasionally move close enough to be captured by Jupiter's gravity, thereby accounting for the planet's unusual outer satellites.

With a million asteroids whirling around the Sun, the main belt was once thought to be a hazard to space travel. In reality, however, the volume of space they occupy is so large that any one asteroid is typically several million kilometers away from its nearest neighbor, and the asteroids are so small that it would be hard to hit one even if you tried to do so. Thus, despite their vast numbers, the asteroids leave plenty of room for space flight. The *Pioneer 10* and *11*, *Voyager 1* and *2*, *Ulysses*, *Galileo* and *Cassini* spacecraft have all passed through the asteroid belt unharmed.

And since the asteroid belt is largely empty space, collisions between asteroids are relatively rare over short timescales. It takes tens of millions of years for shattering collisions to occur, so they do happen but it takes a long time.

Trojan asteroids

Not all asteroids are found in the main belt. An especially interesting type is further away from the Sun than the asteroid belt. They move along Jupiter's orbit, keeping pace with the giant planet. The first known one, 588 Achilles, was discovered photographically by the Heidelberg astronomer Max Wolf (1863–1932) in 1906. Hundreds of them are now known, traveling on both sides of Jupiter in two clouds, one preceding the giant planet and one following it (Fig. 12.1). As with Achilles, they are all named after heroes of the Trojan War and they are therefore collectively known as the Trojan asteroids.

The Trojan asteroids are held captive by the gravity of both Jupiter and the Sun. They are found near two of the five Lagrangian points, named after the Italian-born French mathematician Joseph Louis Lagrange (1736–1813) who predicted their existence 134 years before the discovery of 588 Achilles. At these points, the gravitational force of Jupiter is equal to that of the Sun, which is much more massive than Jupiter but also a lot further away from the asteroids. These Lagrangian points lie in the corners of equilateral triangles that have Jupiter and the Sun at the other corners (Fig. 12.1).

The Trojan asteroids do not stay precisely at a Lagrangian point, but instead oscillate around it. They pace back and forth along Jupiter's orbit in paths that take them toward and away from the planet over a cycle lasting hundreds of years.

Once locked into their haven near the Lagrangian points, the Trojan asteroids move at slow speeds along well-defined paths. The Trojans therefore suffer fewer collisions than their counterparts in the main asteroid belt, and their surfaces have probably gone unchanged for billions of years. Thus, the Trojan asteroids may be pristine remnants of the early solar system.

The Trojan asteroids are in a 1:1 mean-motion resonance with Jupiter, meaning that they have the same orbital period. Other groups of asteroids are locked into other dynamical resonances with Jupiter. These include the Hildas at the 3:2 mean-motion resonances, where the asteroids orbit with two-thirds of Jupiter's orbital period, the Thule group at the 4:3 resonances, and the Hungarias group at the 9:2 resonances.

Near-Earth asteroids

Although the vast majority of asteroids travel in the main belt lying between the orbits of Mars and Jupiter, there are some notable exceptions that reside within the inner solar system. Known as the near-Earth asteroids, they move inward toward our planet as they travel around the Sun.

One of them could hit our planet someday, with devastating consequences. So astronomers are still actively trying to find all of these interlopers and monitor their motions, with the hope of avoiding the collision. Since these asteroids move closer to the Sun than Mars, they travel at faster speeds than asteroids in the main belt, and thus make longer trails on photographic or computerized images taken to discover asteroids.

There are three populations of near-Earth asteroids, called the Atens, Apollos and Amors (Fig. 12.2). Both the Aten and Apollo asteroids move on eccentric orbits that can cross the Earth's path in space. The Atens are always close to the Sun, never moving out as far as the orbit of

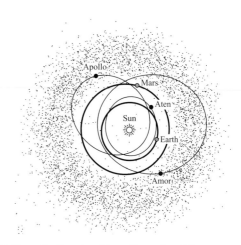

Fig. 12.2 Near-Earth asteroids   The paths of three representative near-Earth asteroids, 1221 Amor, 1862 Apollo and 2062 Aten, all come closer to the Sun than most asteroids, located in the main belt beyond the orbit of Mars. Amor crosses the orbit of Mars, and almost reaches the Earth's orbit. Apollo crosses the orbits of Mars, Earth and Venus (not shown). Aten is always fairly close to the Earth's orbit.

Mars. The elongated orbits of the Apollo objects loop in from the main belt to within the Earth's orbit. The Amors travel around the Sun between the orbits of Mars and the Earth, and often cross the orbit of Mars.

Chaotic orbits of asteroids

Why do some asteroids move near the Earth, while most of them stay in the asteroid belt? Gaps of missing asteroids in the main belt provide some clues to these wandering interlopers. These are the Kirkwood gaps (Fig. 12.3), discovered long ago by the American astronomer Daniel Kirkwood (1814–1895).

The locations of these clearings correspond to orbital resonance with Jupiter, in which the orbital periods are exact fractions of the giant planet's period. Any object that orbits the Sun at the 3:1 resonance, for example, would have exactly one-third, or 1/3, the orbital period of Jupiter, and it would complete three circuits around the Sun for every one that Jupiter completes. Such an asteroid would revolve around our star at a distance of 2.5 AU with a period of 3.95 years, compared with Jupiter's orbital distance of 5.2 AU and orbital period of 11.86 years. An asteroid that happened to stray into this resonance would come close to Jupiter at almost the same part of the asteroid's orbit at regular 11.86-year intervals and the accumulated gravitational interaction with Jupiter could dislodge the asteroid from its orbit.

So the net effect of the resonance is to clear asteroids out of the resonant locations, and some of them might eventually be brought into Earth-crossing orbits. Asteroids

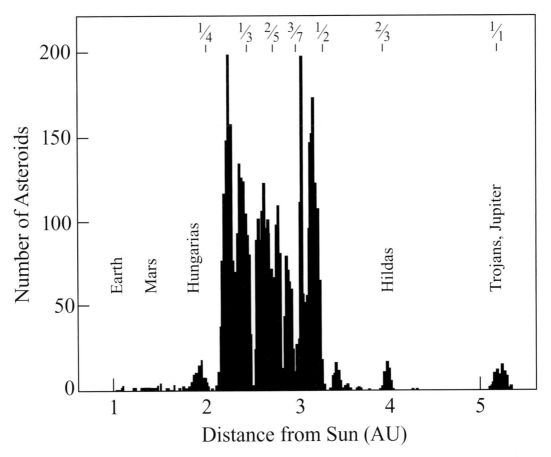

Fig. 12.3 Kirkwood gaps in asteroid orbits Most of the asteroids are found in the asteroid belt that lies between 2.2 and 3.3 AU from the Sun, but there are very few asteroids at certain distances from the Sun. These Kirkwood gaps have orbital periods of 1/4, 1/3, 2/5, 3/7 and 1/2 of Jupiter's orbital period. These fractions are placed above the relevant gap in the figure. The fraction 1/4 indicates, for example, that an asteroid at that distance makes four revolutions around the Sun for each one revolution completed by Jupiter. Asteroids are tossed out of such resonant orbits by Jupiter's repeated gravitational perturbations. In addition, there are several peaks corresponding to groups of asteroids with nearly the same orbital distance, such as the Trojan asteroids that have orbits that are identical in size to the orbit of Jupiter.

that are not in a resonance are affected by Jupiter at completely random time intervals and places along their orbit, so the giant planet's gravitational perturbations tend to cancel each other over long times. It is somewhat analogous to repeated pushes on a swing. If the pushes occur at the same point in each swing, they can amplify the change, but haphazard pushes would produce little net effect.

A satisfactory explanation of the Kirkwood gaps was not achieved until the 1980s when powerful computers were used to study Jupiter's influence on the motion of asteroids. The computer simulations showed that Jupiter induces a chaotic zone in the vicinity of an orbital resonance, and that an asteroid that moves into the resonant orbit will eventually be tossed out of it. An asteroid in the chaotic zone can spend tens of thousands of years in a well-behaved, near-circular orbit. But that ordered placid behavior could be unexpectedly interrupted after 100 000 years or so, when the orbit is suddenly stretched and elongated in a chaotic way (Fig. 12.4).

The increasingly elongated orbit can become so elliptical that it crosses the orbit of Mars or the Earth, and gravitational interaction with these planets can fling the asteroid into a totally different trajectory that removes it from the asteroid belt. Thus, orbits that initially fall within the chaotic zone around a resonance with Jupiter would be gradually cleaned out, creating a gap in the asteroid belt. Asteroids that have been thus redirected could abruptly end their voyage through space in a collision with Earth.

12.2 Origin of the asteroids

Former worlds

In the past, there have been two extreme theories for the origin of asteroids. According to the first, the asteroids represent the fragments of a former planet that has been torn apart. The second theory proposes that the asteroids are the

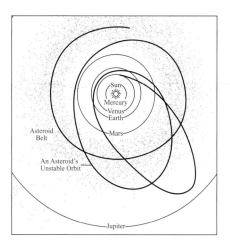

Fig. 12.4 Chaotic asteroid orbit   Asteroid orbits can become chaotic under the gravitational influence of nearby massive Jupiter. Asteroids at certain locations in the main belt follow a trajectory that becomes increasingly off-center over thousands of orbits. They may eventually become interlopers with orbits that cross the Earth's orbit. Some of these near-Earth asteroids may eventually collide with our planet.

pieces of a planet that never formed. Today, astronomers favor a theory that lies between the extremes.

It is now known that the combined mass of all asteroids is far too small to make up a major planet. If all the known asteroids were brought together, they would create a body less than 5 percent the mass of the Earth's Moon. So the first extreme must be discarded. On the other hand, the second extreme can also be excluded because there is strong evidence that many of the asteroids were once collected into a relatively small number of slightly larger parent bodies.

Remnants of a planet that never formed

Why did the asteroids fail to coalesce into a single planet? It is likely that gravitational forces from the rapidly forming and massive Jupiter took charge of its neighborhood, stirring it up and keeping the original asteroids from growing too large. Numerous asteroids in resonant orbits with youthful Jupiter probably permeated the region of the main asteroid belt. Chaotic zones in the vicinity of these resonances would have pumped up the eccentricities of initially circular orbits, flinging the resident bodies into elongated and inclined orbits, accelerating them to high velocities, and causing them to crash into each other. The colliding objects would be moving too fast to stick to each other. Instead, they would break apart into fragments.

Collisions therefore pulverized the early asteroids, grinding them down to the numerous smaller asteroids

we see today. Almost every asteroid we see today must be a fragment of a larger original body.

In contrast, a swarm of small, solid bodies in the inner solar system, with orbits closer to the Sun than the asteroid belt, were located far from Jupiter's gravitational influence. These so-called planetesimals therefore remained in nearly circular orbits, moving at slow velocity around the Sun. This permitted the neighboring planetesimals to merge gently together and form larger ones. They eventually coalesced into the four terrestrial planets – Mercury, Venus, Earth and Mars.

If it wasn't for Jupiter's chaotic interference, a similar terrestrial planet might have formed in the asteroid belt. Gravitational interaction with the giant planet removed objects from this region, throwing them into collisions with the other planets or out of the solar system. These events cleared the asteroid zone of 99.9 percent of its original mass. The missing material could have coalesced to form an Earth-sized planet. What remain today are the relic building blocks of a planet that failed to form.

A lifetime of catastrophic collisions

Billions of years ago, before Jupiter began to disrupt nearby objects, a few large bodies probably inhabited the asteroid belt. The largest of these would-be planets accumulated enough internal heat to differentiate, their dense material sinking to form iron cores and leaving rocky residues in their outer layers. Volcanic flow may have covered the surface of some of them.

Most of the initial asteroids never grew large enough to form cores, so they preserved matter typical of the region in which they formed. The majority of asteroids, both large and small, are undifferentiated asteroids.

Encounters among the earliest asteroids became increasingly violent as Jupiter stretched and twisted their orbits into eccentric and inclined orientations. These orbits crossed each other, sometimes resulting in violent collisions. Instead of continuing to grow, the largest asteroids were chiseled and blasted apart by mutual collisions.

So the original asteroids never grew larger than about 1000 kilometers across, and they never accumulated into a major planet. The pulverized remnants of these former worlds became the present asteroids, often orbiting the Sun in families with common orbital characteristics and spectral properties.

The Japanese astronomer Kiyotsugu Hirayama (1874–1943) discovered asteroid families. He noticed that groups of main-belt asteroids share very similar orbits (Fig. 12.5), suggesting that they are the broken fragments of larger objects. Hirayama called each group a family because he believed the members shared a common origin as the

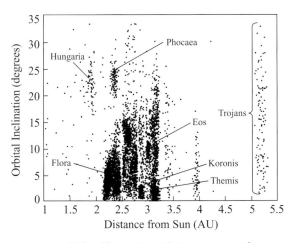

Fig. 12.5 Asteroid families The orbital parameters of many asteroids are very similar, as shown in this diagram of orbital inclination (*vertical axis*) plotted as a function of orbital distance from the Sun (*horizontal axis*). Three families with common orbital elements are the Koronis, Eos and Themis families. The Flora family is sometimes subdivided into several separate ones. The Hungaria and Phocaea group of asteroids at high inclinations are separated from the main belt by secular resonance with Jupiter, that clears asteroids out of certain locations, and they are not true families. The Kirkwood gaps, also cleared by resonance, are noticeable by vertical white spaces; the one at 2.5 AU corresponds to an orbital period of 4 years and the 3 : 1 resonance. Another sharp break is present at 3.3 AU, corresponding to a period of 6 years and the 2 : 1 resonance. [Adapted from Charles T. Kowal's *Asteroids* (New York: John Wiley, 1996).]

children of a bigger parent body. He also named a number of families after their largest member asteroids, such as the Eos, Koronis and Themis families.

Within each of these families, the orbits are so similar that the members must have originated from a single object. Hundreds, and perhaps thousands, of small asteroids making up each family are probably the debris of a collision that disrupted the once-larger parent body. These parents may have been several hundred kilometers in diameter.

And not only are the orbits similar within each family, the colors and surface compositions are also often alike. These similarities imply that the families are real physical groupings. The Koronis family is, for example, composed of bright silicate asteroids and the Themis family of dark carbonaceous ones. Some family members are spectrally diverse, perhaps sampling different parts of the interiors of their parental precursors.

The asteroid belt has been dominated by collisions for billions of years. Major impacts between asteroids larger than one kilometer across occur every 10 to 100 million years. So each one of them must have suffered roughly

a hundred devastating collisions over the past 4 billion years.

Over time, the asteroids crashed into one another and modified their shapes. When a small asteroid hit a larger one, it gouged a crater out of the surface of the larger one. If the two colliding objects were of comparable size, they could have broken each other apart into smaller fragments. As a result, some present-day asteroids are the metal-rich cores of larger former parents, stripped of their rocky mantles by the ongoing collisions. Others are the shattered remains of bodies that have remained homogeneous. Such destructive encounters occur even now, dominating the evolution, shapes and sizes of the asteroids.

12.3 Viewing asteroids from a distance

The size of asteroids

Due to its small size, an asteroid remains an unresolved point of light in even the best telescopes on Earth, just like a faint star. This explains the name *asteroid*, which comes from a Greek word that means "starlike". Although the name describes the visual appearance of these objects in a telescope, it is totally inappropriate to their physical nature. Using our instruments on Earth, we can determine the sizes of asteroids, and they are much smaller than either a star or a major planet.

One method of measuring the size of an asteroid is to watch it pass in front of a star. During such a stellar occultation, the asteroid casts a shadow on the Earth, and the width of this shadow is equal to the asteroid's diameter. As the Earth rotates and the asteroid moves across the sky, the asteroid's shadow is swept by an observer on Earth, and its width can be determined by measuring the length of time the star is invisible. Observations of stellar occultations have established accurate values for the sizes of the largest asteroids, such as 1 Ceres and 2 Pallas (Table 12.1).

Even the biggest asteroids are smaller and less massive than the Moon. Ceres is by far the largest asteroid, having a radius of about 475 kilometers and a mass of 9.43×10^{20} kilograms. That is about a third the radius of the Earth's Moon and only 0.016 the mass of the Moon. The brightest asteroid, Vesta, is about half the size of Ceres and about the same size as Pallas.

There are many more small asteroids than big ones. About 1000 asteroids are larger than 15 kilometers in radius. Surveys of the faintest asteroids suggest that there are about half a million asteroids in the main belt larger than 1.6 kilometers. Yet, despite their vast numbers, the total mass obtained by adding up the contributions of all asteroids, of all sizes, is far less than the mass of any major

Table 12.1 Principal characteristics of the three largest asteroids[a]

Asteroid	Type	Radius (km)	Mass[b] (10^{21} kg)	Bulk density (kg m^{-3})	Rotation period (hours)	Semi-major axis (AU)
1 Ceres	C	475	0.943	2120 ± 40	9.1	2.77
2 Pallas	B	266[c]	0.214	2710 ± 110	7.8	2.77
4 Vesta	V	265	0.267	3440 ± 120	5.3	2.36

[a] The radius is in units of kilometers (km), the mass is in units of kilograms (kg), and the bulk mass density is in units of kilograms per cubic meter (kg m^{-3}).

[b] The mass determinations are from the work of E. M. Standish at the Jet Propulsion Laboratory where he uses observations of Mars to solve for the masses of the largest asteroids.

[c] Pallas is not quite spherical. A tri-axial ellipsoid fit to occultation observations has diameters of 559 km, 525 km, and 532 km with a mean diameter of 538 ± 12 km.

planet. The entire asteroid belt is only 0.05 the mass of our Moon and just 0.0006 the mass of the Earth.

Another way of estimating an asteroid's size is to measure its apparent brightness. Bigger asteroids reflect more sunlight and are therefore brighter. The amount of sunlight reflected from an asteroid nevertheless depends on both its size and reflectivity, or albedo: this is the ratio of light reflected by a surface to the incident light, and therefore is a measure of the efficiency of the reflection process. As an example, 1 Ceres is the biggest asteroid, but 4 Vesta can appear brighter because it has a higher albedo. Vesta reflects 42 percent of the sunlight that strikes it, compared with Ceres' reflectivity of only about 9 percent.

Both the size and the albedo of an asteroid can be determined by observing it in visible and infrared light. When sunlight strikes an asteroid, some of the light is reflected, but most of it is absorbed. The absorbed radiation makes the surface heat up and emit infrared light. By measuring the infrared radiation from the asteroid and comparing it to the amount of reflected visible light, we can determine its albedo. And by knowing the asteroid's apparent brightness and distance as well, we can infer its size.

Shape and form of asteroids

The rotation of a non-spherical asteroid can bring various surfaces into view, reflecting different amounts of sunlight. This produces a brightness variation, with periods of hours, that can be used to infer an asteroid's irregular shape. These rotational modulations of the light reflected from the largest asteroid, 1 Ceres, are very small, indicating that this asteroid is practically spherical in shape.

The round shape of 1 Ceres has been confirmed from images taken with the *Hubble Space Telescope* (Fig. 12.6). Ceres is massive enough that its gravity has crushed the

Fig. 12.6 Dwarf planet–asteroid 1 Ceres   A *Hubble Space Telescope* image of 1 Ceres, the largest object in the asteroid belt and the first object discovered there. The round shape led to its classification in 2006 as a dwarf planet, and suggested that its interior is layered, with a rocky core, icy mantle and thin, dusty outer crust. Ceres is approximately 950 kilometers across. The differences between lighter and darker regions could be impact features or simply due to a variation in surface material. (Courtesy of NASA/ESA/J. Parker, SRI, P. Thomas, Cornell U., L. McFadden, U. Maryland, and M. Mutchler and Z. Levay, STScI.)

object into this spherical form. The International Astronomical Union has therefore added a new designation to Ceres, classifying it as a dwarf planet since it is small, orbits the Sun, and has sufficient mass to attain a nearly round shape.

Asteroids are also known as minor planets, but they usually have irregular shapes and differ noticeably from the familiar major planets. Because of their small size, asteroids have very little mass or gravity. Unlike the larger planets, asteroids are not big enough to hold onto an atmosphere or to retain a spherical shape. The exception is Ceres.

Heat from formation and the decay of radioactive elements most likely elevated the interior temperature of Ceres to the point of differentiation, in which heavier, denser material sank to the center while lighter materials rose to the surface. Scientists therefore think that Ceres has a rocky inner core surrounded by a thick water-ice mantle and a thin, dusty outer crust.

Most, if not all, of the smaller, less massive asteroids have irregular shapes, and retain their asteroid classification. Their self-gravity is not sufficient to overcome internal rigid body forces. Their irregular shapes cause pronounced rotational brightness variations.

The majority of asteroids are not large enough to have ever had a molten core, and lava has never flowed out from the inside of most of them. So, except for ongoing collisions, most asteroids have been unaltered since the formation of the solar system about 4.6 billion years ago. In contrast, the surfaces of large planetary bodies have been destroyed by eons of planetary evolution, including immense volcanoes and extensive lava flows across their surfaces.

The form and amplitude of the periodic light variations have demonstrated that most asteroids are at least slightly elongated, chipped and pummeled into irregular shapes. They are too small to retain a spherical shape during their lifetime of disruptive collisions.

The stretched-out irregular shapes of some asteroids have also been determined from radar observations of near-Earth asteroids that travel close enough for scientists to detect the echoes of radio waves bounced off them. During their close approach, these asteroids speed by the Earth at distances of several hundred thousand kilometers, permitting brief high-resolution radar images before they move on and fade from view.

Radar is an acronym for Radio Detection and Ranging. In this technique, a pulsed radio signal is sent to the asteroid, and the properties of the return "echo" are compared to those of the transmitted signal. This yields information about the composition, surface texture, shape and rotation of the asteroid. Smooth surfaces reflect more radio energy than rough ones, and the faster an asteroid rotates the broader the range of returned wavelengths, which have been Doppler-shifted by the rotation to longer and shorter waves.

The radar data indicate that the overall shape of some asteroids is dominated by two irregular, lumpy components that touch each other, something like a dumbbell. Such an asteroid is a double object, that is, two bodies in contact. The two pieces probably merged after a past catastrophic collision of a larger body; they may have been

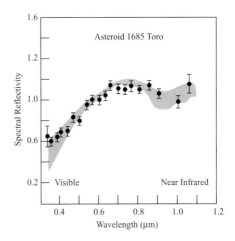

Fig. 12.7 Finding the composition of asteroids A prominent silicate absorption feature is present in the spectrum of sunlight reflected from the S-type asteroid 1685 Toro (*dark dots with error bars*). Asteroids like Toro may be the source of the stony meteorites recovered on Earth. The shaded spectrum is the reflection spectrum of a stony-chondrite meteorite. The wavelength is in units of micrometers (μm), or 10^{-6} meters.

thrown apart and subsequently came together under their mutual gravity. Or they might be two former asteroids that joined in a gentle encounter. Radar images have also revealed at least one triple asteroid, a central body about 700 meters across orbited by two smaller moons.

An asteroid's color

Because asteroids display no visible disk when observed from Earth, their physical characteristics must be inferred from the intensities and spectral properties of their reflected sunlight. By comparing an asteroid's reflected light, wavelength by wavelength, with that of the incident sunlight, it is possible to deduce its surface composition. Astronomers divide the amount of incident sunlight at each wavelength by the amount of reflected sunlight at that wavelength, and the ratio tells them how much light of each color is reflected compared to any other color. Such spectral measurements have revealed the physical diversity of the asteroids, and shown that their compositional differences tend to depend on distance from the Sun.

The bulk of main-belt asteroids can be divided into two broad spectral categories, known as the S, for silicate, and C, for carbonaceous, types. The bright S-types have a reddish color and exhibit spectral dips identified with absorption by silicate minerals (Fig. 12.7). They prevail in the inner part of the asteroid belt, orbiting closer to the Sun than the belt mid-point (Fig. 12.8).

In contrast, the C-type asteroids are darker, bluer and richer in carbon, with relatively flat and featureless spectra

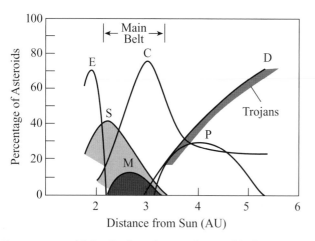

Fig. 12.8 Asteroid distribution of spectral type with distance The color, or surface composition, of the asteroids is correlated with distance from the Sun. In order of increasing distance, there are the white E asteroids, the reddish S or silicate ones, the black C or carbonaceous ones, and the unusually red D asteroids. This systematic change has been attributed to a progressive decrease in temperature with distance from the Sun at the time the asteroids formed. The rare metallic M asteroids are probably the cores of former, larger parent bodies. The P asteroids have spectra that are indistinguishable from the M ones, but the P types are darker.

at visual wavelengths. The C-type asteroids far outnumber all types, possibly composing three-fourths of the main belt. The C-type asteroid 1 Ceres is a representative example; it has a smooth visible spectrum with infrared absorption features attributed to clay-like minerals that include water as a part of their molecular structure. The iron-rich clays may have formed on what was once a wet surface.

The S-type asteroids probably account for up to 15 percent of all asteroids. Some of the less common M-type asteroids reflect sunlight in a way that suggests that their surfaces are composed of nickel and iron, hence the designation M for metallic for at least some of them. These objects could be the metal cores of larger parent bodies, stripped down to their cores by collisions. The M-types are most common in the middle of the asteroid belt. Future space entrepreneurs may want to mine them for valuable metals (Focus 12.1).

There is an intriguing connection between the composition of asteroids and their distance from the Sun. The innermost asteroids, with orbits closest to the Sun, are rocky, siliceous and dry, while the outer ones are carbonaceous with water-rich, clay-like minerals.

There is a related, progressive decrease in an asteroid's reflecting power with increasing distance from the Sun. The brightest asteroids that reflect the most sunlight tend to lie near the inner edge of the main belt, closest to the

Focus 12.1   Mining asteroids

The asteroids are rich storehouses of valuable materials such as iron, nickel and water. The utilization of their minerals could overcome growth limits imposed by dwindling natural resources on Earth.

Prospecting spacecraft might be sent to an asteroid in search of a cosmic El Dorado. Because of its low gravity, valuable metals could be easily removed, and the extracted material might be shipped back by spacecraft from an asteroid that traveled near to the Earth. Imaginative engineers speculate that a more distant asteroid might be brought closer to Earth using a "mass-driver", a device that would chew off pieces and fling them into space, propelling the asteroid like a rocket.

Metals are relatively scarce in the Earth's outer layers because most of them sank inside to the central core where they will remain forever sequestered, buried under our planet's crust and mantle. In contrast, some metallic asteroids may be pure core. The outer mantle and crust of their former parents were probably chipped away by eons of cosmic collisions. So the metals on an asteroid could be relatively abundant and easy to extract. Moreover, the residual gas and dust would be swept from the solar system by the solar wind, thereby avoiding the problems of industrial pollution when recovering valuable metals on Earth.

Sun, while the most distant asteroids are, on the average, the darker ones with the lower reflecting power.

These differences in the composition and reflecting power of asteroids are probably related to conditions in the primeval solar nebula – the interstellar cloud of gas and dust from which the solar system originated. They may be a consequence of a decrease in temperature with increasing distance from the young Sun when the asteroids were formed.

The middle of the main asteroid belt marks the boundary dividing the cold, outer, water-condensing regions and the hot, inner, water-vapor parts of the primeval nebula. Dark material, rich in carbon and water, could condense in the colder regions farther from the Sun, but not in the hot regions near to our star where the early water was probably vaporized. On the other hand, the bright rocky material was less volatile and could remain within the hotter regions closer to the Sun. In this way, the temperature of the solar nebula may have led to the pattern of materials and colors now seen in the asteroids.

Astronomers have identified a plethora of other, less-common classes of asteroids, based upon the shape and slope of their reflectance spectra. In addition to the most common S-, C- and M-types, there are at least 11 other classes denoted by different letters, such as the red, possibly organic D-types found in the outer belt, and the white E-types that are closer to the Sun. In addition, rare individual asteroids exhibit unique well-identified absorption features in their spectra. The brightest and third largest of the asteroids, 4 Vesta, for example, shows the distinct absorption signature of volcanic basalt. Its parent body spawned a little family of smaller V-type asteroids with similar spectra and basaltic composition.

The spin of an asteroid

Asteroids do not shine like a steady beacon with constant brightness. They instead reflect a varying amount of sunlight toward the Earth. The observed brightness variation, also known as a light curve, is periodic, often with two maxima and two minima. The overall repetition is due to rotation, while the double pattern of variability results from alternating side views of an asteroid's elongated shape. When we see the biggest side of an asteroid, with the greatest area, the asteroid is brightest, while the smaller area reflects less sunlight and the asteroid is dimmer.

Almost all asteroids spin about a single axis. The period of rotation is inferred from the amount of time that it takes for the complete pattern of brightness variation to repeat itself. The rotation periods are usually between 2.4 and 24 hours (Fig. 12.9), although a few of them rotate with longer periods, such as 253 Mathilde with a rotation period of 17.4 days, and some have periods of only a few minutes. Frequent oblique collisions can increase the rate of rotation or decrease it, depending on whether the collision is in the direction of rotation or opposite to it.

Some asteroids are probably rotating as fast as they can. If an asteroid is not solid, and is thus bound only by its own gravity, it can only spin at a certain maximum rate before material is whirled off it. Asteroids larger than 200 meters seem to have reached this limit, for most of them do not rotate faster than once every 2.4 hours (Fig. 12.9), suggesting that there is nothing stronger than gravity holding them together. If it lacks the material strength of a solid, an asteroid with a faster rotation rate and a shorter period will throw material off its surface and fly apart.

Asteroids smaller than 200 meters in diameter can rotate at faster rates, some turning once every few minutes. Their rotation is too rapid for these asteroids to consist of multiple components bound together by mutual gravitation. They must instead be rock solid. Some small asteroids rotate so swiftly that their day ends almost as soon as it begins. Asteroid 2008 HJ has one of the fastest

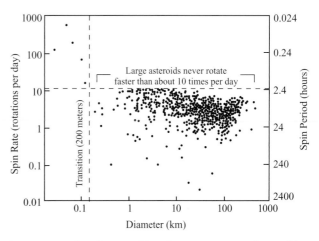

Fig. 12.9 How fast do asteroids spin? Two groups of asteroids are indicated on this plot of their rotation rates (*vertical axis*) versus size (*horizontal axis*). There are the large asteroids that rotate at a wide variety of speeds, except at the fastest rates, and the small asteroids that spin the fastest. No known asteroid larger than 200 meters across rotates faster than once every 2.4 hours, perhaps because these asteroids are piles of rubble that fly apart if spun too fast. Smaller asteroids, which can turn once every few minutes, must be solid rocks.

known rotation periods at just 42.7 seconds; it is thought to be a compact, massive object about 20 meters across.

The direction of an asteroid's rotation axis can be determined by noting the way the brightness variation changes as the asteroid moves about the Sun. A highly elongated asteroid will reveal little or no light variation if its spin axis points directly toward us. As the asteroid's orbital motion carries it away from this alignment, the amount of light reflected toward Earth will increase, reaching a maximum when its broadest equatorial surface is turned toward Earth. Studies of this effect suggest that the rotation axes of the asteroids are haphazardly oriented, pointing in all directions in space, contrary to the orderly rotations of most of the major planets whose rotation axes point more or less north–south. Frequent collisions may have been responsible for the diverse orientations of asteroid rotation axes.

12.4 Spacecraft view asteroids close up

Flyby missions of asteroids

A nearby high-resolution view was required to examine the surface details of asteroids, and this has been accomplished by several passing spacecraft on their way to accomplish a different goal. The first close encounters were provided when the *Galileo* spacecraft flew by two asteroids, 951 Gaspra and 243 Ida, on its way to Jupiter (also see Section 2.1). The images, taken on 29 October 1991 and

Fig. 12.10 Asteroid 21 Lutetia   An instrument aboard the *Rosetta* spacecraft captured his image of the main-belt asteroid 21 Lutetia on 10 July 2010. *Rosetta* is expected to orbit and then land on comet Churyumov–Gerasimenko in 2014. (Courtesy of ESA/*Rosetta* team.)

28 August 1993, respectively, showed irregular shapes peppered by an abundance of small craters, suggesting that asteroids are the battered remains of larger bodies subject to numerous collisions with objects of comparable or much smaller size.

Somewhat blurry images were obtained during the *Deep Space 1* flyby of asteroid 9969 Braille, on 28 July 1999 while testing new technologies, and when the *Stardust* spacecraft passed asteroid 5535 Annefrank on 2 November 2002, en route to its primary target, comet Wild 2. As expected, both asteroids are irregular, elongated bodies, several kilometers across on their longest sides.

On 5 September 2008 the *Rosetta* spacecraft flew by the main-belt asteroid 2867 Steins at a distance of about 800 kilometers. These were the first close-up views of a rare E-type asteroid, classified from its high albedo and infrared spectrum. The *Rosetta* images reveal a conical, oblate diamond-shaped body, with dimensions between 4.5 and 6.7 kilometers, covered with shallow craters. *Rosetta* flew by the large main-belt asteroid 21 Lutetia on 10 July 2010, revealing an irregularly shaped, cratered object about 100 kilometers across (Fig. 12.10). Data obtained during the encounter will help determine if Lutetia is a primitive undifferentiated C-type asteroid or a metallic M-type asteroid from the core of a larger differentiated asteroid. Lutetia is the Latin name for Paris. The *Rosetta* spacecraft

is expected to orbit its primary scientific target, comet Churyumov-Gerasimenko, beginning in 2014.

A sideways glimpse of 253 Mathilde, a dark carbon-rich C-type asteroid, was acquired on 27 July 1997 from the *NEAR Shoemaker* spacecraft on its way to orbit asteroid 433 Eros. Like the other asteroids seen close up, Mathilde has survived blow after blow of cosmic impacts (Fig. 12.11). Its surface is covered with the crater scars of past collisions that have disfigured the asteroid's shape, like the battered and scarred face of a professional boxer who has just lost a fight. Huge pieces have been removed from Mathilde, leaving four enormous craters tens of kilometers across.

Over billions of years, asteroids on intersecting orbits have collided with enough force to shatter and break 253 Mathilde into pieces. Instead of dispersing, the fragments have apparently re-accumulated into a rubble pile. The mass of the asteroid was determined by radio tracking of small perturbations in the trajectory of *NEAR Shoemaker* during its close flyby. A similar technique established an accurate mass for 433 Eros and 25143 Itokawa when spacecraft circled them. These mass determinations were combined with measurements of the asteroid volume to provide their bulk mass density (Table 12.2). *Galileo* flew by too fast and too far away to be affected noticeably by the asteroids it encountered, so masses could not be determined in this way. But a rough estimate for the mass of

Table 12.2 Physical properties of asteroids visited by spacecraft[a]

Asteroid	Type	Overall dimensions (km)	Mass (10^{15} kg)	Bulk mass density (kg m^{-3})	Rotation period
21 Lutetia	M	$132 \times 101 \times 26$	1700	3400	8.17 hours
243 Ida	S	$60 \times 25 \times 19$	42 ± 6	2600 ± 500	4.63 hours
253 Mathilde	C	$66 \times 48 \times 46$	103.3 ± 4.4	1300 ± 300	17.4 days
433 Eros	S	$31 \times 13 \times 13$	6.687 ± 0.003	2670 ± 30	5.27 hours
951 Gaspra	S	$18 \times 11 \times 9$	–	–	7.04 hours
25143 Itokawa	S	$0.535 \times 0.294 \times 0.209$	$3.58 \times 10^{-5} \pm 0.018 \times 10^{-5}$	1950 ± 140	12 hours

[a] The overall dimensions in kilometers (km) are the diameters of a triaxial ellipsoid fit. The mass is in kilograms and the bulk mass density is in units of kilograms per cubic meter (kg m^{-3}).

Fig. 12.11 Asteroid 253 Mathilde An image mosaic of the C-type asteroid 253 Mathilde acquired by the *NEAR Shoemaker* spacecraft on 27 June 1997. The part of the asteroid shown is about 59 kilometers by 47 kilometers across. The angular shape of the upper left edge is a large crater viewed edge on. Mathilde's low mass density indicates that it is a porous rubble pile. (Courtesy of NASA/JHUAPL.)

243 Ida was inferred from the motion of its tiny moon Dactyl.

Mathilde has an exceptionally low bulk mass density of just 1300 kilograms per cubic meter, only slightly higher than water at 1000 in the same units, and about half the average density of the Earth's crust. The asteroid must therefore contain as much empty space as rock in its interior. It is indeed a loose pile of rubble, broken apart and stuck back together again, so pervasively fragmented that no solid bedrock is left.

The porous interior of Mathilde might explain the mysterious absence of visible rims or ejected deposits around its enormous craters. The unusually large craters may have been formed by compression of the surface during impact, like the dents in a beanbag, rather than by excavation of the material. Mathilde might even have swallowed up the colliding objects, like a bullet shot into a sandbag.

We will next turn to the results of two asteroid-orbiting spacecraft, *NEAR Shoemaker* for 433 Eros and *Hayabusa* for 25143 Itokawa. Eros is solid inside, held together by its own material strength. In contrast, Itokawa seems to be a rubble pile, the reassembled debris of previous impacts, consisting of smaller pieces loosely held together by their mutual gravitation.

NEAR embraces asteroid 433 Eros, a solid rock

On Valentine's Day, 14 February 2000, the *Near Earth Asteroid Rendezvous (NEAR)* spacecraft became the first to orbit an asteroid, arriving at 433 Eros, a bright silicate S-type asteroid, after a four-year journey from Earth. It studied the shape and form of the asteroid's surface in high resolution, and determined its composition, mass and bulk density before landing on the surface nearly one year after beginning orbit.

The *NEAR* craft was the first in NASA's Discovery Program of no-frills, scientifically focused, low-cost missions, designed to do quality science in a "faster, better, cheaper" mode. The mission took just 26 months from start to launch, at a bargain total cost of $224 million. The car-sized vehicle circled Eros for a year, landing on the asteroid on 12 February 2001, another historic first. In the meantime, NASA renamed the spacecraft *NEAR Shoemaker* in honor of the astronomer–geologist Eugene M. Shoemaker (1928–1997), a pioneering expert on asteroid and comet impacts.

Discovered in 1898, asteroid 433 Eros was one of the first asteroids found to cross the orbit of Mars and come near the Earth's path in space. When *NEAR Shoemaker* arrived at the asteroid, it was just 1.7 AU from Earth, or less than twice the average distance between Earth and the Sun. Eros is one of the largest of the near-Earth asteroids, about twice the size of Paris and with a mass thousands of times greater than other asteroids that approach the

Fig. 12.12 Asteroid 433 Eros, a solid rock
This global view of the S-type asteroid 433 Eros was obtained by the *NEAR Shoemaker* spacecraft on 29 February 2000 from a distance of 200 kilometers. This perspective highlights the major features of the asteroid's northern hemisphere. The asteroid's largest crater (*top*) measures 5.5 kilometers wide and sits opposite from an even larger 10-kilometer, saddle-shaped depression (*bottom*). Studies of the spacecraft's orbit around the asteroid indicate that 433 Eros is a solid rock. (Courtesy of NASA/JHUAPL.)

Earth. So Eros is big and relatively nearby, two important reasons for a visit. It is also named after the Greek god of love – most appropriate for an encounter on Valentine's Day.

Radio-tracking of the orbiting spacecraft was used to determine the mass of Eros, which weighed in at 6.687×10^{15}, kilograms, about one-billionth the mass of Earth. That means that most adults would weigh less than a few ounces if standing on Eros, about as heavy as a bag of airline peanuts on Earth. And on Eros you could jump thousands of meters high, never to return. The gravity is so slight that *NEAR Shoemaker* had to keep its speed down to about 5 kilometers per hour to stay in orbit, moving about as fast as a casual bicyclist. If it moved at a faster speed, the spacecraft would escape the asteroid's feeble gravity and move into interplanetary space.

Although previous spacecraft passed close to a few asteroids, none had orbited one. In contrast to these previous brief flybys, *NEAR Shoemaker*'s cameras scrutinized 433 Eros for a solid year, sending back 160 000 images of the asteroid and recording its diverse surface from all angles and distances.

Eros is a warped and misshapen world, with heavily cratered expanses abutting relatively smooth areas

(Fig. 12.12). The asteroid's biggest crater measures 5.5 kilometers across, and most of the surface is peppered with smaller craters (Fig. 12.13). The *NEAR* scientists spotted thousands of them.

The craters on Eros have been given names of famous lovers from history and fiction, taken from different cultures. They include Bovary from Gustav Flaubert's (1821–1880) novel *Madame Bovary*, Don Quixote and his Dulcinea from Miguel de Cervantes' (1547–1616) novel *Don Quixote de la Mancha*, Lolita from Vladimir Nabokov's (1899–1977) novel of that name, Don Juan, the legendary Spanish nobleman known for his seduction of women, and both Eurydice and her lover Orpheus, from Greek mythology.

The surface of Eros is saturated with craters. There are as many craters on its surface as there can be, so continued cratering would not change the overall appearance of its surface. It has been completely formed and shaped by the unrelenting bombardment of impacts over billions of years.

There are nevertheless an unusually low number of small craters pockmarking Eros. Scientists think that shaking and vibrations resulting from impacts have obliterated the small craters, filling those smaller than 100 meters across with rock and dust. They call this erasure *seismic*

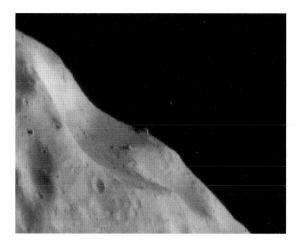

Fig. 12.13 Craters on asteroid 433 Eros   The many craters on the surface of the asteroid 433 Eros are attributed to eons of collisions with other asteroids. Large boulders, perhaps broken off Eros during these impacts, are perched on one of the crater's edge. The largest boulder, on the horizon in the center of the picture, is about 40 meters long. The two overlapping craters shown here were probably formed many millions of years apart. This picture, taken on 7 July 2000 from the *NEAR Shoemaker* spacecraft, is 1.8 kilometers wide. (Courtesy of NASA/JHUAPL.)

Fig. 12.14 Back in the saddle again   This image of the saddle region on 433 Eros was taken on 22 March 2000 by the *NEAR Shoemaker* spacecraft. The saddle is about 10 kilometers across. It may be the scar of an ancient crater, or somehow related to a different large crater on the opposite side of the asteroid. (Courtesy of NASA/JHUAPL.)

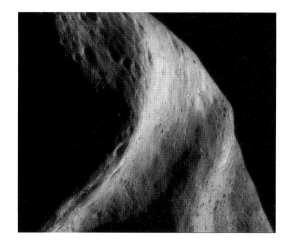

Fig. 12.15 The south saddle of Eros   This mosaic of four images, photographed on 26 September 2000, was taken as the *NEAR Shoemaker* spacecraft looked down on the saddle region of the asteroid 433 Eros. A broad, curved depression stretches vertically across the image, as if scooped out by a cosmic sculptor. A boulder-rich area is seen in the lower right. (Courtesy of NASA/JHUAPL.)

shaping, in analogy with terrestrial earthquakes that shake the ground with seismic waves, but from a totally different cause.

Eros has also been smoothed and rounded by glancing blows during its catastrophic past. This cosmic sculpture rivals the smaller bronze and marble forms of Constantin Brancusi (1876–1957) and Henry Moore (1898–1986). Equally beautiful is the broad, curved depression that connects two mountains on Eros, each thousands of meters high (Figs. 12.14, 12.15).

Far from being a barren lump of rock, Eros has a dusty, boulder-strewn landscape. Despite its weak gravity, the diminutive asteroid has managed to hold on to about 7000 boulders larger than 15 meters, forced out of craters and pulled back to the surface during the relentless bombardment of its past. Some of the isolated stones are as large as a house, and up to 100 meters across.

The positions of the rocks on Eros indicate that at least 3000 of them were ejected from a single impact crater, perhaps a billion years ago. Some of these boulders went straight up and straight down. Most of the remainder traveled as far as two-thirds of the way around the asteroid, in all directions, before finally coming to rest on the surface.

Smaller rocks and a loose layer of dirty debris came into view when the *NEAR Shoemaker* spacecraft moved in to land on the boulder-strewn surface of 433 Eros. It took pictures as close as 250 meters above the surface, showing features as small as a golf ball (Fig. 12.16). They indicate that Eros is something between a very big rock and a planet, large enough to hold on to its pieces yet too small to lose its odd, distorted shape.

Asteroid 433 Eros has a bulk mass density of 2670 kilograms per cubic meter, comparable to that of terrestrial rocks, and Eros is just like an extra big, chipped rock. Millions of range, or distance, measurements with a laser aboard *NEAR Shoemaker* have established the asteroid's complex shape and topography. Comparisons of this

Fig. 12.16 Close-up view of the surface of 433 Eros   This *NEAR Shoemaker* picture of the surface of asteroid 433 Eros was taken from a range of 250 meters on 12 February 2001, just before landing on the asteroid. The image is just 12 meters across, and the cluster of rocks at the upper right measures 1.4 meters across. (Courtesy of NASA/JHUAPL.)

shape with radio-tracking of the gravitational pull on the orbiting spacecraft show that the mass density of Eros' interior must be nearly uniform. Thus, the asteroid 433 Eros is mostly solid throughout, with a uniform composition and homogeneous internal structure, rather than a loosely bound collection of smaller components. It is an undifferentiated object that never separated into a distinct crust, mantle and core. Although 433 Eros has been heavily fractured and whittled away by impact collisions, it was never completely demolished and reformed as a rubble pile.

Most asteroids are probably covered with a blanket of dust, pebbles and rocks that rests on solid bedrock. This layer of loose rock particles is known as the *regolith*, from the Greek word for "rock layer". The regolith formed during repeated bombardment by small meteorites that break apart the surface rock. The Earth's Moon and Mars contain such dusty surface layers. Small asteroids may have a thin coating of dust, while the largest asteroids could have a thicker, powdery veneer, even thousands of meters deep.

The fine, dusty material on Eros has settled downhill, collecting in ponds that are tens of meters wide and a few meters deep. Hundreds of them are found in low-lying hollows, such as the bottoms of craters. The powdered deposits have flat, level surfaces, resembling ponds of water on Earth, but there is no water on a rocky asteroid like Eros, and there hasn't been any for billions of years. Something else is causing the mobile soil to flow down hill. Since the ponds are found in well-lit areas, scientists

speculate that the dust moves when it has been in the Sun too long, but the details of the sorting mechanism are not well understood.

Hayabusa orbits 25143 Itokawa, a rubble pile

On 9 May 2003 the Japanese Institute for Space and Astronautical Sciences (ISAS) launched its *MUSES-C* spacecraft to the near-Earth asteroid 25143 Itokawa. As is the custom, the spacecraft was given another name following successful launch; it is now known as *Hayabusa* – Japanese for "falcon".

Propelled by an ion-drive engine, *Hayabusa* arrived at asteroid Itokawa on 30 September 2005, hovering at an altitude of just 7 kilometers above its surface for about a month. The falcon then landed on the asteroid surface for brief 30-minute stays on 19 and 25 November, like a bird swooping down for fish near the surface of the sea.

Asteroid 25143 Itokawa is a small rocky body, just 535 meters in longest dimension, spinning a couple of times a day and belonging to the S-type of spectroscopic class common in the inner portion of the asteroid belt. It is a nearby Apollo asteroid, discovered in 1998, which follows a 1.5-year orbital path a distance of 0.953 AU from the Sun, just inside the Earth's orbit, to 1.695 AU, just outside the orbit of Mars. The asteroid was thus selected as one of the most accessible targets for the low-energy launch vehicle, as well as providing an opportunity to learn more about small asteroids that regularly move past the Earth and may one day collide with it.

The *Hayabusa* images indicate that 25143 Itokawa has a lumpy, oblong shape that has been likened to a sea otter. It is composed of two parts: a smaller "head" and a larger "body" with a smooth "neck" separating them (Fig. 12.17). The head and body could have been formed separately and later came into contact at a slow relative speed, or the depressed neck region might have been formed by a large impact with a single entity. Shaking and jostling of the asteroid by former impacts may account for the absence of craters, filling them in, as well as the sorting of large boulders and small gravel into different places on the surface.

This asteroid seems to be nothing but pieces, hanging together by its weak gravity, continuously pulverized by impacts and sorted by vibrational shaking. Impact craters seem to be filled as quickly as they form, and smooth, deep gravel beds have been created. Itokawa's vital statistics reflect its small size, with a mass of just $(3.58 \pm 0.18) \times 10^{10}$ kilograms and a low bulk mass density of

Fig. 12.17 Asteroid 25143 Itokawa, a rubble pile The small asteroid 25143 Itokawa is just 500 meters long, and it has an orbit that crosses that of the Earth. This image, taken from the Japanese spacecraft *Hayabusa* in late 2005, strongly suggests that the asteroid is a loose pile of rubble rather than a solid rock. The difference between the smooth and rough terrain might be due to impacts that jostled and shook the asteroid, creating a separation of large and small rocks on the surface. (Courtesy of ISAS/JAXA.)

1950 ± 140 kilograms per cubic meter, indicating a porous interior filled with holes something like sand.

Thus 25143 Itokawa is considered to be a veritable rubble pile because of its low bulk density, high porosity, boulder-rich appearance, shape and absence of craters. It may have been formed by the early collision and breakup of a pre-existing parent asteroid followed by re-agglomeration into a rubble-pile.

Hayabusa returned to Earth from its seven-year, six-billion-kilometer trip on 13 June 2010, burning up in the atmosphere after releasing its small asteroid return capsule that was parachuted to the vast Australian Outback, where it was recovered and transported to Japan.

On 15 April 2010, President Barack Obama (1961–) announced a goal of sending astronauts beyond the Moon and into deep space, to encounter and perhaps land on an asteroid by 2025. The astronaut mission may develop the capacity to mine fuel and metals from an asteroid, as a precursor to a human mission to Mars, or to deflect a near-Earth asteroid headed for collision with Earth. It reminds one of Antoine de Saint-Exupéry's (1900–1944) charming novella about *The Little Prince*, who lives on an asteroid the size of a house and named B-612, making profound observations about life and human nature.

The incredible successes of the *NEAR Shoemaker* and *Hayabusa* missions have provided background data for the *Dawn* mission that will rendezvous with two of the largest bodies in the asteroid belt, 4 Vesta and 1 Ceres, beginning in 2011 and 2015, respectively (Focus 12.2).

12.5 Meteorites

Space rocks

Most meteors, or shooting stars, are produced by tiny fragments of comets: icy dust and pebbles that burn up in the air and never reach the ground, but occasionally a stone will fall from the sky, producing a brilliant trail of light flashing across the night sky. A rumbling sound and what appears to be a great burst of sparks may accompany it. These are fireballs and they are produced by tougher chunks of matter from space, resembling rocks (Fig. 12.18).

Extraterrestrial chunks of rock and metal that survive the fiery descent through the atmosphere and reach the ground have been given the name meteorites. Strictly speaking a *meteoroid* is the solid object in space that appears as a *meteor* when it lights up in the atmosphere and becomes a *meteorite* if it reaches the ground.

Focus 12.2   The *Dawn* mission to asteroids Ceres and Vesta

The *Dawn* spacecraft began a 3-billion kilometer voyage to the main asteroid belt on 27 September 2007, acquiring a gravity assist during an encounter with Mars in February 2009. The mission will be the first to orbit two solar-system bodies in a single voyage, revolving about the asteroid 4 Vesta from September 2011 to April 2012 and orbiting the dwarf planet–asteroid Ceres in 2015. By utilizing the same set of instruments, it can make accurate comparisons of two very different objects, including their mass, bulk density, shape, surface topography, tectonic history, and elemental and mineral composition.

Vesta is a dry, rocky, evolved and differentiated body with basaltic lava flows on its surface, perhaps resembling the magma maria on the Earth's Moon. The asteroid is representative of the building blocks that constructed the terrestrial planets of the inner solar system. The other object, 1 Ceres, has a surface with water-bearing minerals and may well be icy on the outside with water on the inside. It may be representative of the building blocks of the outer planets.

Dawn is expected to answer questions about the formation of the solar system by studying two contrasting protoplanets, one dry and the other wet, that never gathered together with other objects to form a planet. According to current theories, their differing properties are the result of their being formed and evolving in different parts of the solar system, and they are expected to characterize conditions and processes at the earliest moments – the dawn – of the planets.

Dawn is expected to arrive at 1 Ceres just five months prior to the arrival of the *New Horizons* spacecraft at Pluto in the cold outer precincts of the planetary system.

Fig. 12.18 Fireball   A great flash of light, called a fireball, is produced when a large meteoroid streaks through the atmosphere. It is often accompanied by sonic booms and rumbling noises. A camera-chopping shutter used for timing and velocity determinations produced spaces between the luminous segments of the fireball's trajectory. The faint curved lines in the background are star trails caused by the Earth's rotation during the three-hour exposure. (Courtesy of SAU.)

Fig. 12.19 Antarctica   The midnight Sun illuminates the wind-swept ice at the bottom of the world. Numerous meteorites have been found embedded in the ice in this region near Allan Hills, Antarctica. These meteorites are probably fragments of asteroids that once had orbits between those of Mars and Jupiter, but a few of them may have come from the Moon or even Mars. (Courtesy of Ursula Marvin/CfA.)

Meteorites have long been recognized as celestial objects. The Acts of Apostles in the New Testament (Acts 19:35) refers to a temple dedicated to Artemis in which there is a "sacred stone that fell from the sky". These black objects have also been found in the Egyptian pyramids with a hieroglyph meaning "heavenly iron".

Until relatively recently, only about 10 meteorites were recovered each year. They were ones that happened to fall near populated regions of the Earth, and many more must be buried at the bottom of the ocean, lost in the jungles, or buried in desert sand.

The Antarctica lode

In 1969, a group of Japanese scientists discovered a bountiful source of meteorites on the blue ice fields at the bottom of the world (Fig. 12.19), leading to a dramatic increase in the number of recovered meteorites. During the next four decades, tens of thousands of these cosmic rocks have been harvested from the Antarctic ice.

The most productive areas were near the Allan Hills in Victoria Land and the Yamato Mountains in Queen Maud Land.

Many of the objects recovered from Antarctica must be fragments of the same meteorite, but they also represent thousands of other different meteorites, more than doubling the number found on Earth. Before scientists traveled to Antarctica, the world's meteorite collections only had about 2600 different specimens, most of them collected during the past two centuries.

A meteorite landing on the Antarctic ice becomes buried in compressed snow, and is quickly frozen into the thickening ice. The cosmic rock soon sinks to great depths, where it remains preserved against corrosion. The tremendous mass of the ice, which reaches a thickness of 4 kilometers, squeezes the ice downward and pushes it outward toward the edges of the continent. The buried meteorite becomes caught in these huge ice-flows that move like rivers under the surface, creeping along at rates of several meters per year.

Fig. 12.20 Stony meteorite   Fragment of the stony, chondrite meteorite that fell near Johnstown, Colorado. (Courtesy of the American Museum of Natural History.)

Some of the ice and its enclosed meteorites ultimately reaches the sea and breaks off as icebergs. At other locations the ice-flow encounters an obstacle, such as a mountain range. The moving ice-flow then thrusts the meteorite upward and forces the buried rock to the surface. Strong winds corrode and wear away the surface, removing the ice covering and exposing the meteorite. The dark crusts of the meteorites are then easy to spot against the bright icy background.

Meteorites recovered in Antarctica have stayed there for relatively long times. Radioactive dating indicates that most of them have spent about half a million years entombed in the ice, remaining virtually unchanged since the time they struck the Earth. By way of comparison, most of the meteorites found elsewhere on Earth have fallen within the past 200 years.

Typical meteorites

Meteorites, together with rocks returned by astronauts from the Moon and grains of dust collected from the high levels of the Earth's atmosphere or from comet Wild 2 by the *Stardust* mission, are the only samples we know of extraterrestrial material. Although a meteorite's surface is usually coated with dark glassy material, known as a *fusion crust*, that melted during its descent through the atmosphere, the heat of friction did not have time to penetrate deeply into the falling rock. So most recovered meteorites are cold inside, and their interiors are unaffected by their rapid fall though the atmosphere. Meteorites may therefore be cut open and examined with microscopes and subjected to chemical analysis that reveals their original constitution.

Most meteorites that have been seen to fall and then recovered are stones rather than chunks of metal (Fig. 12.20). About 94 percent of the fallen meteorites are stones, 5 percent irons and 1 percent stony-irons (Table 12.3). About 90 percent of the stony meteorites are, in turn, classified as chondrites. So most of the meteorites that fall to Earth are chondrites, which contain millimeter-sized chondrules (Fig. 12.21). The name "chondrite" is derived from the ancient Greek word *chondros*, meaning "grain" or "seed". The other 10 percent of the stony meteorites are achondrites, which show signs of past igneous activity (Fig. 12.22).

The chondrites have been additionally divided into clans with common properties, suggesting that each clan formed in the same region of the solar system. They are the ordinary chondrites, which are the most abundant, the carbonaceous chondrites, and the enstatite

Table 12.3 Classes of fallen meteorites

Name	Composition	Density[a] (kg m^{-3})	Percent[b]
Stones	Silicates (75 to 100 percent) and metallic nickel-iron (0 to 25 percent)	3500 to 3800	94
Irons	Nickel, iron (100 percent)	7600 to 7900	5
Stony-irons	Silicates (50 percent) and metallic nickel-iron (50 percent)	4700	1

[a] For comparison, typical rocks on Earth are largely silicates with densities in the range 3100 to 3300 kilograms per cubic meter (kg m^{-3}), so meteorites are usually denser than other rocks found on the Earth's surface.
[b] Percent of total meteorites.

chondrites, named for their high abundance of the mineral enstatite.

Most meteorites are denser than terrestrial rock. So, if you find a dark, rather smooth rock that you suspect of being a meteorite, it must weigh at least as much as an ordinary Earth-born rock of the same volume if it is to pass muster as a rock from the sky.

But, of course, every rule has an exception – except that rule. There is a rare class of meteorites, with the ponderous name carbonaceous chondrites, that are fragile and have unusually low densities in the range 2200 to 2900 kilograms per cubic meter, making them less dense than an average terrestrial rock. They often contain appreciable amounts of carbon and water and they are considered to be among the most primitive and least altered samples of solids in the solar system.

Chronology of the meteorites

If we ask "How old is a meteorite?" the question can have several meanings. Each meaning refers to the time since a significant event in the history of a meteorite, and there are three of them: the formation, breakup and impact of a meteorite with Earth.

1 Formation

Meteorites are as old as the solar system, dating back to its earliest days. Most of them were formed at about the same time as the planets. They accumulated directly from the primeval solar nebula and they have compositions similar to that of the Sun, except for their lack of hydrogen and helium.

The dates of formation of the meteorites can be determined by radioactive dating, in much the same way that the

ages of lunar rocks were determined. The relative concentrations of the decay products of elements such as rubidium and uranium reveal the time since these rocks were formed. Such measurements indicate that the carbonaceous chondrite meteorites formed 4.566 ± 0.002 billion years ago, and that some stony and iron meteorites are 4.55 ± 0.07 billion years old.

Rounding off the numbers, and allowing for possible systematic errors, we obtain an age of about 4.6 billion years for most meteorites. That is when the mineral grains in the meteorites crystallized, and their radioactive clocks started ticking. The presence of decay products from short-lived radioactive elements additionally indicates that the meteorites formed in just a few million years some 4.6 billion years ago.

The meteorites are the oldest rocks that we can touch. They are hundreds of millions of years older than the oldest rocks on the surface of the Earth. All of the planets and satellites are thought to have originated together with the meteorites 4.6 billion years ago, but erosion and geological processes have destroyed the original rocks on Earth. As a result, it is meteorites that reveal the age of the solar system and give clues to its origin.

2 Breakup and exposure

Another type of radioactivity also occurs in meteorites – radioactivity that is continually being caused by cosmic rays in the solar system. These "rays" are not rays in the usual sense; they are very energetic protons that bombard the meteorites and penetrate their surface for short distances. This cosmic-ray bombardment performs a bit of alchemy and transforms some of the atoms of the meteorite into radioactive nuclei. These radioactive nuclei slowly disintegrate, creating "daughter" nuclei. As the meteorite continues to be exposed to cosmic rays, the daughter nuclei

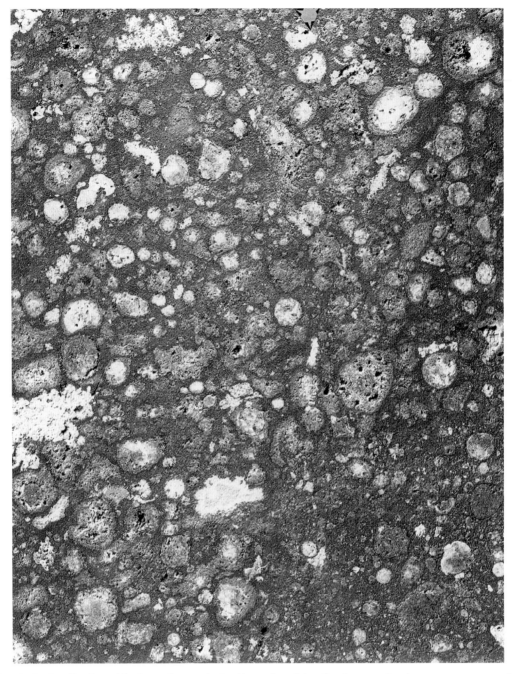

Fig. 12.21 Chondrules in Allende   This photomicrograph of a thin section of the Allende meteorite shows numerous round silicate chondrules together with irregular inclusions. The meteorite section is 0.021 meters across and 0.027 meters high. (Courtesy of the Smithsonian Institution.)

become more and more abundant, and by a careful measurement of the amount of daughter nuclei in a meteorite it is possible to estimate the duration of this exposure interval, or the exposure age of the meteorite.

Now, the exposure ages of the meteorites that have been recovered on Earth are remarkably short in astronomical terms. Typically they are between 5 and 60 million years – just an instant in the life of the solar system and the meteorites. Evidently the meteorites have spent most of their lives shielded from cosmic rays. Astronomers now believe that most meteorites spent a larger portion of their life protected inside a parent body that was much smaller than the Earth but larger than a typical meteorite. According to this view, an important event in the chronology of most meteorites was the breakup of a parent body, exposing smaller fragments to space and to the bombardment by cosmic rays. The exposure ages measure the time that has elapsed since the breakup took place.

Fig. 12.22 Achondrite meteorite A photomicrograph of the achondrite meteorite that fell near Juvinas, France, on 15 June 1821. It contains basalt, resulting from the melting and separation of material inside an asteroid-size parent body. The section shown here is 0.0032 meters across. (Courtesy of Martin Prinz, American Museum of Natural History.)

3 Collision with the Earth

Finally, when a meteorite falls through the blanket of the Earth's atmosphere, it becomes protected from cosmic-ray bombardment. No more radioactive atoms are created, and the ones that already exist inside the meteorite begin to decay, like the slow ticking of a clock.

In this way, the atoms of a meteorite carry a record of their chronology that can be unlocked with radiochemistry.

Rare and exotic finds

The frozen cargo of the Antarctica ice includes at least a dozen, greenish-brown meteorites that are strikingly similar to the welded highland rocks from the Earth's Moon. The abundance of various elements and gases in these meteorites are virtually identical to those found in lunar rocks; at the same time, they are unlike those found in any other known meteorite or terrestrial rock. These small stones were apparently blasted off the Moon by impacting objects.

Out of the thousands of stony meteorites now found in terrestrial collections, roughly a dozen are believed to be pieces of Mars. They were similarly blasted into space by impacting objects, with such force that they escaped the red planet's gravitational pull and eventually reached Earth. One of them, dubbed ALH (for Allan Hills in Antarctica) 84001 was once thought to contain evidence for ancient microscopic life on Mars, but that evidence has subsequently been attributed to other causes.

The meteorites from Mars contain small amounts of water and water-altered minerals. Moreover, gases trapped in bubbles within some of them have the exact, unique composition as the gases in the Martian atmosphere, as measured by the *Viking* landers, and found nowhere else. So there can be little doubt that these meteorites came from Mars.

They are often referred to as the SNC, pronounced "snick", meteorites, short for Shergotty (India), Nakhla (Egypt), and Chassigny (France) – three locations where they were observed to fall from the sky. Radioactive dating indicates that many of the SNCs were molten in the relatively recent history of our solar system. Those named for Nakhla and Chassigny hardened into solid rock 1.3 billion years ago, while the ones found near Shergotty solidified from molten lava just 180 million years ago. In comparison, most other meteorites solidified from molten materials between 4.5 and 4.6 billion years ago.

Organic matter in meteorites

For more than a century, organic molecules have been suspected inside meteorites, although their presence in newly arrived meteorites led to the suggestion that they were the result of terrestrial contamination. But their existence became a certainty when 20 kinds of amino acids were found in the carbonaceous chondrite meteorites from Antarctica. These meteorites had lain in a sterile environment and were collected using sterile procedures. The organic matter in these carbonaceous chondrites apparently formed within the meteorite parent bodies, probably when water was bound into the structure of its clay minerals.

Many organic compounds, such as the amino acids, can come in two versions that are mirror images of each other. They are identified as left- and right-handed, based on their ability to rotate light in one direction or another. All living organisms on Earth use only left-handed amino acids. In contrast, the carbonaceous chondrites contain roughly equal amounts of both types, including the right-handed amino acids that are not found in living systems on Earth. This provides convincing evidence that the organic matter in meteorites did not originate on our planet, and that it is not directly related to life as we know it.

This discovery implies that the amino acids and other organic molecules probably existed in the solar system a billion years before the appearance of living things on Earth. But the molecules found in carbonaceous chondrites are generally thought to be of non-biological origin. So the organic molecules found in meteorites are not in themselves vestiges of extraterrestrial life. But they are certainly primitive, and their cousins may have been the precursors to living matter.

The asteroid–meteorite connection

What is the source of meteorites? The primitive nature of most meteorites, called chondrites, indicates that they came from objects that have not experienced geologic processes. Most of them are not igneous, and did not go through a hot, liquid stage. This is most easily understood if their parent bodies were small asteroids. Only a small fraction of meteorites, known as the achondrites, formed by igneous processes in larger parent bodies, most likely one of the original asteroids.

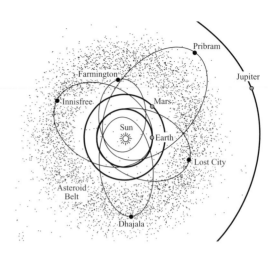

Fig. 12.23 Meteorite orbits   The calculated orbits of five meteorites, inferred from their trajectory before hitting the ground. All of them originated in the asteroid belt, indicating that these meteorites are chips off asteroids.

There is little doubt that most of the meteorites have come from the asteroid belt. They are probably chips off wayward asteroids, and there are three pieces of evidence for this conclusion. They are the orbits, colors and crystalline structure of meteorites.

1 Orbits

Photography of meteorites as they descend through the Earth's atmosphere can be used to determine their precise speed and direction of motion when they encountered the Earth. From these data, their orbits may be inferred, and many of the objects came from space beyond Mars, in the main belt of asteroids (Fig. 12.23).

2 Colors

The surface composition of asteroids has been inferred by breaking down their reflected sunlight into its component colors. Such spectral displays are similar to those of meteorites, suggesting that the meteorites are the debris of colliding asteroids. The relative abundance of asteroid types are not like those of the fallen meteorites, but this may simply reflect the ease or difficulty in sending asteroid fragments to Earth.

The sunlight reflected from C-type asteroids, that far outnumber all other asteroid types, closely resembles that of the relatively scarce carbonaceous chondrite meteorites. But the C-type asteroids reside in the remote, outer part of the main belt, furthest away from the Earth. The most common meteorites, the ordinary chondrites, display spectral colors that resemble those of the S-type asteroids found in the inner half of the asteroid belt. Due to its proximity, most meteorites are expected to originate from this part of the main belt. The light reflected from rare,

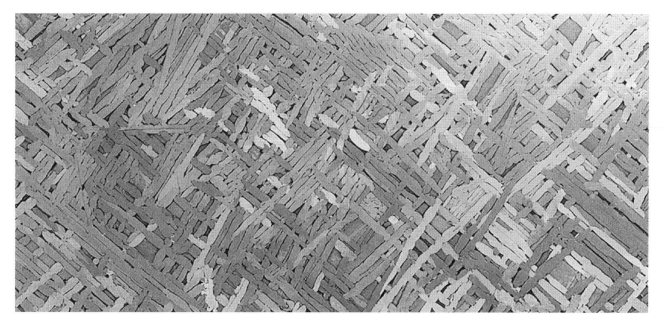

Fig. 12.24 Widmanstätten pattern When polished and etched with acid, an iron meteorite displays this distinctive Widmanstätten pattern produced by crystals of two different iron-nickel alloys. The pattern provides evidence that this meteorite was once buried within a parent body between 50 and 200 kilometers in radius. This sliced specimen is about 0.05 meters across. (Courtesy of the Smithsonian Institution.)

bright, M asteroids, which reside near the middle of the main belt, matches the spectrum of relatively scarce iron meteorites.

Nevertheless, the color coordination is not exact. The S-type asteroids have redder overall colors and subdued light absorption compared with the abundant meteorites. This discrepancy may be explained if solar radiation or small impacting cosmic particles gradually altered the thin outer layers of asteroids, darkening and reddening their surfaces. Similar "space weathering" apparently makes the lunar surface much redder than the color of unexposed rocks returned from the Moon. And the *Galileo* spacecraft showed that material near the sharp-edged, relatively young craters on 951 Gaspra and 243 Ida is slightly bluish, while the low-lying, older areas are slightly reddish. So the evidence now suggests that the color of freshly exposed asteroid surfaces gradually redden with time. The bluish surfaces are recently exposed, while the red areas have been weathered for millions and even billions of years.

Powerful collisions most likely excavate meteorites from deeper layers inside asteroids, which retain their pristine, unweathered color. So the surface appearances of asteroids are slightly misleading. Deep down inside, the most common, inner main-belt asteroids are probably very similar to the most common meteorites.

Instruments aboard *NEAR Shoemaker* revealed that 433 Eros, also an S-type asteroid, has the same basic composition as some ordinary-chondrite meteorites. The composition of the surface of asteroid 1 Ceres is similar to that of a water-rich carbonaceous chondrite meteorite. Asteroid 4 Vesta is an exception, bearing signs of significant heating, differentiation and surface lava flow. Five percent of the found meteoritic samples on Earth, the Howardite Eucrite Diogenite (HED), meteorites, are thought to be the result of a collision or collisions with Vesta.

3 Crystalline structure

When the majority of iron meteorites are cut and polished and then are etched with acid, a delicate and complex pattern emerges (Fig. 12.24). It is produced by regions of crystalline structure, depending on the local orientation of the crystals in the iron. The sizes and shapes of these crystals indicate that they grew very slowly, and that the meteorite must have been hot, almost to the melting point, for tens of million of years. It probably cooled at the rate of a few degrees in a million years.

Such a slow cooling rate is compelling evidence that the meteorites were once inside a sizeable parent body. If a small meteorite had been exposed to space when it was still hot, it would have cooled in a matter of days. Small meteorites cool rapidly because their material is close to the surface, through which the heat can escape. Large crystal patterns would not have grown in such small bodies.

The meteorites that retained their heat for millions of years must have been buried within parent bodies between 50 and 300 kilometers in radius, and this is just the size of large asteroids.

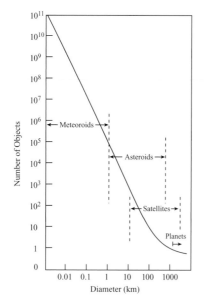

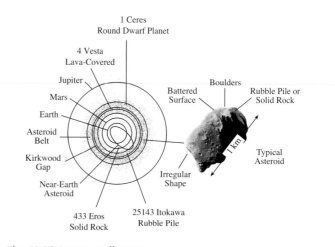

Fig. 12.26 **Summary diagram**

Fig. 12.25 **Interplanetary objects**   Repetitive collisions between interplanetary objects have produced many more small meteoroids than large ones. Some of the largest asteroids are comparable in size to small moons, and ongoing collisions between asteroids have produced numerous smaller meteorites.

Additional suggestive evidence is the relationship among the sizes of asteroids, meteorites, and meteoroids (Fig. 12.25). The classes are not mutually exclusive, and there is considerable overlap. The simplest explanation of all the evidence is that meteorites are the debris of collisions among the asteroids.

Whatever their precise history, the asteroids and meteorites are primitive objects that can act as beacons to the past. They represent a tableau of the ancient objects that littered space 4.6 billion years ago, and they carry messages for our understanding of the formation of the solar system.

13 Colliding worlds

- At least 20 pieces of a comet hit Jupiter on 7 July 1992, producing explosive fireworks and dark scars that fascinated astronomers throughout the world.

- Some comets are on suicide missions to the Sun, diving into our star and being consumed by it.

- Most of the impact craters on the Earth disappeared long ago, but a few of the relatively recent ones have been located from space.

- An asteroid wiped out the dinosaurs when it hit the Earth 65 million years ago.

- If an asteroid or comet of about 10 kilometers in size hit the Earth, the horrific blast could generate overpowering ocean waves, block out the Sun's light and heat, ignite global wildfires, drench the land and sea with acid rain, and produce deadly volcanoes on the other side of the Earth.

- The Earth is immersed within a cosmic shooting gallery of potentially lethal, Earth-approaching asteroids that could collide with our planet and end civilization as we know it.

- The lifetime risk of your dying as the result of an asteroid striking the Earth is about the same as death from an airplane crash, but a lot more people would die with you during the cosmic impact.

- It is estimated that the Earth receives a direct hit by an asteroid about two kilometers in size every million years or so, resulting in a global catastrophe. It could happen tomorrow or it might not occur for hundreds of thousands of years.

- Astronomers are now taking a census of most of the Near-Earth Objects that are big enough and close enough to threaten us with global destruction.

- With enough warning time, we could redirect the course of an asteroid or comet that is headed for collision with the Earth.

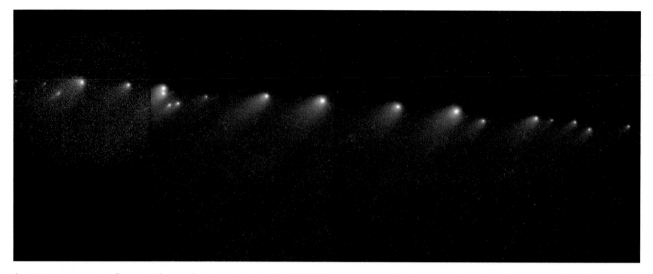

Fig. 13.1 Fragments of comet Shoemaker-Levy 9   In July 1992, this comet passed so close to Jupiter that the icy material of its nucleus was torn apart by the differential gravitational forces of the giant planet. This panoramic image of the comet fragments was taken from the *Hubble Space Telescope* in January 1994, eight months after they were discovered and six months before they dived into the atmosphere of Jupiter. The length of the string of comet pieces is about 1.1 million kilometers, three times the distance from the Earth to the Moon. The largest of the fragments in the string is about two kilometers across. Each fragment mimics a larger comet with a round coma and a dusty tail. (Courtesy of NASA/STScI.)

13.1 A comet hits Jupiter

Comet Shoemaker-Levy 9

Eugene Shoemaker (1928–1997), his wife Carolyn, and the amateur astronomer, David Levy (1948–) were involved in routine observations the cloudy evening of 23 March 1993. They were continuing a ten-year search for comets and asteroids that might be headed toward the Earth using the small 0.46-meter (18-inch) wide-field photographic telescope at Palomar Observatory in California. Two days later, when Carolyn examined the images, taken on fogged film, she saw an elongated feature that looked to her like a "squashed comet". When the discovery was confirmed with better telescopes, the stretched-out blur of comet light was resolved into several little comets aligned along a single straight line projected in the sky, each with a nearly spherical coma and elongated dust tail (Fig. 13.1). In accordance with tradition, it was named comet Shoemaker-Levy 9, after the last names of the discoverers, ninth in a series of objects the trio found traveling around the Sun in short-period orbits.

Comet Shoemaker-Levy 9 (SL9) consisted of the pieces of a former single comet that had been trapped in a two-year orbit around Jupiter for decades. But when it traveled too close to Jupiter in 1992, the comet was torn apart by the difference between the planet's gravitational attraction on the near and far side of the comet, ripping the fragile comet into at least 20 observable pieces.

A few previous comets had been known to orbit Jupiter temporarily, and the disruption wasn't unprecedented. What made SL9 unique was that the broken comet was inexorably hurtling toward a collision with Jupiter. Orbital calculations indicated that the train of comet fragments would plunge into the giant planet in July 1994, two years after the former single comet's disruption and more than one year after the discovery of its pieces (Fig. 13.2).

Such advance knowledge of a collision with any planet was unprecedented in human history. Numerous craters on the terrestrial planets, as well as the Earth's Moon and some of Jupiter's satellites, bear witness to cosmic collisions of the past, but now for the first time astronomers could see the collision happening before their very eyes. The anticipated impact was an incredible opportunity, occurring just once in the lifetime of any astronomer and most likely once in a millennium.

Impact of a comet with Jupiter

The collision of comet Shoemaker-Levy 9 with Jupiter in July 1994 was perhaps the most widely witnessed event in astronomical history. Practically every telescope in the world was trained on Jupiter during impact week, between 16 and 22 July 1994. Infrared heat detectors were placed at the focal point of the Keck Observatory's giant 10-meter telescope atop Mauna Kea in Hawaii, and the *Hubble Space Telescope* was poised to record the event at visual wavelengths. Every other major astronomical observatory

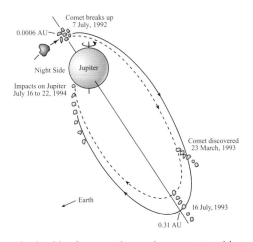

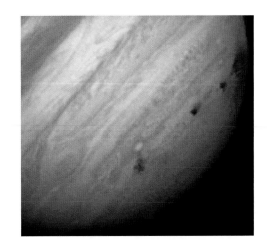

Fig. 13.2 Final orbit of comet Shoemaker-Levy 9 This comet was orbiting Jupiter for more than half a century, until it was ripped apart during a close encounter with the planet and collided with it two years later. The disruption occurred on 7 July 1992 when the comet passed within 0.0006 AU, or 90 000 kilometers, from the planet's center. Since Jupiter has a radius of just over 70 000 kilometers, the comet passed within about 20 000 kilometers of the planet's cloud tops. Jupiter's unequal gravitational pull on the near and far sides of the comet nucleus then tore the object apart. Carolyn and Eugene Shoemaker and David Levy discovered the comet fragments on 23 March 1993, when the broken comet was almost at its farthest distance from Jupiter, at 0.31 AU. One by one the icy fragments exploded in Jupiter's cloud tops during impact week, from 16 to 22 July 1994.

Fig. 13.3 Dark impact scars of comet fragments hitting Jupiter This *Hubble Space Telescope* image shows several dark spots (*right*) on Jupiter, each marking the impact site of a fragment of Comet Shoemaker-Levy 9. The Earth-sized scars remained visible for about five months, until the winds in Jupiter's outer atmosphere pulled them apart. A thin expanding ring of dark material, suggesting waves spreading out from the impact explosion, surrounded some of the dark central spots. Jupiter's Great Red Spot is also prominent, as it has been for centuries. (Courtesy of NASA/STScI.)

participated, as did numerous amateur astronomers from their own backyards.

Detailed calculations indicated that the collisions would be on the dark "back" side of Jupiter, hidden from the Earth's view by the body of the giant planet. So it might be something like watching a World Series ball game from a seat behind a stadium post. The comet fragments would nevertheless strike Jupiter close to the side facing Earth, so astronomers hoped that something would be seen when the planet's rapid rotation, of once every 9 hours 55.5 minutes, brought the impact sites into view. Moreover, the *Galileo* spacecraft, on its way to Jupiter, had a direct view of the actual collisions from its unique position in space.

No one was disappointed! When the comet fragments plowed into Jupiter, each of them exploded with an energy equivalent to the simultaneous explosion of hundreds of thousands of nuclear bombs on Earth. As it penetrated the outer atmosphere of Jupiter, each comet fragment heated and compressed the surrounding gas, producing a violent explosion high in the atmosphere. The resultant fireball punched a hole through the overlying material and sent plumes of hot gas rising into space. Using instruments

on the *Galileo* spacecraft, they were detected at the side of Jupiter by their infrared heat radiation, indicating temperatures of up to 20 000 kelvin. It took 10 or 20 minutes for each plume to rise and fall again, by which time the impact site had rotated into view from Earth.

The comet fragments slammed into Jupiter, one after another, like the cars of a train when its locomotive is derailed. After generating a bright ball of light, each fragment disfigured Jupiter with a black scar that had never been seen before, twice as large as the Earth and spanning tens of thousands of kilometers (Fig. 13.3). Meanwhile, waves swept across the impact site and reverberated deep within the planet, which seemed to shudder from the impacts. The dark marks endured for months, gradually spreading, merging and slowly fading from view as the winds in Jupiter's atmosphere dispersed their material.

13.2 Consumed by the Sun

Some comets plunge deep into the Sun's thin million-degree outer atmosphere, or corona. Instruments aboard the *SOlar and Heliospheric Observatory* (*SOHO*) satellite have recorded their death-defying trip around the Sun. One of its instruments uses an occulting disk to block out the bright light of the visible solar disk, enabling it to detect the comets as they move through the inner corona (Fig. 13.4). A comet often pays a heavy price for this trip, sometimes breaking apart because of the Sun's gravitational forces.

Fig. 13.4 Fatal impact of a comet into the Sun   This composite image records a comet plunging into the Sun on 23 December 1996. The innermost image (*center*) records the bottom of the million-degree solar atmosphere, known as the corona. The electrically charged coronal gas is seen blowing away from the Sun just outside the inner dark circle, which marks the edge of one instrumental occulting disk. Another instrument records the comet (*lower left*), as well as the coronal streamers at more distant regions and the stars of the Milky Way. (Courtesy of the *SOHO* EIT, UVCS and LASCO consortia. *SOHO* is a project of international collaboration between ESA and NASA.)

Other comets are hurtling toward complete meltdown, passing so close to the Sun that the encounter is fatal. Though rarely, if ever, hitting the visible solar disk, or photosphere, these comets can come closer than 50 000 kilometers from it. They are unlikely to survive the Sun's intense heat and gravitational forces at that range.

Most of the comets discovered by *SOHO*, about 90 percent of them, are small comet fragments known as the Kreutz sungrazers, which closely approach the Sun from one direction in space. They are named after the German astronomer Heinrich Kreutz (1854–1907) who found that many of the comets that traveled exceptionally near to the Sun in the 19th century seemed to have a common origin with similar orbits. They are all probably fragments of a single large comet that first broke up when passing very close to the Sun thousands of years ago.

When a member of the Kreutz sungrazer group moves around its orbit and returns to our vicinity, it can dive into the inner corona and disappear forever (Fig. 13.5). Each comet fragment can be very small, just 6 to 12 meters across, despite their spectacular display. Such a tiny object, falling so close to the Sun, would vaporize completely away.

The occulting instrument aboard *SOHO* has unexpectedly watched more than a thousand comets pass near the

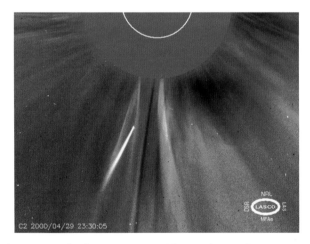

Fig. 13.5 Death of a sungrazer   This *SOHO* image shows a bright sungrazing comet (*bottom center*) headed into the inner atmosphere of the Sun on 22 October 2001. The partial white circle marks the outer edge of the visible solar disk, whose intense glare is hidden by the instrument's occulting disk (*opaque circular region at top center*). The million-degree solar atmosphere can also be seen, steaming away from the Sun. (Courtesy of the *SOHO* LASCO consortium. *SOHO* is a project of international collaboration between ESA and NASA.)

Sun or even into it. Amateur astronomers from all over the world have examined *SOHO*'s real-time images, posted on the Internet at http://sohowww.nascom.nasa.gov/, discovering hundreds of previously unknown comets, including numerous Kreutz sungrazers and many other new comets as well.

The collisions of comets with Jupiter and the Sun have helped raise worldwide awareness of a similar threat to our home planet. Such impacts have happened on Earth in the past and they could happen again, with devastating effects to civilization.

13.3 Impacts of asteroids with the Earth

The solid surfaces of the terrestrial planets, as well as the Earth's Moon, are marked with impact craters, the scars of past collisions with cosmic objects speeding through space. They originated by the coalescence of these objects, and the barrage continues at a lower rate today when many cosmic objects travel across the orbits of the terrestrial planets (Fig. 13.6). So the Earth does not occupy a secure niche in space. Our planet is instead immersed in a cosmic shooting gallery, subject to a steady bombardment by potentially lethal, Earth-approaching objects.

The large majority of asteroids are in orbits between those of Mars and Jupiter where they pose no threat to Earth. Some of these rocky objects, however, follow a more eccentric course that takes them closer to Earth, sometimes crossing its path. And although most short-period comets do not come closer to the Earth than the nearest planets, some of these icy intruders can pass perilously near to us. It is these threatening asteroids or comets, collectively known as Near-Earth Objects (NEOs), which may be on a collision course with the Earth. If one of these cosmic bombs hits our planet, it could explode with a violence that far surpasses the world's entire nuclear arsenal, threatening civilization and possibly making humans extinct.

Explosions in the atmosphere

We have dramatic evidence that intruders from space are constantly bombarding the Earth today. Unknown to the public, the US Department of Defense has been detecting them for decades. Military satellites, designed to watch for enemy rocket firings and nuclear explosions, have detected the explosions produced when speeding, house-sized cosmic objects enter the air. The incoming projectiles are heated to incandescence and then self-destruct in the upper atmosphere, vanishing without a trace on the ground.

Ground-based defense networks, designed to listen to the sounds generated by man-made nuclear explosions, have confirmed the satellite results. They show that one of these cosmic bombs is bursting overhead every month, each with an energy equivalent to a nuclear bomb. These investigations might be aiding world peace by helping to distinguish between natural explosions and those caused by humans, thereby preventing false warnings of clandestine nuclear tests or of nuclear attack by terrorists.

The largest object to strike the Earth in the 20th century wasn't quite big enough to reach the ground. It disintegrated between 5 and 10 kilometers up, over the Podkamennaya Tunguska River in central Siberia. The shock wave generated by the ensuing explosion leveled trees over 2 trillion (2×10^{12}) square meters of the underlying land, an area larger than New York City and surrounding suburbs (Fig. 13.7). The energy produced was equivalent to the aerial explosion of the nuclear bomb that leveled Hiroshima. So much devastation, yet it failed to produce a crater.

A Tunguska-like atmospheric explosion, relatively near the ground, is the usual fate for a stony asteroid fragment roughly 50 meters across. An asteroid of this size would have had the internal strength to penetrate deeply into the Earth's atmosphere before exploding, but a comet of comparable size would have disrupted too high in the atmosphere to cause much damage to the ground.

Scientists estimate that a Tunguska-like explosion should occur every century or two, on average. If it happens above a large city, the results will be disastrous. But then, it is more likely to enter the atmosphere above the oceans or remote land regions, since they cover much more of the planet's surface area. Terrestrial impacts by larger cosmic objects occur less frequently, but they will reach the ground with more powerful consequences.

The terrestrial impact record

Cosmic intruders just tens of meters in size can survive passage through the Earth's atmosphere more or less intact and strike the terrestrial surface at high velocity. Even a relatively small metallic asteroid, just tens of meters across, is tough enough to penetrate the atmosphere and hit the ground. Stony asteroids have to be more than 50 meters in size to survive passage through the atmosphere, and the fragile comets must be at least ten times as big to make it through.

A classic example is the 1.2-kilometer-wide Meteor Crater in northern Arizona, also known as the Barringer Crater (Fig. 13.8). It was formed about 50 000 years ago by the impact of a nickel-iron asteroid just 60 meters in diameter. When it struck the Earth, this relatively small projectile released an estimated 100 million billion (10^{17}) joules of kinetic energy, an amount comparable to the

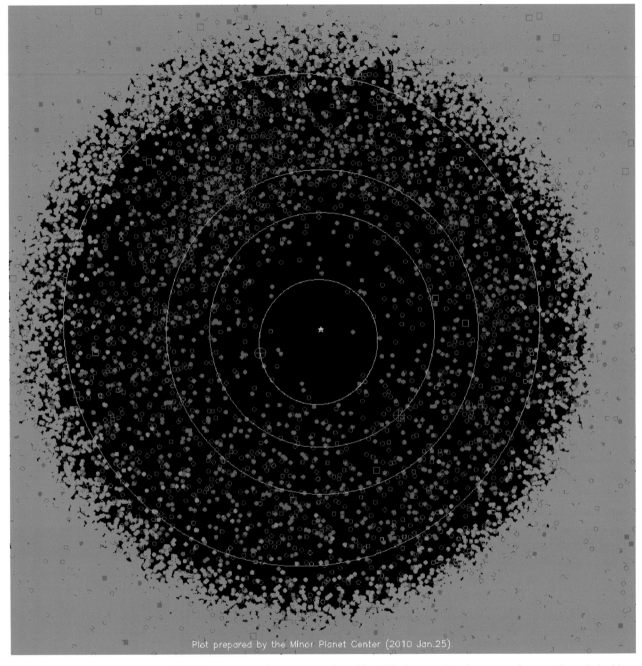

Plot prepared by the Minor Planet Center (2010 Jan.25).

Fig. 13.6 Minor and major planets in the innermost solar system   The orbits of the inner major planets are shown as large light-blue circles, for Mercury, Venus, Earth and Mars from the center out. Small red circles indicate the locations of the minor planets coming within 1.3 AU of the Sun at perihelion, where 1.0 AU is the mean distance between the Earth and the Sun. Small filled circles indicate objects observed at more than one opposition. Outline circles indicate objects seen at only one opposition. Numbered periodic comets are shown as filled light-blue squares. Other comets are shown as unfilled light-blue squares. The plot is for the date 20 January 2010. (Courtesy of Gareth Williams, Minor Planet Center.)

energy released by the explosion of 20 million tons of TNT (trinitrotoluene).

Even the largest craters, produced by the biggest aster-oids, will gradually disappear from sight with the passage of time. The same forces that erode mountains, deposit sediments, eject lava and shift continents are erasing the craters and removing them from sight. If not for these dynamic forces, the craters accumulated over the ages on the Earth would be as densely distributed and prominent as the overlapping craters on our Moon.

Only about 176 terrestrial impact craters have managed to survive the ravages of time. They have been identified

Fig. 13.7 The Tunguska explosion   On 30 June 1908, a giant blue-white ball of fire streaked across the daytime sky above the Tunguska River in Siberia, apparently becoming brighter than the Sun. The Tunguska fireball then exploded in the atmosphere, felling underlying trees like matchsticks. All of the toppled trees point away from a central location that must have been directly below the point of the explosion. Interpreting the extent and orientation of the tree-fall pattern, scientists concluded that the explosion above Tunguska released an energy equivalent to exploding a nuclear bomb between 5 and 10 kilometers above the Earth's surface.

on images taken from space, using airplanes, the *Space Shuttle*, or satellites such as *Landsat*. These craters can be first identified from aerial photographs, by their circular shapes and uplifted and overturned rims (Fig. 13.9). Aerial images of about one quarter of the identified Earth impact craters have been provided by scientists at the Lunar and Planetary Institute on the Internet at http://www.lpi.usra.edu/publications/slidesets/craters/.

But since other processes, such as volcanism and erosion, can also leave circular holes, confirming evidence of an impact origin must be gathered from rocks in and around the crater. How do geologists know that some terrestrial craters are due to the explosions of projectiles coming from space? They look for rocks that have been transformed under the conditions of extreme temperature, pressure, and shock associated with a high-velocity external impact. The most apparent shock effect is the formation of conical structures called shatter cones, which point toward the center of the impact (Fig. 13.10). Other evidence includes glassy, previously molten material formed at high temperature, and minerals with a deformed crystal structure produced by a shattering, high-pressure impact. Roughly 25 percent of the craters also contain meteorites that had to come from space.

Geologists have dated some of the craters by radiometric age determinations of previously melted rock, determining the time that has elapsed since the molten rock cooled and solidified. They find ages from a few thousand to 2 billion years, but most of them are younger than 200 million years old. The lunar crater record suggests that the cratering rate on Earth must have been roughly constant during the past 3 billion years, so erosion and other geological processes have worn most of the older ones away.

Only the largest terrestrial craters have been able to survive the wearing effects of time for more than a few million years. One of the largest and oldest ones, for example, is located near Vredefort, South Africa; it is 300 kilometers in diameter and it was formed about 2 billion years ago.

Fig. 13.8 Meteor crater   One of the best and earliest known impact craters on Earth, located in the northern Arizona desert. It has been named the "Meteor Crater" for the nearby post office named Meteor, but it is also widely known as the Barringer Crater in honor of Daniel Barringer (1860–1929), a mining engineer and business man, who was the first to suggest, in 1903, that it was produced by meteorite impact. The crater, about 1.2 kilometers in diameter, was excavated about 50 000 years ago during the impact of a nickel-iron meteorite about 50 meters across. When first discovered, about 30 tons of meteoritic iron was found scattered about the crater. The meteorite is officially named the Canyon Diablo Meteorite, after the nearby ghost town of Diablo Canyon. (Courtesy of NASA.)

Fig. 13.9 Wolf Creek impact crater   This relatively well-preserved crater near Wolf Creek, Australia, is partly buried under windblown sand. Iron meteorites have been found in the vicinity, as well as some impact glass. The rim diameter of this crater is 850 meters and the impact that created it occurred about 300 000 years ago. (Courtesy of Virgil L. Sharpton and the Lunar and Planetary Institute.)

Table 13.1 The 10 largest identified terrestrial impact craters[a]

Crater name	Location	Diameter (km)	Age (My)	Meteoritic component
Vredefort	South Africa	300	2023 ± 4	Chondrite
Sudbury	Ontario, Canada	250	1850 ± 3	–
Chicxulub	Yucatán, Mexico	170	64.98 ± 0.05	Chondrite
Popigai	Russia	100	35.7 ± 0.2	Chondrite
Manicouagan	Quebec, Canada	100	214 ± 1	–
Acraman	South Australia	90	~590	Chondrite
Chesapeake Bay	Virginia, USA	90	35.3 ± 0.1	
Puchezh-Katunki	Russia	80	167 ± 3	
Morokweng	South Africa	70	145.0 ± 0.8	Chondrite
Kara	Russia	65	70.3 ± 2.2	Chondrite?

[a] The crater diameters are in units of kilometers (km), and the ages are in millions of years (My).

Fig. 13.10 Shatter cones   The shatter cones that are found in the vicinity of many terrestrial craters provide evidence for shocks associated with the impact of a cosmic object. They point towards the direction of impact, like the cone-shaped plugs of glass that are often formed when a bullet strikes a window. The shatter cones shown here are about 0.05 meters in height. They are from the Wells Creek Tennessee Basin, a crater that is about 14 kilometers across and roughly 200 million years old.

An extensive database of known terrestrial impact craters is maintained by the Planetary and Space Science Center, University of New Brunswick, Canada, at http://www.unb.ca/passc/ImpactDatabase, listing them by location, age, diameter or name with other supplemental information including images. Many of the 10 largest ones have been identified with chondrite meteorites and they are between 35 million and 2 billion years old (Table 13.1).

The consequences of these bigger impacts are even more sobering than the small ones. If an asteroid of just 1 kilometer in size hit the Earth, the power of its explosion could not be matched by the world's entire nuclear arsenal. Such an impact is estimated to occur every one million years or so. An asteroid of exceptional size, say 20 kilometers across, might hit the Earth less often, every 100 million years, on average. As we shall next see, such collisions have happened in the past, when they altered the course of biological history.

13.4 Demise of the dinosaurs

Catastrophe from the sky

Collisions by objects from outer space have always been a menace to life on Earth. During the planet's first billion years, the barrage was probably so intense that living things could not exist on the Earth's surface. After those early times, the rate of bombardment slowed down, so impacts of exceptionally large cosmic projectiles became less frequent. But these giant impacts continued every once in a while, with devastating consequences. The most recent death-rock arrived 65 million years ago, wiping out the dinosaurs and many other living things. Such an abrupt destruction of an entire species by a force of nature is known as a mass extinction.

A thin worldwide layer of iridium-rich clay, just 0.01 meters thick, provided the initial evidence that a cosmic collision wiped out the dinosaurs. When a team headed by the American geologist Walter Alvarez (1940–) determined the layer's age, from its position among geologically dated strata, they found that it was deposited about 65 million years ago when the dinosaurs and a variety of plants and animals disappeared.

The clay layer contained unusual amounts of the rare element iridium. It is not found in such quantities in terrestrial rocks since most of the primordial iridium on Earth sank to the planet's core during differentiation early in its history. In contrast, iridium is much more abundant in certain meteorites, so they might have supplied the clay layer with iridium.

Since the cosmic iridium rains steadily down through the atmosphere and settles in the soil, the amount of iridium in a layer of sediment can be used as a cosmic clock. In an average century, a certain amount of iridium will mix with the soil and become part of any new layers that are forming. If a layer requires twice as long to form, it will have twice as much iridium.

But the geologists found that the iridium clock had gone wild for a short interval about 65 million years ago. The amount of iridium they found in this layer of clay was far higher than normally found in the Earth's crust and about 30 times higher than that found in the fossilized limestone above and below the clay. Moreover, the same type of iridium enrichment was found in clay at widely scattered points on the Earth. So the entire globe had been drenched with an unusually large amount of iridium for a short time.

Walter, his father Luis Alvarez (1911–1988), and their colleagues concluded in 1980 that the iridium deluge came from outside the Earth, delivered by a large asteroid that struck the Earth and vaporized about 65 million years ago. According to their hypothesis, the iridium was lofted into the atmosphere along with other debris by the fireball of hot gas created during the collision, and then carried by the winds over much of the globe. The worldwide cloud of iridium-rich dust then slowly filtered back down to the ground where it produced a thin global layer that contained relatively large amounts of iridium. They estimated that a layer 0.01 meters thick covering the entire Earth would be deposited by an asteroid about 10 kilometers in diameter.

Most geologists and biologists must have initially dismissed the idea of a killer asteroid from space. They probably attributed it to spaced-out astronomers or science-fiction enthusiasts. Such an abrupt cataclysm conflicted with the prevalent concept of gradual evolutionary change over the eons. Sudden, short-lived events were just not supposed to affect the course of evolution. Yet the evidence for catastrophic events is found throughout the solar system, including the cratered surfaces of the Earth and its Moon.

Fig. 13.11 Site of impact that wiped out the dinosaurs   The long-sought site of the scar left by a killer asteroid has been found near Chicxulub (*filled circle*), a small village at the tip of the Yucatán Peninsula. Material from the submerged crater has been dated at about 65 million years ago, coinciding with a blast that triggered the eradication of most life on Earth. Thick sedimentary deposits laid down 65 million years ago in Haiti (*big open circle*) contain exceptionally large amounts of iridium, shocked quartz and glassy debris that are thought to be part of the impact across the Caribbean, almost 2000 kilometers away. The small open circles mark the sites of marine wave deposits associated with the same impact. [Adapted from Alan R. Hildebrand and William V. Boynton, *Natural History* **6** 47–53 (1991).]

The idea just would not go away and supporting evidence kept accumulating. Shocked quartz grains were found in the iridium-rich clay layer worldwide. No known terrestrial process, including volcanic flows or explosions, can generate pressure high enough to alter the grains in the observed way – only the sudden shock of an impact can. Still, the skeptics asked: Where is the crater of the impact that occurred 65 million years ago and was big enough to obliterate most of the Earth's life forms?

After years of searching, the telltale crater was found straddling the northern coastline of the Yucatán Peninsula (Fig. 13.11). It is located below the Mayan village of Chicxulub (pronounced Cheek-shoe-lube, a Mayan phrase for "horns of the devil"), and is hence known as the Chicxulub impact basin. The discovery of this crater and the subsequent confirmation of its age at 65 million years led to the widespread acceptance of the impact hypothesis for the demise of the dinosaurs.

At the time of the impact, that part of the Yucatán Peninsula was below sea level, and the center of the crater now lies buried below 1.1 kilometers of limestone laid down in the intervening years. So there is nothing on the surface to betray the crater's existence, and the vast scar cannot be seen directly.

Focus 13.1 The belt of an asteroid

The energy of the collision of an asteroid with the Earth can be calculated from the kinetic energy (K.E.) of the impacting object, which is given by the expression

$$\text{K.E.} = \frac{1}{2} M V^2$$

where M is the mass of the projectile and V is its incoming velocity. The mass of an impacting asteroid can be determined from the mass density and volume. Assuming a mass density, ρ, of about $\rho = 3000$ kilograms per cubic meter, comparable to that of stony asteroids and meteorites, and a radius, R, of $R = 5$ kilometers, a mass of

$$M = \rho \times \frac{4}{3} \pi R^3 = 1.6 \times 10^{15} \text{ kilograms}$$

is obtained. Most near-Earth objects travel with orbital velocities of about 30 kilometers per second, comparable to the Earth's, and using this velocity with the mass we have estimated, we obtain

$$\text{K.E.} = 7.2 \times 10^{23} \text{ joules}$$

This amount of energy is equivalent to the explosion of nearly 100 000 terrestrial bombs, each with a destructive force of 100 megatons (100 million tons or 10^{11} kilograms) of trinitrotoluene (TNT). The destruction produced by the impact would therefore be many orders of magnitude larger than that caused by the simultaneous explosion of the world's entire arsenal of nuclear weapons. It would be equivalent to the detonation of the blast that destroyed Hiroshima, at just 13 000 tons of TNT, every second for 175 years.

The serendipitous discovery of the submerged crater began with a search for oil in the region. The Mexican national petroleum company, Petróles Mexicano or Pemex for short, commissioned an aerial magnetic survey to assess the thickness of sedimentary – and possible oil-bearing – rocks in the region, which revealed a large, buried semicircular structure. Coarse gravity maps showed a similar feature, and exploratory oil-drilling in the area revealed an underground layer of broken, melted rock. Subsequent radioactive dating of the drill-core fragments showed that they were exactly contemporaneous with the clay layer, with an age of 65 million years.

Chicxulub is the biggest thing to hit Earth in the past 1 or 2 billion years, and of just the right size to have been excavated by an asteroid of 10 kilometers in extent. The energy released during the collision of such an intruder is enough to trigger a mass extinction (Focus 13.1).

The day the dinosaurs died

Most scientists are now convinced that the dinosaurs, which had dominated the Earth for over 160 million years, were destroyed when a marauding asteroid dropped out of the sky and struck the Earth 65 million years ago. The explosive impact generated an enormous ball of fire, which incinerated everything in the immediate area, blasted billions of tons of debris into space, and gouged out a crater 170 kilometers across. But that was just the beginning.

The cosmic intruder was completely vaporized during the explosive impact, and a great fireball rose into the stratosphere, carrying with it large amounts of pulverized debris. Vast clouds of dust and ash remained suspended and were circulated by air currents until they encircled the Earth, covering it in total darkness. Since the global shroud blocked the Sun's heat, as well as its light, the surface temperature plummeted and the planet entered a dark chill, lasting for months before the dust eventually settled back to Earth.

Plants would have failed to receive enough sunlight to allow photosynthesis to continue, and a prolonged "winter" of unusual cold would have added to the devastation. As plant life withered and froze, plant-eating animals dependent on them would also die, as would meat-eating animals once their plant-eating prey were gone. By the time that the dust settled, many land plants and animals were no longer there.

The cosmic object that slammed into the water above the current Yucatán Peninsula would have also generated enormous sea waves, great moving walls of water that surged into the East Coast of what is now known as the United States, leveling everything in their path and leaving almost nothing standing.

Other devastating effects most likely included: extensive wildfires, ignited when hot material ejected from the impact fell back down to the ground; acid rain that poisoned the water and destroyed shell-bearing creatures, disrupting the entire food chain; and volcanoes that might have contributed to the mass extinction by spewing carbon dioxide, deadly ash, sulfur and other substances that disrupted the atmosphere and altered the climate. When combined with the dust, sulfur, and hot debris tossed skyward by the original impact, the volcanic activity could have spelled doom for many living things.

When the cosmic blast and its aftermath were over, the dinosaurs were gone, along with most marine animals and many land plants. Altogether roughly half of all animal and plant species were wiped out. That sounds pretty awful, but from catastrophes there arise opportunities. The biological devastation caused by the impact apparently cleared the way for the rise of the relatively small mammals, so your

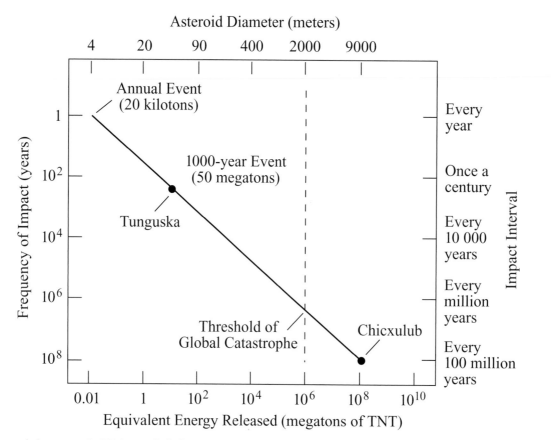

Fig. 13.12 Cosmic impact probabilities   A hail of cosmic objects continually pelts the Earth from space. This plot shows the likelihood (*vertical axis*) that a member of the current population of Earth-crossing asteroids will hit our planet. An object two kilometers in diameter, capable of producing certain worldwide damage, hits the Earth every million years on average. An impact like the one that wiped out the dinosaurs, giving rise to the Chicxulub crater, is estimated to occur every 100 million years. Since smaller asteroids are much more numerous than larger ones, the smaller objects strike our planet more often. An impact like the Tunguska event, which occurred on 30 June 1908, might occur every 1000 years or so. The explosive energy of the impact is also given (*horizontal axis*) in units of megatons of exploded TNT. One hundred megatons (10^{11} kilograms) of TNT is equivalent to about 4×10^{17} joules of energy, the amount released by a typical nuclear bomb on Earth.

very distant ancestors may have benefited. Some scientists have even argued that killer asteroids may periodically sweep a wave of death across the Earth, thereby ending the rule of the dominant species.

We humans have flourished in the past half million years, developing wonderful civilizations, building great cities, generating profound knowledge, and sending spacecraft throughout the solar system. Yet it might suddenly come to an end. Such catastrophes have a small probability of occurring during our relatively short lives, but over astronomical times of millions and billions of years, the exceedingly unlikely becomes a virtual certainty. After all, 99 percent of all species that ever lived have gone extinct.

13.5 Assessing the risk of death from above

Somewhere in space, an asteroid is hurtling toward a future collision with Earth. And if it is large enough, the

impact will severely disrupt terrestrial life. It's only a matter of time.

So when are we going to be hit hard enough to worry about it? Since astronomers have not yet located the doomsday rock that will definitely collide with Earth, we don't know exactly when the next impact will take place. But we can calculate the odds, and they depend on the size of the colliding object. Since there are many more small cosmic objects than large ones, the smaller ones hit our planet more frequently. Bigger asteroids strike the Earth less often, but they can cause vastly greater damage.

Thus, to estimate the risk of being hit in a way that matters, the potential impacting projectiles first have to be sorted according to size (Fig. 13.12). Fragments smaller than a few tens of meters across burn up in the atmosphere and rarely reach the ground. Asteroids a hundred meters in diameter are expected to strike Earth every thousand years on average. They could take out a city and cause

Table 13.2 The dangers of a lifetime[a]

Cause of death	Chance of dying in a 65-year period	Number of people killed
Car accident	1 in 100	5
Murder	1 in 300	1
Gun accident	1 in 2500	1
Accidental electrocution	1 in 5000	1
Asteroid impact (two kilometers in diameter, global catastrophe)	1 in 20 000	100 million
Airplane crash	1 in 20 000	300
Tornado	1 in 60 000	10
Snake bite or bee sting	1 in 100 000	1
Food poisoning	1 in 3 000 000	1

[a] Adapted from Clark R. Chapman and David Morrison, *Nature* **367**, 33–40 (1994). The risks are for a person living in the United States.

severe local damage, but pose no threat to the Earth as a whole.

Asteroids about one kilometer in size pose a greater peril. They are large enough to destroy a large country and produce global consequences. Contemporary surveys indicate that there may be about one thousand of these objects now on paths that come near the Earth's orbit. Most of the time, the Earth will be somewhere else if one of them crosses its path, but occasionally they will arrive almost simultaneously at the intersection. The average time between such impacts is about one million years, an interval vastly longer than the history of civilization.

During your lifetime, the chance that you will be wiped out by the impact of a two-kilometer asteroid is 1 in 20 000, the same as death from an airplane crash (Table 13.2). But it is not just one person that dies when a large cosmic projectile hits the Earth. About 100 million people are expected to perish if the object is two kilometers across, and a bigger one could destroy all of us. The other causes of death usually affect just one person at a time, or perhaps a few of them.

Moreover, to declare that cosmic disaster strikes the Earth, on average, just once in a million years or so does not mean that we are guaranteed such a long interval between catastrophic impacts. The very small chance of such a collision is the same today as it will be millions of years from now. It could be tomorrow, or it might be long after you're gone.

In fact, there are all sorts of other things you can worry about if you are in a morbid mood. The chance of dying before age five is 1 in 8 in South Asia, and 1 in 100 in the United States. The risk of dying in childbirth in the United States is about 1 in 12 000, roughly twice the chance of death by the impact of a two-kilometer cosmic projectile, but thousands of times greater than death by dog bite or by drinking detergent.

But to get back to the dangers of cosmic impact, it takes an even larger, rarer asteroid, 10 kilometers across or larger, to render the human species extinct. The impacting object would blast out a huge crater and eject billions of tons of pulverized rock and dust into the air. The globe-circling pall of dust and other debris would block out the Sun's light and heat, crippling agriculture, producing widespread starvation, and perhaps leading to a worldwide breakdown of our fragile civilization.

Such a mass extinction might occur once in a hundred million years on average, destroying in an instant what it has taken humans millennia to build. The chances of that happening are very low, but they are not zero. The lifetime risk of your being wiped out with the rest of humanity during such a mass extinction is roughly one in a million. So it might happen, and a prudent society should prepare for the possibility.

13.6 Breaking a date with doomsday

Finding the hidden threat

While we know that asteroids and comets have collided with the Earth in the past, and that they will inevitably hit our planet in the future, we do not yet know if any them are now headed for a deathly collision with our solitary outpost of life. Astronomers are therefore taking a census of everything out there that is big enough and close

Fig. 13.13 Asteroid streak   As an asteroid moves along its orbit it will produce an elongated trail in an image taken with a telescope following the background of "fixed" stars, which are not moving in this way. One previously unknown asteroid was discovered in 1998 as a long blue streak (*top center*) against the white background stars in this archival image taken by the *Hubble Space Telescope*. The bright asteroid has an estimated diameter of 2 kilometers, and was located at about 140 million kilometers from the Earth. (Courtesy of NASA/R. Evans and K. Stapelfeldt, JPL.)

enough to threaten us (Figs. 13.13, 13.14, 13.15). Once all of these Near-Earth Objects (NEOs) are located, and their current trajectories known, astronomers can use computers and refined observations to determine their precise future paths and establish whether and when any of them will strike the Earth.

After decades of scanning the skies, ongoing search programs will find and catalog the most threatening asteroids (Focus 13.2). They will provide the exact positions and orbits of about 90 percent of the near-Earth asteroids larger than one kilometer in size. That is the minimum diameter of a space rock that could have global consequences if it hit Earth.

Astronomers have found nearly one thousand near-Earth asteroids with diameters of one kilometer or larger, and they seem to account for the majority of the largest space rocks that might wreak global havoc. None of those discovered so far is on a direct collision course with Earth – at least in the near future. The search is therefore slowly diminishing the chances of our demise, gradually improving the odds of this cosmic Russian roulette.

Just as there are more small fish than large ones in the sea, there are many more small cosmic objects that are now moving about in the space near our planet, and there are also about one thousand of them that are already known to be potentially hazardous. Most of these probably pose

Fig. 13.14 Near-Earth asteroid   The red dot at the center of this image is the first near-Earth asteroid discovered, on 12 January 2010, with the *Wide-Field Infrared Survey* (*WISE*) spacecraft. The asteroid is about 1 kilometer in diameter, and moves on an elliptical orbit that is inclined to the plane of our solar system. This orbit takes the asteroid beyond Mars and as close to the Sun as the Earth. The all-sky *WISE* infrared survey, which began on 14 January 2010, is expected to find about 100 000 previously undiscovered asteroids in the main belt between Mars and Jupiter, and hundreds of near-Earth asteroids. It will reveal the darkest asteroids, which don't reflect enough visible sunlight to be seen with conventional optical telescopes, but heat up and emit the infrared radiation detected from *WISE*. (Courtesy of NASA/JPL-Caltech/UCLA.)

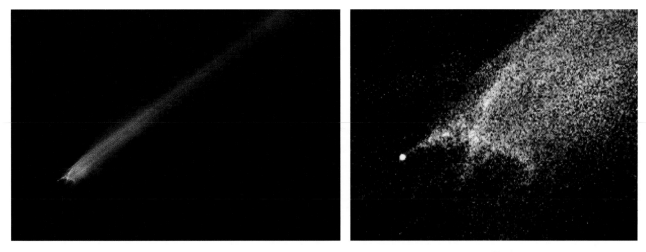

Fig. 13.15 Suspected asteroid collision leaves trailing debris   This unusual object was first discovered by the ground-based Lincoln Near Earth AsteRoid (LINEAR) program, designed to detect asteroids that may collide with the Earth in the future. When the *Hubble Space Telescope* zoomed in to take a look with its sharper vision, it found a comet-like tail of debris (*left*), flowing from an X-shape seen in the close-up view (*right*) of the front center of the tail. The X may mark the spot of an asteroid collision in the main belt of asteroids that created the debris. The remains of one of the asteroids, a nucleus 140-meters in size, are offset from the tail center and possible collision site. (Courtesy of NASA/ESA/D. Jewitt, UCLA.)

a significant threat, but not a global one. They can hit the Earth with an energy equivalent to a large nuclear bomb.

Today it is beyond our technology to detect and defend against the smaller cosmic bombs of several meters across, but they can be seen when coming very near the Earth. Some of these asteroids have come uncomfortably close, passing within less than the mean distance between the Earth and its Moon, which is 384 400 kilometers away. By cosmic standards, that is a close call. About 100 asteroids have passed this close to the Earth in the past several years, most of them less than 200 meters across.

On 13 January 2010, for example, an asteroid of 10 to 20 meters in diameter passed within one-third the distance from the Earth to the Moon. Others have nearly hit our planet, passing within a few Earth radii (Table 13.3), sometimes with just a few hours notice or with no advance warning at all.

Table 13.3 Very close approaches of asteroids to Earth[a]

Closest distance from Earth (R_E)	Date of closest approach	Estimated diameter (meters)
1.0	7 October 2008	2 to 5
2.0	31 March 2004	5 to 12
2.1	9 October 2008	0.5 to 1
3.2	6 November 2009	5 to 12
4.8	20 October 2008	5 to 12
5.3	19 December 2004	1 to 3
7.1	3 November 2008	2 to 5
7.6	8 March 2004	20 to 50

[a] The radius of the Earth is R_E = 6378 kilometers.

Doing something about the threat

The identification and deflection of a near-Earth asteroid that is heading toward a direct collision with Earth is now regarded as a global public safety issue. By 2010 NASA had identified more that 1000 potentially hazardous NEOs, but a full inventory was still incomplete, and the White House has urged the development of evasive action to deflect one when we see it coming, perhaps related to the goal of landing an astronaut on an asteroid by 2025.

How much warning will we have? When the inventory of large NEOs is complete, and one is found on its way to strike the Earth, then warning would probably come decades in advance of a collision. The threatening asteroid would most likely swing near the Earth and loop around the Sun several times before hitting our planet. Since existing surveys do not regularly detect small NEOs, less than 200 meters in size, as efficiently as they discover larger ones, we may not know about them until they are about to collide with us. Thus we will either have a long lead-time or none at all.

And what do we do if we find a large cosmic projectile headed our way? The Earth cannot be moved out of the way, but we could launch an intercept mission to redirect

Focus 13.2   Searching for cosmic bombs headed our way

The collision of a comet with Jupiter, anticipated in 1993 and watched by millions in 1994, raised public consciousness of the impact threat to planet Earth. The US Congress held hearings to study the threat, and asked NASA to formulate plans to deal with the problem. Public awareness of the cosmic bombs was notched up once more with the release, in 1998, of two blockbuster movies, *Deep Impact* which deals with a killer comet and *Armageddon* in which two astronauts save humanity by diverting a rogue asteroid headed toward collision with Earth. In the same year, the US Congress held more hearings about the threat, and NASA initiated the Spaceguard Survey, intended to find, within the next decade, 90 percent of the Near-Earth Objects (NEOs) which might be on a collision course with Earth and are larger than one kilometer.

By the end of the 20th century, several teams of astronomers were surveying the sky with electronic detection equipment and computers to complete the inventory of large NEOs, supported by NASA and the US Air Force. The search involves finding every object that moves against the background stars down to 19.5th magnitude, almost 100 000 times fainter than the detection limit of the human eye.

Exceptionally productive search programs, resulting in the largest number of discoveries of near-Earth asteroids, are the LIncoln Near Earth Asteroid Research (LINEAR) project of MIT's Lincoln Laboratory and the Catalina Sky Survey, based at the University of Arizona. They have together discovered between 100 and 400 near-Earth asteroids every year between 1998 and 2010, and nearly one thousand of these are thought to have diameters of one kilometer or larger.

More recently, the *Wide-field Infrared Survey Explorer* (*WISE*) has turned its infrared eyes to the sky, expecting to detect hundreds of previously unseen near-Earth objects. It will reveal the darkest members of this population, which don't reflect much visible light but do emit detectable infrared heat radiation.

Once an NEO is discovered by the ongoing surveys, previous unsuspecting observations made before the NEO discovery and follow-up observations with powerful radar telescopes can be used to refine knowledge of its future trajectory and push predictions far into the future. By bouncing radio waves off the object, and examining the return echo, the radar technique can be used to establish an exact orbit, providing us with centuries of advance notice of close encounters or impacts by NEOs.

Near-Earth Objects are catalogued at the International Astronomical Union's Minor Planet Center, located on the web at http://www.cfa.harvard.edu/iau/mpc.html. They list more than one thousand of the potentially hazardous objects, with the greatest likelihood of close approaches to Earth. The Near Earth Object website of the Jet Propulsion Laboratory, located at http://neo.jpl.nasa.gov/, carries several tables that can be sorted by close approach, date, object name, past and future approaches, and comets or asteroids. The European Spaceguard Foundation includes a Spaceguard System of observatories that are engaged in Near-Earth Object observations in order to discover them and protect the Earth from the possible threat of their collision.

the asteroid's course. If the impact is many years away and the threatening object relatively far away from us, all we have to do is give it a little nudge. By the time the asteroid reaches the Earth's vicinity, that small change in trajectory will make a big difference, enabling it to bypass the planet.

An orbiting mirror could be used to focus sunlight and vaporize the asteroid's surface, producing a jet of gas and dust that might change its course. Astronauts could rendezvous with it, attaching small explosives, rocket engines, solar sails or mass drivers that could push it into a harmless trajectory. A large solar sail would use the pressure of sunlight to move the body slowly, while a mass driver would scoop up surface material and hurl it away, creating an appropriate recoil reaction just as the small explosives or rocket engines would.

If the warning time is only a matter of months or less, the sole recourse might be to send a high-powered rocket armed with a bomb powerful enough to redirect the object or blow it up. Such a possibility has sparked the interest of bomb designers and some members of the military (Focus 13.3). A conventional nuclear weapon might be used to deflect or destroy a small, solid, rocky asteroid, but a much larger explosion could be needed to divert or pulverize a loosely bound one. Moreover, it might not be a good idea to blow the threatening object up, for the Earth might then be struck by a hail of dangerous projectiles rather than a single blast. Some of the pieces could still head straight at us. If they were not small enough to burn up in the atmosphere, the fragments might cause massive destruction in several places on the Earth.

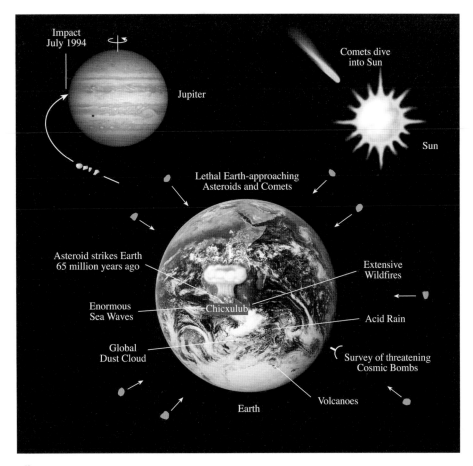

Fig. 13.16 Summary diagram

Focus 13.3 Star Wars in outer space

When it was realized that an asteroid might be hurtling toward collision with the Earth, experts in nuclear weapons and missile defense systems had something other than nuclear war to worry about. They proposed that nuclear-tipped missiles could be used to blow the threatening object up or to deflect it en masse from its earthward trajectory.

Of course, the military would want to test the warheads. With the whole world threatened you wouldn't want to use an untested weapon. In 1996, China even advocated underground tests to prepare nuclear warheads as a possible defense against the cosmic threat.

That brings up the question of who is told about plans to protect civilization from the intruder. A madman, or someone making a stupid error, could use the Star Wars weapons to deflect a previously benign asteroid into a collision course with Earth. After all, we now know how terrorists can redirect airplanes to destroy thousands of innocent lives.

In the absence of early warning and evasive action, advance knowledge of the time and place of an impact would at least allow people to store food and supplies and to evacuate regions near the impact site where damage would be the greatest. It could happen today, or it might not happen for millennia. Some of us will place our trust in God, and many of us have more immediate concerns, but it will not hurt to subscribe to a cosmic insurance policy and fully identify the swarm of cosmic objects that might be headed our way. The policy has relatively small premiums and enormous potential benefits.

14   Comets

- The sudden apparition, changing shapes, and unpredictable movements of comets have puzzled humanity for centuries. To ancient cultures they were harbingers of disaster and portents of great events.

- Comet Halley has returned to fascinate and frighten the world for more than 2000 years.

- Long-period comets, with orbital periods greater than 200 years, have been tossed into the planetary realm from a remote, spherical shell, named the Oort cloud, located about a quarter of the way to the nearest star.

- A million million (10^{12}) invisible comets have been hibernating in the deep freeze of the Oort comet cloud since the formation of the solar system 4.6 billion years ago.

- Many Jupiter family comets, with orbital periods of less than 20 years, probably came from the Kuiper belt, which lies in the outer disk of the planetary system beyond the orbit of Neptune and may contain more than a billion unseen comets.

- Comets light up and become visible for just a few weeks or months when their orbits bring them near the Sun. The solar heat then vaporizes some of the comet water ices, permitting the comets to grow large enough to be seen. The water ice sublimates, or turns directly from solid ice to water vapor.

- The solid comet nucleus is just a gigantic ball of water ice, other ices, dust and rock. Some of the comet nuclei are about the size of Paris or Manhattan and roughly one-billionth the mass of the Earth. Other comets are much smaller.

- No two comets ever look identical, and every comet changes shape and form as it whips around the Sun, but they all develop a glowing spherical cloud of gas and dust, known as the coma, when moving close enough to the Sun.

- The comet coma can be larger than the Earth and as big as the Sun, and around the coma there is an even larger envelope of atomic hydrogen, known as the hydrogen cloud, that shines in ultraviolet light.

- Some comets develop tails that flow away from the Sun, briefly attaining lengths as large as the distance between the Earth and the Sun, but other comets have no tail at all.

- Comets can have two kinds of tails: a long, straight ion tail, that re-emits sunlight with a faint blue fluorescence, and a curved dust tail that shines by reflecting yellow sunlight.

- The *Giotto*, *Deep Space 1*, *Stardust* and *Deep Impact* spacecraft have respectively peered into the icy hearts of four comets – Halley, Borrelly, Wild 2 and Tempel 1 – showing that their nuclei are blacker than coal and reflect just a few percent of the incident sunlight. *Deep Impact* continued on to encounter comet Hartley 2.

- Gas and dust jet out from the sunlit side of the nucleus of comet Halley, from fissures in its dark crust, but nearly 90 percent of the surface of its nucleus was inactive at the time of the *Giotto* encounter.

- Comet Borrelly is covered with a dark, unreflective carbon-rich material, and contains surface features that are most likely supported by solid water ice.

- Comet Wild 2 has been exposed just a few times to the Sun's intense heat, and it has a dark, pockmarked surface with pits, craters and jets of gas and dust.

- The *Stardust* spacecraft gathered dust from the coma of comet Wild 2 in January 2004, returning the dust in a capsule that was parachuted to Earth two years later. The returned comet dust contains a mix of minerals formed at both cold and high temperatures, two types of nitrogen-rich organic molecules, and the amino acid glycine.

- The *Deep Impact* spacecraft collided with comet Tempel 1 on 4 July 2005; spectroscopic examination of the ejected cloud of dust revealed fine porous material, water vapor, water ice, carbon dioxide, hydrocarbons and silicates or sand.

- The *Deep Impact* spacecraft flew past comet Hartley 2 on 4 November 2010, revealing a small active nucleus composed of two rough parts joined at a smooth waist.

- When a bright comet nears the Sun, it turns on its celestial fountains, spurting out about a million tons of water each day.

- The recoil effect of jets of matter ejected from a comet's spinning, icy nucleus can push a comet along in its orbit or oppose its motion, causing the comet to arrive closest to the Sun earlier or later than expected.

- Most of the comets seen during recorded history will vanish from sight in less than a million years, either vaporizing into nothing or leaving a black rock behind.

- About 40 000 tons of small, cosmic dust particles fall onto the Earth in a typical year, wafting gently through the atmosphere to the ground.

- Visible comets are in their death throes, but they may carry the residues of creation in their ice and dust.

- Meteor showers, commonly known as shooting stars, are produced when sand-sized or pebble-sized pieces of an icy comet burn up in the atmosphere, never reaching the ground.

- Comets strew particles along their orbital path as they loop around the Sun, and when the Earth passes through one of these meteoric streams a meteor shower occurs, recurring at the same time every year.

Fig. 14.1 The Great Comet of 1577 This drawing by a Turkish astronomer appeared in the book *Tarcuma-I Cifr al-Cami* by Mohammed b. Kamaladdin written in the 16th century. The yellow Moon, stars and comet are shown against a light blue sky. (Courtesy of Erol Pakin, Director, Istanbul Universitesi Rektorlugu.)

14.1 Unexpected appearance of comets

Every few years, on average, an unusually bright comet will blaze forth in the night sky, becoming visible to the unaided eye and sporting a graceful tail resembling long hair blowing in the wind (Figs. 14.1, 14.2). In fact, the word *comet* is derived from the Greek name *aster kometes*, meaning "long-haired stars". But a comet is not anything like a star. Their dramatic display emanates from a relatively small, blackened chunk of ice and dust, comparable to a large city in size.

Unlike the planets, the comets can appear almost anywhere in the sky, remain visible for a few weeks or months, and then vanish into the darkness. Astronomers call this period of visibility an "apparition". During its apparition, a comet changes its shape, often from night to night.

Many of the enigmatic comets travel far outside the paths of the planets and move in every possible direction around the Sun. Their orbits are inclined at all possible angles to the ecliptic, the plane of Earth's orbit, and different comets move in either the same direction around the Sun as the planets or in the opposite retrograde direction. Other comets move in tighter orbits, circling the Sun within the bounds of the outer planets.

Comets inspired awe and fear in ancient cultures. By their unexpected arrivals, these celestial intruders seemed to upset the natural order of the otherwise placid firmament, and to presage changes in the order of things on Earth, such as the death of rulers, wars and other disasters (Fig. 14.3). One comet appeared in 44 BC, the year that Julius Caesar was assassinated. The fallen emperor's adopted son declared the comet to be Caesar's soul rising to heaven, and used the apparition to gain control over the entire Roman Empire as Augustus Caesar (63 BC–14 AD). William Shakespeare (1564–1616) wrote about the comet's link to Julius Caesar's death 15 centuries later, with:

> When beggars die, there are not comets seen;
> The heavens themselves blaze forth the death of
>   princes.
> Cowards die many times before their deaths;
> The valiant never taste death but once.

> *Julius Caesar*, Act II, Scene ii, line 30

Another famous example was the Norman conquest of England in 1066, which was coincident with the appearance of what is now called comet Halley. This comet was also seen in 1456 when the Turks conquered Constantinople, and some in Europe prayed for protection from "the Devil, the Turk and the Comet". More recently, the Great Comet of 1811 was supposed "to portend all kinds of woes and the end of the world" in Leo Tolstoy's (1828–1910) *War and Peace*.

The unexpected appearance of a bright comet was also once taken as an omen of doom and the harbinger of disaster. In *Paradise Lost*, John Milton (1608–1674) imagined a Satan-like comet that "from his horrid hair shakes pestilence and war." Even in 1910 there were speculations that comet Halley would impregnate the air with poisonous vapors such as cyanogen, and wipe out life on Earth, but there were no noticeable effects on humans or other living things when the Earth passed near the comet's tail.

Awe-inspiring comets still arrive without warning today, becoming brighter than the most brilliant stars. They are usually named after the last names of their discoverers, unless a spacecraft is involved (Focus 14.1). The brightest, most spectacular apparitions are also known as Great Comets.

At night, the Great Comets can remain visible to the unaided eye for months, and they sometimes become visible in daylight. Some of them have enormous tails (Fig. 14.4). Great Comet Hyakutake, for example, came within 0.1 AU of the Earth in 1996, with a tail that stretched one-quarter the way across the sky.

As illustrated in Table 14.1, the Great Comets all travel closer to the Sun than the Earth, which orbits at a mean distance of 1 AU, and sometimes pass quite near the Earth itself. But a decade or more can pass between the unanticipated discoveries of truly Great Comets.

Fig. 14.2 Comet Kohoutek A modern photograph of a comet's flowing tail. It was taken on 12 January 1974 with the 1.2-meter (48-inch) Schmidt telescope of the Hale Observatories with a 3-minute exposure in blue light. (Courtesy of the Hale Observatories.)

14.2 The return of comet Halley

Many of the brightest comets seem to come out of nowhere, suddenly moving past the Sun, and are never seen again. Yet, one famous comet has come back for repeat performances, fascinating the world for more than 2000 years. It is now known as comet Halley, named for the British astronomer Sir Edmond Halley (1656–1742).

Halley demystified comets by showing that at least one of them travels in an elongated orbit around the Sun. He found that the orbit of the bright comet of 1682 was similar to those of comets observed in 1607 (seen by Johannes Kepler, 1571–1630) and in 1531 (observed by Petrus Apianus, 1495–1552). All three comets moved around the Sun in retrograde orbits with a similar orientation. Halley also

Fig. 14.3 The eve of the deluge   People believed for centuries that the unexpected appearance of comets was a premonition of war, death and other disasters. Here the arrival of a comet foretells the great flood at the time of Noah. The 1835 apparition of comet Halley may have influenced the artist, John Martin (1784–1854), for he finished this painting a few years later in 1840. (Collection of Her Majesty the Queen.)

Focus 14.1   Naming comets

Amateur astronomers often discover the brightest comets, diligently searching for them with small telescopes or even large binoculars. Professional astronomers sometimes accidentally come across one while using a large telescope for another purpose. In accordance with a tradition that has gone on since the time of the French comet hunter Charles Messier (1730–1813), new comets are now given the last name of their discoverer, the last names of their independent discoverers, or the acronym of a spacecraft used in the discovery, such as *SOHO*.

A prefix "P/" is now used for a periodic comet, defined to have a revolution period of less than 200 years with confirmed observations at more than one perihelion passage, and "C/" for a comet that is not periodic in this sense. A number is also added before the prefix P to designate the order of discovery. For instance, 1P/Halley was the first periodic comet known and 2P/Encke the second. Comets are also now designated by the year of their discovery, the upper-case letter identifying the half-month of the observation during that year, and a consecutive numeral to indicate the order of discovery announcement during that half-month. The letters "I" and "Z" are not used to make a total of 24 half-months. For example, the third comet reported as discovered during the second half of February 1995 would be designated 1995 D3.

knew that the Great Comet of 1456 had also traveled in the retrograde direction, and he concluded that all four apparitions were the same comet traveling along an identical elongated orbit around the Sun and appearing at 76-year intervals when coming close to our star. Halley confidently predicted its return in 1758, noting that he would not live to see it.

Halley also pioneered our understanding of cartography, diving bells, mortality tables, naval navigation, stellar proper motions, tides and trade winds, and helped with the publication of Isaac Newton's (1643–1727) powerful description of gravity and orbital motion. But Halley is best known for the comet that now bears his name, comet Halley (Fig. 14.5). It was re-discovered, on Christmas night of the predicted year.

Halley's is the most famous of the comets because it was the first to arrive on schedule. Its fame is deserved

on other counts as well. It displays a complete range of comet fireworks including an exceptionally long tail, a bright head, and jets, rays, streamers and halos. Moreover, we now know that it has been observed for a longer period of time than any other comet in recorded history. The earliest apparition established with confidence from Chinese chronicles dates back to 240 BC; since then, all its perihelion passages have been retraced in the ancient or modern records of astronomers (Table 14.2).

After its 1910 apparition (Fig. 14.6), comet Halley moved away from the Sun into the outer darkness, arriving in 1948 at the remotest part of its orbit at 35 AU. The comet then turned the direction of its course, and began falling back toward the heart of the solar system with ever-increasing speed. It reached perihelion, or its closest distance from the Sun, on 9 February 1986. Comet Halley and the Earth were then on opposite sides of the Sun, so this was among the least favorable apparitions for observing the comet with the unaided

Fig. 14.4 A Great Comet lights up When a comet travels close to the Sun, the solar heat vaporizes ice from the comet's surface, and solar forces bend the liberated material into comet tails that always point away from the Sun rather than toward it. The long tail of this Great Comet stretches 120 million (1.2×10^8) kilometers, or nearly the mean distance between the Earth and the Sun, at 1 AU $= 1.496 \times 10^8$ kilometers. It is also named comet Ikeya–Seki (1965 S1) after the last names of its discoverers, Kaoru Ikeya (1943–) and Tsutomu Seki (1930–), and the year and order of its discovery. (Courtesy of the Lick Observatory.)

eye. Nevertheless, it still became one of the most thoroughly studied apparitions in the history of comet research (Fig. 14.7), including visits by several spacecraft. After these visits, Halley's comet headed for the cold reaches of the solar system, to return in the Sun's neighborhood in 2061.

14.3 Where do comets come from?

Comets are primitive bodies that formed at the same time as the Sun and planets about 4.6 billion years ago. But once they come close enough to be seen, comets begin to fall apart and they must eventually vanish from sight, often in

Table 14.1 Some Great Comets of the 19th and 20th centuries[a]

Name	Perihelion date	Days visible[b]	Perhelion distance (AU)	Brightest apparent magnitude
Great Comet of 1807	19 Sept. 1807	90	0.65	1 to 2
Great Comet of 1811	12 Sept. 1811	260	1.04	0
Great March Comet of 1843	27 Feb. 1843	48	0.006	1
Comet Donati	30 Sept. 1858	80	0.58	0 to 1
Great Comet of 1861	12 June 1861	90	0.82	0 (or −2?)
Great Comet of 1865	14 Jan. 1865	36	0.03	1
Comet Coggia	09 July 1874	70	0.68	0 to 1
Great September Comet	17 Sept. 1882	135	0.008	−2
Great Comet of 1901	24 Apr. 1901	38	0.24	1
Great January Comet	17 Jan. 1910	17	0.13	1 to 2
Comet Halley (in 1910)	20 Apr. 1910	80	0.59	0 to 1
Comet Skjellerup-Maristany	18 Dec. 1927	32	0.18	1
Comet Ikeya-Seki	21 Oct. 1965	30	0.008	2
Comet Bennett	20 Mar. 1970	80	0.54	0 to 1
Comet West	25 Feb. 1976	55	0.20	0
Comet Hyakutake	01 May 1996	30	0.23	1 to 2
Comet Hale-Bopp	01 Apr. 1997	215	0.91	−0.7

[a] The perihelion distance is the distance from the Sun at the closest approach to the star, given in astronomical units (AU), roughly the mean distance between the Earth and the Sun. The apparent magnitude is a measure of the apparent brightness of a celestial object, in which brighter objects have smaller magnitudes. Sirius A, the brightest star other than the Sun, has an apparent visual magnitude of −1.5. The nearest star other than the Sun is about 0 on the magnitude scale, while Venus has an apparent magnitude of −4 when brightest, and at its brightest Jupiter appears at magnitude −2.7. Adapted from Donald K. Yeomans' *Great Comets in History*, at the website http://ssd.jpl.nasa.gov/great_comets.html.

[b] Days visible to the naked eye unaided by binoculars or a telescope.

less than a million years after first sighting. So comets are very old, but once they swing near the Sun they do not last very long.

Something must be furnishing the inner solar system with new comets, and they come from two reservoirs. One is very far away, at the fringe of the outer solar system, and another nearer one is at the edge of the planetary realm. We distinguish between these source regions on the basis of the orbital periods of the comets. Both types of small icy worlds have been hibernating in the cold outer reaches of space ever since the formation of the solar system.

The long-period and short-period comets

Traditionally, comet orbits have been classified as short period or long period with the dividing line at orbital periods of 200 years. Most newly discovered comets are long-period comets, with orbital periods larger than 200 years. They have arrived near the Sun from distant regions far beyond the major planets, traveling along very elongated trajectories that take them back to the distant regions

they came from. The long-period comets are observed in the inner part of the solar system just once, arriving unannounced and unpredicted. As you might expect, they come from very far away, at the outer fringes of the solar system.

Nowadays, astronomers divide the former short-period designation into the Jupiter family, with orbital periods less than 20 years, and the Halley-type comets that have orbital periods between 20 and 200 years. The Jupiter-family comets move about the Sun with a period comparable to Jupiter's 10-year period, and have orbits affected by the gravitational influence of the giant planets.

The Jupiter-family comets are seen time and again, trapped in tight orbits within the planetary realm. It is these comets whose regular returns we are able to predict, and which we can examine in detail with spacecraft (Table 14.3). They travel on direct orbits, in the same direction as the planets orbit the Sun, and most of them have low orbital inclinations near the plane of the Earth's orbit, with mean distances from the Sun of just a few times that of the Earth.

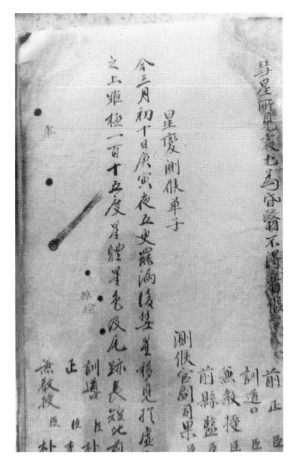

Fig. 14.5 Comet Halley in 1759 AD   This Korean record of comet Halley was made during the comet's first predicted return in 1759. The Korean astronomers have been recording the appearance of comets and other unusual celestial objects for more than 3000 years. (Courtesy of Il-Seong Na, Yonsei University, Seoul.)

The Oort cloud of comets

The size and orientation of the trajectories of long-period comets can be explained if they come from a remote, spherical shell belonging to the outer parts of the solar system (Fig. 14.8). This vast comet reservoir is known as the Oort cloud, named after the Dutch astronomer Jan H. Oort (1900–1992) who first postulated its existence in 1950. His careful examination of the trajectories of observed long-period comets, which became visible when they entered the inner parts of the solar system, could only be explained if these comets came from a distant reservoir, which he located at between 50 000 and 150 000 AU. At greater distances the stars in the neighborhood of our solar system compete for gravitational control of the comets.

And because long-period comets enter the planetary realm at all possible angles, with every inclination to the Earth's orbital plane, they must come from a spherical shell. This would also explain the fact that long-period comets move in all directions. Roughly half of them move along their trajectories in the retrograde direction, opposite to the orbital motion of the planets.

Modern calculations suggest that there is an inner Oort cloud with an inner edge at around 3000 AU and a density falling off with greater distance. The outer Oort cloud is continuous with this, but is defined to be those objects at distances greater than 20 000 AU. The cloud fades away with increasing distance, and its tenuous outer edge is dynamically limited to about 200 000 AU by the galactic gravity field.

Table 14.2 Thirty-two perihelion passages of comet Halley[a]

240 BC	25 May
164	13 November
87	6 August
12 BC	11 October
66 AD	26 January
141	22 March
218	18 May
295	20 April
374	16 February
451	28 June
530	27 September
607	15 March
684	3 October
760	21 May
837	28 February
912	19 July
989	6 September
1066	21 March
1145	19 April
1222	29 September
1301	26 October
1378	11 November
1456	10 June
1531	26 August
1607	28 October
1682	15 September
1759	13 March
1835	16 November
1910	20 April
1986	9 February
2061[a]	28 July
2134 AD[a]	27 March

[a] The perihelion of a comet is the point in its orbit that is closest to the Sun. The future two perihelion passages of Halley's comet are predicted dates; all of the others have been recorded.

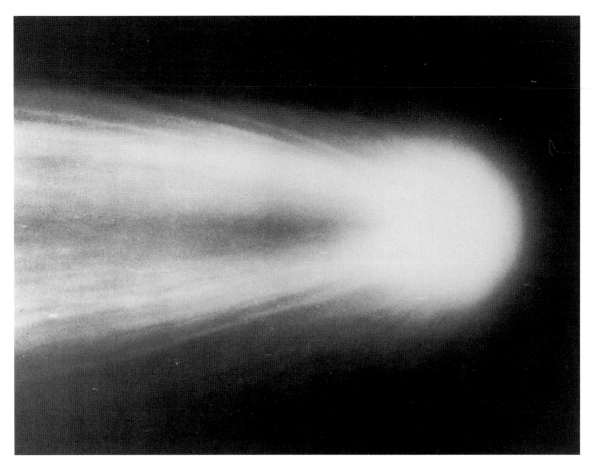

Fig. 14.6 Apparition of comet Halley in 1910   The head region or coma of comet Halley observed on 8 May 1910 with the 1.5-meter (60-inch) telescope on Mount Wilson. The comet's tail flows to the left, away from the Sun. (Courtesy of the Hale Observatories.)

The Oort cloud has also been divided into two components: the spherical outer cloud discussed by Oort and a more flattened inner cloud. The inner cloud is probably the source of the Halley-type comets; they require a closer origin to be captured into stable orbits around the Sun with periods between 20 and 200 years.

Formation of the Oort comet cloud

Where did the Oort-cloud comets originally come from? They could not have formed in their current position, because the material at such large distances from the young Sun would have been too sparse to coalesce. They instead originally condensed and agglomerated into comet-sized bodies between the orbits of Jupiter and Neptune, as the leftover bits and pieces from the formation of the solar system. Once formed, the kilometer-size comet bodies were swept out of the region by the newborn giant planets. Gravitational perturbations by the newly formed giant planets sent some of the comets to large distances from the Sun and some into the inner solar system where they faded long ago. Thus, the nascent giants acted like

cosmic street-cleaners, either hurling the newborn comets into distant regions or consuming them.

Many of the comets sent to large distances escaped from the solar system. Some remained barely bound to the Sun by its weakened gravity at large distances; in a roughly spherical cloud about 10 000 to 200 000 AU in radius, the Oort cloud. Out there it takes roughly 10 million years to complete one orbit around the Sun.

Jupiter and Saturn, the two most massive planets, might have ejected some of them into interstellar space, but they also placed a significant fraction of comets into the Oort cloud. Jupiter may have additionally tossed nearby comets into a collision course with our planet, perhaps supplying some of early Earth's water and organic compounds. Uranus and Neptune, with lower masses, could not easily throw the primitive comets into the space between the stars, but they should have tossed about the same number of comets into the Oort cloud as Jupiter and Saturn did. The outer giants would have also pulled nearby comets into themselves, helping them to grow.

The comets that belong to the Oort cloud are therefore mementos of creation, frozen into the deep freeze of outer

Table 14.3 Selected short-period comets[a]

Name	Orbit period (years)	Perihelion date[b] (year)	Perihelion distance[b] (AU)	Orbital inclination (degrees)	Absolute magnitude
2P Encke	3.30	2003	0.34	11.8	9.8
6P d'Arrest	6.51	2002	1.35	19.5	8.5
9P Tempel 1[c]	5.51	2000	1.50	10.5	12.0
19P Borrelly[c]	6.88	2001	1.36	30.3	11.9
21P Giacobini–Zinner	6.61	2005	1.00	31.9	9.0
26P Grigg–Sjkellerup	5.11	2002	0.99	21.1	12.5
46P Wirtanen[c]	5.46	2002	1.06	11.7	9.0
67P Churyumov–Gerasimenko	6.57	2009	1.29	7.1	15.4
73P Schwassmann–Wachmann 3	5.34	2001	0.94	11.4	11.7
81P Wild 2[c]	6.39	2003	1.58	3.2	6.5
103P Hartley 2	6.46	2004	1.05	13.6	16.6

[a] Adapted from Kenneth R. Lang, *Astrophysical Data: Planets and Stars* (New York, Springer Verlag, 1992) and Gary M. Kronk's comet website http://cometography.com.

[b] The given perihelion date is the first to occur in the 21st century, and the perihelion distance is in AU, the mean distance between the Earth and the Sun. The absolute magnitude is a measure of the intrinsic brightness of a comet, and a smaller magnitude indicater a brighter comet.

[c] Comets that have either been visited by spacecraft in the past or have been considered for encounters in the future, listed in the order of their recognition, or periodic comet number.

[d] Also asteroid (4015).

Fig. 14.7 The return of comet Halley in 1986 Rays, streamers and kinks can be seen in the ion tail of comet Halley during its 1986 return to the inner solar system. The broad, fan-shaped dust tail can also be seen. The radio galaxy known as Centaurus A, or NGC 5128, can be seen in the bottom left corner. It is about ten trillion, or 10^{13}, times further away from the Earth than the comet is. Photograph taken by Arturo Gomez on 15 April 1986 with the Curtis Schmidt telescope of Cerro Tololo. (Courtesy of NOAO.)

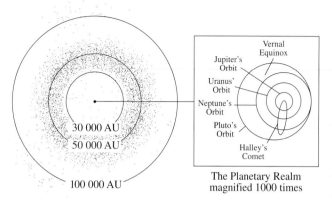

Fig. 14.8 The Oort comet cloud   More than 200 billion comets hibernate in the remote Oort comet cloud, shown here in cross-section. It is located in the outer fringes of the solar system, at distances of about 100 000 AU from the Sun. By comparison, the distance to the nearest star, Proxima Centauri, is 270 000 AU, while Neptune orbits the Sun at a mere 30 AU. The planetary realm therefore appears as an insignificant dot when compared to the comet cloud, and has to be magnified by a factor of 1000 in order to be seen. This comet reservoir is named after the Dutch astronomer Jan H. Oort (1900–1992) who first postulated its existence in 1950.

space, tumbling unseen in the remote blackness for billions of years. They are much too small and too far away to be seen. So they will remain forever invisible, and will never be directly detected. As the Stoic philosopher Seneca (4 BC to 65 AD) put it:

> How many bodies besides these comets move in secret, never rising before the eyes of men? For God has not made all things for man.
>
> *Natural Questions, Book 7, Comets*

A few of them occasionally return as comets that we see.

Dislodging comets from the Oort cloud

But how do comets fall from the Oort comet cloud to the heart of the solar system? At such enormous distances, comets in the Oort cloud are loosely bound by the Sun's gravity, and are easily disturbed by massive objects passing by, such as nearby stars or interstellar molecular clouds, which throw some of the comets back into the planetary system. The random gravitational jostling of individual stars passing nearby, for example, knocks some of the comets in the Oort cloud from their stable orbits, either ejecting them into interstellar space or gradually deflecting their paths toward the Sun. Every one million years, about a dozen stars pass close enough to stir up the comets, sending a steady trickle of comets into the inner solar system on very long elliptical orbits. A giant interstellar

molecular cloud can also impart a gravitational tug when it moves past the comet cloud, helping to jostle some of them out of their remote resting-place. Tidal forces generated in the cloud by the disk of our Galaxy, the Milky Way, also help to feed new long-period comets into the planetary region.

As time goes on, the accumulated effects of these tugs will send a few comets in toward the Sun – or outward to interstellar space. If the several hundred new comets observed during recorded history have been shuffled into view by the perturbing action of nearby stars or molecular clouds, then there are at least 100 billion, or 10^{11}, comets in the Oort cloud. There may be a trillion, 10^{12}, or even 10 trillion, 10^{13}, of them. This large population of unseen comets can sustain the visible long-period comets and persist without serious depletion for many billions of years, until long after the Sun expands to consume Mercury and boil the Earth's oceans away.

The Kuiper belt of comets

The Jupiter family of comets, with orbital periods less than 20 years, cannot come from the Oort cloud. These comets have relatively small orbits tilted only slightly from the orbital plane of the Earth, and they usually move in the same prograde direction as the planets. Unlike their longer-period cousins, the motions of the Jupiter-family comets resemble those of the planets, and their origin requires a close-in, flattened source.

This is now thought to be the Kuiper belt (Fig. 14.9), a ring of small icy objects at the outer edge of the planetary realm, just beyond the orbit of Neptune and a thousand times closer than the Oort cloud. It is named after the Dutch-American astronomer Gerard P. Kuiper (1905–1973) who predicted its existence in 1951. The name Edgeworth–Kuiper belt is used in the United Kingdom, acknowledging Kenneth E. Edgeworth's (1880–1972) proposal of the belt's existence in 1943.

Kuiper believed that comets must have formed throughout the early solar system. Although the giant planets cleared out any comets that were in their vicinity, the comets that formed beyond these planets should still be there. The density in this outer region of the primeval planetary disk was so low that the small objects did not coalesce into a single larger planet. They instead formed the flattened Kuiper belt of 100 million to 10 billion, or 10^8 to 10^{10}, small frozen worlds that have remained there for billions of years.

Armed with sensitive electronic detectors and powerful telescopes, astronomers have shown that the planetary system does not end abruptly at Neptune. They have discovered a substantial population of small, previously unseen

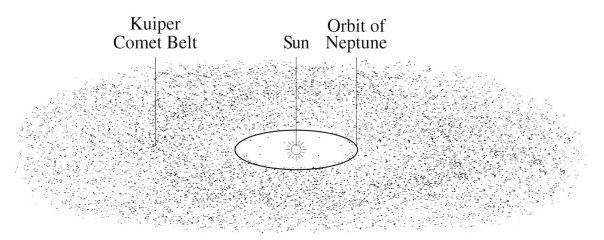

Kuiper Comet Belt   Sun   Orbit of Neptune

Fig. 14.9 The Kuiper belt of comets A repository of frozen, comet-sized worlds resides in the outer precincts of the planetary system, just beyond the orbit of Neptune and near the orbital plane of the planets. Known as the Kuiper belt, it is thought to contain 100 million to 10 billion comets. Many Jupiter-family comets are tossed into the inner solar system from the Kuiper belt. It is named after the Dutch-American astronomer Gerard Kuiper (1905–1973).

bodies in the Kuiper belt, each millions of times fainter than can be seen with the unaided eye (Chapter 15). All of these newly found dwarf planets travel in trans-Neptunian orbits that are only slightly tilted from the ecliptic, encircling the planetary system somewhat like the ring that wraps around Saturn. In fact, Pluto is probably one of them. Based on these detections, scientists estimate that the Kuiper belt contains tens of thousands of objects larger than 100 kilometers in size.

All of the objects observed in the Kuiper belt are larger than comets, which have a nucleus of about 10 kilometers across, but this is an observational selection effect. Smaller objects cannot be directly seen in the Kuiper belt with existing telescopes and detectors. The belt ought to contain a large population of small comet-sized bodies, just a few kilometers across. There are probably at least a billion pristine comets located just beyond the orbit of Neptune, each so faint and distant that they cannot be seen.

Unlike the comets in the Oort cloud, those in the Kuiper reservoir formed at their current locations at the dim horizon of the planetary realm, and have not been significantly perturbed since the origin of the solar system about 4.6 billion years ago. But Neptune's gravity slowly erodes the inner edge of the Kuiper belt, within about 45 AU from the Sun, launching comets from that zone into the inner solar system.

A few of the small Kuiper-belt objects are routed from their reservoir by the gravitational influence of Neptune and the other giant planets, and become regularly appearing Jupiter-family comets of shorter orbital periods. Like the comets in the Oort cloud, however, most of the comets in the Kuiper belt never come anywhere near the Sun.

14.4 Anatomy of a comet

All of the comets in the Oort cloud, and most of those in the Kuiper belt, are invisible. They are the nuclei of comets that can only be seen if they are dislodged and sent near the Sun. Each nucleus is the solid, enduring part of a comet. It is just a gigantic ball of frozen water ice and other ices laced with darker dust and pieces of rock, only a few kilometers in size. Light from the distant Sun is much too feeble to warm the comet ices, which remain frozen solid at the low temperatures in the remote comet reservoirs, so the comet never changes in size out there.

Unlike the planets, a comet lights up and becomes visible for a brief, fleeting interval during its long journey through space. It can often be detected only when it moves into the inner solar system, within the orbits of the terrestrial planets (Focus 14.2). Then the comet loops around the Sun and heads outward in more or less the same direction that it came from. As it moves away from the Sun, a comet receives less solar heat, becoming cold and inert and fading into darkness.

No two comets ever look identical (Fig. 14.10), just as no two snowflakes are alike, but most comets have basic features in common. When they emerge from the deep freeze of outer space and move toward the Sun, the comets then become visible as an enormous moving patch of light. This glowing, misty ball of light is called the *coma*, the Latin word for "hair". One or more tails can eventually stream from the coma, in a direction away from the Sun.

Comet gas and dust are initially ejected primarily in the general direction of the Sun; solar forces push them into tails that flow in the opposite direction. As a result, a

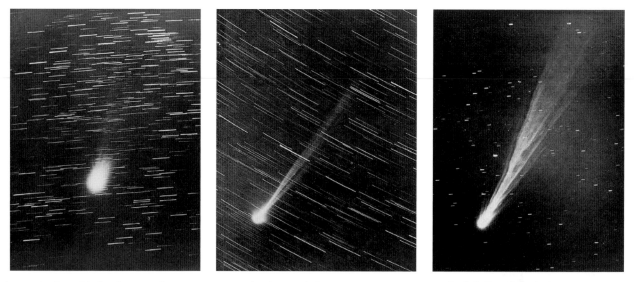

Fig. 14.10 Three kinds of comet shapes   Comet Perrine (1902 III) shows a transparent coma and tail (*left*), comet Finsler (1937 V) exhibits a coma and tail that are asymmetrical (*center*), and comet Morehouse (1908 III) is remarkable for the rapid variations in the structure of its tail (*right*). When a telescope follows a comet, stars move across the field of view, producing numerous short star trails. [Courtesy of the Royal Observatory Greenwich (*left and right*) and the Norman Lockyer Observatory (*center*).]

Focus 14.2   What turns a comet on?

When a comet nucleus emerges from the deep freeze of outer space and moves toward the Sun, the increased solar heat causes the comet's surface material to sublimate, with gases escaping through fissures in the crust of the nucleus. At the low-pressure conditions of space, the solid ice goes directly into gas without passing through a liquid state, in a process called sublimation, just as dry ice does on Earth. The escaping gases also carry along dust particles. The gas and dust make the comet grow in size, enabling it to be seen.

The distance at which an invisible comet nucleus turns on, and grows large enough to be seen, varies from comet to comet. Some Jupiter-family comets are first seen at several astronomical units from the Sun, but new long-period comets that are traversing the planetary system for the first time can be detected at greater distances. For instance, many first-time visitors to the solar neighborhood become unusually bright and extensive at distances of 5 AU or more.

The outer layers of old comets, which have made many passages close to the Sun, have been "cooked" and partially stripped off, leaving behind a dark insulating crust composed largely of dust. Solar radiation has a more difficult time penetrating this material than the fresh, icy surface of a new comet. This explains the limited loss of material from periodic comets that have been repeatedly exposed to the Sun.

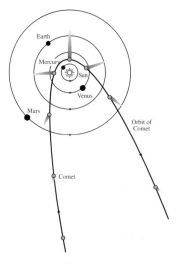

Fig. 14.11 Trajectory and tails of a comet   The path and changing shape of a typical comet as it enters the inner solar system. Note that the tail of the comet is oriented away from the Sun, independent of the direction of travel of the comet.

comet travels headfirst when approaching the Sun and tail first when moving away from it (Fig. 14.11).

A comet exerts very weak gravity, so the gas and dust that are blown off the surface of the central nucleus easily escape to interplanetary space, forming the coma and comet tails. All this material is lost to the comet forever, and must be continuously replenished from the solid comet nucleus.

The visible coma, or head, is a spherical cloud of gas and dust that has emerged from the nucleus, which it surrounds like an extended atmosphere. The central body, the

Table 14.4 Structural features of a comet[a]

Feature	Size	Composition	Appearance
Nucleus	1 to 10 kilometers	Dust, ice and rock	Very dark
Coma	Up to 0.01 AU	Neutral (un-ionized) molecules and dust	Slightly yellow
Hydrogen cloud	Up to 0.1 AU	Hydrogen atoms	Ultraviolet radiation
Dust tail	Up to 0.1 AU	Dust particles	Yellow, curved
Ion tail	Up to 1 AU	Ionized molecules	Blue, straight

[a] One AU, roughly the average distance between the Earth and the Sun, is about 149.6 billion, or 1.496×10^8, kilometers.

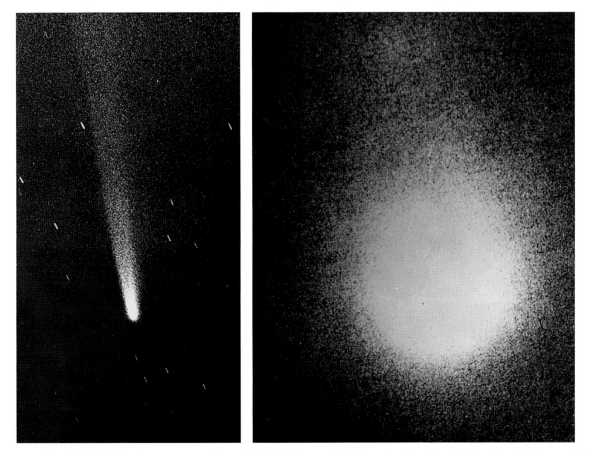

Fig. 14.12 Hydrogen cloud of a comet A comparison of the visible image (*left*) of comet Kohoutek with a far ultraviolet image (*right*) on the same scale, taken from *Aerobee* rocket flights on 4 and 7 January 1974. The ultraviolet image shows a gigantic cloud of hydrogen nearly 10 million kilometers in size, or about 10 times bigger than the Sun. It is being fed by the comet nucleus at the rate of 500 billion billion billion, or 5×10^{29}, atoms of hydrogen every second. The large size of the hydrogen cloud is due to the fact that hydrogen atoms are much lighter, and move into space faster, than the other atoms, ions, molecules and dust particles, which produce the visible light of the coma. (Courtesy of Chet B. Opal, NRL.)

nucleus, has typical dimensions of 1 to 10 kilometers, and the bulk of its composition is water ice and dust. The coma sometimes reaches a million kilometers in size, which is about as large as the Sun, and they usually become bigger than the Earth (Table 14.4).

A vast cloud containing hydrogen atoms that emit ultraviolet radiation envelops the coma and nucleus. Observations of this glow – invisible to the eye – indicate that the hydrogen halo can be 10 million kilometers across (Fig. 14.12), or about 10 times bigger than the Sun. The atomic hydrogen is produced when water molecules, released from the comet nucleus, are torn apart by energetic sunlight. The relatively light hydrogen atoms travel at high speeds to great distances before they are also

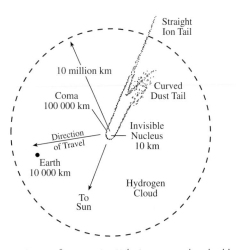

Fig. 14.13 Anatomy of a comet   What you see when looking at a comet depends on how you look at it. The nucleus of a comet is usually invisible, unless a spacecraft is sent in to take a glimpse. A comet first becomes visible when it develops a coma of gas and dust. When the comet passes closer to the Sun, long ion and dust tails become visible, streaming out of the coma in the direction opposite to the Sun. When looking at a comet in ultraviolet light, the hydrogen atoms in its huge hydrogen cloud are detected.

ionized by the Sun's energetic light and swept away by its winds.

If the comet travels closer to the Sun than about 1.5 AU, it usually develops a gossamer tail; but many comets that stay outside this distance have no tail. The wispy, ghost-like tails are paler and more tenuous than the coma. The long flowing tails sweep across the sky in regal splendor, attaining lengths of 10 million, or 10^7, kilometers, and even 100 million, or 10^8, kilometers, about 1 AU. Thus, the tails of comets can briefly become the largest structures in the solar system.

Yet the comet tails appear much more substantial than they really are. You can sometimes see stars shining through the tails as if they were not there at all. It is therefore no wonder that the Earth has passed through many comet tails unscathed. So comet tails look awesome, but they contain so little matter that they are very close to being nothing at all.

To sum up, a comet's anatomy consists of a concealed nucleus, an Earth-sized or Sun-sized coma, a vast hydrogen cloud, and two types of tails, the dust and ion tails (Fig. 14.13). But a comet's anatomy is not a static thing, for comets are always changing shape. All of the comet tails grow when the comet approaches the Sun, and shrink when the comet moves away from the Sun. There is no such thing as a typical comet tail. They differ in shape, size and structure. Some comets have multiple tails, some have only one tail, and others have no tail.

14.5   Two comet tails

Some comets show two types of tails at the same time. They are the long, straight, blue ion tails, attributed to gas escaping from the comet, and the shorter, curved yellow dust tails, composed of small solid particles driven away from the comet as its ices sublimate and the gases expand. Most of the dust is comparable in size to the width of a human hair, but larger solid particles are also released from the comet, with sizes comparable to sand or even pebbles.

The gases liberated by a comet nucleus become ionized by the action of solar ultraviolet radiation and emit the faint blue light of the ion tail by fluorescence. The dust tail shines only by reflecting yellow sunlight. Since the individual dust particles enter slightly different orbits of their own, the dust tail often spreads out into a fan shape (Fig. 14.14). A comet may have a dust tail, an ion tail, both types of tail, or no tail at all.

But what are the solar forces that blow the gas and dust into comet tails? The gentle pressure of sunlight pushes the tiny, solid dust grains along curved paths as the comet moves through space. When the Sun's light bounces off the dust particles, it gives them a little outward push, called radiation pressure, and this forces them into the dust tails. For larger solid particles, comparable in size to sand or pebbles, the Sun's gravitational pull overcomes the radiation pressure, and so these particles stay near the orbital path of the comet and they do not enter the dust tails.

A solar wind of electrically charged particles and magnetic fields propels and constrains the ions on straight paths away from the Sun. The solar wind, which continuously flows away from the Sun, also accelerates the ions to high velocities.

Thus, the ion tail acts like a windsock and, in fact, the existence of the solar wind was hypothesized from observations of comet ion tails before the age of space exploration. The gas lost from a comet is ionized by ultraviolet sunlight, producing an ionosphere that envelops the comet nucleus. Magnetic fields carried by the solar wind are unable to penetrate the ionosphere, so they pile up in front of it and drape around it to form nearly parallel, adjacent magnetic field lines that point toward and away from the Sun. Guided and constrained by these folded magnetic field lines, the comet ions are pushed away from the Sun by the much faster solar wind particles, forming a straight, blue ion tail.

But the interplanetary magnetism extending from the Sun is divided into sectors that point in opposite directions, toward and away from the star. When a comet crosses from one sector to another, the magnetism that envelops its ion tail becomes pinched and the comet loses the tail,

Fig. 14.14 Dust tail of a comet   This photograph of comet West (1976 VI) shows a broad, curved, pearly-hued dust tail. Because dust particles scatter sunlight, the dust tail has a slightly yellow color. It has a delicate lacy structure, created by countless dust particles shed from the comet nucleus over many days. (Courtesy of Stephen Larson, LPL/U. Arizona.)

somewhat like a tadpole. Unlike a tadpole, the comet soon grows another ion tail.

14.6 Spacecraft glimpse the comet nucleus

There is an invisible source of everything we can see when a comet passes near the Sun. It is the central nucleus of a comet, a small, icy object, typically 1 to 10 kilometers across, which is hidden within the brilliant glare of the coma's fluorescing gases and reflected sunlight.

A comet nucleus cannot be resolved using conventional telescopes, even when the comet passes very near to the Earth. A comet nucleus at a distance of only 0.2 AU would subtend an angle of about 0.002 seconds of arc. Ground-based telescopes are limited by atmospheric turbulence to a resolution of about 1.0 seconds of arc, and even outside our atmosphere the *Hubble Space Telescope* has a resolution of about 0.1 seconds of arc. So the only way to see the normally invisible comet is to send a spacecraft in close, to see through the coma and down to the detailed surface of the nucleus.

Table 14.5 Imaging missions to comets

Spacecraft	Comet	Encounter date
VEGA 1	Halley	6 March 1986
VEGA 2	Halley	9 March 1986
Giotto	Halley	14 March 1986
Deep Space 1	Borrelly	22 September 2001
Stardust	Wild 2	2 January 2004
	Tempel 1	14 February 2011
Deep Impact	Tempel 1	4 July 2005
	Hartley 2	4 November 2010
Rosetta[a]	Churyumov–Gerasimenko	August 2014

[a] The *Rosetta* spacecraft is on its way to a future encounter with a comet in 2014.

Imaging missions to comets are summarized in Table 14.5. They began by encountering comet Halley in March 1986, and continued with three Jupiter-family ones, comet Borrelly in September 2001, comet Wild 2 in

January 2004 and comet Tempel 1 in July 2005. The imaging is expected to continue in 2014 when the *Rosetta* spacecraft encounters comet Churyumov-Gerasimenko.

As we shall see, all of the comet space missions confirm the view that the nucleus of a comet is a single, solid frozen mixture of ices laced with rocky dust. The nucleus is made up mostly of water ice, but other ices such as carbon dioxide and methane are also present. As it approaches close to the Sun, energy supplied by solar radiation raises the temperature of the near-surface layers facing the Sun, and sublimation of ices, mostly water ice, takes place. The water ice goes directly from solid to the gaseous state, or to water vapor, without passing through the liquid state. This sublimation of ices produces the gas molecules that form the gas coma and subsequently the ion tail. When the ices sublimate, the embedded dust particles are also released to form the dust tail. The dust particles that are not carried away and fall back onto the nucleus form a dark insulating crust on the surface.

Observing the nucleus of comet Halley

An international flotilla of six spacecraft, belonging to four space agencies, flew by comet Halley in March 1986, to examine the gas and dust in the vicinity of the comet and to photograph its nucleus. Japan launched the *Sakigake* and *Suisei* spacecraft, meaning "pioneer" and "comet", which observed the comet from a safe distance and measured the interaction of the solar wind with the comet's atmosphere. The American probe *International Cometary Explorer* (*ICE*) also examined the solar wind upstream from the comet. *ICE* had already flown through the tail of the short-period comet 22P/Giacobini-Zinner on 11 September 1985. The two Soviet probes, named *VEGA 1* and *VEGA 2*, penetrated to within 9000 kilometers of the sunlit side of the nucleus, and the European *Giotto* approached to within 596 kilometers.

The major objective of the cameras on board both the two *VEGAs* and *Giotto* was to image the bare surface of a comet nucleus, which no one had ever seen. All three spacecraft penetrated the coma of comet Halley and detected its nucleus, despite damage by the hail of comet particles.

Giotto obtained the best images, with the highest resolution (Fig. 14.15), showing that the nucleus of comet Halley has an elongated, non-spherical and irregular shape with dimensions of $16 \times 8.5 \times 8.2$ kilometers, about the size of Paris or Manhattan. For a mass density about that of water, this volume corresponds to a mass of about 10^{15} kilograms or 1000 billion tons. A varied, lumpy topography was seen, with craters, valleys, hills, and mountains. Bright jets were spewing gas and dust from the comet's sunlit side, but there wasn't much white ice in site.

The ice had been evaporated away from the outer layers of the nucleus, to leave a dark tar-like crust all over the surface, reflecting only about 4 percent of the sunlight falling on it. When Halley passes close to the Sun, and the solar heat makes the comet active, gas jetting out from cracks and holes in its surface carries dust with it. But when the comet is far away from the Sun at the other end of its orbit, some of the black dust settles back down over the surface to form a dark crust, rich in carbon compounds.

Gas and dust can now only get out in vents where the crust has broken to expose the underlying ice, rather than from the whole sunlit hemisphere. The sunward jets are emitted from roughly 10 percent of the total surface area. Nearly 90 percent of the surface was inactive at the time of observation.

When the nucleus of comet Halley was at its maximum rate of gas emission, near its closest approach to the Sun, it released up to 1.7×10^{30} water molecules per second, or up to 56 tons of water every second. On average, the comet nucleus was emitting 20 tons of gas every second during the *Giotto* encounter, and about 80 percent of that was water. This is comparable to a loss of about 1.4 million tons of water every day, the approximate amount of water needed to supply the hydrogen halos observed around other comets. The nucleus of comet Halley was also ejecting about 10 tons of dust every second. This fully confirmed the icy conglomerate model in which a comet nucleus consists mainly of water ice and dust particles.

If the comet suffers such a high rate of water loss during the few months when it is close to the Sun, then it must lose on the order of 100 million tons of water ice during the course of each orbit. With a total mass of 100 billion tons, the comet can survive for about 1000 orbits, or 76 000 years, before it wastes away.

Deep Space 1 encounters comet Borrelly

Comet Borrelly also has a tar-black surface, as unreflective as that of comet Halley. The *Deep Space 1* spacecraft was directed toward this comet after completing its primary mission of flight-testing an ion engine and other advanced technologies. On 22 September 2001, the spacecraft whizzed by comet Borrelly at a distance of just 2200 kilometers, revealing an irregular chunk of rock and ice, about 8 kilometers long and perhaps 4 kilometers wide (Fig. 14.16). It is covered with a dark carbon-rich slag that reflects only about 3 percent of the incident sunlight, on average; comparable to the reflectivity of the powdered toner used in laser printers.

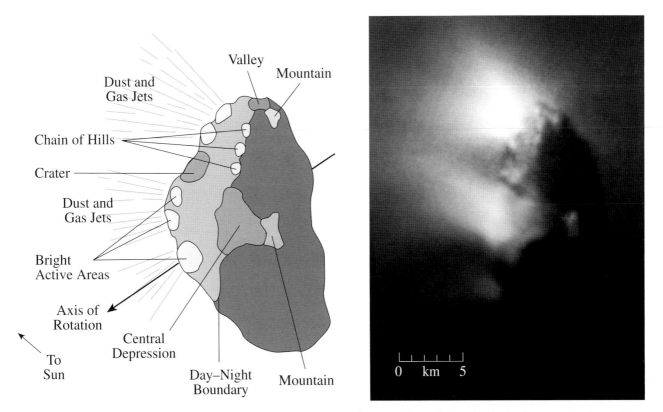

Fig. 14.15 Nucleus of comet Halley   A composite image of the nucleus of comet Halley (*right*) obtained using images taken in March 1986 with the camera on board the *Giotto* spacecraft, from a distance of 6500 kilometers before comet dust destroyed the camera. It is compared with a schematic drawing (*left*) that highlights the major features recognizable in the photographs. The nucleus is about 16 kilometers long and 8 kilometers wide. Dust and gas geyser out of narrow jets from the sunlit side of the nucleus, but about 90 percent of the surface is dark and inactive. The gas is mainly water vapor sublimated from ice in the nucleus, while a significant fraction of the dust may be dark carbon-rich matter. A dark surface crust, which insulates most of the underlying ice, is blacker than coal, reflecting about 4 percent of the incident sunlight. "Mountains" rise about 500 meters above the surrounding terrain, while a broad "crater" is depressed about 100 meters. (Image courtesy of Harold Reitsema of the Ball Aerospace Corporation and Horst Uwe Keller.)

The surface of this comet nucleus has no water ice in sight, for it has all been covered up during the comet's successive passages near the Sun, leaving the dark crust behind like a dirty street a few days after a snowstorm. The sublimation has also produced a variety of surface features likened to ridges, hills, depressions, mesas and deep fractures. Beneath the rugged terrain and insulating dust there must be a supporting nucleus of solid water ice.

Stardust brings some of comet Wild 2 back home and is retargeted to comet Tempel 1

The *Stardust* spacecraft was launched on 7 February 1999, for an encounter with comet Wild 2 in early 2004. This comet is a fairly recent arrival to the inner solar system, captured into its current orbit by a close encounter with Jupiter in 1974. So it has traveled near the Sun only a few times, and its material has probably not

been significantly altered by solar heat and extensive sublimation, when compared to comet Halley that has made hundreds and perhaps thousands of close passes by the Sun.

The flyby of comet Wild 2 on 2 January 2004 came within 236 kilometers of the nucleus, and resulted in detailed high-resolution images. They revealed a rugged diverse surface terrain on a dark, round body about 4 kilometers across and reflecting only about 3 percent of the incident sunlight (Fig. 14.17). Steep-walled pits and craters with flat floors were found beneath violent jets of gas and dust, shooting out from the internal, solid water ice required to feed the jets and support the surface.

The main goal of the *Stardust* mission was to collect samples of comet dust particles during the flyby and bring them home to Earth. The pioneering spacecraft caught the pristine comet dust by extending a paddle covered with an exotic substance called aerogel. It has one of the lowest

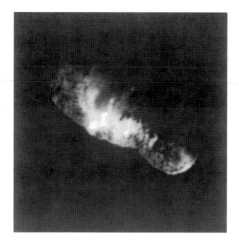

Fig. 14.16 Nucleus of comet Borrelly   A camera on board the *Deep Space 1* spacecraft peered into the icy heart of comet Borrelly on 22 September 2001, taking this image from a distance of 3400 kilometers. The nucleus is shaped like a gigantic bowling pin, with a length of about 8 kilometers and a width of roughly half that size. A dark veneer of material covers most of the nucleus, reflecting only about 3 percent of the incident sunlight on average. Rugged terrain is found on both ends of the nucleus, while bright smooth plains are present in the middle. Jets of gas and dust shot out from all sides of the comet's nucleus as it rotated, producing a flow of ions that was not centered on the nucleus. (Courtesy of NASA/JPL.)

mass densities known, essentially because it is filled with holes, and was thus able to capture the dust intact. The dust particles would have evaporated when striking a denser, metallic surface.

The collected dust samples were placed within a capsule that was carried back to Earth and parachuted to the desert salt flats in Utah on 15 January 2006, while the rest of the spacecraft remained in space. The sample return capsule exceeded all expectations, containing thousands of coma particles that have been analyzed and scrutinized in terrestrial laboratories (Focus 14.3).

The *Stardust* spacecraft has been reused to encounter comet Tempel 1 on 14 February 2011, providing unique observations of a comet nucleus after its close approach to the Sun. This is a second visit to the comet, which was previously struck by the *Deep Impact* spacecraft on 4 July 2005.

Deep Impact strikes comet Tempel 1 and encounters comet Hartley 2

The *Deep Impact* spacecraft was launched on 12 January 2005 on a six-month journey to comet Tempel 1. The mission consisted of two parts, a flyby spacecraft, which

Fig. 14.17 Nucleus of comet Wild 2 An image of the dark, cratered nucleus of comet Wild 2 taken from the *Stardust* spacecraft during its comet flyby on 2 January 2004. The nucleus is about 5 kilometers in diameter. This is a composite of a short exposure image showing surface details, and a long exposure image, taken just 10 seconds later, showing the active surface jetting dust and gas streams into space. During this flyby *Stardust* gathered comet dust and subsequently returned the comet sample to Earth. (Courtesy of NAS/JPL-Caltech.)

Focus 14.3 High-temperature and organic materials found in comet dust return

Comet particles returned from comet Wild 2 by the *Stardust* spacecraft originated in the frozen nucleus of the comet, which has hibernated for billions of years in one of the coldest places in the solar system and has not been significantly altered by repeated passages near the Sun. Yet the returned comet dust included high-temperature minerals, and contained a mix of material that must have formed at some of the highest and lowest temperatures that existed in the early solar system. These discoveries might alter views of comet formation.

Two kinds of nitrogen-rich organic molecules and the amino acid glycine have also been discovered in samples of comet Wild 2. These results indicate that comets could have delivered similar organic compounds to the Earth along with water, especially during the planet's early history when comet impacts were more frequent than now. The rain of comet dust may have even brought the self-replicating, carbon-rich molecules necessary for the origin of life, but this remains a speculation without hard scientific evidence.

Fig. 14.18 Impact with nucleus of comet Tempel 1 The initial material ejected when the *Deep Impact probe* collided with the nucleus of comet Tempel 1 on 4 July 2005. The dark, cratered nucleus is an oblong body between 4.9 and 7.6 kilometers in size. Spectroscopic examination of the ejected material showed that the comet nucleus is mainly composed of water ice and dust. (Courtesy of NASA/JPL-Caltech/UMD.)

Fig. 14.19 The nucleus of comet Hartley 2 Jets stream out of the sunlit side (*right*) of the nucleus of the small, active comet Hartley 2, whose long axis spans about 2 kilometers. A narrower, smooth waist joins rough areas at both ends of the elongated nucleus. (Courtesy of NASA/JPL-Caltech/UMD.)

would image the comet, and a smaller impactor that would collide with the comet and find out what is inside it. The collision occurred as scheduled on 4 July 2005, in celebration of the American Independence Day.

Moving at a speed of 10 kilometers per second, the 820-pound impactor vaporized deep below the comet's surface and kicked up a cloud of dust (Fig. 14.18). The opacity of the plume that the impactor created, and the light it gave off, suggested that the dust excavated from comet Tempel 1 is very fine and porous, like snow rather than sand. It might cover an interior that has remained unchanged since the comet formed 4.6 billion years ago.

Telescopes trained on the ejecta detected spectra of water vaporized by the heat of impact, but also the absorption spectra of ice particles ejected from the surface a few seconds after the water vapor was initially released. This was the first time direct evidence had been obtained for water ice in a comet and on its surface – not melted, vaporized or sublimated but solid ice. Most of this escaping water was contained in ice particles from below the surface.

Instruments aboard the *Spitzer Space Telescope* detected a huge increase in the amount of carbon-containing molecules, including carbon dioxide and hydrocarbons, detected in the spectral analysis of the ejection plume,

indicating that comets contain substantial amounts of organic material.

After releasing its impact probe, the *Deep Impact* spacecraft was retargeted to fly past comet Hartley 2, as part of the extended *EPOXI* mission. The closest approach, at a distance of 700 kilometers, revealed a small, oblong nucleus with activity on the sunlit side and two rough areas joined at a smoother middle (Fig. 14.19).

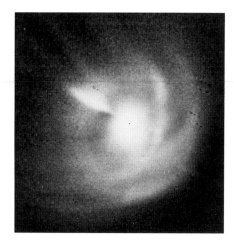

Fig. 14.20 Rotating comet This photograph of comet Halley shows jets of dust ejected from a rotating nucleus. (Courtesy of Stephen Larson, LPL/U. Arizona.)

A *Rosetta* future

On 2 March 2004, the European Space Agency launched the *Rosetta* spacecraft on a complex 10-year journey to comet Churyumov-Gerasimenko, using four gravity assists, three from Earth and one from Mars. Beginning in August 2014 the spacecraft is expected to spend approximately 2 years orbiting and observing the comet as it approaches the Sun. *Rosetta* is also expected to place a small lander on the comet's surface in November 2014.

14.7 Rotating comet nucleus

A comet's nucleus also rotates (Fig. 14.20). Typical rotation periods are a few hours to a few days. Observations of comet Halley, for example, indicate that it rotates around its longest axis once every 7.4 days, and that it wobbles about its shortest axis once every 2.2 days, or 53 hours. As the nucleus rotates, new regions turn to face the Sun, heat up and become active, while others face away from the Sun and momentarily turn off their activity.

Jets from a rotating comet nucleus can help explain comets that seem to defy gravity by arriving at perihelion a few days before or after the time calculated using Newton's theory of gravitation. The gas and dust streaming off the sunlit side of a comet nucleus initially heads toward the Sun, before being swept back into the comet tails. The expelled material pushes the comet in the opposite direction, making it arrive sooner or later than expected (Fig. 14.21). A similar recoil effect explains the darting action of a small balloon when it is released, as well as the forward thrust of a rocket engine. The thrust of the rocket-like jets either pushes the comet along in its orbit or slows it down.

14.8 Comet decay and meteor showers

Comets fall apart

Once a Jupiter-family comet enters the inner solar system, it returns again and again on a relentless voyage of continual disintegration, eventually being consumed by their own emissions. They are caught in a life of continual decay, sublimating and blowing part of themselves away each time they come near the Sun.

Sooner or later most Halley-type and Jupiter-family comets will either fall apart or turn into a dark rocky corpse that looks like an asteroid. It's just part of the aging process. Once you can see one of these comets, they are doomed to disappear.

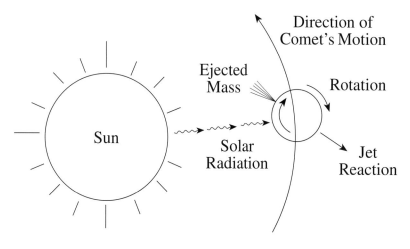

Fig. 14.21 How to make a jet engine out of a dirty ball of comet ice Unexpected cometary motions are attributed to non-gravitational forces caused by jets of matter ejected from a comet's spinning, icy nucleus. In this illustration, the ejected material pushes the comet in the opposite direction to its motion, causing the comet to arrive closest to the Sun at a time later than expected. If the comet had been rotating in the opposite direction, the jets would have pushed the comet along in its original direction, resulting in an early arrival time.

Fig. 14.22 Comet clones   The splitting of the nucleus of comet West (1976 VI) photographed (*top to bottom*) on 8, 12, 14, 18 and 24 March 1976, in yellow-green light using a 0.60-meter (23.6-inch) Cassegrain reflector. On 18 March, the diameter of the four features was about 10 000 kilometers. (Courtesy of C. Knuckles and S. Murrell, New Mexico State University Observatory.)

Comets are very fragile, with little internal strength and a very low mass density. Some comet nuclei are crumbly, fluffy structures with mass densities less than that of solid ice and far less than solid rock. The central pressure of their nucleus is probably comparable to that under a thick layer of blankets. So it is little wonder that some comets have been observed to break up as the result of tidal forces from either the Sun or Jupiter. They pull on the near side of the comet a little more than the far side, tearing the comet apart. The nucleus of comet West was, for example, split into pieces when it passed near the Sun in 1976 (Fig. 14.22).

In other instances, a comet has spontaneously split apart, with no apparent gravitational or tidal disruption from Jupiter or the Sun. Sometimes a comet just blows off its fragile outer parts with no apparent reason (Fig. 14.23).

The pieces of the split nucleus have too little mass to pull themselves together gravitationally. So once a nucleus splits, its pieces remain forever separated. The jets of escaping gas kick them away from each other, and they continue to drift farther apart.

Any Jupiter-family comet is slowly wasting away from the outside in. Each time it approaches the Sun, a small percentage of a comet's icy surface will sublimate and escape into space, dragging along grains of dust. An example is comet Encke that revolves about the Sun with a 3.3-year orbital period; it loses about a meter of ice and dust each time it passes near the Sun. If its nucleus were 3 kilometers across, then this comet would disappear in just 10 000 years. Comet Halley is not expected to last more than about 76 000 years, and most of the comets seen during recorded history will probably vanish from sight in less than a million years, either sublimating away into nothing or leaving behind a black, rocky, burned-out corpse.

Nights of the shooting stars

Spectacular meteor showers, spawned by passing comets, have periodically returned for at least a thousand years, inspiring fear, wonder and admiration. William Blake (1757–1827) caught some of the excitement of the August shower, the Perseids, in his poem "The Tyger":

> When the stars threw down their spears
> And water'd heaven with their tears . . .

Meteor showers are commonly called shooting stars (Fig. 14.24), but they are not stars. They are fragile material from comets. In addition to spewing off small dust particles, which can be captured by Earth, drift down to the ground, and enter your hair, comets also expel larger particles ranging in size from sand grains to pebbles. This debris burns up when it enters our atmosphere, producing visible meteors.

The meteors associated with comets vaporize completely in flight. In fact, no meteor associated with a comet has ever reached the ground. As a cometary meteoroid enters the Earth's atmosphere, it will decelerate and heat up as the result of friction, producing a luminous trail until the object becomes hot enough to vaporize completely away. Thus, there is a great difference between the fragile meteoric material associated with comets and the tough meteorites that survive the atmospheric flight to the ground.

When just one of the comet particles rubs against the air, it vaporizes in a streak of light, producing the luminous

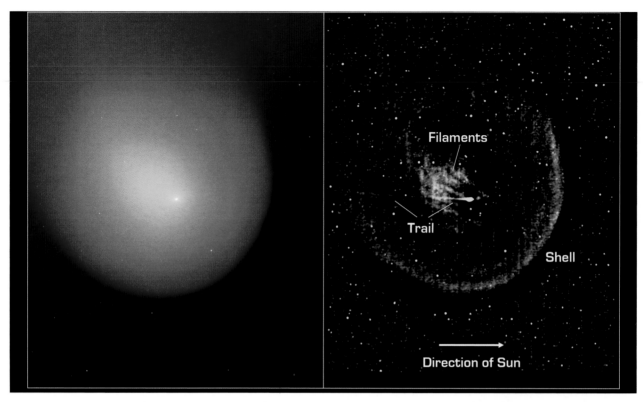

Fig. 14.23 Comet explosion   Comet Holmes has exploded just twice in 118 years, in November 1892 and October 2007, while approaching the asteroid belt and moving from the vicinity of Jupiter toward the Sun. This infrared image was obtained from the *Spitzer Space Telescope* in March 2008; five months after the comet suddenly erupted and brightened a million-fold overnight. The infrared picture (*left*) reveals dust particles that make up the outer shell, which envelops solid particles (*yellow*) blown from the exploding comet and the comet nucleus (*white center*). The contrast-enhanced image (*right*) shows the outer shell and filaments or streamers of dust. (Courtesy of NSA/JPL-Caltech.)

Fig. 14.24 The shooting stars   Two couples portray meteor showers or falling stars in this picture painted by Jean-Francois Millet (1814–1875) in 1847. They soar through the skies, perhaps illustrating the transcendental nature of erotic love. (Courtesy of the National Museum of Wales, Cardiff.)

Fig. 14.25 Meteor trail A glowing meteor streaks across the stars near the constellation Cygnus. The straight trail was produced by a sand- or pebble-sized piece of a comet, burned up by friction as it entered the Earth's atmosphere. The curved structure (*left center*), known as the Cygnus Loop, is an expanding shell of material thrown off during the supernova explosion of a massive, dying star. (Courtesy of the Yerkes Observatory.)

Table 14.6 Comets associated with meteor showers[a]

Meteor shower[b]	Comet
Lyrids	Thatcher C/1861 G1
Eta Aquarids	1P/Halley
Scorpiids-Sagittariids	2P/Encke
Bootids	7P/Pons-Winnecke
Perseids	109P/Swift-Tuttle
Aurigids	Kiess C/1911 N1
Draconids	22P Giacobini-Zinner
Orionids	1P/Halley
Epsilon Geminids	Ikeya C/1964 N1
Taurids	2P/Encke
Adromedids	3P/Biela
Leonids	55P/Temple-Tuttle
Geminids	3200 Phaëthon[c]
Monocerotids	Mellish C/1917 F1
Ursids	8P/Tuttle

[a] Adapted from Kenneth R. Lang, *Astrophysical Data: Planets and Stars*, New York, Springer Verlag (1992).
[b] The visibility dates of these showers are given in Table 14.7.
[c] 3200 Phaëthon is cataloged as an asteroid, but it may be an inactive comet nucleus.

trail of a meteor (Fig. 14.25). And when many fall into the dark night sky, they produce meteor showers.

From the luminous path of a meteor, it is possible to determine the incoming particle's orbital path around the Sun, and in most cases the orbits are similar to those of comets (Table 14.6). A comet ejects the particles along its orbital path as it loops around the Sun, and this material continues to revolve around our star. The swarm of comet material is called a meteoroid stream, which is spread out all along the orbit of the comet (Fig. 14.26). And when the Earth passes though one of these streams, it intercepts some of the orbiting particles that enter our atmosphere, and creates a meteor shower (Fig. 14.27).

When a meteor shower includes large numbers of shooting stars, the trails appear to intersect and emanate from a distant point called the *radiant* (Fig. 14.28). But meteors that appear to diverge from a point are actually moving on parallel paths, just as parallel railroad tracks seem to come from a point on the distant horizon. Meteor showers are named after the constellation in which their radiant appears (Table 14.7).

Because the Earth often passes through a comet's orbit just once a year, a meteor shower usually appears at yearly intervals. The Lyrids, for instance, appear in April, the Perseids are seen every August, and the Leonids light up the night sky in November. Since the distribution of material along a comet orbit is generally non-uniform, the rates of meteors in a particular shower can vary from year to year.

The Earth has been passing through some comet orbits for hundreds of years. The Lyrids were known in 687 BC, the Perseid stream was first recorded in 36 AD and the Leonids were first recorded in 902 AD. Nevertheless, the comet that produces a meteor shower will eventually disintegrate completely, and that shower will not be seen again. But this may not happen for a million years from now.

Burned-out comets that look like asteroids

Although comets are distinguished by their ability to emit gas and dust when near the Sun, some comets may be either dormant or extinct. In fact, there are a few small

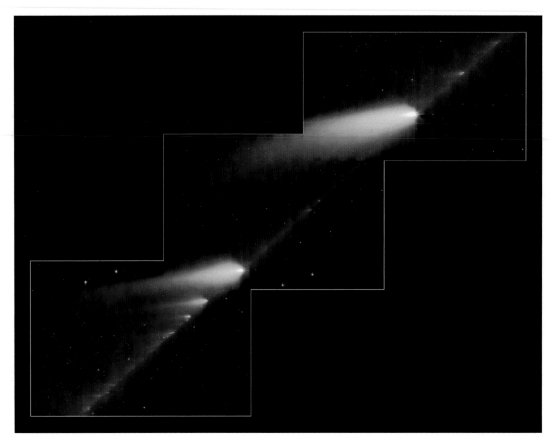

Fig. 14.26 Meteoroid stream from disintegrating comet   The broken comet Schwassman-Wachmann 3 moves along a trail of debris formed during its multiple trips around the Sun, which the comet circles every 5.4 years. In 1995 the comet broke apart, and two of the biggest fragments are shown here, as miniature comets, together with about 36 smaller fragments. The line bridging the large comet fragments is the meteor stream of dust, pebbles and rocks left in the comet's wake during numerous previous journeys. This infrared image was taken on 4 to 6 May 2006 from the *Spitzer Space Telescope*. (Courtesy of NASA/JPL-Caltech.)

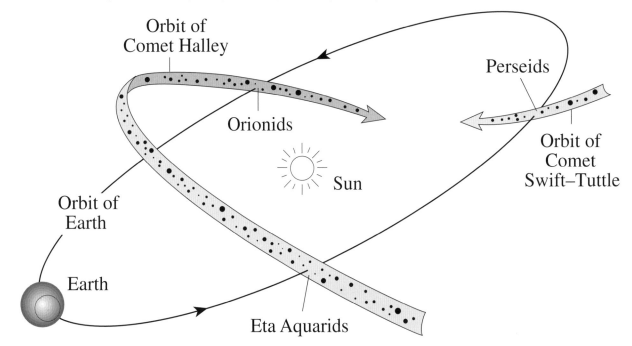

Fig. 14.27 Comets produce meteor showers   The Earth's orbit intersects a stream of meteoric material left along the orbit of comet Halley, producing two meteor showers, the Eta Aquarids in May and the Orionids in October. Other comets intersect the Earth's orbit just once during their trip around the Sun. Annual meteor showers are created when the Earth enters the intersection point, such as the August Perseids produced by debris from comet Swift-Tuttle. The orbit of comet Halley is inclined by 162 degrees with respect to the ecliptic, the plane of the Earth's orbit, while the orbit of comet Swift-Tuttle has an inclination of 114 degrees.

Table 14.7 The principal annual nighttime meteor showers[a]

Shower	Maximum date	Radiant position Right ascension	Declination	Visibility dates	Meteors per hour
Quadrantids	3–4 January	15h 28m	+50°	1–6 January	110
Alpha Aurigids	6–9 February	04h 56m	+43°	Jan.–Feb.	10
Virginids	12 April	14h 04m	−09°	March–April	5
		13h 36m	−11°		5
Lyrids	21–22 April	18h 08m	+32°	19–25 April	15
Eta Aquarids	5 May	22h 20m	−01°	24 April–20 May	35
Alpha Scorpiids	28 April	16h 32m	−24°	20 April–19 May	5
	13 May	16h 04m	−24°		
Ophiuchids	9 June	17h 56m	−23°	May–June	5
	19 June	17h 20m	−20°		
Alpha Cygnids	21 July	21h 00m	+48°	June–August	5
Capricornids	8–15 July	20h 44m	−15°	5 July–20 August	5
Alpha Capricornids	2 August	20h 36m	−10°	15 July–20 August	5
Delta Aquarids	29 July	22h 36m	−17°	15 July–20 August	25
	6 August	23h 04m	+02°		10
Iota Aquarids	6 August	22h 10m	−15°	July–August	10
Piscis Australids	31 July	22h 40m	−30°	July–August	5
Perseids	12 August	03h 04m	+58°	25 July–20 August	80
Alpha Aurigids	28 August	04h 56m	+43°	August–October	10
Piscids	9 September	00h 36m	+07°	Sept.–Oct.	10
Orionids	21 October	06h 24m	+15°	15 Oct.–2 Nov.	30
Taurids	3 November	03h 44m	+14°	15 Oct.–25 Nov.	10
Leonids	17 November	10h 08m	+22°	15–20 November	45
Geminids	13–14 December	07h 28m	+32°	7–15 December	70
Ursids	22–23 December	14h 28m	+78°	19–24 December	10

[a] The celestial coordinates of the radiant are right ascension (RA), in hours (h) and minutes (m), and declination (dec.), in degrees. The maximum hourly frequency of meteors assumes the radiant is at the zenith, but this rate can vary from year to year because of the non-uniform distribution of the relevant meteoroid stream. [Adapted from N. Bone, *Meteors*, Cambridge, MA, Sky Publishing Co. (1993), and Kenneth R. Lang, *Astrophysical Data: Planets and Stars*, New York, Springer Verlag (1992).]

objects that move like comets, but emit no gas and dust and display neither a coma nor a tail. Some of them have eccentric, elongated comet-like orbits that stretch into the vast outer reaches of the solar system, beyond the most distant major planets. Yet they show no trace of comet activity when approaching the warmth of the Sun, sometimes passing nearer to it than the Earth.

They may be inert comets, which have turned into inactive objects. These comets either exhausted all the volatile ices they once had to feed a coma and tail, or their ice might be completely shrouded in a thick, insulating cover of dust and dirt, preventing the ice from sublimating into luminous material.

The near-Earth asteroid 3200 Phaëthon is a well-known example. It moves in a highly eccentric orbit – once thought to be the hallmark of comets. In fact, it follows the orbit of the meteoroid stream that produces the Geminids meteor shower. Since the large majority of meteor showers are caused by debris scattered along a comet's orbit, it is likely that this object is a defunct comet that has now lost its ability to emit gas and dust.

These anomalous bodies account for a very modest fraction of the thousands of known comets and asteroids.

We now turn our full attention to other larger objects in the Kuiper belt beyond Neptune.

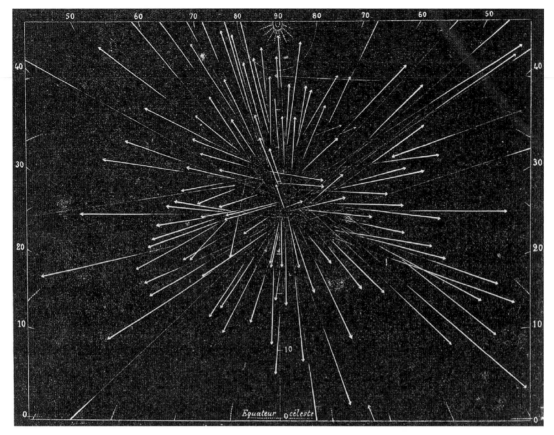

Fig. 14.28 Radiant meteors The apparent paths of shooting stars on 27 November 1872. Meteor showers are named after the constellation in which their radiant appears. This meteor shower is called the Andromedids meteor shower because its radiant appears in the constellation Andromeda. The shower occurs every November when the Earth intersects the debris that has been scattered along the orbit of Biela's comet. [Adapted from Amedee Guillemin's *Le Ciel*, Librairie Hachette, Paris (1877).]

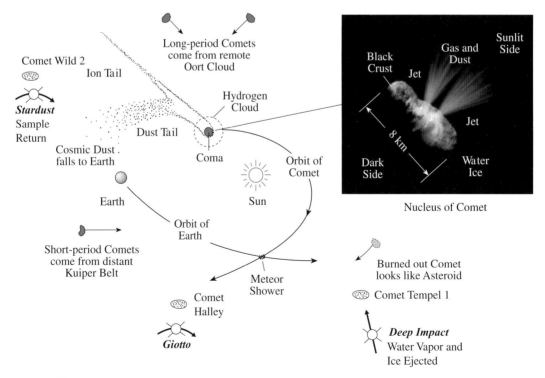

Fig. 14.29 Summary diagram

15   **Beyond Neptune**

- The discovery of Pluto was serendipitous, the result of a long careful search based on incorrect predictions.

- Pluto moves between 29.7 and 49.3 AU from the Sun, and just crosses within the orbit of Neptune, at 30 AU, without ever colliding with it.

- Pluto has a thin atmosphere of nitrogen, methane and carbon dioxide, and large variable dark and light surface makings.

- Pluto has three companion moons, named Charon, Nix and Hydra. Charon is an oversized satellite, about half the size of Pluto.

- The *New Horizons* spacecraft will encounter Pluto in 2015, to study its atmosphere, surface and interior, then travel into the Kuiper belt.

- Centaurs are small bodies that orbit the Sun between Jupiter and Neptune and cross the orbit of one or more giant planets. They are thought to originate further away from the Sun in the Kuiper belt. Some centaurs display comet comas, and most of them are expected to become comets.

- Several trans-Neptunian objects orbit the Sun beyond the orbit of Neptune, including Eris, Makemake and Haumea. They are either larger than Pluto or comparable to it in size.

- Pluto has been demoted from the ninth major planet to a dwarf planet, and then reclassified as a plutoid.

- A dwarf planet orbits the Sun and has a rounded shape, but it does not clear out the orbit around it. The asteroid 1 Ceres has been designated a dwarf planet.

- Plutoids are trans-Neptunian dwarf planets, and they include Eris, Haumea, Makemake and Pluto.

- The *Voyager 1* and *2* spacecraft have crossed the termination shock of the solar wind at 94 AU and 84 AU respectively.

- The outer Oort cloud lies between 20 000 and 200 000 AU, far beyond the termination of the solar wind, but still within the Sun's gravitational control.

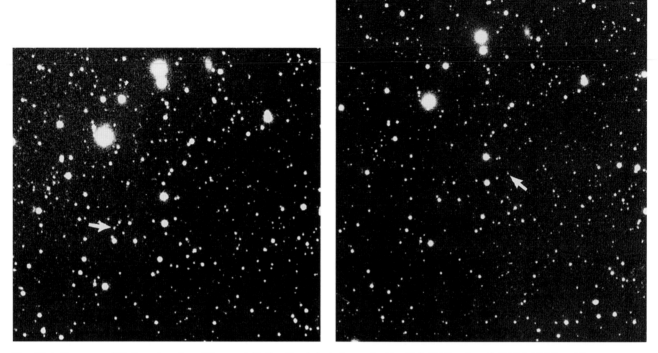

Fig. 15.1 Discovery of Pluto   A region of the constellation Gemini, photographed by Clyde W. Tombaugh (1906–1997) on 23 January 1930 (*left*), and the same region photographed six days later (*right*). When comparing the two plates on 18 February 1930 with a blink microscope, Tombaugh noticed an object (*arrows*) on the second plate that had changed its location with respect to the background stars since the first plate was taken. This was a previously unknown object that had to belong to the solar system. Because of its slow apparent motion across the sky, the planet images were separated by just 3.5 millimeters on the two photographs. (Courtesy of the Lowell Observatory.)

15.1 Pluto: a small frozen world with companions

The serendipitous discovery of Pluto

The discovery of Neptune in 1846 resulted from a mathematical study of the differences between the predicted and observed positions of Uranus, attributed to the gravitational pull of the then unknown planet. Astronomers hoped that similar irregularities in Neptune's motion would lead to the discovery of another remote planet; but because of Neptune's long 165-year orbit there were insufficient observations. Prediction of another unknown planet therefore had to be based upon perturbations in Uranus' motion, after corrections for the gravitational effects of Neptune.

Two astronomers used the corrected Uranus data to predict an undiscovered planet beyond Neptune. The first such prediction was made in 1909 when William Henry Pickering (1858–1938) argued that both Neptune and a remote Planet O were producing gravitational tugs on Uranus. Percival Lowell (1855–1916) made the next attempt in 1915; he called his unknown object Planet X.

Lowell directed the most ambitious search for the trans-Neptunian planet, at his observatory in Flagstaff, Arizona, but no new planet was found at a variety of predicted locations between 1905 and 1919. The search from the Lowell Observatory continued a decade later using a new 0.33-meter (13-inch) photographic telescope. Once three photographs had been taken at intervals of several days, they were set in pairs in a blink microscope that would show the apparent motion of a planet, asteroid or comet against a background of nearly half a million stars on each photograph.

After months of painstaking work, Clyde William Tombaugh (1906–1997) discovered, on 18 February 1930, the sharp, faint, moving image of the elusive quarry (Fig. 15.1). The new object was named Pluto, for the Roman god of the underworld. It is a small frozen world at the outer fringe of the planetary system, traveling along an elongated, 248-year path that is inclined by 17 degrees to the orbital plane of the major planets.

Pluto's orbit around the Sun is so far from circular that it moves between 29.7 and 49.3 AU from the Sun, and it crosses the orbit of Neptune at 30.06 AU. Pluto is, however, protected from possible collision with the large planet by a special orbital relationship, called a

Table 15.1 Physical characteristics of Pluto

Mean radius	1153 kilometers
Mass	1.309×10^{22} kilograms
Bulk density	2050 ± 40 kilograms per cubic meter
Sidereal rotation period	-6.3872 Earth days
Sidereal orbital period	247.92 years
Average distance from Sun	39.48 AU

resonance. Simply put, for every three orbits of Neptune around the Sun, Pluto completes two. This means that although Pluto's orbital plane crosses that of Neptune, the two bodies never come closer than 17 AU from each other.

Pluto is an anomaly. It is much smaller than the giant planets, and is comparable in size to some of their satellites (Table 15.1). Pluto is smaller than Saturn's satellite Titan and all four of Jupiter's largest moons. In many ways, Pluto is the twin of Neptune's largest satellite Triton. They have almost the same size and mean mass density. Moreover, many other small worlds have now been discovered just beyond the orbit of Neptune. Pluto is more akin to this group of objects than to the families of terrestrial or giant planets.

The discovery of an object that orbits Pluto led to a determination of its mass, at about only 1.3×10^{22} kilograms, or just 0.2 percent (0.002) of the Earth's mass and only about one-sixth the mass of the Earth's Moon. This means that Pluto was not found because it was correctly predicted. Its mass is far too small to have noticeably influenced the past motions of Uranus. In fact, the distant Earth exerts a larger gravitational influence on Uranus than Pluto does. The discovery of Pluto was the result of a meticulous and systematic search that was guided by an incorrect prediction, which merely happened to point in the general direction of Pluto.

After Pluto's discovery, astronomers continued to speculate that some unknown, massive and remote planet was responsible for the apparent perturbations in the motion of Uranus. However, after accounting for the gravitational effects of Neptune, using a precise mass obtained when *Voyager 2* encountered the planet in 1989, the small unexplained differences between the predicted and observed locations of Uranus simply disappeared. This means that there is no massive trans-Neptunian planet, and that all of the sizeable planets in our solar system have been discovered. Neptune is effectively the outermost major planet, serving as a lonely distant sentinel to outer space.

Pluto's atmosphere and surface

In 1998 Pluto passed in front of a star, revealing a thin, extended atmosphere that caused a gradual reduction in the star's light. If there were no atmosphere, the starlight would vanish abruptly. But Pluto does not have much of an atmosphere, for its surface pressure is about one-millionth that of Earth's.

The atmosphere's three main gases – nitrogen, methane and carbon monoxide – will partially freeze onto Pluto's surface during the 100-year long period when it is farthest from the Sun. When closest to the Sun, the gases will slowly escape into space because of Pluto's low gravity.

Spectroscopic evidence, first obtained in 1976, shows that methane frost covers much of the surface of Pluto. Nitrogen frost has also been detected on Pluto. A comparison of *Hubble Space Telescope* (*HST*) images of Pluto taken in 1994 with those taken in 2002–03 indicates that its surface is gradually changing with the seasons. *HST* images also show that large-scale dark and light markings are distributed around Pluto's globe (Fig. 15.2). They may be attributed to topographic features or to the distribution of frosts.

Pluto's companions

Pluto has an oversized companion that is half as big as Pluto is. This discovery was an accidental by-product of observations made for another purpose. In 1978, astronomers at the United States Naval Observatory were obtaining a series of photographs to improve the accuracy of Pluto's orbit, when several of the images appeared slightly distorted, from a round to oblong shape. The elongation seemed to disappear every few days, and careful examination showed that it is caused by another small world that orbits Pluto. The two objects are so close to each other and so far away that they remain blurred together when viewed with even the best telescopes on the ground, but they can be clearly resolved with the *Hubble Space Telescope* that orbits the Earth above its obscuring atmosphere (Fig. 15.3).

Pluto's companion is named Charon, after the boatman who ferried new arrivals across the river Styx at the entrance to Pluto's underworld, Hades. Penniless ghosts are said to have waited endlessly because Charon gave no free rides.

The announcement of this remarkable doubling was a happy surprise, for it permitted determining the mass of Pluto. Charon orbits Pluto at a distance of 19 640 kilometers once every 6.387 Earth days. For comparison, if our Moon were that close to Earth it would orbit in 7 hours.

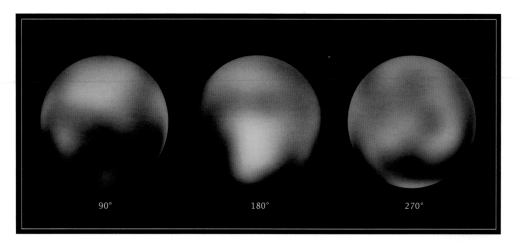

Fig. 15.2 Pluto's changing surface   Dark-orange and charcoal-black terrain is seen on three sides of Pluto. Ultraviolet radiation from the Sun is thought to break up methane that is on the dwarf planet's surface, leaving behind a dark carbon-rich residue. The center image, at 180 degrees, has a bright spot that is unusually rich in carbon-monoxide frost. These images were constructed from multiple *Hubble Space Telescope* (*HST*) observations in 2002–03. When compared with *HST* images taken in 1994, seasonal changes in color and brightness are detected, probably created when ices melt and refreeze and the tenuous atmosphere changes. Although these views are not sharp enough to resolve mountains or craters, they provide background for closer scrutiny when the *New Horizons* spacecraft encounters Pluto in 2015. (Courtesy of NASA/ESA/Marc Buie, SRI.)

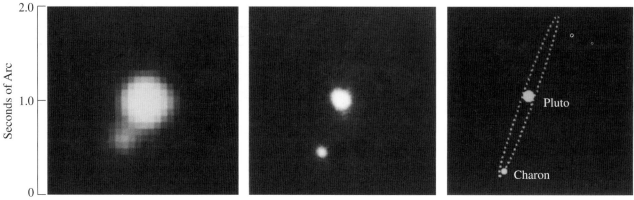
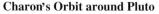

| Ground-based Telescope | Hubble Space Telescope | Charon's Orbit around Pluto |

Fig. 15.3 A double object   The *Hubble Space Telescope* (*HST*) distinguishes between Pluto, the bright object at the middle of the central image, and its companion Charon, which is the fainter object in the lower left of the central image. Observations from telescopes on Earth (*left*) were unable to clearly resolve the pair because of atmospheric distortions. At the time of the *HST* photograph, Charon's orbit around Pluto (*right*) was seen nearly edge-on, and Charon was near its maximum angular separation from Pluto, a mere 0.9 seconds of arc. Pluto's mean radius is 1153 kilometers, Charon is half that size, and the two objects are just 19 640 kilometers apart. The *HST*'s ability to distinguish Pluto's disk at its distance of 4.4 billion kilometers is equivalent to seeing a baseball at a distance of about 100 kilometers. (Courtesy of NASA.)

Charon's slow revolution about Pluto is a result of Pluto's small mass.

When Pluto and Charon pirouette into an edge-on view from Earth, we see a series of mutual eclipses as the two objects alternately pass directly in front of each other. Timing the starts and ends of such occultations permit an accurate determination of their size. Pluto has a mean radius of just 1153 kilometers, or about two-thirds that of the Earth's

Moon. Charon is nearly half the size of the Pluto, with a mean radius of 603.5 kilometers.

In mid-2005, almost three decades after Charon's discovery, a team of astronomers found two more moons of Pluto as the result of a dedicated search with the *HST* (Fig. 15.4). The new moons have been named Nix, for the mother of Charon and the goddess of darkness, and Hydra, for the nine-headed serpent of the underworld. The

Fig. 15.4 Pluto's three moons Pluto is the brightest object in this image, centered in the diffraction cross. Its large, nearby moon Charon is seen just below and to the right of Pluto. The other two moons, Nix and Hydra, which appear as small points of light to the right of Charon, are about 5000 times fainter than Pluto. The oversized moon Charon was discovered in 1978, while Nix and Hydra were discovered in mid-2005. Compared to Pluto and its large moon Charon, at 2360 and 1210 kilometers in diameter respectively, Nix (*inner moon*) and Hydra (*outer moon*) are estimated to be only 40 and 160 kilometers across. They are about two to three times farther from Pluto than Charon is. (Courtesy of NASA/ESA/Harold Weaver, JHUAPL, Alan Stern, SwRI, and the HST Pluto Companion Search Team.)

two moons are small, each about 80 kilometers across. They move in circular orbits in Pluto's equatorial plane, as Charon does, but at 48 700 and 65 000 kilometers, compared with the 19 640-kilometer separation between the centers of Pluto and Charon.

Perhaps it is no coincidence that the first letters of Nix and Hydra coincide with those of the spacecraft *New Horizons*, which will arrive at Pluto in 2015 and travel on to study other trans-Neptunian objects that most likely resemble Pluto (Focus 15.1).

15.2 Small cold worlds in the outer precincts of the planetary system

As it turned out, Pluto is only one of a host of small icy worlds revolving about the Sun beyond Neptune's orbit (Fig. 15.5). Perhaps a billion of them are comets, too small to be seen with any telescope. But an estimated 70 000 trans-Neptunian objects are over 100 kilometers in diameter and may be potentially observable. Many of them are located close to the orbital plane of the major planets and concentrated between 38 and 48 AU. Others are scattered into more-inclined orbits and extend to greater distances of up to 200 AU.

Focus 15.1   *New Horizons* mission to Pluto and the Kuiper belt

The *New Horizons* spacecraft was launched on 19 January 2006 toward an encounter with Pluto that is expected in July 2015. Since Pluto is about 9.5 billion kilometers away, *New Horizons* must travel at a speed of about 43 000 kilometers per hour to reach Pluto in 9.5 years, which means that the spacecraft will be traveling too fast to be inserted into orbit around Pluto. It will nevertheless move on to explore other parts of the Kuiper belt in which Pluto is located.

The spacecraft's instruments will study the atmosphere, surface and interior of Pluto. They include an ultraviolet spectrometer to measure gas ingredients, an infrared spectrometer to determine surface composition and structure, a radio instrument to study the composition and temperature of the atmosphere, an optical or visual light telescope for high-resolution images of the surface, and energetic particle and dust detectors.

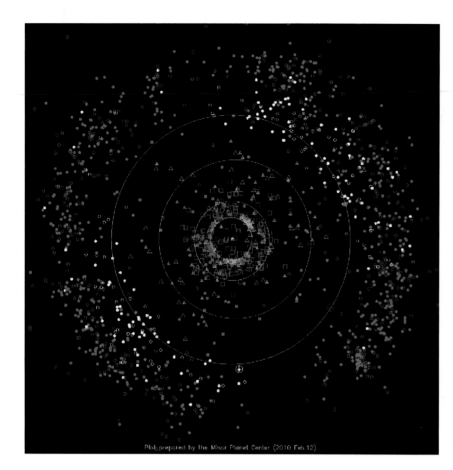

Plot prepared by the Minor Planet Center (2010 Feb 12)

Fig. 15.5 Outer fringes of our planetary realm The orbits of major planets in our solar system are shown in large light blue circles in this plot, prepared on 12 February 2010. The outermost circle denotes the orbit of Neptune and the current location of each planet is marked by a large dark-blue symbol. The current locations of the minor bodies of the outer solar system are shown in different colors to denote different classes of objects. Unusual high-eccentricity objects are shown as cyan triangles, centaur objects as orange triangles, plutoids, a subcategory of dwarf planets, as white circles, and Pluto itself as the large white symbol. Scattered-disk objects are denoted as small magenta circles, and "classical" or "main-belt" objects as small red circles. Objects observed at only one opposition are denoted by open symbols, objects with multiple-opposition orbits are denoted by filled symbols. Filled light-blue squares denote numbered periodic comets; other comets are shown as unfilled light-blue squares. (Courtesy of Gareth Williams, Minor Planet Center.)

The distant, flat ring just outside of Neptune's orbit, at 30 AU, is known as the Kuiper belt in recognition of Gerard P. Kuiper's (1905–1973) prediction of its existence in 1951. He argued that the dark outer edge of the planetary realm is not empty, but is instead full of small, unseen bodies created from the leftover debris of the formation of the giant planets. The low-density material in these distant regions would have been spread out into such a large volume, and moving in such slow, ponderous orbits around the Sun, that it could not gather or coalesce into a body much larger than Pluto.

Jane X. Luu (1963–) and David C. Jewitt (1958–) discovered the first trans-Neptunian, Kuiper-belt object, other than Pluto, on 30 August 1992 using the University of Hawaii's 2.2-meter (87-inch) reflector on Mauna Kea. An electronic detector attached to a large telescope was required to detect the meager amount of sunlight reflected back from such a small object at such great distances. Within a decade of the first discovery, hundreds of these objects were identified (Fig. 15.5). Every one of them is millions of times fainter than can be seen with the unaided eye.

The dynamic orbits of the scattered disk objects occasionally force them into the inner solar system, becoming first centaurs and then short-period comets. The centaurs are small bodies that orbit the Sun between Jupiter and Neptune, and cross the orbit of one or more giant planets (also see Fig. 15.5). The name of the mythological creatures, part human and part horse, was chosen because the centaurs behave as half asteroid and half comet.

The first object to be recognized as a member of this population was discovered by Charles Kowal (1940–) in 1977. It was designated as 2060 Chiron, the two-thousand-and-sixtieth numbered asteroid. Chiron has a diameter of about 200 kilometers, and moves between the orbits of Neptune and Saturn. But then it was discovered that Chiron has a dual personality, developing a luminous shroud of gas and dust like a comet. At least three of the dozens of currently known centaurs display detectable comet comas. Any centaur that is perturbed close enough to the Sun is expected become a comet.

The important thing is that an entirely new class of small worlds has been discovered orbiting the Sun in the outer fringes of the planetary disk, making Pluto much less unique than had been thought previously. Since the pioneering work of Luu and Jewitt, several large trans-Neptunian objects have been found (Figs. 15.6, 15.7). There is Eris, discovered in 2003 by Michael E. Brown (1965–) and colleagues, with a diameter of about

Fig. 15.6 Large trans-Neptunian objects Comparison of the disk sizes of the eight largest trans-Neptunian objects known in 2009. Five of them are known to have moons. The top four were officially designated as dwarf planets, and they have now been re-designated as plutoids. The other four are candidates for this classification. (Courtesy of Wikimedia Commons.)

Fig. 15.7 Trans-Neptunian object In this artist's visualization, a small, cold trans-Neptunian object dubbed Sedna is shown where it resides at the outer edges of our solar system, beyond the orbit of Neptune. The object is so far away that the Sun appears as an extremely bright star instead of the large, warm disk observed from Earth. A distant, hypothetical small moon is shown above the object and nearly in the direction of the Sun. (Courtesy of NASA/JPL.)

2400 kilometers. It is larger than Pluto and travels in an eccentric orbit that takes it up to 97 AU from the Sun and well beyond the classical Kuiper belt. Eris is named after the Greek goddess of strife, perhaps as the result of discord over the naming of the object.

Two years later Makemake was found, and named for a Polynesian fertility god. The object is roughly half the diameter of Pluto and orbits the Sun at an average distance of 46 AU. Another large trans-Neptunian object Haumea, named for the Hawaiian goddess of childbirth, has the oblong shape of an egg, and is significantly larger than Pluto in its longest dimension. Haumea has two small moons, and rotates end over end as it moves closer to the Sun than Pluto and then further away.

Since Eris is larger than Pluto, it was initially described as the "tenth planet", but a committee of the International Astronomical Union (IAU) demoted both Pluto and Eris to dwarf planets, different from the eight major ones. They defined a dwarf planet as a celestial body that is in orbit around the Sun and has sufficient self-gravity to overcome rigid body forces and become crushed into a rounded shape. All the major planets meet these criteria, but a dwarf planet has not cleared the neighborhood around its orbit as the major planets have. By this definition, Eris, Haumea, Makemake and Pluto, as well as the largest asteroid, 1 Ceres, are all dwarf planets.

Almost two years after the introduction of this category, the IAU decided to rename all trans-Neptunian dwarf planets as plutoids. So Ceres became the only known dwarf planet that was not also a plutoid, and many of us would prefer to keep on calling it an asteroid.

In any event, the trans-Neptunian objects mark the outer fringes of the planetary system, and this is about where the influence of the Sun's winds stops.

15.3   Edge of the solar system

The wide-open spaces between the planets, once thought to be a tranquil empty void, is swarming with hot, charged invisible pieces of the Sun. They expand and flow away from the Sun at supersonic speeds, faster than 300 kilometers per second, forming a perpetual solar wind. The solar wind is an exceedingly rarefied mixture of electrons and protons set free from the Sun's abundant hydrogen atoms.

The tenuous solar gale moves past the planets and surrounds them, carrying the Sun's magnetic fields and outer atmosphere into the space between the stars. It thereby creates a teardrop-shaped bubble in interstellar space, which is inflated by the solar wind, and threaded by open magnetic fields that have one end attached to the Sun (Fig. 15.8).

Focus 15.2   The heliosphere's outer boundary

The solar wind carves out a cavity in the interstellar medium known as the heliosphere. The radius of the heliosphere can be estimated by determining the stand-off distance, or stagnation point, in which the ram pressure, P_W, of the solar wind falls to a value comparable to the interstellar pressure, P_I. As the wind flows outward, its velocity remains nearly constant, while its density decreases as the inverse square of the distance. The dynamic pressure of the solar wind therefore also falls off as the square of the distance, and we can use the solar-wind properties at the Earth's distance of 1 AU to infer the pressure, P_{WS}, at the stagnation-point distance, R_S. Equating this to the interstellar pressure we have:

$$P_{WS} = P_{1AU} \times \left(\frac{1 AU}{R_S} \right)^2 = (N_{1AU} V_{1AU}^2) \times \left(\frac{1 AU}{R_S} \right)^2$$
$$= P_I$$

where the number density of the solar wind near the Earth is about $N_{1AU} = 5$ million particles per cubic meter and the velocity there is about $V_{1AU} = 500$ kilometers per second.

To determine the distance to the edge of the solar system, R_S, we also need to know the interstellar pressure, and that is the sum of the thermal pressure, the dynamic pressure, and the magnetic pressure in the local interstellar medium. Its estimated value results in $R_S = 100$ AU or more, well beyond the orbits of the major planets. However, the estimates by different authors give a broad range for the distance to the edge of the heliosphere, depending on the uncertain values of various components of the interstellar pressure, and the termination shock at the edge of the solar wind has now been measured from the *Voyager 1* and *2* spacecraft.

How far does the solar wind extend, and where does its influence end? Since the solar wind is weakened by expansion, thinning out as it moves into a greater volume, it eventually becomes too dispersed to repel interstellar forces. The winds are no longer dense or powerful enough to withstand the pressure of gas and magnetic fields coursing between the stars. The radius of this celestial standoff distance, in which the pressure of the solar wind falls to a value comparable to the interstellar pressure, has been estimated at about 100 AU, or one hundred times the mean distance between the Earth and the Sun (Focus 15.2).

Instruments aboard the twin *Voyager 1* and *2* spacecraft, launched in 1977 and now cruising far beyond the

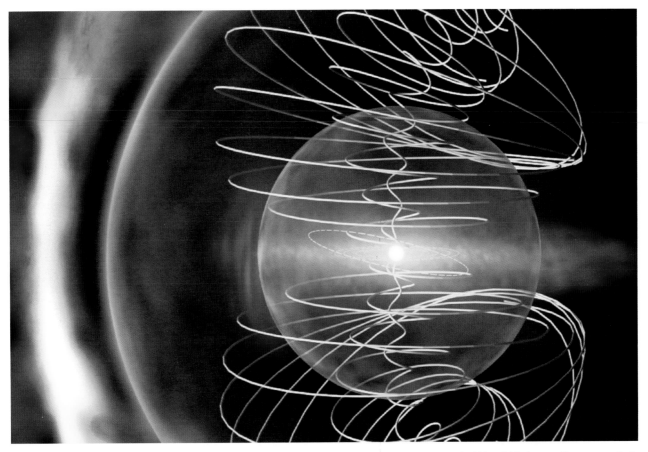

Fig. 15.8 The Sun's domain With its solar wind going out in all directions, the Sun blows a huge bubble within interstellar space called the heliosphere, with the Sun at its center and the planets inside. It is threaded by open magnetic fields that are connected to the Sun at one end, carried into space by the relentless solar wind on the other end, and twisted by the rotating Sun. Interstellar winds mold the heliosphere into a non-spherical teardrop shape, creating a bow shock (*left*) where interstellar forces encounter the solar wind. The heliosphere extends to about 100 times the distance between the Earth and the Sun. (Courtesy of Thomas H. Zurbuchen.)

outermost planets, have approached this edge of the solar system from different directions, *Voyager 1* moving in the northern hemisphere of the heliosphere and *Voyager 2* in the southern hemisphere. *Voyager 1* crossed the termination shock of the supersonic flow of the solar wind on 16 December 2004 at a distance at 94 AU from the Sun. At this distance, the spacecraft's instruments recorded a sudden increase in the strength of the magnetic field carried by the solar wind, as expected when the solar wind slows down and its particles pile up at the termination shock.

Voyager 2 crossed the termination shock on 30 August 2007 at a distance of 84 AU from the Sun. It appears that there is a significant north/south asymmetry in the heliosphere, likely due to the direction of the local interstellar magnetic field.

Both *Voyager 1* and *2* have therefore now crossed into the vast, turbulent heliosheath, the region where the interstellar gas and solar wind interact, due to the reflection and deflection of the solar-wind ions by the magnetized wind beyond the heliosheath. In technical terms, the solar-wind ions in the heliosheath are deflected by magnetosonic waves reflecting off of the heliopause, causing the ions to flow parallel to the termination shock toward the heliotail.

But the Sun's winds are not alone in the dark, cold outer fringes of the solar system. About a million million (10^{12}) unseen comets have been hibernating out there in the Oort comet cloud ever since the solar system formed. Modern calculations suggest that there is an inner Oort cloud with an inner edge at around 3000 AU and a density falling off with greater distance. The outer Oort cloud is continuous with this, but is defined to be those objects at distances greater than 20 000 AU and less than about 200 000 AU. So far, the two *Voyager* spacecraft have traveled nowhere near the more remote comet reservoir. Moreover, the comets are so small and widely spaced that it is exceedingly unlikely that the *Voyager 1* or *2* spacecraft will ever encounter even one of them, just as the *Voyagers* passed through the asteroid belt unaffected by the billions of asteroids there.

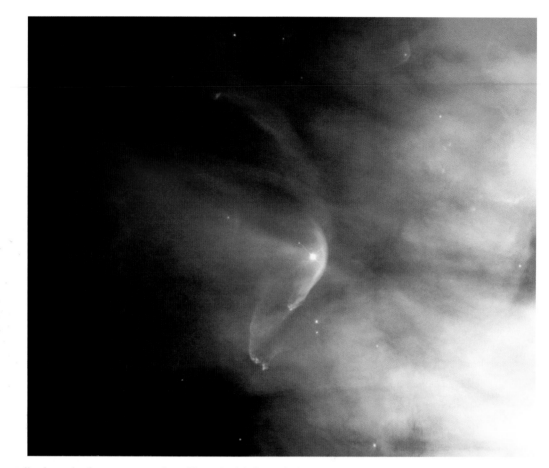

Fig. 15.9 Stellar bow shock   A crescent-shaped bow shock is formed when the material in the fast wind from the bright, very young star, LL Ori (*center*) collides with the slow-moving gas in its vicinity, coming from the lower right. The stellar wind is a stream of charged particles moving rapidly outward from the star. It is a less energetic version of the solar wind that flows from the Sun. A second, fainter bow shock can be seen around a star near the upper right-hand corner of this image, taken from the *Hubble Space Telescope*. Both stars are located in the Orion Nebula; a bright star-forming region located about 1500 light-years from the Earth. (Courtesy of NASA, the Hubble Heritage Team, STScI, and AURA.)

Both *Voyager* spacecraft are equipped with plutonium power sources expected to last until at least 2020 and perhaps 2025. So they ought to eventually measure the heliopause, at the outer edge of the heliosheath. It is the place where interstellar space begins.

In the meantime, the *Interstellar Boundary EXplorer* (*IBEX*) was launched on 19 October 2009. Instruments on this spacecraft, which operates in Earth orbit, detect neutral, or un-ionized, atoms coming from the termination shock and the boundary between the solar wind and interstellar space.

The motion of the interstellar gas, with its own wind, compresses the heliosphere on one side, producing a teardrop-like, non-spherical shape with an extended tail. A bow shock is formed when the interstellar wind first encounters the heliosphere; just as a bow shock is created when the solar wind strikes the Earth's magnetosphere. The graceful arc of a bow shock, created by an interstellar wind, has been detected around the young star LL Orionis (Fig. 15.9), and this brings us to the captivating recent discoveries of planetary systems around nearby stars other than the Sun.

16   **Brave new worlds**

- According to the nebular hypothesis, the Sun and planets formed out of a single collapsing, rotating cloud of interstellar gas and dust called the solar nebula. This hypothesis provides a natural explanation for the highly regular pattern of the planet and satellite orbits.

- Conservation of angular momentum in gravitational collapse suggests that the Sun initially rotated much more rapidly than it does now.

- Spiral nebulae were once thought to be young stars enveloped by nascent planetary systems, but they are now known to be distant galaxies, each containing roughly 100 billion stars.

- The youngest stars in our Milky Way Galaxy are surrounded by dusty planet-forming disks, initially discovered by their infrared radiation, and detected in large numbers and great detail by the *Hubble Space Telescope* and the *Spitzer Space Telescope*.

- Vast interstellar clouds of gas and dust are the incubators of large numbers of newborn stars, many of them embedded in the material from which they arose and surrounded by flattened, rotating planet-forming disks.

- At least two planets with a mass comparable to that of the Earth were discovered orbiting a cold, dark pulsar.

- The first unseen planets circling ordinary Sun-like stars were inferred from the tiny, periodic Doppler wavelength shifts of their parent star's spectral lines, caused by the motion of the orbiting planet. They became known as "hot Jupiters", since they revolve unexpectedly close to their star and have masses comparable to that of Jupiter.

- Hundreds of planets have been found circling nearby stars, including many multi-planet systems. These newfound planets are known as extrasolar planets, or exoplanets for short.

- Most of the exoplanets have been indirectly inferred from the miniscule, periodic velocity changes they create in their parent star, but many have also been inferred from the brief, periodic dimming of starlight when they pass in front of, or transit, their star as viewed from Earth.

- The orbital motions of a few exoplanets have been confirmed by direct observation of the moving planets.

- The *Kepler* mission is searching for Earth-size planets that reside in the habitable zone.

- The *COROT* spacecraft and the world's best ground-based telescopes have detected several super-Earths with a mass between those of the Earth and Jupiter.

- The atmospheres of transiting exoplanets are being investigated using the *Hubble Space Telescope*, the *Spitzer Space Telescope* and ground-based telescopes.

- Water vapor, carbon dioxide and methane have been detected in the atmosphere of one hot Jupiter, named HD 189733b.

- Velocity observations with ground-based telescopes have resulted in the discovery of six planets orbiting a nearby Sun-like star, Gliese 581. One of the exoplanets, designated GL 581g, is located in the potentially habitable zone where liquid water could exist on its surface, and has the right mass, of about three times that of the Earth, to hold an atmosphere.

- There is most likely a very large number of habitable, Earth-size planets in the Milky Way galaxy.

16.1 How the solar system came into being

Any successful theory for the origin of the planets and satellites in our solar system must account for the regular arrangement of their orbits. The planets all move in a narrow band across the sky, the zodiac, which implies that the orbits lie nearly in a plane, and they all orbit the Sun in the same direction that our star rotates. These orbital paths are nearly circular, and the equator of the Sun's rotation nearly coincides with the plane of the planetary orbits. The orbits of most of the satellites imitate the planets in being confined to the planet's equatorial plane and revolving about their planet in the same direction that it orbits the Sun.

This regular orbital arrangement of the planets and satellites is not accidental. Even if a million million million (10^{18}) solar systems were made haphazardly and the planets and moons were thrown into randomly oriented orbits, only one of these solar systems would be expected to look like our own. So it is exceedingly unlikely that the planets became aligned by chance.

Although gravity and motion describe the present behavior of the solar system, they cannot explain the remarkable arrangement. Some additional constraints are required, which describe the state of affairs before the planets were formed and set in motion. These initial conditions are provided by the nebular hypothesis, in which the Sun and planets formed out of a single collapsing, rotating cloud of interstellar gas and dust, called the solar nebula. This hypothesis provides a natural explanation for the highly regular pattern of the planet and satellite orbits.

The German philosopher Immanuel Kant (1724–1804) proposed the nebular hypothesis in his book *Allgemeine Naturgeschichte und Theorie des Himmels*, or *Universal Natural History and the Theory of the Heavens*, published in 1755. He pictured an early Universe filled with thin gas that collected into dense, rotating gaseous clumps. One of these primordial concentrations was the spinning solar nebula. Attracted by its own gravity, the nebula fell in on itself, getting denser and denser, until the middle became so packed, so tight and hot, that the Sun began to shine. Meanwhile, the rotation spun the surrounding material out into a flattened disk revolving about the central Sun. The planets formed from swirling condensations in this circumstellar material, which explains qualitatively why all the planets revolve in the plane that coincides with the equator of the rotating Sun (Fig. 16.1).

The French astronomer and writer Pierre Simon Marquis de Laplace (1749–1827) popularized this nebular hypothesis for the origin of the solar system in 1796, extending it to the formation of rings and moons around the planets. According to Laplace's modification, the shrinking Sun shed a succession of gaseous rings, and each ring condensed into a planet. Then each planet, in turn, became a small rotating nebula in which its own family of rings and satellites was born.

Modern versions of the nebular hypothesis provide additional caveats, but the basic tenants of the original idea are still valid. Billions of years ago, a dense interstellar cloud, the spinning solar nebula, collapsed until the Sun began to shine at its center. The planets formed at the same time, within a flattened rotating disk centered on the contracting proto-Sun.

This is the essence of the original nebular hypothesis, which explained qualitatively the fact that the planets and their moons all revolve in the plane that coincides with the equator of the rotating Sun. The highly regular pattern,

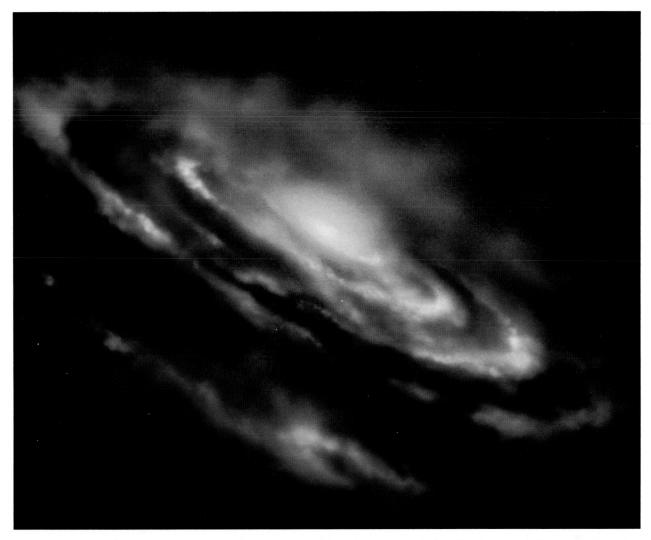

Fig. 16.1 Formation of the solar system An artist's impression of the nebular hypothesis, in which the Sun and planets were formed at the same time during the collapse of a rotating interstellar cloud of gas and dust that is called the solar nebula. The center collapsed to ignite the nuclear fires of the nascent Sun, while the surrounding material was whirled into a spinning disk where the planets coalesced. (Courtesy of Helmut K. Wimmer, Hayden Planetarium, American Museum of Natural History.)

which cannot be accidental, is a natural consequence of the rotation and collapse of a solar nebula composed of gas and dust from which the planets were produced.

There are a few details that need to be explained if the nebular hypothesis is correct, such as the present distribution of mass and angular momentum in the solar system. According to the law of conservation of angular momentum, the rotation of a shrinking object will speed up as the radius decreases, so the young Sun must have been spinning much more rapidly than it does now (Focus 16.1).

Also, if the nebular hypothesis is correct, and the whole solar system originated at the same time, then you might expect the planets to have a similar chemical composition to the Sun. The abundance of the elements in the giant planet Jupiter does indeed mimic that of the Sun, with a predominance of the lightest element hydrogen. Unlike the Sun, the Earth is mainly composed of heavier elements,

perhaps because the volatile gases near the young Sun were driven away by its powerful winds. In the inner regions of the solar nebula, the higher temperatures would also vaporize icy material that could not condense, leaving only rocky substances to coalesce and merge together to form the terrestrial planets. The modest masses of the terrestrial planets and their proximity to the Sun did not allow them to capture and retain the abundant lighter gases, hydrogen and helium, directly from the solar nebula.

At larger distances, where the solar nebula was cooler, icy substances could condense and combine with heavier ones to form the large, massive cores of the giant planets. These cores eventually became sufficiently massive to gravitationally accrete, accumulate and capture the surrounding hydrogen and helium. The low temperatures at remote distances from the Sun thus enabled the giant planets to retain the abundant light gases and grow even

Focus 16.1   Conservation of angular momentum in the early solar system

During the gravitational collapse of the rotating interstellar cloud, called the solar nebula, the angular momentum is conserved. For a body of mass M, radius R and rotation period P, this means that

$$\text{Angular momentum} = \frac{2\pi\, M\, R^2}{P} = MVR = \text{constant}$$

where the rotation velocity V is

$$V = \frac{2\pi\, R}{P}$$

Since the Sun contains 99.87 percent of the mass of the solar system, we can assume that the mass remains constant during the collapse of the solar nebula to form the Sun. We might propose, for argument's sake, that the solar nebula had an initial radius R_{sn}, located at about the current edge of the solar system, where interstellar forces are comparable to those within the solar system. It is located at a radius of about 100 AU, or at about 1.5×10^{13} meters. Assuming that the initial rotation velocity of the solar nebula was comparable to the velocities of interstellar matter, or that $V_{sn} \approx 100$ kilometers per second, the initial rotation period P_{sn} would be $P_{sn} = 2\pi R_{sn}/V_{sn} \approx 1.9 \times 10^9$ seconds. Since angular momentum is conserved during collapse, the rotation period of the solar nebula when it had collapsed to the present size of the Sun, with radius $R_\odot = 6.96 \times 10^8$ meters, is

estimated to be

$$\text{Rotation period of the young sun} = P_{sn}\left(\frac{R}{R_{sn}}\right)^2$$
$$= 4.1 \text{ seconds}$$

And even if the initial solar nebula was rotating at a relatively slow speed of just 1 kilometer per second, the young Sun would be rotating with a period of just 4.1×10^4 seconds or 11.4 hours, comparable to Jupiter's rotation period of 9.9 hours.

However, the Sun now has an equatorial rotation period of 25.7 days, or 2.22×10^6 seconds, many times longer than expected. Some process other than gravitational contraction must have slowed the Sun's spin after its birth. One possibility is the action of magnetic fields that could connect the Sun to the distant, slowly rotating material in the surrounding disk and act like brakes to slow the solar rotation. Another possibility is that frictional forces caused mass to move inwards from the disk to the central Sun while transporting angular momentum outward. Whatever the exact explanation, it seems to apply to other newborn stars, for it is the youngest stars that rotate at the fastest speeds, while older ones spin at a slower rate.

To put the problem in another way, most of the mass of the solar system is in the central Sun, while there is about 10 times as much orbital angular momentum in Jupiter as there is rotational angular momentum in the spinning Sun. So a very small fraction of the mass of the solar system has significant angular momentum, while most of the mass has relatively little angular momentum.

---

bigger, with large masses and low mass densities. But this scenario does not seem to apply to giant planets recently discovered circling nearby stars in close, hot orbits.

16.2   Newborn stars with planet-forming disks

Twentieth-century astronomers have long been on the lookout for planetary systems around stars other than the Sun, and this has led to some happy surprises. It was once thought, for example that the spiral nebulae (Fig. 16.2) represented an early stage in the evolution of stars, with planets forming around their bright centers. Observations of these objects, with the hope of understanding planetary formation, led to the unanticipated discovery of their enormous velocities and eventually to the discovery of the expanding Universe (Focus 16.2).

It wasn't until 1983 that astronomers used instruments aboard the *InfraRed Astronomical Satellite* (*IRAS*) to unexpectedly obtain the first evidence for planet-forming disks. Using technology pioneered by the military to detect the infrared heat of the enemy, the satellite was designed to detect cosmic infrared radiation, which is mainly inaccessible from the ground owing to absorption in the atmosphere. Because of their low temperature, dust particles emit most of their radiation at infrared wavelengths, while radiating no detectable visible light. It is the other way around for the hot stars, which shine brightly in visible light and emit relatively little infrared.

The *IRAS* instruments found the bright infrared glow of dusty clouds, disks and rings circling bright, massive stars such as the brilliant blue-white Vega, as well as less-massive, solar-type stars. In fact, the youngest nearby stars are usually found embedded in the dense clouds of interstellar gas and dust that spawned them.

Focus 16.2 Spiral nebulae and the discovery of the expanding Universe

In the early 20th century, Vesto Slipher (1875–1969) unexpectedly helped us move beyond the stars into the expanding Universe. Working at the Lowell Observatory in Flagstaff, Arizona, he was measuring the rotations of spiral nebulae, whose bright centers were thought to be newborn stars – the surrounding spiral arms had been interpreted as nascent planetary systems. Using a spectrograph camera with a modest 0.61-meter (24-inch) refractor telescope, he found, in 1917, that the outward velocities of 25 spiral nebulae were well in excess of the velocity of any known cosmic object. Almost all of the spiral nebulae were moving away from the Earth, at astonishingly high velocities, up to 1100 kilometers per second. This suggested to Slipher that the spiral nebulae were stellar systems, or "island universes", seen at great distance rather than nearby stars attended by planet formation.

By 1929 Edwin Hubble (1889–1953) showed that the measured distances, established by him using the superb light-gathering power of the 2.5-meter (100-inch) Hooker telescope on Mount Wilson, were roughly correlated with Slipher's velocities. This relationship is now attributed to the expanding Universe, which no one had anticipated at the time Slipher made his measurements. The spiral nebulae are now known as spiral galaxies, each containing roughly 100 billion stars.

Fig. 16.2 The Andromeda nebula At one time, spiral nebulae like Andromeda were thought to be young stars enveloped by protoplanetary material. We now know that the Andromeda nebula, also known as M31, is the nearest large galaxy to our own, located at a distance of about 2.6 million light-years. Both the Andromeda nebula and our Galaxy are spiral galaxies with similar sizes and total masses, containing roughly 100 billion (10^{11}) visible stars. This photograph was taken in visible light with the 5.0-meter (200-inch) telescope at the Palomar Observatory near Pasadena, California. The two smaller elliptical galaxies are at about the same distance as M31, but with only about one-hundredth of its mass. (Courtesy of the Palomar Observatory.)

The *Spitzer Space Telescope* has recently used its powerful infrared vision to detect hundreds of stars with excess infrared radiation, suggesting that they harbor planet-forming disks. The closest disk system to our own, surrounding the star Epsilon Eridani, contains two infrared-emitting belts, one at approximately the same position as the asteroid belt in our solar system, and the second, denser belt between the first one and a more remote ring of icy comets similar to our own Kuiper belt.

The *Hubble Space Telescope* (*HST*) has discovered flattened disks of dust swirling around at least half the young stars in the Orion nebula, shining in reflected visible light. The high resolution and sensitivity of the *HST* have also been used to obtain detailed images of dusty, planet-forming disks surrounding other Sun-like stars, providing insights to the beginnings of our solar system (Fig. 16.3). The flattened, rotating disks encircling other star suggests that the nebular hypothesis applies to them. This material is expected to coalesce into full-blown planets if it hasn't already done so.

16.3 The plurality of worlds

The ancient Greeks imagined that all matter consists of tiny moving particles, both indivisible and invisible, which they called atoms, and that all material things can be created by the coming together of a sufficient number of atoms. The Greek philosopher Epicurus of Samos (276–194 BC) proposed that the chance conglomerations of innumerable atoms, in an infinite Universe, should result in the formation of a multitude of unseen Earth-like worlds.

The Roman poet Lucretius (99–55 BC) also wrote about the plurality of worlds within a Universe without end, declaring that seeds innumerable in number are rushing on countless courses through an unfathomable Universe, making it highly unlikely that our Earth is the only one to have been created and that all those particles outside are accomplishing nothing.

Astronomers eventually showed that the planets in our solar system are revolving around the Sun, and that the Sun is itself one of innumerable stars. These discoveries opened up the possibility that there might be planets orbiting other stars, some possibly inhabited. Such a

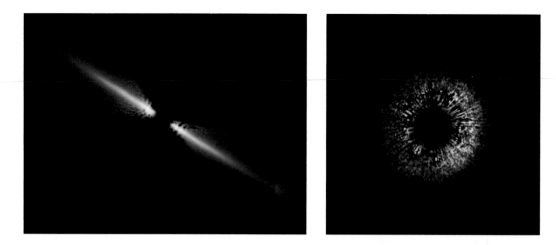

Fig. 16.3 Dusty disks around Sun-like stars   Instruments aboard the *Hubble Space Telescope* have obtained these images of the visible starlight reflected from thick disks of dust around two young stars that might still be in the process of forming planets. Viewed nearly face on, the debris disk surrounding the Sun-like star known as HD 107146 (*right*) has an empty center large enough to contain the orbits of the planets in our solar system. Seen edge-on, the dust disk around the reddish dwarf star known as AU Microscopii (*left*) has a similar cleared-out space in the middle. HD 107146 is 88 light-years away, and is thought to be between 50 million and 250 million years old, while AU Microscopii is located 32 light-years away and is estimated to be just 12 million years old. [Courtesy of NASA, ESA, STScI, JPL, David Ardila – JHU (*right*), and John Krist – STScI/JPL (*left*).]

belief dates back at least as far as the Italian philosopher and priest Giordano Bruno (1548–1600), who reasoned in 1584 that these other planets should be orbiting bright stars, and would remain invisible to us because they are much smaller and non-luminous.

Bruno spent the last eight years of his life in the prisons of the Inquisition. He was eventually tried by the Catholic Church, bound to a stake, and burned alive in Rome in 1600, perhaps more for his heretical religious views, such as his doubts about the Immaculate Conception, the Holy Trinity, and Christ's divinity, than for his belief in an infinite Universe filled with countless habitable planets circling other stars.

The invention of the telescope, and the construction of increasingly large ones during ensuing centuries, have enabled astronomers to detect signs of some of these innumerable worlds that were once only imagined. They have caught the light of vast, interstellar clouds of gas and dust, the future incubators of newborn stars (Fig. 16.4).

Some of the giant clouds, with a mass of about a million Suns, are even now in the process of creating stars, falling in on themselves due to the mutual gravitation of their parts. Once this gravitational collapse is underway, the giant cloud fragments into smaller components that eventually collapse to become stars like the Sun.

As the force of gravity pulls the local concentrations of material together, the core of each cloud fragment becomes more compressed and the temperature rises. After about a million years, the central core becomes hot enough to ignite nuclear fusion, and make a star. The most massive stars turn on in a shorter time and shine with greater brightness, lighting up the surrounding material and forming colorful regions within the dark clouds.

A rotating disk is created around a newborn star, the future home of planetary systems. Observations with the *Hubble Space Telescope* and the *Spitzer Space Telescope* indicate that the youngest nearby stars are often embedded in the gas and dust that spawned them, encircled by flattened, rotating planet-forming disks. Instruments aboard the *HST* have discovered flattened disks swirling around at least half the young stars in the Orion Nebula (Fig. 16.5).

But not every interstellar cloud is now in the process of stellar formation. For the most part, the gas is too tenuous and agitated to collapse under its own weight. These dark, stable interstellar clouds contain roughly as much material as is found in visible stars, so there is still plenty of material around to create new stars for billions of years in the future, whenever outside forces might trigger their collapse.

Individual planets shine by reflecting light that is much fainter than the light of the star that illuminates them. The visible light reflected by Jupiter is, for example, about a billion times dimmer than the light emitted by the Sun, and that reflected by Earth is 10 billion times fainter still. As a result, distant planets are almost always too small and too faint to be seen directly in the bright glare of their nearby star. Their presence has only recently been inferred from their miniscule gravitational effects on the motions of the

Fig. 16.4 Mountains of creation   The infrared heat radiation of hundreds of embryonic stars (*white/yellow*) and windblown, star-forming clouds (*red*), detected from the *Spitzer Space Telescope*. The intense radiation and winds of a nearby massive star, located just above the image frame, probably triggered the star formation and sculpted the cool gas and dust into towering pillars. (Courtesy of NASA, JPL-Caltech, Harvard-Smithsonian CfA, ESA, and STScI.)

star they revolve around, or when they chance to pass in front of a star, momentarily blocking the star's light when viewed from Earth. Such extrasolar planets, which orbit around stars other than the Sun, are called exoplanets for short.

So the hunt is on, and the prize will be an Earth-size planet with an atmosphere, orbiting a Sun-like star at just the right distance to retain liquid water on its surface. This suggests that we might find companionship in the vast and lonely Universe (Fig. 16.6).

16.4 The first discoveries of exoplanets

Pulsar planets

The first planets to be found outside the solar system were unexpectedly discovered in 1992 by two American radio astronomers, Aleksander Wolszczan (1946–) and Dale A. Frail (1961–). Wolszczan wasn't looking for planets. He was using the giant radio telescope at Arecibo, Puerto Rico, to search for pulsars that spin very rapidly, at the rate of several hundred times a second. Since the telescope is always in high demand, the search only became possible when it was shut down for repairs and pointing in just one direction in the sky. As luck would have it, a previously unknown, fast-spinning pulsar, designated PSR 1257+12 for its position in the sky, happened to pass through the immobile antenna beam. As with other pulsars, it is a tiny, superdense neutron star that emits precisely periodic radio radiation as it rotates.

Once the telescope was repaired, Wolszczan used large computers and comparisons to atomic clocks to measure the arrival of millions and then billions of the rapid, uniform pulses, obtaining a very accurate repetition period.

Fig. 16.5 The great nebula in Orion   This nebulosity, the brightest in the sky and designated M 42 or NGC 1976, forms the middle star of Orion's sword. It is 1500 light-years away, a relatively short distance compared with the 100 000 light-year width of our Milky Way Galaxy. Gravity has already pulled some of the interstellar gas into dense concentrations, igniting the celestial fires of newborn stars. The massive stars in the center of the nebula have blown out most of the gas and dust in which they formed, providing a clearer view of other young stars, some of them still embedded in protoplanetary disks in which future planetary systems can form. Hot, massive stars also ionize the nearby interstellar gas, causing this debris of other long-dead stars to fluoresce with red light. More than 3000 stars of various sizes appear in this crisp, detailed image that was taken from the *Hubble Space Telescope* in 2005. (Courtesy of NASA/ESA/Massimo Robberto, STScI and the Hubble Space Telescope Orion Treasury Project Team.)

The best results indicated that the pulsar was spinning once every 0.0062 seconds, or 6.2 milliseconds, rotating on it axis 161 times every second. But there seemed to be something wrong with the data, for the repeating pulses did not match a single, well-defined periodicity, and the computer could not predict exactly when each pulse would arrive at Earth.

The position of the pulsar could be slightly in error, causing a mismatch in the computer analysis of the pulsar timing, which corrects for the motion of the Earth towards

So astronomers were disappointed. They had hoped to discover planets around a perfectly ordinary star like the Sun, whose steady light and heat would at least be compatible with the notion of extraterrestrial life.

Discovery of an unseen planet circling an ordinary star

If any planet in our solar system were placed in orbit around any other star, it would vanish from sight, lost in the star's glare. The presence of the unseen planet has to be instead deduced by indirect means, by recording the way its gravity pulls at the star it orbits. The more massive the planet, and the closer it is to the star, the stronger the planet's gravitational pull on the star and the more the planet perturbs it.

Like two linked, rotating dancers, the planet and star tumble through space, pulling each other in circles. They both orbit a common center of mass where their gravitational forces are equal, somewhat like the equilibrium point of a seesaw, where the forces of two people balance. This fulcrum is closest to the heavier person, or to the massive star in the stellar case. So the star moves in a much smaller circle, a miniature version of the planet's larger path.

To detect this tumbling motion, astronomers had to look for the subtle compressing and stretching of starlight as the unseen planet tugged on the star, pulling it first toward the Earth and then away, causing a periodic shift of the stellar radiation to shorter and then longer wavelengths (Fig. 16.7).

The effect is entirely analogous to the well-known Doppler effect, discovered by the Austrian physicist Christian Doppler (1803–1853), who in 1842 described how sound depends on the relative motion of the source and listener. If the source is moving toward us, the motion compresses the sound waves, pushing them at us and shortening their wavelength. Sounds emitted at a given wavelength then arrive at a shorter wavelength than that emitted by a stationary source; more crests strike the ear each second and a higher frequency or pitch is heard. The sound waves of a receding source are pulled away from us and stretched out to longer wavelengths, with a lower pitch than would be emitted by a non-moving source (see also Section 6.4 and Focus 6.1).

Just as a source of sound can vary in pitch or wavelength, depending on its motion, the wavelength of electromagnetic radiation shifts when the emitting source moves with respect to the observer. If the motion is toward the observer, the Doppler shift is to shorter wavelengths, and when the motion is away the wavelength becomes longer. The amount of the wavelength shift, $\Delta\lambda$, at wavelength λ

Fig. 16.6 Astrologers of life   Two silhouetted figures search the heavens in this 1947 painting by Rufino Tamayo (1889–1991). It may represent our modern attempts to understand the Universe and our search for habitable planets within it. A comet and full Moon illuminate the azure blue sky. The geometric diagrams in the foreground could portray stellar or planetary configurations. A red radio tower in the background sends out signals, perhaps to civilizations on other worlds. (Courtesy of Sotheby Parke Bernet Inc., New York, 1985.)

any point in space. So Dale Frail used the Very Large Array in New Mexico to obtain an accurate location. With the new position, the timing data made sense. The pulse arrival times were being affected by at least two unseen planets.

It was a momentous occasion, for the first planets had been found outside our solar system. But the response of the astronomical community was at best lukewarm, for the planets are orbiting the wrong kind of star. Without any thermonuclear fuel to make it shine, the pulsar PSR 1257+12 is a cold, dark star, emitting no light to warm the newfound planets. In addition, the intense, spinning magnetic fields of the pulsar must accelerate and send lethal high-energy particles and radiation to the planets.

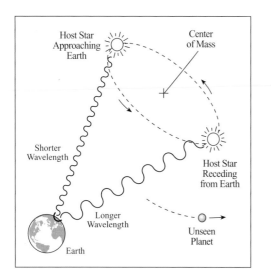

Fig. 16.7 Starlight shift reveals invisible planet   An unseen planet exerts a gravitational force on its visible host star. This force tugs the star in a circular or oval path, which mirrors in miniature the planet's orbit. As the star moves through space, it approaches and recedes from Earth, changing the wavelength of the starlight seen from Earth through the Doppler effect. When the planet pulls the star toward us, its light waves pile up in front of it slightly, shortening or "blueshifting" the wavelength we detect. When the planet pulls the star away from us, we detect light waves that are stretched or redshifted. During successive planet orbits, the star's spectral lines are periodically shortened and lengthened, revealing the presence of the planet orbiting the star, even though we cannot see the planet directly.

can be used to infer the velocity of motion along the line of sight, known as the radial velocity V_r, by the equation

$$\frac{\Delta \lambda}{\lambda} = \frac{V_r}{c},$$

where $c = 2.9979 \times 10^8$ meters per second is the velocity of light.

But an orbiting planet produces an exceedingly small variation in the wavelength of spectral lines emitted from its star. Massive Jupiter, for example, makes the Sun wobble at a speed of only about 12 meters per second. To detect the Doppler effect of a star moving with this speed, astronomers would have to measure the wavelengths with an unheard-of accuracy of at least one part in 30 million.

So the effect could only be detected once very sensitive spectrographs were constructed to precisely spread out the light rays. The enhanced light-collecting powers of electronic charge-coupled detectors were also needed to record the dispersed starlight. And since no single line shift is significant enough to be seen, computer software had to be written to add up all the star's spectral lines, which shift together, combining them over and over again

at all possible regularities, or orbital periods, with continued comparison to non-moving laboratory spectral lines.

It took decades for astronomers to develop these complex and precise instruments. Then, in the 1990s, the time was ripe, and two Swiss astronomers from the Geneva Observatory in Switzerland, Michel Mayor (1947–) and Didier Queloz (1966–), accomplished the seemingly impossible. In April 1994 they attached a new, exquisitely precise, computerized spectrograph to the 1.93-meter (76-inch) telescope at the Observatoire de Haute-Provence in the south of France, and within a year and a half they had found the first planet that orbits an ordinary star, the faintly visible, Sun-like star 51 Pegasi, only 50.9 light-years away from Earth in the constellation Pegasus, the Winged Horse.

Hints of the planet were found a year before the announcement of its existence, but the Swiss team had to be very careful. There had been many false planetary discoveries before, and they seemed to have found a giant planet with an unexpectedly short orbital period of just 4.23 days. By way of comparison, Jupiter orbits the Sun once every 11.86 years. Observations were stopped in March 1995 because the star moved too close to the Sun to be seen, and renewed in the first week of July when the two astronomers returned with their families. Armed with a precise prediction of what the spectrograph would show if the unseen planet really existed, they pointed the telescope at 51 Pegasi, and saw exactly what they had hoped for. As Mayor described it, the occasion happened like a dream, a spiritual moment.

Mayor and Queloz submitted a discovery paper to *Nature* magazine the following month, and announced it at a professional meeting in Firenze, Italy, on 6 October 1995. They had detected the back-and-forth Doppler shift of the star's light with a regular 4.23-day period, measured by a periodic change of the star's radial velocity of up to 50 meters per second (Fig. 16.8). To produce such a quick and relatively pronounced wobble, the newfound planet had to be large, with a mass comparable to that of Jupiter, which is 318 times heftier than Earth, and moving in a tight, close orbit around 51 Pegasi (Focus 16.3).

Planets that are closer to a star move around it with greater speed and take less time to complete an orbit, all in accordance with Kepler's third law. Thus, the Earth takes a year or 365 days to travel once around the Sun at a mean distance of one astronomical unit (1 AU), while Mercury, the closest planet to the Sun, orbits our star with a period of 88 days at 0.387 AU. A short orbital period of only 4.23 days meant that the newfound planet is located at a distance of just 0.05 AU from its parent star, or about one-eighth the distance between Mercury and the Sun. Thus, a completely unanticipated planet had been found, rivaling

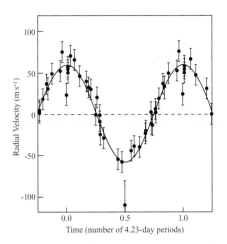

Fig. 16.8 Unseen planet orbits the star 51 Pegasi Discovery data for the first planet found orbiting a normal star other than the Sun. The giant, unseen planet is a revolving around the solar-type star 51 Pegasi, located 50.9 light-years away. The radial velocity of the star, in units of meters per second, has been measured from the Doppler shift of the star's spectral lines. The velocity exhibits a sinusoidal variation with a 4.23-day period, caused by the invisible planetary companion that orbits 51 Pegasi with this period. The observational data (*solid dots*) are fit with the solid line, whose amplitude implies that the mass of the companion is roughly 0.46 times the mass of Jupiter. The 4.23-day period indicates that the unseen planet is orbiting 51 Pegasi at a distance of 0.05 AU, where 1.00 AU is the mean distance between the Earth and the Sun. [Adapted from Michael Mayor and Didier Queloz, "A Jupiter-mass companion to a solar-type star", *Nature* **378**, 355–359 (1995).]

Jupiter in size and revolving around 51 Pegasi in an orbit smaller than Mercury. No one expected a giant planet to be revolving so close to its star.

Less than two weeks after the announcement of a giant planet circling 51 Pegasi, two American astronomers Geoffrey W. Marcy (1955–) and R. Paul Butler (1962–) confirmed the result using the 3-meter (120-inch) telescope at Lick Observatory near Santa Cruz, California. On 17 October 1995, Marcy and Butler issued a press release containing the confirmation, which hit the front-page headlines of newspapers throughout the world. And now that they knew that giant planets could revolve unexpectedly near a star, with short orbital periods, they used powerful computers to re-examine their observations of other nearby stars accumulated during previous years, announcing in January 1996 the discovery of two more Jupiter-sized companions of Sun-like stars.

These were astounding discoveries. In just a few months, astronomers had detected the first planets circling ordinary stars just like our Sun. Other worlds were no longer limited to philosophical musings, scientific speculations, or artists' imaginations. After two millennia, a long dream has come true. We can now look up at the night sky and say that there are definitely invisible planets out there, orbiting perfectly ordinary stars that are now shining brightly in the night sky.

16.5 Hundreds of new worlds circling nearby stars

Scientists had spent decades looking for giant planets far from their central star, only to find that they are easy to detect once you look close in. After scientists realized that a large planet could be so near to its star, they knew where and how to look. And by monitoring thousands of nearby Sun-like stars for years, American and European teams have found more than 400 planets revolving about other nearby stars, most of them massive, Jupiter-size planets.

Some of the newfound worlds travel in circular orbits, like those in the solar system, but much closer to their parent stars than Mercury is to the Sun. Dubbed "hot Jupiters" because of their size and proximity to the intense stellar heat, they are much too hot for life to survive or water to exist. Their temperatures can soar to more than 1000 kelvin, far hotter than the surface of any planet in our solar system. Other newfound planets follow eccentric, oval-shaped orbits that deviate from a circular path, so they venture both near and far from their stars. Many multi-planet systems have also been found as the result of longer and improved observations.

Most of these worlds have been discovered by the wobble they create in the motion of their parent star, but some of them were discovered when they passed in front of the star, causing it to dim, or blink. We haven't mentioned this transit method yet, and it works this way. If the planet happens to have a near edge-on orbit, as seen from Earth, it will periodically cross directly in front of, or transit, its host star. Such a transit can only be seen if the orbit of the distant planet crosses the line of sight from Earth, blocking a tiny fraction of the star's observed light and causing it to periodically dim, over and over again during the planet's endless journey around the star. The size of the planet can be derived from the size of the dip. The planet's temperature can be estimated from the characteristics of the star it orbits and the planet's orbital period.

Some of the transiting hot Jupiters are orbiting in the opposite direction to the rotation of their host star, and others have been found with orbits that are steeply tilted with respect to the star's equatorial plane. In contrast, all eight major planets in our solar system orbit the Sun in the same direction as its rotation and in roughly the same plane, which extends from the Sun's equator. The process that pulled the giant exoplanets so unexpectedly close to

Focus 16.3 Determining the mass and orbital distance of an exoplanet

Planet hunters record the spectral lines of a nearby star, and look for periodic variations in the line-of-sight velocities, V_{obs}, detected from the measured Doppler shifts of the lines. Because the orbital plane is normally inclined to the line of sight, the true orbital velocity, V, is related to the observed velocity by:

$$V_{obs} = V \sin i$$

where i is the inclination angle between the perpendicular to the orbital plane and the line of sight.

The period, P, of the velocity variations is given by Kepler's third law

$$P^2 = \frac{4\pi^2 a^3}{G(M_1 + M_2)}$$

where M_1 and M_2 respectively denote the mass of the star and its planet, a is their separation, and the Newtonian gravitational constant G is 6.673×10^{-11} m^3 kg^{-1} s^{-1}. If r_1 and r_2 denote their respective distances from a common center of mass, and we assume circular orbits, then

$$r_1 M_1 = r_2 M_2$$

with

$$a = r_1 + r_2 = \frac{r_1(M_1 + M_2)}{M_2}$$

Since the orbital velocity is

$$V = \frac{2\pi r_1}{P} = \frac{V_{obs}}{\sin i}$$

we can combine this expression with the equation for P and the expression for a to obtain

$$a = \frac{P V_{obs}}{2\pi \sin i}\left(\frac{M_1 + M_2}{M_2}\right)$$

Substituting into Kepler's third law, given by the equation for P, gives

$$M_2^3 \sin^3 i = \frac{P V_{obs}^3}{2\pi G}(M_1 + M_2)$$

and since the mass of the star will greatly exceed the mass of the planet, or $M_1 > M_2$,

$$M_2 \sin i \approx \left(\frac{P}{2\pi G}\right)^{1/3} V_{obs} M_1^{2/3}.$$

For the first exoplanet to be discovered (Fig. 16.8), we have $P = 4.23$ days $= 3.655 \times 10^5$ seconds and $V_{obs} = 50$ meters per second. Under the assumption that $\sin i = 1$ and the star's mass is comparable to the Sun, with $M_1 \approx M_\odot \approx 1.989 \times 10^{30}$ kilograms, we obtain a planet mass of $M_2 = 7.55 \times 10^{26}$ kilograms, which is comparable to the mass of Jupiter, $M_J = 1.9 \times 10^{27}$ kilograms. But the exoplanet is nowhere near as far away from its star as Jupiter is from the Sun, at 7.78×10^{11} m $= 5.2$ AU. The separation, a, is given by

$$a = r_1 + r_2 = r_1\left(1 + \frac{M_1}{M_2}\right) \approx \frac{r_1 M_1}{M_2} = \frac{M_1}{M_2}\frac{P V_{obs}}{2\pi \sin i}$$

$$\approx 7.66 \times 10^9 \text{ meters} = 0.05 \text{ AU}$$

which is even closer to the star than Mercury is from the Sun, at 5.79×10^{10} meters or 0.387 AU.

their star may also be responsible for their unanticipated backwards or tilted orbits.

You can keep track of the accelerating pace of discovery at the extrasolar planets encyclopedia at http://exoplanet. eu/ or at http://planetquest.jpl.nasa.gov/. In February 2010, for example, 400 candidate planets and 41 multiple-planet systems had been detected by the radial velocity method and 69 candidate transiting planets had been located.

These have all been indirect detections of exoplanets. The important direct confirmation of a planet circling another star was obtained using the *Hubble Space Telescope* to examine the extensive debris disk of dust surrounding the bright star Fomalhaut; the protoplanetary disk had been discovered in the 1980s using the *InfraRed Astronomy Satellite*. In 2004, an occulting disk was used to block out the star's bright light and enable the space

telescope to resolve the visible-light image of a ring of protoplanetary debris, analogous to the Kuiper belt in our solar system. The sharp inner edge of the ring suggested that a nearby planet was clearing out the material beyond it, and by 2008 the light of the Jupiter-size world had been observed. The planet's host star, Fomalhaut, is believed to be a relatively young star; it is only 100 to 300 million years old compared with the Sun's age of 4.6 billion years. Fomalhaut has 2.1 times the Sun's mass and 18 times its luminosity.

Also in 2008, astronomers used the Keck I telescope in Hawaii to directly confirm the orbital motion of three planets around the star HR 8799, using adaptive optics at infrared wavelengths (Fig. 16.9). The host star is roughly 1.5 times as massive as the Sun, and about 5 times as luminous; but it is much younger, with an estimated age of 60 million years. The planets, designated HR 8799 b,

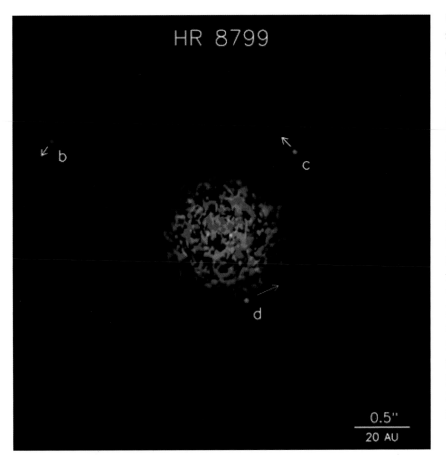

0.5"

20 AU

Fig. 16.9 Three planets orbiting a nearby star The speckled residual infrared light of a host star HD 8799 (*center*) and three orbiting planets (*red dots*) are found in this image taken in 2008 from the 10-meter Keck I telescope in Hawaii. The planets all orbit their star in the counter-clockwise direction, as indicated by the arrows, which show their positional displacement between 2004 and 2008. The planets, labeled b, c and d, are located at distances of 70, 40 and 25 AU from the star. The distance of the inner planet is comparable to Neptune's orbital distance at 30 AU. The star HR 8799 has a mass of about 1.5 times that of our own Sun, and is located about 130 light-years away. The planets most likely formed inside a protoplanetary disk about 60 million years ago. The planet masses are estimated as 7, 10 and 10 times that of Jupiter, for b, c and d respectively. (Courtesy of National Research Council, Canada, Christian Marois, Bruce Mcintosh and Keck Observatory.)

c and d, orbit inside a massive dusty disk at distances of roughly twice those of Neptune, Uranus and Saturn from the Sun. Their masses lie between 8 and 10 times the mass of Jupiter. The three planets have subsequently been imaged in infrared light using the relatively small, 1.9-meter diameter portion of the Hale telescope.

Another exoplanet, with a mass of about 9 Jupiter masses, has been observed moving around the star Beta Pictoris, from one side of the star, behind it, and on to the other side (Fig. 16.10). These observations, taken in 2003, 2008 and 2009, were performed using an infrared adaptive optics instrument attached to the Very Large Telescope located in Chile. The relatively cold giant planet, designated Beta Pictoris b, is located at a distance from its host star of between 8 and 15 times the Earth–Sun distance of 1 AU, or at about the same distance as Saturn is from the Sun at 9.539 AU. A dusty disk surrounding Beta Pictoris, extending up to 1000 times the Earth–Sun distance (also see Fig. 16.10), was also discovered during pioneering infrared observations. The presence of a giant planet was subsequently proposed to account for the gap in the dust disk and to explain the observed warp of its inner parts. The planet's host star is just 75 percent more massive than the Sun, but with an estimated age of only 12 million years. Because Beta Pictoris is so young, the exoplanet had

to form relatively quickly, in a timespan as short as a few million years.

16.6 Searching for habitable planets

The discovery of hundreds of planets orbiting other stars has created intense excitement and popular interest. From a human perspective, the most interesting planets will be those as small as the Earth, in circular orbits at just the right distance from the heat of a Sun-like star to provide a haven for life.

The orbital size can be calculated from the period of the repeated eclipse, and the planet's temperature estimated. This information would tell us if the planet resides within the warm habitable zone, the range of distances from a star where liquid water can exist on the planet's surface and life might exist. At closer distances, the water would all be boiled away, and at more remote distances it would be frozen solid.

Two missions observe planetary transits from space, attempting to find Earth-size planets in a habitable zone. They are the European *COnvection ROtation and planetary Transits (COROT)*, mission, launched on 27 December 2006, and NASA's *Kepler* mission, launched on 7 March 2009.

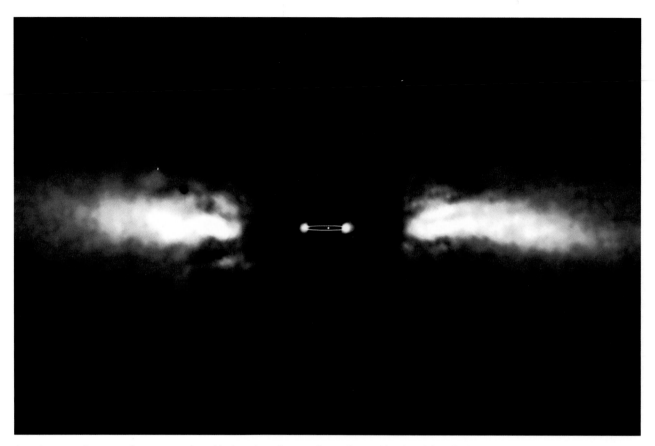

Fig. 16.10 Exoplanet on the move   The orbital motion of an exoplanet, denoted by the white elliptical line, was imaged from an adaptive optics instrument attached to the Very Large Telescope in Chile. The small white spot at the center shows the location of the host star, Beta Pictoris. Observations in 2003 are at the left side of the ellipse and those in 2009 are on the right side. The larger dust disc surrounding the host star is also shown by the large flattened blue image at the left and right. (Courtesy of ESO/A.M. Lagrange.)

The *COROT* spacecraft is capable of detecting extrasolar planets with short orbital periods and large terrestrial size. In February 2009, the mission announced its seventh planet discovery named COROT-7b, which has a radius of 1.7 times that of the Earth and is about 4.8 times as massive. The newfound planet's orbital period is nevertheless just 20.5 hours, implying a distance of just 0.017 AU from the host star, or only four times the star's radius and far too hot for comfort.

The *Kepler* mission is specifically designed to detect planets comparable to the Earth in size or smaller, and located at or near the habitable zone. By continuously measuring the brightness of 100 000 stars for four years, it will detect the periodic dimming of starlight produced when the planets pass in front of the star. A transit by an Earth-size planet will produce a small change in the star's brightness of about one ten-thousandth lasting for 2 to 16 hours.

Once detected, the planet's orbital size can be calculated from the period of the repeated eclipse and the mass of the star using Kepler's third law of planetary motion. From the orbital size and the brightness of the star, the planet's temperature can be calculated. This information would tell us if the planet resides in or near the habitable zone. In addition, the size and probable mass of the planet can be found from the depth of the transit, or how much the brightness of the star drops during the transit. The fractional change in brightness, or transit depth, is equal to the ratio of the area of the planet to the area of the star. For the Earth and Sun, as an example, the transit depth is 0.000 084.

Kepler began science operations on 12 May 2009, and within a year had identified more than 700 planet candidates, including at least five candidate systems that appear to exhibit more than one transit. Follow-up observations with ground-based telescopes are required to confirm which candidates are really planets.

Kepler will continue to search for smaller planets at least until November 2012. Since transits of planets in the habitable zone of Sun-like stars occur about once a year and require three transits for verification, it is expected to take at least thee years to locate and verify such a world. Ground-based observations will also be required to confirm the discoveries.

Fig. 16.11 Alien world with an atmosphere An artist's portrayal of a giant Jupiter-size planet passing in front of the star HD 189733b, which lies about 63 light-years away. It is so close to its parent star that it takes just over two days to complete an orbit. In 2007, the spectral signatures of water vapor were observed in the atmosphere of the hot, transiting exoplanet using an instrument aboard the *Spitzer Space Telescope*, and confirmed by observations from the *Hubble Space Telescope*, which also revealed molecules of methane and carbon dioxide. Also in 2007, a spectrograph on the small ground-based InfraRed Telescope Facility, with a mirror diameter of just 3.0 meters, was used to detect carbon dioxide and methane in the atmosphere of HD 189733b. Although these organic molecules are also found in living things on Earth, the planet is so massive and so hot that it is considered an unlikely habitat for life. (Courtesy of ESA/C. Carreau.)

In the meantime, the world's best telescopes are being employed to find new exoplanets using the velocity method. The European Southern Observatory's 3.6-meter telescope in La Silla, Chile, has discovered many new ones, including several super-Earths, and the 10-meter Keck I telescope atop Mauna Kea in Hawaii has been used to discover many more, including a super-Earth with about four times the mass of the Earth. Exoplanets with a mass lying between that of the Earth and Jupiter have been dubbed "super-Earths".

The atmospheres of transiting exoplanets are also being investigated using the *Hubble Space Telescope*, the *Spitzer Space Telescope*, and ground-based infrared telescopes. As the planet passes in front and behind its star, astronomers can subtract the light of the star alone, when the planet is blocked, from the light of the star and planet together prior to eclipse. That isolates the emission of the planet and makes possible the detection of the infrared spectral signatures of gases in the planet's atmosphere. Water vapor, carbon dioxide and methane have, for example, been found in the atmosphere of HD 189733b, a hot Jupiter-size planet that orbits its star in just 2.2 days and is nearly 63 light-years away from the Earth (Fig. 16.11).

In 2010 Steven Vogt (1949–) of the University of California, Santa Cruz, and R. Paul Butler (1962–) of the Carnegie Institution of Washington, and their colleagues used 11 years of radial velocity observations of the nearby Sun-like star Gliese 581 (pronounced GLEE-za) to confirm four previously discovered planets orbiting this star, and to discover two more. The sixth planet to be found, designated GL 581g, is within the potentially habitable zone, at the right distance to harbor liquid water on its surface. It has a mass of about three times the Earth's mass, and is thus capable of holding an atmosphere. The dim, red parent star is about 20 light-years away from Earth, about one-third the mass of the Sun, and only about one-hundredth as bright, which means that the habitable zone is relatively close to the star and that a planet in it has relatively short orbital period.

This detection of a potentially habitable planet orbiting a nearby star, coupled with radial velocity surveys of Sun-like stars in the immediate solar neighborhood, suggests that Earth-size exoplanets are common, outnumbering large exoplanets just as sand and small pebbles are more frequent than rocks at the ocean shore. Our Milky Way galaxy is probably teeming with potentially habitable planets located at the right distance to have water and with the right mass to hold an atmosphere. That doesn't mean that they are inhabited with any plant or animal life, but they could be.

Author index

Subject index

51 Pegasi, planet, 454, 455

A Midsummer Night's Dream, 22
A ring, Saturn, 328, 330
aberration: chromatic, 21; spherical, 19
absorption lines, 29, 31
accretion hypothesis, origin of Moon, 197, 198
active region, Sun, 153, 157
active volcanoes, Io, 65, 68, 299–303
Adastea, Jupiter ring moon, 315
aerial refractor, 21
aerogel, 425
age
    meteorites, 188, 385
    Moon, 53
    Moon rocks, 188, 189
    oldest Earth rocks, 188
    solar system, 188
    terrestrial impact craters, 399
    Venus surface, 59, 63, 236
albedo, Moon, 163
ALH 84001, meteorite from Mars, 278, 279
Allgemeine Naturgeschichte und Theorie des Himmels, 446
Almagest, 5
ALSEP, 176
Amalthea, Jupiter ring moon, 315
amino acid, comet Wild 2, 427
ammonia ice clouds: Jupiter, 95, 289, 290; Saturn, 322, 324
Amor asteroids, 368
anatomy: comets, 419–423; crater, 56
Andromeda nebula, 449
angular diameter, planets, 34
angular resolution, telescope, 21
annual parallax, 27–29
anomalous precession of Mercury's perihelion, 217
anorthosites, Moon, 180
Antarctica, meteorites, 383
anticyclone, 9
aperture, 19
aphelion, 13
Aphrodite, 223
Aphrodite Terra, 233, 234
Apollo asteroids, 368

Apollo Lunar Surface Experiments Package, 176
Apollo missions, 169–176
    Apollo 8, 170, 171
    Apollo 11, 39, 171, 172
    Apollo 13, 174
    Apollo 15, 175
    landing sites, 173
Apollo program, to land men on the Moon, 39, 170
apparition, comets, 410
Appenine Mountains, 56, 57
Aqua spacecraft, 148, 149
arachnoids, 240, 242
Arecibo Observatory, 74, 205
Ares Vallis, Mars, 76
Armageddon, 406
asteroids, 365–380
    1 Ceres (dwarf planet), 367, 371–373, 381, 382, 389
    4 Vesta, 367, 372, 375, 381, 382, 389
    21 Lutetia, 376
    243 Ida, 48, 375
    253 Mathilde, 376, 377
    433 Eros, 377–380
    951 Gaspra, 48, 375
    2867 Steins, 376
    25143 Itokawa, 380, 381
    Amor, 368
    Apollo, 368
    astronauts visit, 381
    Aten, 368
    belt, 17, 19
    C-type, 373, 374
    carbonaceous, 373, 374
    chaotic orbits, 368, 369
    close approaches to Earth, 405
    collisions, 370; with Earth, 395–407
    color, 373
    composition, 373, 374
    discovery, 17
    families, 370, 371
    Galileo spacecraft, 48
    Kirkwood gaps, 368, 369
    M-type, 374
    main belt, 367
    mass, 17, 372, 377

mass density, 372, 377
metallic, 374
mining, 374
names, 17, 367
near-Earth, 404
number, 17
orbits, 367–369
origin, 369–371
parent bodies, 370
physical properties, 372, 377
potentially hazardous, 404, 406
radar, 373
radius, 371, 372
regolith, 380
rotation period, 372, 375, 377
rubble pile, 376, 381
seismic shaking, 378, 381
shape, 372
silicate, 373, 374
size, 372, 377
solid rock, 377, 380
S-type, 373, 374
total mass, 371
asthenosphere, 122
astronauts: *Apollo 11*, 171; asteroid, 381; solar threat to, 153
astronomical unit (AU), 13, 26–27
astronomy: from Moon, 188, 189; optical, 19
Aten asteroids, 368
Atla Regio, 237, 239
atmosphere, 80–97
    carbon dioxide concentration, Earth, 144, 148
    comets, 83
    Earth, 87, 91, 138–150
    escape, 84–87
    evolution, 92, 93
    exoplanets, 459
    giant planets, 93–95
    loss, 84–87
    Mars, 80, 91, 253–258
    methane concentration, Earth, 145, 149
    origin, 83
    Pluto, 437
    rise in carbon dioxide, 144
    secondary, 82
    terrestrial planets, 87–92